Fibre Metal Laminates

An Introduction

Edited by

AD VLOT

Delft University of Technology,
Faculty of Aerospace Engineering Delft,
The Netherlands

and

JAN WILLEM GUNNINK

Fibre Metal Laminates Centre of Competence,
Delft, The Netherlands

KLUWER ACADEMIC PUBLISHERS

DORDRECHT / BOSTON / LONDON

A C.I.P. Catalogue record for this book is available from the Library of Congress.

ISBN 1-4020-0038-3

Published by Kluwer Academic Publishers,
P.O. Box 17, 3300 AA Dordrecht, The Netherlands.

Sold and distributed in North, Central and South America
by Kluwer Academic Publishers,
101 Philip Drive, Norwell, MA 02061, U.S.A.

In all other countries, sold and distributed
by Kluwer Academic Publishers,
P.O. Box 322, 3300 AH Dordrecht, The Netherlands.

Printed on acid-free paper

Printed in the Netherlands.

Table of Contents

Preface

This is the first monograph on Fibre Metal Laminates, a material that has been developed over the last decades at Delft University of Technology, faculty of Aerospace Engineering in the Netherlands. This book was compiled for an international conference at 24-26 September 2001 on the occasion of the formal retirement of Professor Boud Vogelesang. He was the 'founding father' and the driving force during all the years of research. His enthusiasm, team spirit and perseverance were vital for the continuity and the success of the development.

This is one of the three books that were published for the conference. This book covers the technical afternoon sessions. Only a fraction of the chapters in this book were presented at the conference. The other chapters were added to provide a complete picture of the various material properties. The book mainly focuses on Glare, the Fibre Metal Laminate with glass fibres. A second book with the title *Around Glare – the context of a new aircraft material*, covers the broader subject of the morning sessions on the context of material development in the aviation industry, and also includes the farewell lecture of Boud Vogelesang. The third book, *Glare – history of the development of a new aircraft material*, provides the inside story of the development activities in Delft. All three books are published by Kluwer academic publishers b.v.. Arno Schouwenburg of Kluwer did a great job in transforming our electronic data into real material!

I also want to thank the sponsors of the conference. Their names are included at the end of the book. Without their help it would have been impossible to realise this event.

The scientific board helped to assure the quality of the papers. The board members were:

Prof.dr. Peter Beaumont, Cambridge University, United Kingdom
Prof.dr. Alan Baker, DSTO and RMI, Melbourne, Australia
Prof.dr.ir. Andre de Boer, University of Twente, the Netherlands
Dr. Bob Bucci, ALCOA, USA
Dr. Jim Greer, Lt.Col. USAF, Center of Aircraft Structural Life Extension, US Air Force Academy, USA
Dr. Peter Poole, DERA, United Kingdom

Thank you all very much for your time and assistance!

The layout and corrections of the papers were taken care of by a team of students, see Figure 0.1, who did excellent work. The members of this team were:

Ronald van der Meijs
Dort Daandels
Lex Dop
Geoff Morris

It was a joy to work with you!

The cover design is based on a painting by Willemien Veldhoven. Together with Frans Oostrum she did a great job in creating an artistic cover for this technical book.

Prof.dr.ir. Ad Vlot
Delft, June 2001

Figure 0.1 *Stress distribution around LRSMLGTO20 in Glarebook HQ, June 26, 2001.*

A. Material Properties

1

Historical overview

A. Vlot

Abstract - In this chapter a brief overview of the history of Fibre Metal Laminates Arall and Glare is given as background information for the other, technical chapters of this book. This is a summary of a more complete description that can be found in Vlot [1].

1.1 Arall

1.1.1 The development of Arall

The origins of our story can be traced back to the devastated Fokker facilities of 1945. The Fokker engineers that had to start from scratch grasped the opportunity to introduce new technologies. One of those engineers was Rob Schliekelmann, who graduated at Delft University. He did some practical training at the English aircraft manufacturer De Havilland. This was the first company to bond metal parts together. Schliekelmann took the metal bonding technology to Fokker where he optimised the pretreatment of the aluminium layers and the production process of bonded structures, for example by introducing autoclaves. As no expensive milling equipment was available, metal bonding was applied to create a built-up, laminated structure for the Fokker F-27 tailored to its local strength.

Later it was discovered that the bonded, laminated structure also appeared to have good fatigue properties. The team of Schijve at the National Aerospace Laboratory NLR carried out different types of fatigue tests on F-27 centre wings. A major finding was the favourable resistance to fatigue crack growth of laminated sheet metal reinforcements around a large wing joint. The crack growth was extremely slow because cracks started to grow in a single layer. The other intact layers effectively bridged the crack and in this way considerably slowed down crack growth.

In 1973 Schijve came to Delft to become a professor. Boud Vogelesang was already working there in the materials lab. Vogelesang had started his scientific career with test work on etched aluminium and on adhesive bonding. One of the projects of the lab was associated with laminated sheet material, still without fibres. The remarkable result was that the crack growth in the outer layer was slowing down after some growth had taken place. Because the crack did not yet penetrate into the subsurface layers, crack opening of the fatigue crack in the surface layer was effectively restrained, reducing the stress intensity at the crack tip in the outer layer. The inner layers were bridging the fatigue crack of the outer layer, which resembles the fibre bridging mechanism in Arall and Glare.

Composites were widely studied in those days but were very expensive. Against this background, combinations of metal and composites were studied in Great Britain and in the U.S. as a kind of intermediate between the metallic and composite structure. British researchers bonded composites as reinforcements onto rods, tubes and beams in the early seventies and determined the efficiency of these combinations compared to aluminium. Important for them were the lower costs of the combinations compared with full composites. The NASA Langley Research Center also emphasised the lower cost and treated the reinforcement as a way to reduce the weight while limiting the risk of applying a completely new material. NASA researchers were especially interested in local reinforcements bonded to aluminium structures in Space Shuttle components.

At Fokker, co-workers of Schliekelmann saw these tests during a visit to the U.S. at the end of the seventies. Once home, Schliekelmann's group realised that the concept came close to their bonded laminated structure if fibres were added to the adhesive of their laminates. Nylon fibres were applied as a weave, as was carbon in a single direction (unidirectional) embedded in bondlines between one millimetre thick sheets. The fatigue crack growth properties of those fibre-reinforced laminates were determined. Under constant-amplitude fatigue loading, some reduction of the crack growth rates was obtained, which was more significant for the carbon fibres than for the nylon fibres. The strength of the material after inflicting significant damage in the form of saw cuts was measured to determine the damage tolerance of the laminates. The crack growth of the fibre-reinforced laminates was two to three times slower than in aluminium but this was not spectacular in view of the high cost of the material. The concept of fibre reinforcement was technically very interesting but the results were not spectacular and the many inherent problems related to durability and quality control which Schliekelmann had managed to solve for application of bonding in the F-27 seemed to be endless and would lead to expensive products.

Almost from the start, Delft was involved in these developments and the discussions that took place. Schijve and Vogelesang gave student Lex ten Have the task to carry out flight simulation tests in 1978 on carbon and aramid fibre reinforced laminates. The unidirectional carbon fibres remained intact under these conditions, whereas the aramid fibres, which were applied as a weave, broke at some distance from the crack tip. The completely intact carbon fibres, all being in the loading direction, slowed down the crack growth considerably, much more than the only partially intact aramid fibres of which only half were oriented in the loading direction.

While the activities were stopped at Fokker, Delft University started to optimise the material. It had become clear that fatigue in the fibre-reinforced metal laminates was a delicate process. Evidently, loads from the cracking metal layers were transmitted via the adhesive to the fibres, thus unloading the metal layers and slowing down crack growth in these layers. This is called 'fibre bridging'. It was subsequently found that the adhesive, which is loaded in fatigue, started to come apart from the metal. This process is called 'de-lamination', since the various layers separate when the adhesive fails. This release of the bond would appear to be a bad thing. However, without this delamination the fibres would be stretched too much when the crack opens and would therefore break. If, on the other hand, the delamination is too large then the crack will open too much and will consequently grow too fast. Therefore there is a balance between delamination and crack growth. This balance is assured by the strength of the adhesive and its resistance to delamination. This balance must be established over the whole length of the crack, since if the loads in the fibres are higher at a certain location than elsewhere, more delamination will occur there and this will cause lower loads in the fibres at that location. The intact fibres in the wake of the crack considerably restrain the opening of the fatigue crack, which has a most favourable effect on the stress intensity at its tip. As a result, fatigue crack growth is effectively slowed down, and even full arrest of crack growth can occur. In this way insight into the mechanism slowly evolved, making the optimisation of Fibre Metal Laminates possible.

Delft student Marissen developed an analytical model for the prediction of fatigue crack growth in Fibre Metal Laminates, based on fracture mechanics principles. This model was used to calculate the stress intensity factor of the fatigue crack, which accounted for the fatigue cracks in the metal layers and the crack-bridging load of the unbroken fibres in the wake of the crack. Marissen had to make some simplifying assumptions in his model, but it was based on the physical understanding of the observations on the mechanism of crack growth in Fibre Metal Laminates. It turned out that the agreement between predicted crack growth rates and observed crack growth rates was encouraging. It substantiated the basic concepts of a new class of Fibre Metal Laminates characterised by:

1. bridging of fatigue cracks by high strength fibres
2. thinner sheet-metal layers in order to allow more fibre layers to reduce the shear stress in the adhesive between the fibres and the metal layers and thus to limit delamination

Vogelesang baptised the optimised material with the acronym Arall (Aramid Reinforced ALuminium Laminates). Later, when working in a German aeronautical laboratory, Marissen wrote a doctoral thesis [2] on an improved and more detailed model for fatigue crack growth in Fibre Metal Laminates.

From the beginning of the research on Fibre Metal Laminates in Delft, it was clear that the production of the laminates required partnerships with material suppliers. The obvious needs were associated with strong fibres and thin aluminium alloy sheets. The large Dutch chemical company AKZO manufactured aramid fibres suitable for the laminate. ALCOA produced the thin sheets. Contacts were also established with 3M, which made good adhesives and was also able to manufacture the prepregs.

During 1980 other students were involved to further optimise the Fibre Metal Laminates. Finally, a layout was selected with aluminium layers of 0.3 or 0.4 mm thickness. Two aluminium alloys were tested: type 7475, an aluminium alloy with a high strength but relatively poor fatigue properties, and 2024, an aluminium alloy with a lower strength but with superior fatigue properties. For the 7475 variant stretching after curing was necessary to get good properties, while for the 2024 laminate curing alone appeared to suffice. The more detailed contours of the new material now began to take shape. For a laminate with aluminium, fibres and adhesive, many variants were theoretically possible. Research, however, soon focused on specific variants with AKZO's aramid fibre, two aluminium alloys (2024 and 7475) from ALCOA, and an epoxy adhesive from 3M.

The bonded laminates used on the Fokker F-27 could not be characterised as a material. It was a structure; an assembly of different parts by bonding. With the industrial partners ALCOA, AKZO and 3M - all material manufacturers - the fibre reinforced laminates were no longer a structure but a material: a so-called 'semi-finished product', similar to a regular sheet of aluminium. This represented a definite change of perspective. Instead of producing the laminates themselves, future aircraft manufacturers like Fokker had to buy the fibre reinforced laminates from ALCOA, for example.

The patent on Fibre Metal Laminates was filed in the US on 9 January 1981. As inventors of the new material it mentioned Schijve, Vogelesang and Marissen. Actually there were two patents: a general one covering a combination of metal sheet and fibres and one on the stretching process. The name of the applicant is Delft but AKZO holds the rights.

1.1.2 Application of Arall in wing structures

In the beginning the studies for the application of Arall focussed on wing structures. The oldest F-27 aircraft reached their design life and Fokker did a lot of test work on a critical detail of the wing, i.e. a panel with a tank inspection hatch. Because a lot of background information was known, this wing was a perfect playing field to study the application of Arall. Fokker, Delft and ALCOA met in 1984 to discuss the possibilities of the design and testing of a part of the wing of the F-27 made from Arall. ALCOA committed itself to produce the required panels of 2 metres long and 1,5 metre wide, see Figure 1.1. Up to that moment it only had produced tiny panels of roughly A4-size. In 1982/83 the first commercial product under the trade name

Arall was launched by ALCOA. The grades Arall 1 and Arall 2 were standardised. Arall 1 was a variant with aluminium 7075 layers and was stretched after curing, and Arall 2 had aluminium 2024 layers and was in an as-cured condition. Of the fibre/adhesive layers 41% of the weight consisted of aramid fibres. The production started in 1984 and the official introduction took place at the Paris Air Show in 1985. Later, in 1987, two other grades, Arall 3 (7475 aluminium, identical to Arall 1) and Arall 4 (with a different, high temperature adhesive for military applications) were also standardised. The produced panels had a size of 8 feet by 4 feet. A production plant was set up for 1 million square feet per year. Due to the various treatments the material had to undergo in the production of Arall, the price was high: in the order of ten times that of aluminium per kilogram. In Europe the chromic acid anodising process was chosen for the aluminium layers, whereas in the US a more environmentally friendly process using phosphoric acid was applied. Cladding (adding a pure aluminium layer at the outside) was possible for customers that preferred a polished, shiny outside skin of their aircraft.

Figure 1.1 *Arall panels produced by ALCOA.*

The pretreatment, layup and curing of Arall were performed in building G at the ALCOA Technical Center near Pittsburgh. Because the ALCOA division itself was set up in New Kensington, Pennsylvania, the crew would travel to ALCOA Technical Center to manufacture the laminates. Only the ultrasonic inspection of the produced laminates was undertaken in New Kensington itself.

For the application of Arall in the design and construction of an actual aircraft, Fokker was the obvious candidate as far as the Delft group was concerned. But Fokker's problem in applying Arall was that after the F-27 in the mid-1950s, and the F-28 at the end of the 1960s, no new aircraft types had been developed. For the application of Arall this meant that extra costs of changing an existing structure had to be overcome, which would counterbalance the operational cost savings obtained from a lighter structure with a new material. The application of Arall, and later Glare, would therefore only make sense within the context of the development of a completely new aircraft. For years Delft had to keep the material alive, waiting for a window of opportunity through which it could target a new aircraft design. Similar problems were encountered at Airbus and Boeing.

The Netherlands Agency for Aerospace Programmes NIVR stepped in to fund further research on Arall. The work that had been started in Delft on the F-27 wing panels could now really be executed. Two Arall wing panels would finally be built and tested. Flying demonstrators, i.e. tank covers, were also designed, manufactured and mounted by Fokker to fly on F-27's. The F-27 wing-panel project received official approval in 1984 and was started in November 1985. Up to then, many small specimens were tested in Delft which had created a basic confidence in Arall for the different partners, but it was now time to study a realistic aircraft component.

Figure 1.2 *Production of the Arall F-27 panel with access hole.*

The dimensions of the panel would approximately be two by one metres and it would contain one access hole, used in service for internal inspection of the wing, see Figure 1.2. The same loading would be applied as for the aluminium panel, and because the Arall was lighter and thinner, the mean stress level in Arall (mean stress in flight) would be 86 to 100 MPa, whereas the stress in the aluminium panel had been 77 MPa. The target to be achieved was a weight reduction of 25%.

It became evident from the tests that Arall was quite sensitive to strength reductions caused by holes drilled in the material (so-called 'blunt notch strength') and that although the material had excellent fatigue crack growth properties, thickness steps (doublers bonded on the structure to increase strength) were susceptible to premature fatigue cracks. Bending of the material at these locations in particular caused cracks to grow at the outside layer of the laminates since the fibres are less effective during this type of loading by bending. The production of parts at the Fokker plant did not create special problems when compared with aluminium. The large wing panel was assembled from these parts and subjected to three times the design life of the F-27, i.e. 270,000 flights. Only minor cracks occurred at the finger tips, i.e. the thickness steps of the doublers, and only in the outer layer of Arall at that. This meant only a minor surface scratch in Arall whereas the aluminium equivalent would have failed dramatically. The residual strength in a cracked condition was high. On top of this increased safety level a weight saving of 33% compared to the original design in aluminium was achieved. The blunt notch strength appeared to be critical, but the wing panel was a big success.

1.1.3 Application of Arall in different structures

Fuselage

Up to that moment research had concentrated on wing applications and this culminated in the successful F-27 wing panel project. This concentration on the wing had a historical explanation as we saw. However, MBB in Hamburg had also started to become involved in the material and for them the fuselage would be a more likely application. Tests for the application of Arall in the construction of aircraft fuselages had showed, however, that under loading conditions that resemble those of the fuselage of an aircraft, the aramid fibres around a fatigue crack would break. With the aramid broken, crack growth would no longer be slowed down. Of course this was a very serious issue, since without intact fibres the whole concept of Arall would not work. Roebroeks [3] found an explanation. He thought that the aramid fibres hardly had a connection with the adhesive. The aramid fibres lay loose in a tube in the adhesive and when pulled out they therefore slipped out of the tube without deformation of the adhesive. When this was followed by compression, the aramid fibres behaved weakly under compression and would be damaged and break.

Ballistic Arall

But the Fibre Metal Laminate development did not stop there. Already a new variant was being developed in Delft that did not show the fibre failure behaviour at all: a version of Arall with glass fibres, developed by Roebroeks and Vogelesang. Apart from the variant with glass fibres, Delft also developed other new Arall types. In 1986 a patent was filed on a special variant of Arall meant as armour plating, with Vogelesang, Paalvast and Verbruggen as inventors. This Arall type had ceramic tiles at the outside to break up a bullet that would hit the material, while the Arall backing layer would absorb the remainder of its kinetic energy and would in this way stop the bullet. In the same year, government support was obtained to develop a new type of armour made from Arall under the name 'ballistic Arall'. The aramid fibres were widely used in armour material at that time and it was felt that such a ballistic material could be superior. After an extensive test programme carried out by the Dutch military and TNO it was concluded that the material was not viable.

Space structure and tubes

Other developments turned out to have more potential. In 1988 the undergraduate Coen Vermeeren tried to come up with an Arall type which would be able to withstand the high temperatures occurring in space structures and supersonic transports: an Arall with carbon instead of aramid fibres and titanium instead of aluminium layers. The possibilities of Arall with its different components seemed to be endless. New variants were introduced and tested. A patent was filed on a manufacturing process for Arall tubes with Harald Bersee and Vogelesang as inventors. The tubes were promising for applications in the chemical and nuclear industry because they remain leak-proof even if damaged. In the same patent a bike made of those tubes was protected, see Figure 1.3. During the same period an Arall tennis racket which was ten percent lighter than a conventional racket was designed and made. In 1987 a further patent on Arall with a thermoplastic adhesive was filed, mentioning Vogelesang, Meijers and Van Velze as inventors. This material would become soft at high temperatures and therefore was easier to deform than the original Arall.

1.1.4 The breakthrough of Arall

The first flying Arall parts would be a number of wing hatches of an access hole of the Fokker 50, which was given certification by the Dutch Airworthiness authorities RLD. For that reason it was also an important milestone.

Figure 1.3 *Strain survey on a bicycle made from Arall tubes.*

At that time Arall also gained international momentum and was about to take off. In 1988, one year after the first Arall conference in Seven Springs (USA) a second conference was held in Delft. During this conference the French company Aerospatiale presented their study on a fuselage shell of an Airbus A320. However, the results were disappointing. Arall was clearly not suited for fuselages. In the same year, in October 1988, the first contract for the application of Arall was won by the ALCOA marketing department for the C-17 military transport aircraft. The C-17 suffered from heavy weight problems in the back of the aircraft and therefore the large cargo door situated there would be manufactured from Arall by McDonell Douglas. Several other locations for the use of Arall on this new aircraft, which finally opened possibilities for the material to be applied, were also studied in Delft, such as the wing and fuselage. Only the cargo door with dimensions of 5.6 by 9.7 metres was successful. It was manufactured from Arall 3 and led to 26% weight savings. However, it also turned out to be a highly complex product when manufactured from Arall: stretch forming, roll forming, routing, bonding and riveting of many different components were necessary. The Arall panels that could be produced by ALCOA were too small and had to be connected together with expensive titanium straps. This evidently revealed the great Achilles' heel of Arall: per kilogram the laminate was already about eight to ten times as expensive as aluminium. And with the complex production steps that were necessary, this cost factor even became worse. Only about thirty C-17's were built with Arall cargo doors.

1.2 Glare

1.2.1 The development of Glare

A patent on Glare was filed on the 14th of October 1987 by AKZO with Roebroeks and Vogelesang named as inventors. A partnership between AKZO and ALCOA started to operate in 1991 to produce and commercialise Glare. A formal agreement to found the Structural Laminates Company (SLC), a joint venture of AKZO (1/3 owner) and ALCOA (2/3 owner), was signed on the 1st of June 1991. The agreement provided for production to be concentrated in New Kensington, Pennsylvania, while the research, development and marketing would be done in Delft. For this reason a separate company, a subsidiary of SLC with the name Structural Laminates-bv was founded in the Netherlands, headed by Gunnink.

SLC assisted in design studies at Boeing in the New Large Airplane group, which was studying a successor for the Boeing 747 Jumbo Jet. SLC did not have enough data to study the application of Glare in this large aircraft and instead the application of Glare in the Boeing 777 was investigated, a new Boeing aircraft which was then in its final stage of development.

1.2.2 Impact properties of Glare

It was, however, too late to consider Glare for the vital parts of the aircraft, but the material properties of Glare could be used for other parts. Whereas Arall had poor resistance against objects which hit the structure, the so-called 'impact resistance' research, done by Vlot [4] in Delft, had shown that at low velocities Glare is as good as aluminium and superior to carbon fibre composites, while at higher speeds the glass fibre in the laminate becomes relatively much stronger and the impact properties much better than that of aluminium. It showed that Glare is not only more resistant to impact than other materials, but also that the damage – permanent deformation and the denting of the aluminium layers in the laminate – can be found easily with the naked eye due to plastic deformation. Composites do not deform plastically; hardly any dent is created and therefore damage is much more difficult to detect in this type of material. The mechanical properties of Glare after impact damage was created also appeared to be better. In 1990 the excellent impact properties of Glare were implemented in the cargo floor in the Boeing 777. Floor structures are particularly prone to impact damage. This was the first commercial application of a Glare product. A special variant, Glare 5, was later developed and specially optimised for impact properties. It therefore consisted of more glass fibre layers than the regular Glare types.

Apart from its excellent impact properties it was also found that Glare had good burn-through resistance. Boeing carried out fire resistance tests up to temperatures of 1200 degrees Celsius and showed that Glare could prevent fire from penetrating for more than fifteen minutes. Although the outer aluminium layer would melt, the glass fibre was proven to withstand the flame temperature. While the adhesive carbonised, the layers separated by which insulation was improved and the inside temperature remained in the order of only one hundred degrees Celsius. Both the impact and fire resistance properties were demonstrated in a blast resistant container that was tested by the FAA in 1995 as an answer to the Lockerbie crash over Scotland. This container was named ECOS-3 and was developed by Galaxy Scientific Corporation. It later went into production.

1.2.3 Glare in the fuselage

Up until the end of the eighties the design studies on Arall had concentrated on wing applications. Design studies initiated around this time for the C-17 horizontal stabiliser revealed that for such large and therefore thick structures Arall would become a very expensive material with too many layers that were difficult to produce and inspect. Although cracks grow very slowly in Arall, inspection to find them would still be needed, but finding such cracks would be difficult in the hidden aluminium layers inside the laminate. Special techniques such as Eddy current inspection become more difficult if the laminate is thicker. The first decade of Arall and Glare had been spent on wing studies while the future for the material now appeared to depend on the fuselage skin. It was under fuselage fatigue loading that the fibre failure mechanism for Arall was discovered, to which Glare appeared to be the answer. A lot of attention in Delft was soon given to fuselage applications. The

PhD-subject of Dong Chen [5] was on fuselage loading conditions on Arall and Glare.

However, for this area where aluminium was unrivalled, new aluminium alloys were developed. For the Boeing 777, ALCOA developed a special alloy designated as 'C188' and later given the systematic code 2524, which meant that the alloy is a derivative of the familiar 2024 alloy already applied in the DC-2. The C188/2524 alloy had excellent damage tolerance capabilities, which meant that an aircraft fuselage constructed from this material could withstand a large crack, caused by fatigue or impact of engine fragments. Therefore the fuselage of the Boeing 777 did not need the titanium crack stoppers which were still required in the Airbus aircraft to give the fuselage the capacity to carry internal pressure with large cracks. Glare also appeared to be ideally suited for fuselage applications. Apart from Glare 1 and Glare 2, which were unidirectional and resembled Arall 1 and Arall 2 respectively, in 1990 'cross-plied' variants were defined for fuselage applications: Glare 3 with 50% of the fibres in one direction and 50% in the direction perpendicular to it and Glare 4 with twice as many fibres in one direction as in the other direction. Glare 4 was especially suited for locations in the fuselage where the load in one direction was twice as high as in the other direction. Glare 3 on the other hand was defined for the most upper part of the fuselage where this ratio is one, due to the combination of pressure loads and the bending of the fuselage under its own weight.

1.3 Splices in Glare

Boeing considered Glare to be too expensive. As a response to this, a method to produce bigger Glare sheets was developed. The available autoclaves are big enough to bond the aluminium sheets with a limited width of approximately 2.5 metres together to form one bigger panel. SLC quality manager Carl Garesché was the main inventor of this 'splice' concept. Delft extended the idea to a viable solution. Although a lot of work still had to be done to prove the viability of this concept, a patent was filed for 'spliced Glare' with Garesché and Delft engineers as inventors. The main problem was that the strength reduction caused by the interruptions of the aluminium sheets had to be restored. Initially this was achieved by adding material in the form of doublers to the outside of the laminate as a kind of bridge. However, this was not adequate. It was found that the sheets could be internally bridged by adding aluminium strips to the inside to restore the local strength, the so-called 'internal doublers'. Later Delft came to the conclusion that the best option would be not to create gaps between the aluminium sheets but to create an overlap of them by adding some adhesive film in between. Spliced Glare would make wider sheet sizes possible. It also opened the possibility to produce double curved sheets without the need for expensive stretch forming. The splice concept paved the way to cost reduction and large scale application of Glare. However, it was too late for Boeing. From 1993 onward, Glare could only breakthrough for use in primary aircraft structures in Europe.

In 1997 SLC stopped all commercial activity and licensed the technology out. ALCOA wanted to keep control of Glare in the US and therefore SLC was kept alive as a 'paper' company in control of the patent rights in the U.S. and as a joint venture with AKZO. SLC gave out a license for secondary applications such as floors and cargo containers to the company Aviation Equipment, Inc. AKZO got the license in Europe in order to co-operate with Airbus, assumed responsibility for Structural Laminates BV and restructured it as Structural Laminates Industrie (SLI) for this purpose.

1.4 Glare at Airbus

In 1990, in the middle of the change from Arall to Glare a meeting was held at Aerospatiale, the French partner in Airbus Industrie, to discuss the results of fuselage studies with Glare. Johannes Koshorst from the headquarters of Airbus Industrie in Toulouse, who was responsible for co-ordination and stimulation of material developments at Airbus Industrie had initiated these studies. In March 1989 he granted a contract to Delft, and to the French manufacturer and Airbus partner Aerospatiale to carry out this work on 'R-glass Arall', as Glare was still called in these days. The project on R-glass Arall was a follow-up of the project on Arall of which the results were shown at the second specialist conference in Delft in 1988. The results of the fuselage study for aramid Arall had been disappointing. It only indicated an 8% weight reduction against an eight to tenfold price increase. This was certainly not acceptable. Fortunately Glare was introduced soon after and this material was immediately considered to be more promising. In 1989 the results of the new R-glass Arall studies were presented and discussed. The possibility was discussed for the production of the material in Europe and it was reasonable to look for a production site at Fokker. In the meeting on June 1990 of Airbus Industrie and all its European partners to discuss the outcome of the fuselage study, the fuselage in Glare of the Airbus A320 appeared to lead to a 25.9% weight reduction for 280 dollar per kilogram saved weight. This was very promising compared to the disappointing study on the Arall fuselage.

Aerospatiale in Toulouse studied the feasibility of applying Glare in the Aerospatiale parts of Airbus aircraft. DASA in Germany became very active in Glare research. This dated back to 1988; at that time a segment of the Airbus A330 and A340 aircraft fuselage was planned for fatigue testing for certification in Hamburg. It was decided to study the application of new materials such as two lithium-aluminium alloys and the 6013 alloy using the common 2024 fuselage alloy as a reference. Two upper shells, the 'crown section' of the fuselage, the part of the fuselage with the heaviest load, were available for Glare. Six sheets were produced for the barrel at Fokker's Schiphol plant. In 1990 the fatigue test on the barrel was started, see Figure 1.4. It continued until 1994 and was finished after 100,000 flight cycles. Each week one thousand cycles could be applied, after which the barrel was inspected for cracks for a week. After the fatigue tests, artificial damage was applied and the residual strength test was performed. The material was tortured with hammers to produce dents, saws were set in the material to create cracks and milling machines were used to remove the outer aluminium layer to simulate corrosion damage. The residual strength in the presence of these types of damages was a big concern at that time since it was found to be lower than that of the 2024 aluminium alloy. Glare survived all this while all the other materials on the barrel had to be repaired after significant crack growth.

Apart from the barrel, DASA also did other test work. In the full-scale fatigue test on the A320, cracks were found in the circumferential strips ('butt straps') that connected the fuselage sections. Since this strip is located inside the aircraft it is difficult to inspect for cracks. Because the design of the aircraft was already finished, a material was needed that could replace the original strip without necessitating changes in the rest of the structure. Successful tests were done in Hamburg and in Delft on realistic specimens with a butt strap connecting two pieces of skin.

Figure 1.4 *A330/340 barrel test in Hamburg.*

For the A340 a bulkhead section at the rear of the fuselage was designed and manufactured in Glare and tested in the full-scale test. The Glare section produced for the bulkhead was the first Glare part to be manufactured and tested with a curvature in two directions. The aluminium was stretch-formed to the right contours before bonding it together by laying the intermediate prepreg layers in between the formed aluminium parts. At another location of the bulkhead, three radial crack stoppers in titanium were replaced by stoppers made from Glare. Later, in 1996, double curved Glare was produced by laying up aluminium sheets in a double-curved mould without pre-forming and by using splices to prevent excessive wrinkling of the sheets and autoclave pressure to press the sheets in the right contour. Student Van Oostrom did the first production trials in this way in 1996.

The U.S. Air Force officer Rob Fredell [6] began his PhD-study in Delft in 1991. Fredell started a project on repair and managed to gather a large group of undergraduate students in his team. His research focused on the possibility to repair existing aluminium aircraft structures with bonded Glare patches, see Figure 1.5. He compared the results with the conventional composite patches that are applied for these purposes. He discovered that because of the smaller thermal mismatch between aluminium and Glare compared to the combination of aluminium and composite, the Glare patches behave better. This initiated a lot of work on bonded patches for years afterwards in Delft and when Fredell returned to the U.S. Air Force Academy he continued his research there by hiring Dutch graduates, Kees Guijt and Stephan Verhoeven. This repair project even led to the fact that the first flying Glare application on a primary structures was achieved in the U.S.. In October 1995 a bonded Glare repair patch was installed over cracks in the fuselage of a gigantic C-5A Galaxy transport aircraft. Another American project was for the U.S. Navy, in co-operation with Drexel University. From 1992 to 1994, research was carried out on special high-temperature laminates for supersonic transports, which suffer from aerodynamic heating. On the Dutch side, Coen Vermeeren did the work on this type of Fibre Metal Laminates. As a follow-up of this project the NIVR sponsored a project, carried out by Kees Guijt, on high-temperature Fibre Metal Laminates for space applications consisting of thin layers of titanium and special adhesives reinforced by carbon fibres.

Figure 1.5 Repair patch on fuselage skin of a C-5A.

From 1992 to 1996, a project involving a number of aircraft companies in Europe got under way, sponsored by the European Union under the name Brite Euram project BE 2040. Participants in this project were: Fokker, Deutsche Aerospace Airbus, British Aerospace, Aerospatiale, Dassault, Dornier, DASA, Structural Laminates BV, NLR, DLR, TU Delft, TU Braunschweig, Alenia and CASA. In this way other European manufacturers also gained experience with the material. Although the project did not really develop Glare further and repeated much of the work that was already done in Delft and Hamburg, it did help to spread knowledge of and experience with the material across Europe.

In 1994/1995 Roebroeks carried out tests which showed significant strength reductions – in the order of 30% – after exposure of Glare to a high humidity environment. Roebroeks found out that this effect is characteristic for *small* coupon specimens were the moisture can penetrate through the whole width. In contrast he found that experience with *large* glass fibre reinforced structures of the U.S. Navy, when examined after twenty years of service, showed that no degradation had taken place. Roebroeks reached the conclusion that in real structures only the edges of the laminate are affected and that periods of not only high humidity will occur but also dry periods in which the moisture is dried out of the structure. This indicated that the accelerated lab test in small pieces of material were not relevant.

PhD-student Richard Müller [7] extensively studied the behaviour of riveted joints in aluminium and Glare fuselages. This was very important work since those riveted joints are usually the critical locations where fatigue cracks start to develop. Müller established a database of test work on joints, which proved to be essential for the understanding of the behaviour of Glare joints. He showed that the rivet squeeze force is a very important parameter for the fatigue life of a riveted joint.

1.5 Cost estimation of Glare parts

A study of SLC and Aerospatiale on the A330 fuselage as a follow-up on the previously successful A320 study was carried out. The outcome of this A330 work was quite positive and indicated a 20% weight saving. In Germany, design studies were done on upper shell panels of the A340 fuselage that indicated a possible weight saving of 14-17%. It was important that Aerospatiale carried out the first serious cost study on the application of Glare in the A330 aircraft. As a material, Glare was 5 to 10 times more expensive per kilogram than a traditional aluminium alloy and this was an important hurdle to cross. Because Glare led to weight savings, less material is necessary but this will not close the gap between the aluminium and the Glare structure. The remainder has to be balanced by the reduced operating cost because of the reduction of fuel burn due to the lighter structure. However, cost reduction during manufacture became a prime issue. For the material to be viable, the cost of the Glare structure had to come very close to the aluminium structure. This is possible because the excellent fatigue properties of Glare can make an aircraft structure in Glare simpler than in aluminium. Reinforcements have to be added around joints in an aluminium structure in order to reduce the stress levels and prevent fatigue cracking, while this would not be necessary in Glare. Also, the crack stoppers that are necessary in aluminium because significant fatigue cracks may occur and adequate residual strength has to be assured are not necessary in Glare.

In Glare, the reduction of parts reduced the production costs of the structure. At the same time the use of the so-called ‘splices’, which were invented in that time to produce large panels, was favourable to Glare, as larger panels imply less expensive joints. Because the maximum sheet width of the aluminium was 2.5 metres the Glare sheets had to originally be connected by riveted joints every 2.5 metres. Splicing meant that the sheets would be internally connected in the Glare by bridging fibre layers and by adding extra aluminium sheets. The limit of the size was now the size of the autoclave, but they were huge. However, all these advantages were never quantified and the cost of the Glare material was still considered by many Airbus engineers to be the main obstacle to its application. The extensive cost study that was performed indicated that although the material costs of Glare are high, the total, finished cost of a Glare product comes very close to an aluminium product. This definitely proved that Glare should not be produced as sheet material which has to be shaped and machined into a product as is done for aluminium, but that Glare should be produced as a component. It must be laid up and cured in a curved mould such that after cure a product comes out of the autoclave with the right shape for the specific aircraft, including the right local fibre orientations and reinforcements which are necessary. In this way, the number of production steps could be reduced significantly.

Figure 1.6 Glare stiffeners in cargo bay area in F28.

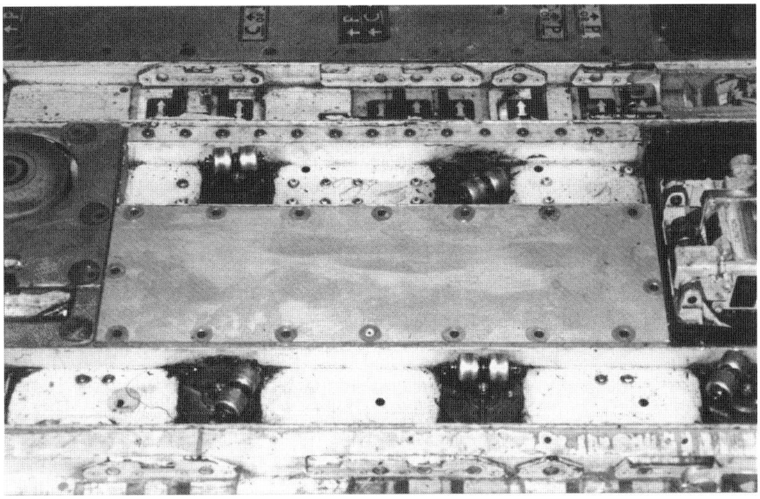

Figure 1.7 *Glare floor panel in DC-10.*

Another important milestone was the first regular application of Glare in a primary structure of an aircraft. In 1996 Shorts of Northern Ireland, one of the companies that made up Bombardier Aerospace, decided to use Glare for the forward bulkhead of the Learjet 45 business-jet. Although this was a small structure, it certainly was an important one. Another flying Glare structure accomplished in this period was a couple of Glare stringers installed by the Indonesian maintenance engineer Soerjanto [8]. He had been sent to Delft by his company for his PhD-study on the maintenance aspects of Glare. The stiffeners had to replace corroded aluminium stiffeners in the cargo bay area of an aeroplane used to transport seafood (which was frequently dripping salt water onto the structure), see Figure 1.6. Since Glare is layered, only the outer aluminium will corrode, after which the corrosion is stopped by the prepreg layer that functions as a corrosion barrier. Soerjanto also used a DC-10 as a test setup for his research by installing a Glare floor panel near the door of a cargo bay at a location that is prone to impact damage, see Figure 1.7.

1.6 GTO and A3XX

A completely new and ambitious very-large aircraft was considered by Airbus which required all the available technological knowledge and capabilities to be possible. This speeded up the development of Glare enormously. Just before Christmas 1996 a meeting was held in Toulouse to discuss what was still necessary for Glare to be applied as the material in this aircraft. The result was a document in which more than a hundred projects were carefully defined in a highly structured way in terms of technical content and required man-hours. These were necessary for Glare to achieve maturity.

The A3XX also meant an impulse for Delft for defining projects for students. This was very successful. Many students began their thesis work in these days and continued to work on Glare after finishing their master thesis and are authors in this book.

The work on the technology readiness of Glare was possible because of extensive financial support by the Dutch Ministry for Economic Affairs through the Netherlands Agency for Aerospace Programmes NIVR. This project was not specifically for the A3XX but aimed at the technology readiness of Glare in general. In July 1997 the project to establish technology readiness of the material was started in The Netherlands. The NIVR controlled the project, which was baptised 'GTO' (Glare Technologie Ontwikkeling, Glare Technology Development) and SLC, NLR and Delft University would carry out the required sub-projects. The roughly hundred sub-projects were divided over seven working groups: 1. Material and qualification, to determine material properties and carry out qualification of the material, 2. Methods, to generate and verify analytical methods to predict the mechanical properties of Glare structures, 3. Design concepts, to define concepts of what a Glare structure should look like, 4. Manufacturing, to study the manufacturing possibilities of Glare products on an industrial scale, 5. Specialities, for special properties such as burn-though resistance, 6. Spin-off, which was required by the Dutch government to generate applications of the material other than in aircraft, and 7. Maintenance, to study possible damage, inspection and repair methods.

The co-operation between the National Aerospace Laboratory NLR, SLC, TU Delft and NIVR in GTO was fruitful. The same is true for co-operation between these Dutch partners with the Airbus partners and especially EADS in Germany and the Large Aircraft Division of Airbus Industrie. After long negotiations, SLI was bought by Stork in December 1998 and joined with Fokker Aerostructures. Fokker Aerostructures became strongly involved in GTO and performed a lot of work on the design and manufacture of Glare panels on an industrial scale. This company will deliver Glare panels for the Airbus A380. GTO participants wrote most of the contents of this book and it reflects the current technology-readiness of Glare.

References

[1] A. Vlot, *Glare – History of the Development of a New Aircraft Material*, Dordrecht: Kluwer, 2001.

[2] R. Marissen, *Fatigue crack growth in Arall a hybrid aluminium-aramid composite material*, dissertation Delft University of Technology, 1988.

[3] G. Roebroeks, *Towards Glare – The development of a fatigue insensitive and damage tolerant aircraft material*, dissertation Delft University of Technology, 1991.

[4] A. Vlot, *Low-velocity impact loading on fibre reinforced aluminium laminates (ARALL) and other aircraft sheet materials,* dissertation Delft University of Technology, 1991.

[5] D. Chen, *Bulging of fatigue cracks in a pressurized aircraft fuselage*, dissertation Delft University of Technology, 1991.

[6] R.S. Fredell, *Damage Tolerant Repair techniques for Pressrized Aircraft Fuselages*, dissertation Delft University of Technology, 1994.

[7] R.P.G. Müller, *An Experimental and Analytical Investigation on the Fatigue Behaviour of Fuselage Riveted Lap Joints*, dissertation Delft University of Technology, 1996.

[8] Soerjanto Tjahjono, *Consequences and Challenges of Glare for Structural Repair and Newly Designed Fuselage Structure*, dissertation Delft University of Technology, 1998.

2

Glare features

G.H.J.J. Roebroeks

Abstract - This chapter gives a definition of Glare as a material-type. Following discussion of the various Glare grades and special material features, a step is made to the value of this material for use in aircraft. Several established yet still unexplored fields of application are mentioned and briefly discussed.

2.1 What is Glare? Constituents, codes and standardisation

When Glare is mentioned, it is often forgotten that this material is a member of the family of Fibre Metal Laminates; laminated materials consisting of thin layers of metal sheet and unidirectional fibre layers embedded in an adhesive system. Glare was optimised for aircraft fuselage skins at a moment in time that the first member of the FML family (Arall) appeared not to be able to cover any aircraft application. The occurrence of failure of the aramid fibres in some (fuselage) loading conditions, as well as their low strength preventing cross-ply applications from being competitive with the strength properties of monolithic aluminium initiated the development of a second FML. After some reluctance within the TU Delft team working on Arall, initially not fully accepting that Glare could be a better candidate for some aircraft applications, the idea of a family of Fibre Metal Laminates was initiated: a set of laminated materials combining the advantages (and obviously also the disadvantages, see below) of metal (aluminium alloys, steel, titanium, etc.) and composites (based on aramid fibres, high strength glass, carbon, PE, M5, etc.), which can provide benefits over conventional materials for a wide area of applications inside and outside the aircraft industry.

Today Glare has six different standard grades. They are all based on unidirectional S-glass fibres embedded with FM 94 adhesive resulting in a 0.127 mm thick prepreg with a nominal fibre volume fraction of 59%. This prepreg is laid-up in different orientations in between the aluminium alloy sheet, resulting in the different standard Glare grades. Table 2.1 shows these grades, including the most important material advantages.

For Glare 1, Glare 2, Glare 4 and Glare 5 the prepregs in each fibre layer are stacked symmetrically. This is not the case for the cross-ply prepreg layers in Glare 3 and Glare 6, which therefore need further definition. In Glare 3 the prepreg of a cross-ply fibre layer closest to the nearest outer aluminium layer of the laminate is in the rolling direction of the aluminium (0°). In Glare 3 laminates with an uneven amount of fibre layers, the first prepreg in the centre fibre layer laid-up during production is in aluminium rolling direction (0°). For Glare 6 the situation is comparable; replace 0° in the definition for Glare 3 by +45° for Glare 6A and replace 0° in the definition for Glare 3 by -45° for Glare 6B (+45° is the angle between 0° and 90°). A laminate coding system is used to comprehensively define laminates from the table above. An example of this is:

Glare 4B-4/3-0.4

Defined as:

- a Glare laminate with fibre orientation according to the Glare 4B definition in the table above,
- 4 layers of aluminium and 3 fibre layers
- each aluminium layer is 0.4 mm thick

Table 2.1 Standard Glare grades.

Glare grade	sub	Metal sheet thickness [mm] & alloy	Prepreg orientation* in each fibre layer**	Main beneficial characteristics
Glare1	-	0.3-0.4 7475-T761	0/0	fatigue, strength, yield stress
Glare 2	Glare 2A	0.2–0.5 2024-T3	0/0	fatigue, strength
	Glare 2B	0.2-0.5 2024-T3	90/90	fatigue, strength
Glare 3	-	0.2-0.5 2024-T3	0/90	fatigue, impact
Glare 4	Glare 4A	0.2-0.5 2024-T3	0/90/0	fatigue, strength in 0° direction
	Glare 4B	0.2-0.5 2024-T3	90/0/90	fatigue, strength in 90° direction
Glare 5	-	0.2-0.5 2024-T3	0/90/90/0	impact
Glare 6	Glare 6A	0.2-0.5 2024-T3	+45/-45	shear, off-axis properties
	Glare 6B	0.2-0.5 2024-T3	-45/+45	shear, off-axis properties

* All aluminium rolling directions in standard laminates are in the same orientation; the rolling direction is defined as 0°, the transverse rolling direction is defined as 90°.

** The number of orientations in this column is equal to the number of prepregs (each nominally 0.127 mm thick) in each fibre layer.

In some instances a different aluminium layer thickness is used in one laminate. In that case the individual layer thickness is fully written down in the order of aluminium layup:

Glare 4B-4/3-0.4/0.3/0.3/0.4

The centre two metal layers are 0.3 mm thick and the outer two are 0.4 mm thick.

The continuous S-glass fibres in Glare are approximately 10 μm thick, have a strength of 4000 MPa, a stiffness of 88 GPa and a strain at failure of 4.45%. The mechanical properties of the adhesive system in which the fibres are embedded are negligible (strength ± 50 MPa, stiffness ± 1.7 GPa, strain to failure 5–10% depending on strain rate). However, the adhesive system very significantly contributes to the performance of the laminate. It determines the bond-strength between the fibre layers and metal layers. These metal layers have pre-treated surfaces (Chromic Acid Anodised or Phosphoric Acid Anodised and primed with BR-127 corrosion inhibiting bond primer), bonded to the adhesive which is also used to impregnate the glass fibres. In many failure modes the cohesive failure of the adhesive nearby the metal sheet / fibre layer interface, is one of the governing parameters. The adhesion between the FM 94 system and pre-treated metal surface and between FM 94 and S-glass fibres is so high that these bond lines often remain intact until cohesive adhesive failure occurs. In order to guarantee this behaviour, and with that guarantee a wide range of mechanical performances, the quality of the bond strength has to be assured. The recommended method and procedures for this are presented in chapter 24.

Figure 2.1 *The manufacture of a flat sheet of Glare.*

2.2 Special Glare features

Glare was initially manufactured as flat sheets, see Figure 2.1. It was believed that the aircraft manufacturer would use these flat sheets in order to manufacture his shells by applying the curvature, thickness steps and joints himself, using conventional methods developed for metal structures (forming, bonding, riveting, etc.). Several studies showed the benefits in performance and weight of these Glare shells but they also indicated the high cost of these parts in comparison with conventional aluminium structures.

When one of the main potential customers of Glare indicated that Glare could be a valuable material for some smaller problem areas only because of the high material cost, a series of developments started, changing the Glare design completely. Through this, the large disadvantage of Glare (high cost of Glare sheets compared to conventional aircraft materials) has turned into one of its main advantages (moderate cost of Glare parts compared to modern aircraft structures). The main development steps are indicated in Figure 2.2, which shows a significant effort on manufacturing development in order to reduce cost. The main technology steps shown in the figure are discussed below.

2.2.1 The splicing concept

In the years 1990–1995 Glare laminates were produced only as flat sheets. It was clear from the beginning that the size of Glare sheet was not limited by any of the material production steps. However, the dimensions of metal sheet with the required 0.3–0.4 mm thickness was limited to a width of 1.65 metres, while fuselage skin sheet sizes up to 2.0 metres or wider were expected to be required.

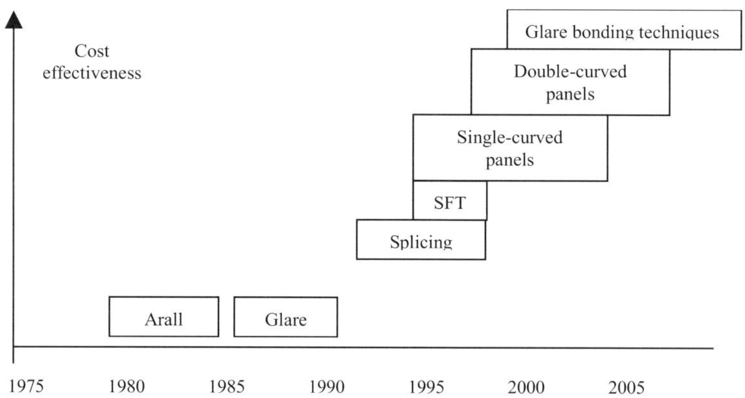

Figure 2.2 The main stages during Glare's development.

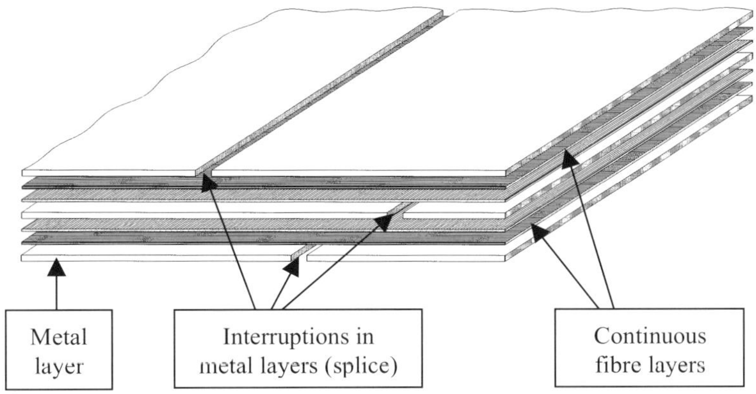

Figure 2.3 *A schematic representation of the splicing technique.*

This was the inspiration for a concept where aluminium sheets would be positioned side by side with a narrow gap in between (1 mm, for example), in order to obtain 3 or even 4.5 metre wide Glare panels. The metal sheet gaps are positioned slightly staggered throughout the thickness of the laminate to prevent too much strength loss in a specific cross-section. The fibre layers between the metal sheets simply cross the gaps, providing load transfer. This method for manufacturing very wide Glare sheets is referred to as the Glare splicing concept and is shown in Figure 2.3.

It was found that the splicing concept described above resulted in sudden delamination, especially of the outer aluminium layers of the laminate when the stress transverse to the splice direction exceeded approximately 400 MPa. (Glare) doublers were therefore bonded over the splice area in a second autoclave cycle, locally reducing the stress in this area to resist high loads without material damage. This concept was applied in one of the full size test panels manufactured at the time, see Figure 2.4. A clear disadvantage at this stage was that spliced panels had to be manufactured in at least two cure cycles; one to manufacture the laminate and one to bond the doubler over the splice area.

2.2.2 The Self-Forming Technique

Because of the difficulties of manufacturing spliced Glare in two bonding cycles, a solution was found allowing for the manufacture of a complete splice including doublers in only one cure cycle. The autoclave pressure forms the laminate over the external and internal doublers because the stiffness of the package of thin aluminium layers and still uncured adhesive and fibre layers is low. This is called the 'Self-Forming Technique' (SFT).

Figure 2.4 *The first full-size, spliced Glare panel.*

Additional adhesive of the same type as that used to impregnate the glass fibres in the prepregs is also added at certain locations. This is to adhere interrupted metal sheets to each other, to adhere thin aluminium internal or external doublers to the aluminium layers of the laminate and to fill gaps in the laminate which would otherwise remain unfilled.

With this method, the initial solution to reduce the material stress below the stress at which delamination could occur in the splice area is changed. Now load transfer over the interrupted metal sheet is mainly provided by thin additional aluminium sheets, adhered to the interrupted (spliced) aluminium sheet with adhesive. The Self-Forming Technique also allows the inclusion of internal doublers of metal and/or fibres to locally optimise the material properties and thickness to the panel loads. Figure 2.5 shows some schematic drawings of configurations using this technique. Figure 2.6 is an example of a Glare aircraft panel manufactured using self-forming and incorporating an internal doubler within the splice.

The additional adhesive in the laminate in the configurations above has much higher shear strength when compared to the prepreg layers. This means that the splices manufactured using the SFT, where adhesive is added at those locations where load transfer from one metal layer to another occurs, are no longer critical regarding delamination during static or fatigue loading. The splice is no longer the weakest link in the panel strength. The design values relevant for material sizing have also become valid for the splice area in the same material.

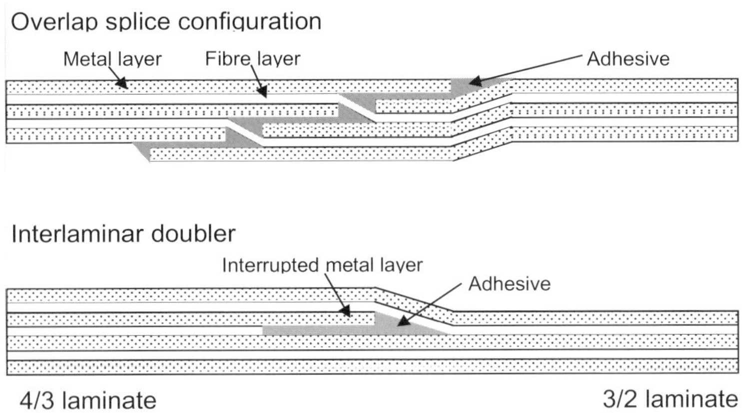

Figure 2.5 *Various self-forming configurations.*

Figure 2.6 *Glare aircraft panel manufactured using self-forming and incorporating an internal doubler within the splice.*

2.2.3 Single-curved panels

Aluminium skin panels are generally roll-formed to the required curvature. This additional production step can be avoided for Glare through layup of the thin metal sheet and prepreg layers in a single curved mould. It has been shown that this does not add to the cost of Glare production beyond the cost of the layup mould.

2.2.4 Double-curved panels

It also proved possible to form flat aluminium sheets to a double curvature under the pressure of the autoclave. This is used to laminate all metal layers and prepreg of a Glare sheet in a double curved mould and form it to the required contour in a single autoclave cure cycle. Although the aluminium sheet width is reduced using this method (depending on the curvature of the panel), it again eliminates the need for a difficult (in the case of Glare) and costly (stretch) forming operation.

2.2.5 Glare bonding techniques

Generally speaking, the bonding of Glare is similar to aluminium bonding techniques. However, Glare allows the option of tooling simplification. Although the procedures for this are still under development, it has already been shown that laminates are not necessarily manufactured to the same contour as that which is required, for instance a doubler on another panel. When a finished Glare laminate with a certain contour is heated up to the cure temperature in a second bond cycle, it can be somewhat formed to another contour under the pressure of the autoclave (3 bars = 30,000 kg/m^2). Some shear deformation of the fibre/adhesive layers at the higher temperature, most probably also causing some adhesive-molecule stretching, delivers similar deformation potential as autoclave age forming does for metals. This increases the efficiency of double curved doubler production.

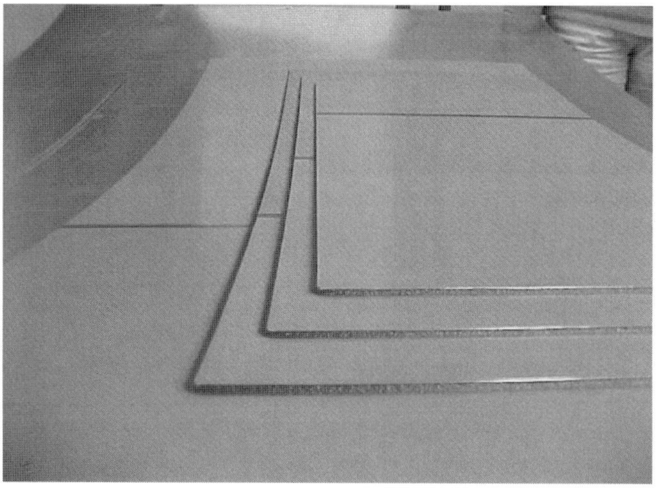

Figure 2.7 *Autoclave forming of Glare.*

It is no longer necessary to laminate and cure the doubler on a tool with exactly the same (gradually changing) contour as it will have when bonded to the skin. Because of this, large doubler panels can be produced with slightly different but constant curvature in the two main directions. From these panels, doublers can be machined and formed to the final curvature in the second cure cycle when the panels are bonded to the Glare skin. These techniques are not possible for aluminium since it will create relatively large internal stress in the sheet material.

Another feature specific to Glare is the ability to bond a skin and doubler, both with thickness steps in it, together in a second bond cycle without adjusting the surface contours to each other. Again, the adhesive in the laminates will become slightly rubbery at the cure temperature, allowing sufficient material deformation to adjust the contours of skin and doublers to each other. An example is shown in Figure 2.7. A skin and three doublers, all containing a splice (0.5 mm thickness step) and all manufactured on a smooth single curved bond tool, are laminated over each other such that the splices in the four laminates are adjacent to each other. In the final product, the inside of the doubler package in the panel (which generally shows a thickness step of the splice) has been formed to a flat surface in the second cure cycle. In that second cycle the four laminates are bonded to a thick package (approximately 10 mm) while "draping" the spliced areas over each other.

2.2.6 Other features

The list of possible features for Glare will grow in the future with new items being added as standard material solutions appear insufficient at a certain point. Some of these options have already been identified:

- Glare is often designed using the blunt notch strength as one of the main criteria. It results in solutions where the material thickness under the stringers and frames governs the thickness of the skin. In aluminium, similar reasons result in pocketed skin; in between frames and stringers the skin thickness is reduced using chemical milling. In Glare this could be applied if the thickness of (one of) the outer aluminium layers is enough to mill it to smaller thickness and obtain the required result. However, a larger thickness of one of the outer aluminium skins of the laminate significantly reduces the fatigue performance of a riveted lap joint. This can be solved by placing a pocketed aluminium sheet at the centre of the laminate. Figure 2.8 shows a Glare 3-5/4-0.4 laminate in which the centre 0.4 mm aluminium sheet is replaced by a 1 mm thick sheet that is reduced to 0.2 mm at three positions. The SFT was obviously used during production, filling the circumferences of the pockets with additional adhesive to prevent local porosity in the fibre layers. It results in a pocketed Glare laminate without the disadvantage of one thick aluminium layer at the outside of the sheet.

Figure 2.8 *A pocketed Glare panel.*

- Other laminate constituent combinations will add new members to the FML family, resulting in new applications of the material. The obtainable property changes may be relatively small, as is the case when a need for somewhat higher yield properties results in the use of aluminium 2024-T81 or 7475-T73 instead of 2024-T3. Complete new members may also be added to the FML family, using Titanium, carbon fibres and/or thermoplastic resins.

2.3 Characteristic properties

It is often claimed for an FML that the material combines the advantages of its constituents (aluminium alloy and composite), not the disadvantages. Although this is not true as a general rule, for some properties it can indeed be found that the properties of Glare are significantly better than would initially be expected when considering the combined properties of the aluminium alloy and composite layer:

- The impact properties of several Glare grades with biaxial fibre layers are better than those of aluminium, while the impact behaviour of glass fibre composites is significantly below that of aluminium. This is combined with the further improved impact behaviour of Glare when high strain rates are applied (hail strike, bird impact). Glare also has a dent in the surface after impact, similar to aluminium (unlike composites), resulting in simple visual inspection for such damage.
- It was initially believed that the FML build up would result in beneficial corrosion behaviour since the fibre layers would inhibit through-the-thickness corrosion. However, it was also found that the corrosion behaviour of the thin 2024-T3 alloy sheet was significantly better compared to that of thicker sheet often used in aircraft. The much faster quench of the thin sheet after rolling results in less alloy elements at the crystal boundaries of the material, resulting in improved corrosion behaviour.

- The damage tolerance of Glare is better when compared to aluminium and composites considering a wide range of possible damage. The strength of Glare structures remains high for damage conditions that cause significant reductions for aluminium alloys or composites. Fatigue damage in many adjacent rivet holes (MSD) causes significant strength loss for 2024-T3 while for Glare the reduction in strength is only small. The effect on (compressive) strength of impacted composite structures can be large while for Glare the typical insensitive aluminium behaviour is found.
- The fire resistance of aluminium is poor. In up to 60 seconds the thickest aluminium skins burn away in a kerosene fire. Glass composites have much better behaviour. Although the matrix of the composite melts or burns off, the (woven) glass fibres stay intact at 1000 °C to 1100 °C, preventing the fire passing this barrier. Glare has shown to have an additional benefit in firewall testing. The heat of the fire makes thicker laminates delaminate, creating a large improvement in insulation. The metal layers behind the first cross ply fibre layer do not melt away and the temperature at the inside of such thicker laminates remain at acceptable levels. The intact metal layers also provide some coherence of the structure, preventing deterioration of the frame / stringer structure and subsequent collapse of interior parts. This also shows an example where the behaviour of Glare is better than could be expected based on the properties of its constituents.

For other properties, Glare just combines the behaviour – including disadvantages – of aluminium alloy sheet and glass composites.

- It has the fatigue behaviour of an aluminium alloy, although the fibre layers of the material significantly improve this behaviour.
- Glare is affected by hot-wet ageing under accelerated conditions, as with composites. However, moisture absorption (through the milled edges only) and property reductions are less when compared to glass composites due to the large amount of unaffected aluminium in the material.

Other typical material properties are:

- The specific weight of Glare is approximately 10% lower than that of aluminium.
- Glare is formable to many single and double curved shapes.
- Glare tensile strength is significantly higher than 1.5 × yield stress.
- Glare blunt notch strength is often one of the governing design values.
- Glare can be machined using similar tools (hard metal steel) and procedures as for aluminium alloys.
- Glare can be repaired using aluminium type repair configurations and procedures.

The above properties are extensively discussed in this book, as are several of the various applications that are possible with this material, see for example Figure 2.9. Some of the applications are briefly mentioned in the next section.

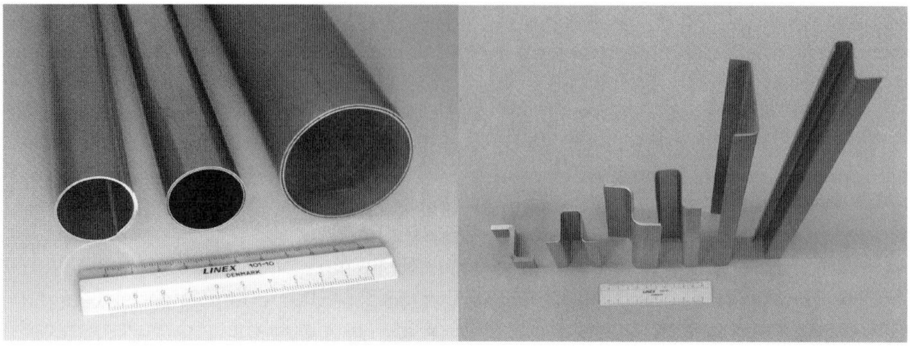

Figure 2.9 *Glare can be formed and manufactured in a wide variety of shapes.*

2.4 Possible applications

Only a few of the applications mentioned in this section have been fully evaluated and can be precisely judged at this stage. Nevertheless, the aforementioned list of typical Glare properties allows us to list those applications that seem possible using Glare.

2.4.1 Fuselage skin

Glare has been optimised for fuselage skin applications. The fatigue properties of the material and its damage tolerance make it especially suitable for this part of the primary structure.

2.4.2 Upper and lower wing skins

The material properties of Glare, such as fatigue properties, strength and damage tolerance, combined with the latest developments in the manufacturing of large double-curved panels make it the ideal material for aircraft wing skins. Significant cost savings are expected in the area of manufacturing, especially for these applications. Tailoring the material to the load conditions of the wing will result in new material combinations including

- "tapered laminates" (material with a centre tapered aluminium sheet from say 2 mm thick at the wing centre section to approximately 0.4 mm thick at the wing tip).
- 7000 series aluminium alloys.
- the Self-Forming Technique in order to splice the skin sheets manufacture ply-drop-offs and the slight double curvature.
- autoclave forming to locally reinforce the laminate with Glare doublers at limited additional tooling cost.

2.4.3 Stringers and frames

The extremely high strength and fatigue properties of the unidirectional Glare grades in particular allows the application of these materials in structural parts with mainly one load direction. A better tailoring of allowable stress in (Glare) skin and attached profiles should allow further weight savings. Obviously, some challenges to reduce manufacturing cost of series production have to be further evaluated.

2.4.4 Floors in passenger and cargo areas

Those parts of the aircraft interior that come into contact with humans prove to be sensitive to impact. Cargo handling often damages low weight composite floors, causing high repair costs. The impact properties of Glare are significantly better than those of composites. Glare type 5 in particular, which was optimised for impact resistance, outperforms other available solutions such as aluminium alloys. For this reason a growing number of aircraft operators fly with Glare floor material, providing much positive in-flight experience.

The impact resistance of Glare improves better than that of composites or aluminium as the strain rate increases. This results in extremely good bird-impact behaviour. For this reason, Glare was applied in the forward bulkhead of a small aircraft. Glare seems to be able to provide advantages in all applications potentially exposed to (high-speed) impact, however. Applications such as the leading edge of tail planes and cockpit roof require bird-impact resistance. Use of Glare as the lower flap skin, see Figure 2.10, on wheel doors or as belly fairings, make use of the material's impact resistance and reparability.

Figure 2.10 Arall C-130 lower flap skin inspection after flight evaluation.

Aircraft doors also frequently suffer from impact damage to the skin material. The use of Glare in this application will improve this property considerably. The large rear cargo door of the C-17 has been manufactured out of Arall for a number of years (this door also suffers from impact during cargo drops using the parachute extraction system). Today, the manufacturing techniques developed for Glare fuselage skins make the application cost competitive with the present aluminium structures, when we also include the improved material performance of Glare compared to Arall these applications seem even more suitable for FML.

2.4.5 Firewalls

Several Glare grades have shown fire resistance in a number of qualification tests. This is combined with the significantly better impact properties in comparison with the glass composite firewalls, its superior bearing properties (fastening the panels to the aircraft structure) and its formability. Some firewall applications are under flight evaluation.

2.4.6 Bulkheads

Glare can be beneficially applied in flat or double-curved pressure bulkheads. The combination of fatigue, strength and damage tolerance properties make it as attractive for this application as for fuselage skins

2.4.7 Cargo barriers

Cargo barriers are heavy structures separating the cargo from the pilot area. When crash conditions occur, the cargo should be prevented from crushing the cockpit. These cargo barriers are designed as stiffened sheets attached to the fuselage skin stringers and frames. The typical advantage of Glare under impact conditions is that after material yielding, the glass fibres in the yielded area can still continue to carry loads even if these loads increase further (bilinear stress-strain relation with considerable stiffness and additional elongation in the second part of the curve). This makes the material attractive for applications such as cargo barriers. This stress-strain behaviour typical of Glare, is one of the main reasons for the material's superior impact performance. It also provides the bomb resistance of cargo containers manufactured out of Glare.

2.4.8 Real-life applications

So far, only the use of Glare in fuselage skins, floors and forward bulkheads has resulted in series applications. It appears that still only limited experience with Glare has been generated. The conservatism in the aircraft industry requires a long path of analysis and evaluation before an application for a new material such as Glare is realised.

None of the above applications is "designed for Glare". Each has other solutions using different materials. In order to realise new applications for Glare; four main conditions have to be fulfilled:

- The manufacturer must have confidence in the material concept, combining thin metal sheet, glass composite and metal bonding technology. This confidence can only grow through an increasing number of Glare applications providing more relevant (than laboratory testing) experience.
- Glare must provide clear benefits over other materials (properties) against acceptable costs (or cost savings).
- The material knowledge required in order to cost-effectively realise the application must be available for all aspects (properties, qualification, calculation methods, design principles, manufacturing, Quality Assurance techniques, and maintenance).
- The material must be sufficiently available, filling the needs of the aircraft manufacturers and operators dealing with costly aircraft development and manufacturing schedules and costly downtime of damaged aircraft taken temporarily out of service for repair.

All this has been sufficiently accomplished for the existing applications, providing many disappointments, moments of doubt, difficult discussions and "challenges" for manufacturing. Each time a new application is developed, part of this will be experienced again so long as Glare remains a "small" material between the giants: aluminium alloys and full composites. During the development of Glare applications for a wide range of structures, the establishment of a network of qualified manufacturers to provide the material (parts) will require the largest effort, rather than the development of the material details (type of aluminium, fibres, adhesive, splicing, etc.) themselves. On the other hand, without the brain-waves and research modifying an existing material into a better solution for a certain application (see the development from flat FML sheet to complex Glare parts as presented above) production facilities for FML applications will not be needed. For this reason we must treasure both our research teams at the Delft University of Technology, headed for many years with great spirit, enthusiasm and workmanship by Boud Vogelesang and the early FML manufacturers, Fokker Aerostructures, Aviation Equipment Inc and Structural Laminates Company.

3

Next Generation Fibre Metal Laminates

T. de Boer

Abstract – As the name Fibre Metal Laminates (FML) already suggests, they are a mixture of sheet material and fibres, bonded by an adhesive. Arall and Glare are the first and second generation FMLs respectively. However, other potential variants have been developed with success as well. In the early nineties a carbon titanium laminate was developed for operating temperatures up to 300 °C, while at this moment, a glass aluminium laminate is under development for elevated temperatures up to 180 °C. As well as developing other laminates, different manufacturing processes have also been investigated. For example, a special manufacturing process for seamless tubes was successfully developed.

3.1 Introduction

Glare has proven to be a promising family of structural materials for aerospace applications. Developed for aircraft structures as a fatigue insensitive material with a low density, Glare has proven to have a range of other promising properties. For example, the excellent fire resistance of Glare has generated ideas for using Fibre Metal Laminates in Thermal Protection Systems for space vehicles. The concept of Fibre Metal Laminates (FML) can easily be extended to other components. (For example, aluminium can be exchanged for more temperature resistant titanium and carbon fibres can easily replace glass fibres.) This gives us the opportunity to generate a large family of structural materials. By choosing the right components and laminate build-up, an FML can be tailored to any desired application.

Besides Glare and Arall, Delft University also investigated many other laminates. This chapter will deal with just some of these potential next generation laminates. In the first part of the chapter, a list of the components that could be used for FML applications is given. In the following sections various laminates will be discussed, such as aluminium/glass, titanium/glass and titanium/carbon laminates. Section 3.6 will discuss special manufacturing processes. The final section contains a summary.

3.2 Different components for FMLs

In theory, many components could be used and mixed freely. However, reality is more complicated. Extreme internal stress, galvanic corrosion, voids and volatiles are just some of the problems that might occur when the components are not tuned well. As well as technological difficulties, availability and costs are also important. A laminate will not be applied when the price is exceptionally high or components are too difficult to acquire. Before discussing some of the examined laminates, this section will give a global summary of the possible components.

Fibre types

The number of potential types of fibre is rather large. In the past, aramid fibres were used in Arall. Nowadays, glass fibres and carbon fibres are commonly applied. Due to the superior bonding properties of epoxy with glass fibres, this combination shows excellent mechanical properties and acceptable environmental durability. Carbon fibres have a relatively low density, a high Young's modulus of elasticity and a high strength, but in combination with aluminium can lead to galvanic corrosion. Carbon fibres combined with titanium results in a high stiffness laminate. Other fibres such as ceramics, synthetics, organics and silicon carbide fibres can be applied as well [1].

Sheet material

Metal sheets are normally applied as a cover material, although ceramics can also be used for ballistic protection material, for example. The most commonly used materials are aluminium and titanium, although more exotic metals such as stainless steel, high strength steel, or niobium can be applied as well.

Resin types

Good shear and toughness properties are required for the adhesive in FML applications. Other properties such as bonding properties with the fibres and sheet materials, temperature influences, sensitivity to moisture, flow properties of the adhesive and glass transition temperatures directly affect FML properties. There are many different adhesive types are available such as epoxy, bismaleide, phenolics, cyanate esters and thermoplastics (PEEK and PEI). So far, different epoxies and thermoplastic have been applied in FMLs.

3.3 Elevated temperature glass/aluminium laminates

Glare was developed for relatively cold structures. It therefore consists of a 120 °C curing epoxy, 2024-T3 aluminium and S2-glass fibres. For extended temperature usage, neither epoxy nor aluminium fulfils the requirements any more. By replacing the normal epoxy with a 177 °C curing epoxy and the aluminium with 2024-T81, a laminate is created which can be used up to 180 °C. The increasing demand for structural materials in this temperature range has also initiated the investigation of an elevated temperature laminate.

The goal of this FML-ET laminate is to cover the temperature area between the cold structures and hot structures. For example, it can be used for engine covers or fighter applications. As a result of aerodynamic heating, supersonic aviation is also one of the possible applications for this laminate.

3.3.1 Mechanical properties of an elevated temperature glass/aluminium laminate

Although this laminate is still under investigation, this section will discuss some of the most important properties. It gives an indication of its capabilities. The properties discussed in this section are preliminary results based on De Boer [2].

Tensile properties

Because of the fibres and 2024 aluminium, the stiffness and ultimate strength of this elevated temperature FML-ET is comparable to Glare. The yield stress is approximately 25% higher, which is caused by the temper of the aluminium. The influence of the elevated temperature is limited for this laminate. The ultimate strength drops by 10% while the yield strength decreases as much as 20% when heated to 150 °C. The reason for this can be found in the fact that both the adhesive and the aluminium lose some of their strength during heating.

Blunt notch properties

The elevated temperature FML has blunt notch properties comparable with those of Glare, i.e. a net blunt notch stress of 490 MPa instead of 500 MPa for Glare. The ratio of blunt notch to ultimate strength ratio $\sigma_{net}/\sigma_{TUS}$ is 70% at room temperature. At 150 °C the blunt notch properties drop by as little as 5%, while the ultimate tensile strength drops by about 10%. Because of this, the ratio $\sigma_{net}/\sigma_{TUS}$ at 150 °C passes 80%. This makes this elevated temperature FML less sensitive to holes at elevated temperatures.

Fatigue properties

Only constant amplitude fatigue was determined during this investigation. For aluminium alone, 2024-T8 has lower fatigue properties than 2024-T3. Therefore it was expected that the fatigue properties would not be as good as those of Glare. Figure 3.1 shows crack lengths using the elevated temperature FML-ET and Glare. At room temperature, the crack length of the elevated temperature variant is about 40% higher than that of Glare. With respect to the fatigue behaviour of aluminium alone, these results are satisfactory. The temperature effect is significant. Compared to room temperature, the crack length doubles after 150,000 cycles at 150 °C.

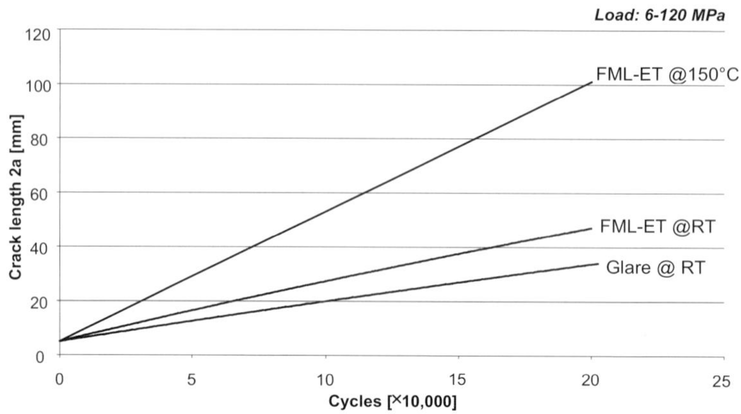

Figure 3.1 *Fatigue results of glass/aluminium laminates.*

3.3.2 Difficulties with the prepreg

Investigating new laminates always leads to difficulties and unforeseen problems. For this FML-ET laminate, the prepreg seemed to be the weak link in the chain. The epoxy used for this FML contained aluminium particles. Normally, these particles are applied to improve the properties at elevated temperature and to control the thickness. However, the aluminium particles in combination with fibres caused large quantities of voids in this FML. A cross-section of this laminate is shown in Figure 3.2. (Scanning Electron Microscope photo) As this figure reveals, many voids are present. Voids are undesirable since they are weak spots in a panel where the components are not bonded. An even more important reason is durability. Voids make it easier for moisture to diffuse into the panel, which affects the durability.

Even though the mechanical properties were satisfactory, the quality of this laminate was insufficient. Changing the pressure or adding an extra dwell time to the curing cycle did not improve the quality. Finally, in co-operation with the manufacturer of this prepreg, it was decided to modify the prepreg by omitting the aluminium particles. After this modification new laminates were cured with better results. No voids were found and the C-scan results were comparable to, or even better than Glare. A cross-section of this laminate is shown in Figure 3.3. As one can see, the quality of this laminate is a lot better than that shown in Figure 3.2.

3.3.3 Current status

Although the most important properties have been determined, many more properties still have to be investigated. The modification of the prepreg made new tests necessary, while other properties like impact, burn-through, sonic fatigue and exposure still have to be determined. When these investigations are finished at the end of 2001, a new Glare variant will be ready for consideration as regards application. The interest in aircraft materials with application temperatures up to 150 °C is also growing as a result of the increasing interest in supersonic (civil) aviation.

3.4 Carbon/aluminium laminates

Developments in carbon fibre production resulted in a large variety of fibres with many different mechanical properties. Strength and stiffness dominated structural applications can be covered in this way. The high stiffness of carbon fibres allows for extremely efficient crack bridging and therefore very low crack growth rates. At the same time, the presence of a metal layer is very favourable for the impact properties.

The combination of high stiffness and strength with good impact properties gives carbon/aluminium laminates a great advantage for space applications. Other applications for this laminate are impact absorbers for helicopter struts and aircraft seats. Because of the demand for laminated tubes, a special production process has been developed. The process of manufacturing seamless tubes is explained in section 3.6.1.

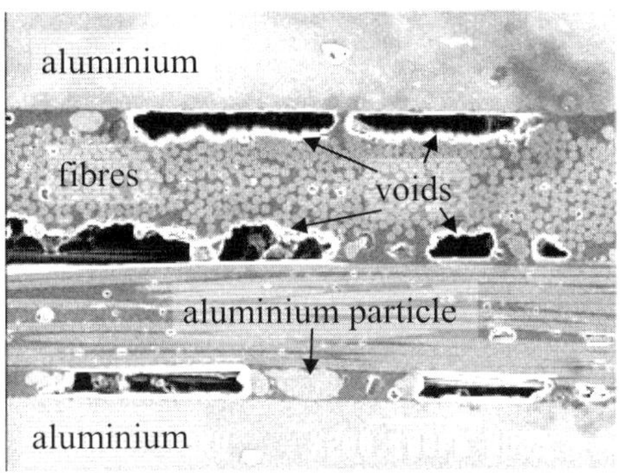

Figure 3.2 *Glass/aluminium laminate with voids.*

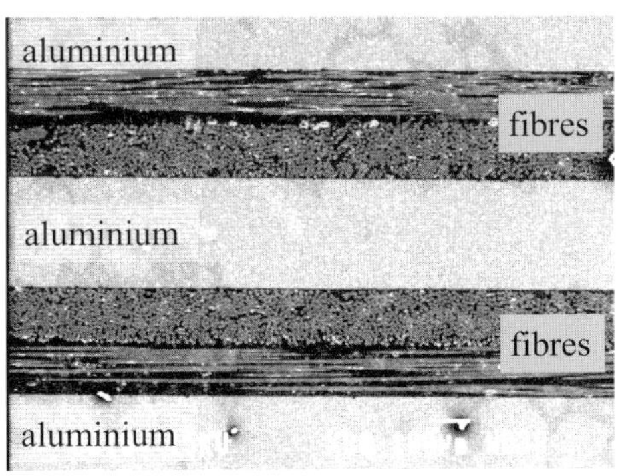

Figure 3.3 *Glass/aluminium laminate without voids.*

Various components were used for this investigation. Primarily 2024-T3 and 7075 aluminium alloys were tested. The list of tested carbon fibres is somewhat longer. HM35 Tenax, T300 and T800 Toray and FT700 Tonen are just an arbitrary example of the carbon fibres used, all with their specific properties. The use of aluminium and carbon in contact with each other is often thought of as a major galvanic corrosion problem. A simple but effective solution was found: placing proper isolation between the carbon fibres and the aluminium sheets. A more detailed explanation of this investigation can be found in section 3.4.2 [3].

3.4.1 Mechanical properties of a carbon aluminium laminate

Various carbon fibres and aluminium alloys have been tested, too numerous to discuss all the possible combinations here. In this section we shall only discuss the results of a laminate composed of T800 carbon fibres with 0.3 mm 2024-T3 aluminium (a 3/2 layup with a cross-ply). The adhesive used for this laminate is AF-163 manufactured by 3M. According to Vermeeren [4], only tests at room temperature were executed.

Table 3.1 Blunt notch results of a carbon/aluminium laminates.

Shape; dimension	K_t [-]	Net stress [MPa]	σ_{net}/σ_u [-]
Round hole; diameter 25 mm	2.74	505	0.69
Slot; length 25 mm, tip radius ρ=5 mm	3.32	469	0.64
Slot; length 25 mm, tip radius ρ=2 mm	4.74	456	0.63
Saw cut; length 25 mm	∞	406	0.65

2024-T3 0.3 mm, epoxy, T 800 Carbon (cross-ply); specimen width W=100 mm

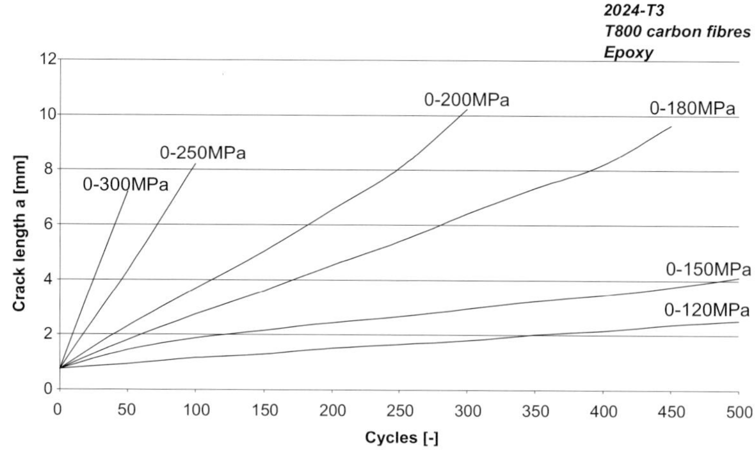

Figure 3.4 Fatigue results of Carbon Aluminium laminate.

Tensile properties

In this laminate, the carbon fibres contribute to a higher tensile ultimate strength and tensile modulus. The tensile modulus for these carbon aluminium laminates are therefore high (near 100 GPa), while a comparable Glare laminate has a modulus of about 55 GPa. For Glare, the ultimate strength is approximately 40% higher. For very strong and stiff carbon fibres, such as Tenax IM600 and Toray T800, the yield criterion is meaningless. The material cannot yield because of the low strain of the fibres. The ultra high strength fibres will exhibit elastic behaviour until failure at approximately 0.45% elastic strain.

Blunt notch properties

The tensile strength of Fibre Metal Laminates is dominated by the tensile strength of the applied fibre. The blunt notch properties are also highly dependent on this fibre strength. It is found that Glare can show static delamination during static failure, resulting in even higher blunt notch strength. This phenomenon is observed for laminates with high strength carbon fibres as well. The limited failure strain of the carbon fibre, when applied in a FML, was thought to be a disadvantage compared to both aramid and glass fibres. In general, it should be mentioned that a lower fibre failure strain would limit, to some extent, the stress carrying capabilities of the aluminium due to the limited strain hardening. This effect, however, is small for 2024-T3.

As far as static strength properties are concerned, the absolute strength of FML is dependent on the failure strength of the fibre and not on its strain. The higher strength of the carbon fibres results in a blunt notch strength that is even higher than for Glare. In a relative comparison, the blunt notch ratio is important. (Ratio of the blunt notch and the tensile strength: σ_{net}/σ_u.) The ratio of Glare is approximately 68% (for K_t=2.74), while the carbon aluminium laminate has slightly higher ratio, See Table 3.1.

Fatigue properties

The efficiency of crack bridging strongly depends on the stiffness of the crack bridging prepreg. For this reason, a strong stiffness of the fibres is desired. Carbon fibres have stiffness in the range of 250 to 700 GPa or more, which is three to eight times stiffer than glass fibres. So it is to be expected that good fatigue results will be achieved using carbon/aluminium laminates. In Figure 3.4, the crack lengths of constant amplitude fatigue test results are shown. Loads up to 300 MPa have been applied and it should be emphasised that loads of these magnitudes are in the order of 50% of the tensile strength of this material.

3.4.2 Galvanic corrosion

The use of aluminium and carbon in contact with each other is often thought of as a major galvanic corrosion problem. Isolating the aluminium from the carbon fibre eliminates this problem. Research has shown that coating the aluminium surface with a thermoplastic coating like PolyEtherImide (PEI) gives good isolation. Care should be taken when making notches in this material, because it enlarges the change of making contact again. Another possibility is isolating the carbon prepreg with glass prepreg on both sides. Though only present for isolation purposes, the glass prepreg contributes to the laminate strength but also reduces the stiffness of the laminate. For both solutions tests have showed that the mechanical properties of the isolated laminates differ from those of the non-isolated laminates [4].

3.4.3 Current status

By solving the galvanic corrosion problems, this laminate can be used freely. The high stiffness of the carbon fibres means that the tensile and fatigue properties are promising. In the early nineties, this laminate was developed for space applications. Other purposes for this laminate included tubular applications. A special manufacturing process was devised for producing seamless tubes. Detailed information of this special manufacturing process can be found in section 3.6.1.

3.5 Carbon/titanium laminates

In the early nineties, a large investigation was carried out into carbon/titanium laminates. An advantage of these laminates is their combination of high stiffness, high yield strength, good fatigue and good impact properties at both room and elevated temperatures. Titanium and carbon fibres can easily be exposed to up to 300 °C, which makes this combination perfect for high temperature applications.

The good mechanical and thermal properties, combined with low density, damage tolerance aspects and manufacturing possibilities means that these laminates can be considered for applications where metals or composites are used. Titanium laminates prove to be very attractive structural materials for several applications in space structures.

Purely commercial Titanium and Ti-6AL-4V were used for the investigations. Poly Ether Ether Ketone (PEEK), Epoxy and Phenolic Triazine (PT) are just some of the tested adhesives. PEEK can be used at temperatures up to 250 °C, while PT can even be exposed up to 315 °C [3].

3.5.1 Mechanical properties of a carbon titanium laminate

Many properties for even more exotic laminates have been determined through extensive investigation throughout the years. A discussion of all the properties is not appropriate here; therefore only the Ti-6AL-4V laminates will be reviewed.

Tensile properties

For a laminate made from Ti-6Al-4V, IM600 carbon fibres and PT adhesive, the tensile ultimate stress is almost twice that of Glare. Compared to Glare, the yield stress is a factor three higher. The stiffness is over 100 GPa, which is two times stiffer than Glare. This laminate was tested up to 315 °C, without losing much of its tensile property. Extrapolating from room temperature, at 315 °C, the yield strength and tensile strengths were both approximately 15 MPa lower than expected [5].

Blunt notch properties

For laminates made from Ti-6Al-4V and APC-2 prepreg (PEEK with AS4 low modulus carbon fibres) the blunt notch values are very satisfactory. For a K_t=2.39, the blunt notch strength is over 1000MPa, which is almost 90% of the ultimate tensile strength. Glare, on the other hand, has a value of 63%. This leads us to conclude that this material is relatively notch insensitive. Unfortunately no information is available about the blunt notch behaviour at elevated temperatures [6].

Fatigue properties

The fatigue properties for Carbon Titanium laminates show good results. For laminates made of commercial titanium, the crack growth rate is 40 times lower than that of monolithic titanium (S_{max} =150 MPa, R=0.05). For Ti-6Al-4V laminates, the crack growth rates are 10-25 times lower than for monolithic titanium, even for temperatures up to 280 °C (S_{max} =200 MPa, R=0.05).

The high modulus of elasticity of the fibres results in low crack growth rates. With titanium, the crack initiation period is longer than for aluminium. Furthermore, the fatigue properties at elevated temperatures are better for titanium than for aluminium [7].

3.5.2 Conclusions

When a high modulus of elasticity is desired, titanium should be combined with carbon fibres. This combination shows good properties at room temperatures and high temperatures up to 315 °C. Compared to Glare, it has a higher strength, stiffness and better fatigue properties. The density of this laminate is even lower than that of Glare, due to the lower density of the carbon fibres, which means that a greater weight reduction is possible.

3.6 Special manufacturing processes

The production of Glare and FML in general seems rather optimised. Normally metal sheets and pre-impregnated fibres are stacked, after which the package is cured in an autoclave, see Figure 3.5. Throughout the development of Fibre Metal Laminates, different manufacturing processes have been experimented with. In this section two different processes will be illustrated. Firstly, the production of seamless Fibre Metal Laminate tubes will be explained, after which we shall discuss Resin Transfer Moulding (RTM) in combination with Fibre Metal Laminates.

Figure 3.5 *Taking Glare panels out of the autoclave.*

3.6.1 Seamless Fibre Metal Laminate tubes

The purpose of this investigation was to produce seamless tubes consisting of aluminium inner and outer layers with a carbon fibre prepreg in between. The stiffness over weight ratio of the tubes can be dramatically improved as a result of the fibres. In comparison with steel tubes, a weight reduction of 15% is possible while retaining the original stiffness. Alternatively, a 40% higher stiffness can be achieved for tubes of the original weight. Obviously a large area of possibility lies in between, so that each tube can be tailored to meet the requirements of any application.

FML tubes were originally intended as impact absorbers for helicopter struts and aircraft seats, but are also of interest as construction elements in bicycle frames and space stations, as floor-struts in aircraft or even as pipe-lines in the chemical industries. In the latter application, the resistance to corrosion of Fibre Metal Laminates is of great benefit.

The manufacturing method is based on the differences of strain in two coaxial tubes when stretched radially. The internal stress this creates will cause the prepreg, in between the tubes, to exert pressure on the adhesive when cured in an oven. Consequently, the use of an expensive autoclave will not be necessary.

The main difficulty in producing seamless Fibre Metal Laminate tubes is to achieve a high enough radial pressure on the fibre/adhesive layer. Contrary to this is the requirement that the prepreg layer has to be inserted in between the tubes. For this a relatively large space in needed, since the fibres must be prevented from being damaged either through abrasion or buckling. This sets a limitation on the maximum pressure that can be exerted on the adhesive.

Production

The production process starts by inserting a prepreg layer in between two coaxial metal tubes. The inner radius of the outer tube is sufficiently larger than the outer radius of the inner tube, so that insertion is possible. After this both the tubes are expanded, so that the prepreg comes into contact with the metal surfaces. Investigation has proven that using a mandrel to stretch the two metal tubes with prepreg inserted is a simple and inexpensive way to produce good quality seamless Fibre Metal Laminate tubes. After this, the seamless tube is ready for curing in a normal oven [8]. Figure 3.6 shows three such seamless tubes.

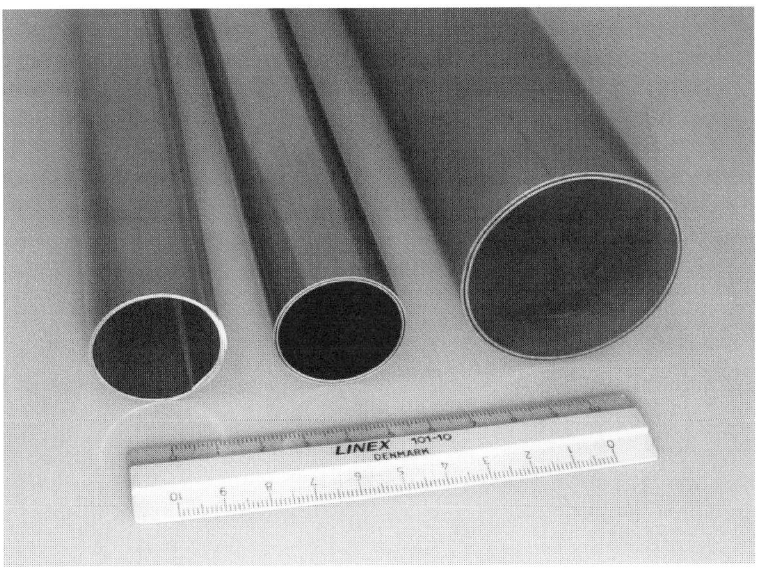

Figure 3.6 *Seamless tubes.*

3.6.2 Resin Transfer Moulding

Recent investigation has shown that manufacturing Fibre Metal Laminates by Resin Transfer Moulding (RTM) could also be a possibility. This manufacturing process is a family of closed-mould low-pressure processes that allow the fabrication of composites ranging in complexity from simple, low performance to complex high-performance articles and in size from small to very large. The common feature of all the resin transfer moulding processes is the flow of resin material through the un-wetted fibres. During the injection of the resin a pressure difference is applied in the closed mould, which forces the resin to flow through the reinforcement. Therefore the permeability of the reinforcement is an important factor. The process is typically used with low viscosity fast-curing resins, such as polyester, epoxy and chopped or continuous mat reinforcement at low fibre volume content. However, the process has been demonstrated with higher fibre volume contents (50 to 60%) [1].

Production

For Fibre Metal Laminates, metal sheets, which may be pre-shaped, and fibres are stacked and placed into a mould, which is then closed. A tube connects the closed tool cavity with a supply of liquid resin that is pumped or transferred into the tool to impregnate the reinforcement, which is subsequently cured. A stringer laminate has been produced with success using this method, see Figure 3.7. As the picture shows, sharp curves can be achieved.

3.7 Fibre Metal Laminates in the future

The freedom of mixing components and layup according to the requirements makes the FML concept competitive for many applications. For almost every requirement a matching laminate can be composed. Aircraft materials are normally expensive but by applying cheaper materials, the FML concept can also become attractive for non-aviation applications. Besides the normal manufacturing process, other manufacturing methods have been investigated and are still under investigation. Reasons for ongoing investigation include the need to fulfil special requirements, such as producing tubes, and to obtain cheaper products.

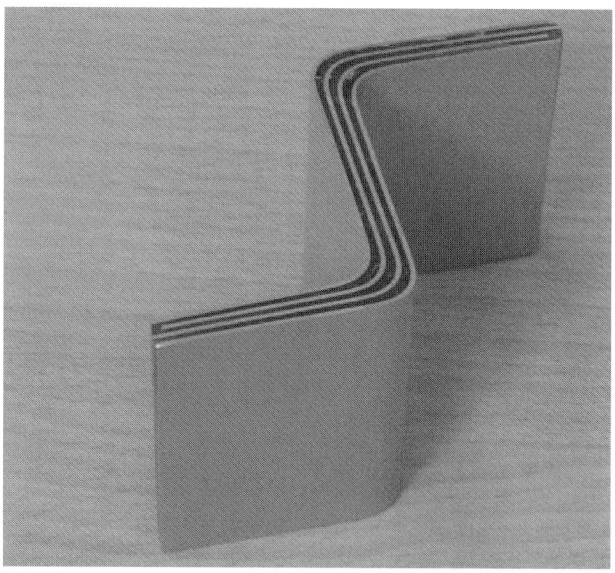

Figure 3.7 Glare stringer manufactured with RTM.

References

[1] M.J.L. van Tooren, J. Sinke, H.E.N. Bersee, *Composties: Materials, Structures & Manufacturing Processes*, Dictaat 32, Delft (NL), May 1993

[2] T. de Boer, *The development of an elevated temperature FML*. Master's thesis, Delft (NL), May 2000

[3] J. L.M. Mensink, *Thermal properties and possible space applications of Fibre Metal Laminates*, Delft (NL), April 1994

[4] C.A.J.R. Vermeeren, *The application of carbon fibres in Arall Laminates.* Report LR 658, Delft (NL), September 1991

[5] E.W. Medenblik, *Enhancement of Fibre Metal Laminates for use ate elevated temperatures*, Drexel University

[6] E.W. Medenblik, *Titanium fibre metal laminates*, Master's thesis, Delft (NL), January 1994

[7] M.G. de Koos, *Peek carbon fibre reinforced titanium laminates*, Delft (NL), October 1990

[8] C. Verseput, *The production of seamless Fibre Metal Laminate tubes.* Delft (NL), May 1994

4

Long-term behaviour

B. Borgonje and W. van der Hoeven

Abstract – Aircraft are operated in a wide variety of environments, for periods of up to 30 years. In order to maintain safe and economical travel, it is necessary to have a good knowledge of the long-term behaviour of the aircraft materials. For longer periods of time, one of the main threats to materials is the influence of the environment, especially that of moisture combined with temperature. The combination of materials used in Glare complicates the investigation of the influence of moisture. However, by investigating environmental influences on each of the different components of Glare, a good understanding of the effects on the laminate as a whole can be obtained. In most cases, the laminated build-up has a positive effect on the environmental durability of Glare, but care needs to be taken in areas of combined moisture absorption and stress concentrations.

4.1 Introduction

Aircraft operate in a wide variety of environments, ranging from cold, dry air conditions at cruise altitude to hot, humid air conditions in tropical environments. At the same time, commercial aircraft are used very intensively for periods up to 30 years. It is therefore important that the materials from which aircraft are built maintain their properties during their entire lifetime, even in the most extreme environments.

The main safety philosophy that is currently practised in aircraft design is the concept of 'damage tolerance'. A damage tolerant structure must be able to sustain regulatory service loads in the presence of damage, until the damage is detected by inspection. Three principal forms of damage must be evaluated: environmental deterioration, accidental damage and fatigue damage. The issue of environmental deterioration will be discussed in this chapter.

The main problem that has been observed in aluminium structures is that of fatigue. Besides this, corrosion is a problem, especially when aircraft operate in humid environments. Fortunately, sufficient measures have been taken to protect aluminium structures against environmental degradation during the years.

In the last few decades, composite materials or fibre-reinforced plastics have found more and more applications in the aircraft industry due to their high strength-to-weight ratio. It is well known that most fibre-reinforced plastics are sensitive to environmental influences, such as solar radiation and moisture. The absorption of moisture is especially undesirable because of the weight increase and its deteriorating effect on the mechanical properties of the composite.

Glare, being a combination of aluminium and fibre-reinforced plastic, is likely to show similar degradation effects as these two materials. However, the combination of the two materials in Glare does lead to some differences that often turn out to be beneficial to its resistance to environmental influences. The influence of the environment on each of the components of Glare will first be described separately in section 4.2. Environmental effects on the laminate as a whole will subsequently be described in section 4.3. In section 4.4 we explain how the influence of environment on Glare is investigated and several results of these investigations are given.

4.2 Environmental effects on the different components of Glare

In an outdoor environment, material degradation may take place due to several processes, of which ultraviolet (UV) radiation, thermal degradation, moisture absorption and exposure to chemicals are most common. With Glare, basically only the outer aluminium layers are exposed as a result of its layup. The prepreg layers are only exposed through free edges of the laminate and holes. UV radiation will therefore not be a problem, since it does not deteriorate the aluminium.
Moisture and chemicals however can penetrate the prepreg. Thermal degradation due to elevated temperature exposure is important, since if it occurs it will affect the entire laminate.
Unlike fluids (such as fuel and hydraulic oils) moisture is always present and can strongly affect material properties. Therefore this chapter will concentrate on the influence of moisture absorption on the properties of Glare. The laminate is 'split' into its three components: aluminium, glass fibres and epoxy. The effects of moisture on each of those components are described in the next sections. As the interfaces between these materials have their own characteristics, these will be described in a separate section.

4.2.1 Environmental effects on aluminium

The combination of moisture and metals is known to lead to corrosion. Pure aluminium is a very reactive metal, which quickly builds up an oxide layer on its surface. This layer protects the aluminium from degradation due to environmental exposure. In order to obtain higher strength and stiffness properties, aluminium is often alloyed with other elements, such as copper, magnesium, zinc, silicon, etc. The presence of these elements, particularly copper, can make the alloy much more sensitive to corrosion.

In order to eliminate or decrease the amount of corrosion, a protective layer can be applied onto the surface of the material. This can be done by anodisation, applying a clad layer of pure aluminium and/or a paint system (primer + top coat). These protection methods have proved to be very effective in decreasing the occurrence of corrosion.

The issue of aluminium corrosion and methods to prevent this are more extensively described in chapter 28.

4.2.2 Environmental effects on glass fibres

In Glare, S2-glass fibres are used, which are made of magnesium aluminosilicate glass and are especially designed for applications that require high strength and modulus, and stability under extreme temperature and corrosive conditions.

Glass by itself is chemically inert and therefore not likely to be subject to environmental degradation. It has been shown however, that exposure to acid solutions can lead to damage (cracks) at the surface of E-glass fibres, due to the presence of metal oxides [1]. As a result of local stress concentrations around these surface cracks, the strength of the fibres can severely decrease. It is not known to what extent this will also affect S2-fibres.

Temperature also influences the fibre strength. S2-fibres show a drop in strength of approximately 5% when temperature is raised from room temperature to 70 °C [2].

4.2.3 Environmental effects on epoxy

The influence of moisture is most notably present in the epoxy. Like all other polymers, epoxies absorb moisture when exposed to humid environments. This takes place through a diffusion process, in which water molecules are transported from areas of high concentration to areas of lower moisture concentration.

Most types of moisture diffusion can be described by Fick's law:

$$\frac{\partial c}{\partial t} = D_x \frac{\partial^2 c}{\partial x^2} + D_y \frac{\partial^2 c}{\partial y^2} + D_z \frac{\partial^2 c}{\partial z^2} \qquad (4.1)$$

with: c = free water molecule concentration
$D_{x,y,z}$ = diffusion coefficient in x, y and z-direction respectively

In the case of Fickian diffusion, the amount of moisture absorbed is linearly dependent on the square root of time at the initial stage of the diffusion process. When the material is exposed to a constant humidity environment, the process continues until the concentration is uniform throughout the material. This concentration level is known as the equilibrium moisture concentration. The diffusion coefficients and the equilibrium moisture concentration together describe the moisture diffusion behaviour of a material.

Both the diffusion coefficients and the equilibrium concentration level are dependent on several internal and external parameters. The main parameters influencing moisture absorption of a polymer are:

- *Temperature*
  The mobility of molecules increases with temperature. Therefore an increase in temperature will result in an increase of the diffusion coefficients and thus in an increase of the diffusion rate.

- *Relative humidity*
  For most materials, the relative humidity rate does not have an influence on the moisture diffusion rate. However, it does determine the equilibrium moisture content. The largest value of moisture content for a given material under humid conditions occurs at 100% relative humidity and is often called the saturation content.

- *Stress level*
  Diffusion coefficients increase when tensile stresses are applied to the material. However, tensile stresses did not affect the value of equilibrium moisture content for a carbon/epoxy laminate [3].

- *Type of material*
  Each material has its own moisture absorption characteristics. It can be stated that all polymers absorb moisture, but the rates of absorption can vary enormously. Even within the group of epoxies there are significant differences in absorption characteristics. Analysis of the moisture absorption data of various epoxy resin compositions shows that the weight gain due to moisture absorption may differ by a factor of 10 or more between different resin chemical structures. For the same epoxy resin cured with different curing agents a factor of 3 or more can be found [3].

- *Quality of the material*
  For polymers, the presence of voids or (micro) cracks can have a very large influence on both the diffusion rate and the equilibrium moisture content. In the case of micro cracks, moisture absorption is accelerated due to capillary action at the micro-cracks and an enlarged exposed area at the surface of the material.

Desorption

Absorption of moisture by polymers is generally a (partially) reversible process. This depends on the way moisture is present in the material. Usually, moisture is present in the form of free molecules, but sometimes water molecules are chemically bonded to polymer molecules. When only free molecules are present, the diffusion process is generally reversible, i.e. desorption will occur when the humidity of the environment decreases. The presence of chemical bonds leads to a more complex desorption behaviour. Some of these bonds can be so strong that only at temperatures beyond the glass transition temperature the water molecules become mobile. In that case, full reversibility is usually not obtained.

When the ambient humidity decreases, the direction of moisture transport is not always towards the outside of the material only. If the material is not fully saturated, part of the moisture will penetrate deeper into the material, due to lower concentration levels.

Effects on physical properties

Several physical properties of the epoxy are changed when moisture is present. The effects can be reversible or irreversible. Reversible effects will disappear when the moisture is removed, whereas irreversible effects will remain. The main processes that can take place are:

Reversible processes

- *Swelling*
  The bound water molecules disrupt the interchain hydrogen bonds in the epoxy. The molecular structure is thus altered to conform to the presence of moisture, resulting in dimensional changes. Local differences in moisture concentrations might subsequently introduce internal stresses to the material.

- *Plasticising*
  Water acts as a plasticiser, which means that it reduces the glass transition temperature T_g. At the glass transition temperature a polymer changes from a glassy to a rubbery state. This is associated with a significant decrease of the modulus of the material caused by an increased mobility of polymer chain fragments. When water molecules enter the polymer, they tend to form hydrogen bonds with certain elements of the polymer. This way, they limit the number of cross-links, leading to a lower T_g.

Irreversible processes

- *Hydrolysis*
  Water absorption has also been shown to induce chemical degradation, especially in polyester resins, by irreversible hydrolysis of ester linkages. Epoxies also have the ability to chemically react with water. In this process, the epoxy structure is irreversibly split into fragments.

- *Physical ageing*
  When polymeric materials are cooled below their glass transition temperature (T_g) a new thermodynamic equilibrium has to be established. This process takes place very slowly and during this time several material properties, such as specific volume and modulus, change continuously. As a result, crazes or cracks may develop. This phenomenon is known as physical ageing. It can only be reversed by heating the material above its T_g. Since in-service temperatures generally remain below the T_g, this process must be regarded as irreversible.

4.3 Environmental effects on Glare

Although the materials that Glare consists of have a relatively well-known long-term behaviour, this knowledge alone is not sufficient to determine the influence of the environment on the laminate. As will be discussed in this section, the build-up of thin layers has a significant impact on the environmental durability properties.

4.3.1 Corrosion

Although Fibre Metal Laminates (FMLs) seem to have similar corrosion characteristics as monolithic aluminium, there are several aspects that distinguish Fibre Metal Laminates from conventional aluminium alloys, which could lead to different corrosion properties. These Fibre Metal Laminate specific aspects are briefly discussed below. A more elaborate description of the corrosion properties of Glare is given in chapter 28.

- *Electrical conductivity of the fibres*
  The use of electrically conductive fibres could lead to the creation of electrical potential differences and therefore to an increase in corrosion sensitivity. This is the main reason why carbon fibres have not been used in Fibre Metal Laminates yet. Since glass fibres are electrical insulators this is not a problem for Glare.

- *Heat treatment during production*
  In aerospace structures, aluminium 2024 is usually applied in the T3 heat-treated condition. This gives the alloy high ductility, excellent fatigue crack propagation resistance and high fracture toughness.
  During the production of Glare the aluminium sheets undergo an additional heat treatment as a result of autoclave bonding. Bonding is carried out at a temperature of 120 °C for about one hour. The effect of this second heat treatment on the properties of Al 2024-T3 is most probably small, due to the relatively low temperature. However, some authors claim that it could have an effect [4].

- *Thickness of the metal layers*
  The aluminium sheets in Glare have a thickness ranging from 0.3 to 0.5 mm, which is much thinner than that of most aluminium sheets used in aircraft structures. The higher quenching rates for thin sheets after the solution heat treatment most probably result in improved corrosion properties compared to that of thicker sheets.

- *Prepreg layers acting as corrosion barriers*
  A major advantage that Fibre Metal Laminates have over monolithic aluminium, is the fact that corrosion damage is limited to the outer aluminium layer, since the fibre layers act as corrosion barriers.

- *Bondline corrosion*
  Corrosion can occur at the interface between aluminium and prepreg. Usually this happens after debonding has taken place, but when a clad layer is present the corrosion itself can be the cause of debonding.

4.3.2 Moisture absorption

Two factors distinguish the diffusion of moisture in Fibre Metal Laminates from that of diffusion in a polymer:

- *Influence of layup*
  As a result of the presence of aluminium layers on the Glare surfaces, moisture can only penetrate the epoxy from free edges of the laminate or through bore holes. This leads to a significantly lower rate of moisture absorption compared to fibre reinforced composites.

- *Influence of fibre*
  As is also the case in fibre reinforced plastics, the presence of fibres influences the moisture absorption behaviour. Moisture absorption by glass fibres is negligible, therefore they act as moisture barriers, see Figure 4.1. This barrier function is strongest in the direction perpendicular to the fibres. Combined with capillary forces acting in the direction of the fibres this causes a much higher moisture absorption rate in the direction of the fibres than perpendicular to this direction.

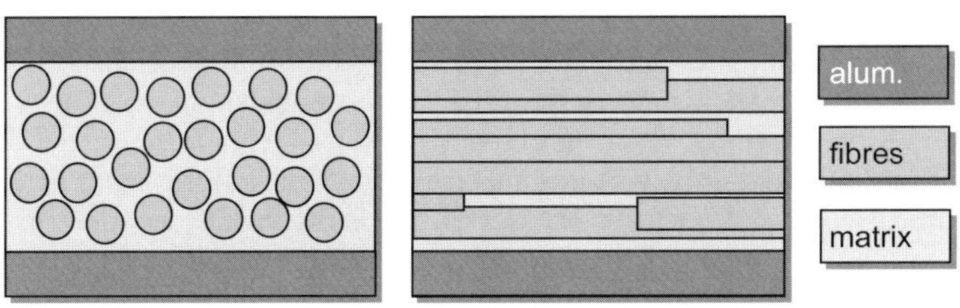

Figure 4.1 *Aluminium and glass fibres acting as moisture barriers in the direction of the fibres (left) and perpendicular to fibre direction (right).*

4.3.3 Environmental effects on interfaces

In Glare, two different interfaces can be distinguished: the aluminium-epoxy interface and the fibre-epoxy interface. These interfaces play an important role in the transfer of loads in the laminate. The influence of moisture on the properties of both interfaces will be discussed next.

Aluminium-epoxy interface

Aircraft structures bonded with epoxy adhesives have been in use for a long time. Since environmental durability of adhesive bonded joints is considered as one of the most important aspects, extensive research has been carried out into the effect of moisture on the properties and behaviour of the aluminium-epoxy interface [5, 6].

The bond between aluminium and epoxy can be degraded by several mechanisms. As has been indicated in the previous sections, the aluminium and especially the epoxy can be degraded by the effect of moisture. The aluminium oxide layer absorbs moisture both from the environment and the adhesive, thereby hydrating the surface of the oxide. This is thought to induce cracks between aluminium and adhesive, further exposing the oxide to moisture. This results in the formation of aluminium hydroxide, which has poor adherence to the aluminium substrate. More importantly, it causes a significant increase in volume leading to higher stresses around the crack tip and subsequently to higher crack growth rates.
Gross corrosion of the aluminium can also weaken the interface bond, but this is usually found in clad aluminium alloys only. Because the clad layer is anodic to the core, it will be prone to corrosion attack.

The epoxy is weakened by processes such as were already described in section 4.2.3. Also, hydrogen bonding in the epoxy will limit its ability to bond to the aluminium substrate.

Fibre-epoxy interface

The bond between the glass fibres and the epoxy matrix plays a very important role in the transfer of stresses in a composite. Unfortunately, the fibre-matrix interfaces are sensitive to the chemical effects of water: glass contains alkaline oxides which can be hydrolysed, possibly leading to permanent changes in the material. The water hydrates these oxides and thus reduces the mechanical properties of the glass [7].

Coupling agents are generally applied to the fibre surface to facilitate good interfacial bonding, which is essential for the transfer of loads between the matrix and reinforcing fibres. In general, epoxy resin adheres to untreated glass fibres via hydrogen bonding only, whereas surface treated glass fibres form both hydrogen and chemical bonds with epoxy. Hydrogen bonds are easily debonded by water, which creates voids at the interface. The debonding of chemical linkages is slower and only occurs under severe conditions [8].

4.4 Experimental work

The physical effects that moisture has on the epoxy in Glare, could lead to a change in the mechanical properties of the laminate. The complicated build-up of Glare makes it very hard to predict this influence, requiring large-scale testing to investigate this.

Different configurations will lead to different moisture absorption behaviour, therefore tests with basic specimens do not provide all the necessary data. Special attention must be paid to the edges of specimens that are used to investigate environmental effects. If for instance the effect of moisture absorption around holes is to be investigated, the uptake of moisture through the edges of the specimen must be prevented.

4.4.1 Test methods

Exposure

The best way to investigate the effect of moisture uptake during the service life of an aircraft would be to expose specimens to a realistic environment. Outdoor exposure tests in a tropical environment (high temperature and humidity) are then one option. However, because the absorption of moisture is a slow process, artificial exposure conditions are frequently used to accelerate the diffusion process. By increasing the temperature, moisture diffusion is accelerated and high humidity levels ensure a worst case scenario. A controlled environment is obviously a requirement to study the kinetics of the diffusion process. To simulate the effects of environmental exposure during the service life of an aircraft, epoxy-based composite materials are frequently tested after exposure. This usually takes place in a climate chamber at a temperature of 70 °C and a relative humidity (RH) level of 85% for a given length of time (often 3000 hours).
As discussed before, some effects due to moisture can be reversible. Therefore continuous exposure to a high temperature and a high relative humidity level in a climate chamber may represent a worst case scenario.

In the past, FML specimens have been immersed in water to investigate their absorption properties [9, 10]. Because the nature of absorption due to immersion is different from that of absorption due to humidity, more realistic test data is obtained by accelerated exposure tests using the standard hot/wet environment (70 °C and 85% RH) [11].
Outdoor exposure remains the most realistic type of investigation, therefore Glare specimens are currently being subjected to such exposure. The tests following this exposure should provide the most realistic test data on the influence of the atmospheric environment on Glare so far.

Temperature effects

As was noted before, the presence of moisture has a plasticising effect on the epoxy. This means that the glass transition temperature will be lowered, which can lead to problems at elevated temperatures. As a result of the presence of moisture, the mechanical properties of an FML can be reduced. For FM 94 epoxy, the T_g is 103 °C in a dry condition. When saturated with moisture at 85% RH this value decreases to approximately 80 °C [12]. As a result, the rubberising effect takes place at temperatures that are approximately 20 °C lower than in Glare that has not absorbed moisture.

4.4.2 Experimental results

Several tests have been performed to investigate the influence of moisture on Glare. A few important types of tests will now be discussed to give the reader an impression of the influence of moisture on different properties of Glare.
It has to be emphasised that in the past different epoxy systems have been used for testing. Earlier versions of Glare contained an AF-163 epoxy adhesive made by 3M. Nowadays, the standard epoxy system used in Glare is FM 94, which is manufactured by Cytec. Besides the epoxy, different sizings have been used. In this section, Glare systems containing AF-163 are marked by an asterisk (*) behind the type number.

Moisture absorption measurements

The rate of moisture absorption into the prepreg layers can be measured by performing weight measurements during the exposure period. The increase in weight is then assumed to represent the amount of moisture that has been absorbed by the prepreg.

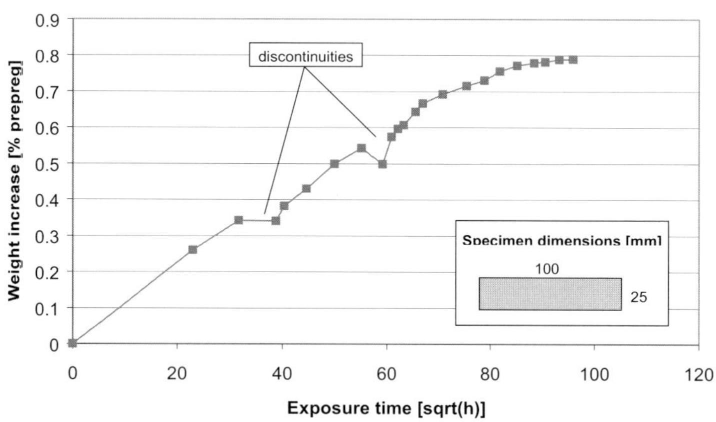

Figure 4.2 Weight increase of Glare specimens exposed at 70 °C, 85 % RH [13].

Even under the relatively harsh conditions of a climate chamber, moisture absorption in Glare is still extremely slow, because of the barrier function of the aluminium outer layers. Specimens measuring 100 × 25 mm have shown to saturate after approximately 8000 hours of exposure to a 70 °C, 85% RH environment, see Figure 4.2. The two discontinuities that can be seen in the moisture absorption curve are due to incidental drops of the humidity level of the climate chamber. Apart from these interruptions, the initial moisture uptake curve shows a linear relationship with the square root of time, indicating Fickian diffusion [13].

Despite the low absorption rate, the problem of moisture should not be underestimated. The longer lifetimes demanded for aircraft will result in higher moisture levels in the material.

Tensile strength

Tensile strength tests on small, moisture-saturated specimens have shown a pronounced effect of moisture on the tensile strength of Glare. Strength values decrease significantly after exposure to a high humidity environment or after immersion in water. For example, after 90 days of immersion in distilled water, strength of Glare* is reduced by up to 36% [10].

The reduction in static strength can be attributed to plasticising of the epoxy (especially noticeable at elevated temperature) and the decrease in strength of fibre-matrix and aluminium-matrix adhesion, which have been discussed in the previous section. Both the epoxy and the interfaces play an important role in the distribution of stresses in a fibre-metal laminate. Degradation of these components can therefore lead to a decrease in tensile strength.

It has to be noted that static strength tests on fully moisture-saturated specimens do not provide realistic strength data. This is due to the fact that in the Glare sheets used in structural applications, moisture absorption will only occur in a relatively small zone close to the free edges (if these edges are not protected by a sealant). Further away from the edges no effect of environmental exposure is expected, simply because there is no moisture. This has been confirmed by exposing sheets of Glare and cutting out the specimens afterwards. The specimens from the middle of the sheet did not show any degradation [14]. Therefore large test specimens show less effect of moisture than small coupon specimens.

Fatigue

Fatigue has been one of the most important problems in aluminium aircraft structures. For Glare, fatigue is much less of a problem. However, it is still important to investigate the effect of moisture and temperature on the fatigue crack growth behaviour of Glare, since these influences could decrease its resistance to fatigue.

Tests have indicated that crack growth in Glare is slightly higher in distilled water than in air [15]. At the same time, an increase in temperature causes an increase in

crack growth rate. This can be explained by a larger absolute amount of moisture present in the air at higher temperatures and its effect on crack growth resistance of the aluminium [16].
The frequency used during fatigue testing plays an important role when investigating the influence of moisture and temperature on materials. A low frequency leads to a longer exposure time per cycle, giving oxygen and moisture more time to react with the aluminium. As a result, crack initiation will start sooner.

Tests have been performed that combine fatigue and exposure. One of the methods is to divide exposure and fatigue time in three parts. One part can consist of 1000 hours of exposure followed by 60,000 cycles. This is repeated three times after which the residual strength is measured. These kinds of tests have led to higher crack growth rates compared to those of specimens that have been exposed for 3000 hours, followed by an 180,000-cycle fatigue test [14]. At all times the exposure conditions were 70° C and 85 % RH.

The influence of moisture and temperature on the fatigue behaviour of Glare is further described in chapter 11.

Blunt notch behaviour

In an aircraft structure, Glare panels will most likely be joined together by means of rivets or bolts. This creates thousands of holes in the structure. When moisture enters through the free edges in the hole, the area of highest stress concentration will be immediately affected. It is therefore important that the influence of moisture around holes is studied thoroughly.
Usually, these holes are filled with a rivet or bolt, but in order to investigate a worst-case scenario, blunt notch specimens were tested with open holes. This will of course speed up moisture absorption significantly.

Tests were carried out after exposing the specimens for 3000 hours in a 70 °C, 85% RH environment. Blunt notch strength showed a decrease of 10 to 15% after this period. Tests performed at different (elevated) temperatures did not show any influence on blunt notch strength [13].

Remarkably, blunt notch tests with filled holes showed similar reductions in strength after exposure. As only a small amount of moisture could have penetrated through the filled hole, this indicates that a degradation process near the edge of the hole is responsible for the strength reduction [14].
Because blunt notch strength of Glare is mainly determined by the strength of the fibres, the most probable explanation for the decrease in blunt notch strength of Glare is the deterioration of the glass fibres due to moisture [13]. This is supported by the fact that temperature plays a minor role in the blunt notch strength of Glare. The properties of the epoxy are largely influenced by temperature, especially after the absorption of moisture, while the strength of the fibres is hardly affected by it.

It was found that the fibre-epoxy interface has a significant influence on the blunt notch strength of Fibre Metal Laminates [17]. When the fibre sizing, which enhances fibre-matrix bonding, is omitted, blunt notch strength increases. The weaker fibre-matrix adhesion of the untreated fibres allows cracks to propagate along the fibre (splitting), rather than to break it. Deterioration of fibre-matrix bonds is therefore not thought to be the cause of the decrease in blunt notch strength after exposure.

Shear strength

Shear strength can be measured in a number of ways, e.g. by testing in-plane shear or interlaminar shear (ILS). The latter is used to check the quality of the bonding cycle during the production of Glare. Interlaminar shear strength is governed by the adhesion between fibres and epoxy and between aluminium and epoxy. The values of ILS strength do not represent specific material properties, but the test provides decent data to be used for qualitative analysis.

Results of ILS tests with Glare show that the presence of moisture has a significant influence on the strength values that are obtained, see Figure 4.3. The effects of moisture on the epoxy and the interfaces, as discussed in section 4.2 can explain this. Tests with different fibre sizings showed significant differences in the results, confirming the important role of the fibre-matrix interface in the ILS properties of Glare [10].

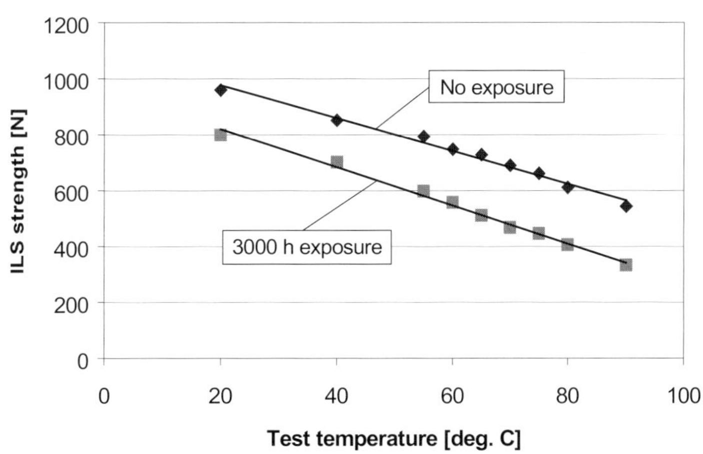

Figure 4.3 *ILS strength of Glare 3-3/2-0.3 specimens.*

Temperature also plays an important role in ILS strength. Tests have shown a linear decrease of ILS strength with increasing temperatures, see Figure 4.3. The presence of moisture leads to a consequent drop in ILS strength throughout the entire temperature range between 20 and 90 °C [13]. This can be attributed to the influence of temperature on the resin; a higher temperature will lead to higher volatility of the polymer chains and thus a lower stiffness. This in turn makes the epoxy less capable in the transfer of stresses, lowering the strength of the laminate.

Other tests to measure shear strength are the floating roller peel test and the cut shear test. Tests with Glare* showed a decrease of 9.9% of floating roller peel strength and a decrease in cut shear strength of 31.1% after exposure for 7 months to a 70 °C, 85% RH environment [4].

Rivet joint strength

The testing of large riveted joint specimens provides a much more realistic form of testing, because these specimens resemble (parts of) real life aircraft structures. Results of strength tests after exposure to a hot/wet environment have shown that the influence of moisture is less than with the more basic specimens, such as blunt notch specimens. This can be easily explained by the fact that the joint specimens are larger, which reduces the influence of moisture entering through the free edges. It can therefore be concluded that tests with basic specimens do not provide realistic data on the influence of moisture on Glare structures.

Previous research investigated the influence of immersion in distilled water and salt spray on Glare* riveted joints. Immersion in water of 60 °C for 90 days led to a strength reduction of 15%. Exposure to salt spray for 90 days led to a reduction in strength of only 5%. The large influence of specimen geometry was demonstrated by increasing the rivet pitch of the water-immersed specimen from 20 to 24 mm. The strength decrease became only 6% after 90 days immersion. This can be explained by the fact that a relatively smaller area has been affected by moisture [10].

A standard exposure period of 3000 hours in a 70 °C, 85% RH environment generally leads to strength decreases of riveted joint specimens of 5 to 10% [14].

4.5 Conclusions

In general, the influence of moisture on Glare is relatively minor, especially when compared to conventional composites. However, some properties still show strength reductions of up to 15%. Care needs to be taken in situations where moisture can reach areas of stress concentrations. More investigation is still needed to obtain a better understanding of the influence of moisture on Fibre Metal Laminates. The influence of realistic outdoor exposure is one of the main research items that are currently studied. With the results of this research a better interpretation of accelerated exposure tests should be made possible.

References

[1] J.M. Hale and A.G. Gibson, *Coupon tests of fibre reinforced plastics at elevated temperatures in offshore processing environments*. Journal of Composite Materials, Vol. 32, No. 6, 1998.

[2] D.R. Hartman, M.E. Greenwood and D.M. Miller, *High strength glass fibers*, Proceedings of the 10th Annual ASM/ESD Advanced Composites Conference, Dearborn, Michigan, USA, 1994, p. 415-426.

[3] P.K. Mallick, *Fiber-reinforced composites: Materials, manufacturing and design.* , New York: Marcel Dekker Inc., 1988.

[4] H. Bommer, *Environmental effects on the material Glare 3-3/2.* F4K/K-95-035 Technischer Bericht, Daimler-Benz, Ottobrunn, 1995.

[5] A.J. Kinloch, *Durability of structural adhesives.* Applied Science Publications, London, 1983.

[6] R. Oosting, *Toward a new durable and environmentally compliant adhesive bonding process for aluminium alloys*. Ph.D. thesis report, Faculty of Aerospace Engineering, Delft University of Technology, 1995.

[7] C. Suri and D. Perreux, *The effects of mechanical damage in a glass fibre/epoxy composite on the absorption rate*. Composites Engineering, Vol. 5, No. 4, Elsevier Science, 1995.

[8] W. Noobut and J.L. Koenig, *Interfacial behavior of epoxy/E-glass fiber composites under wet-dry cycles by Fourier Transform Infrared Microscopy*, Polymer Composites, Vol. 20, No. 1, February 1999.

[9] O.J. Tensen, *The moisture absorption of Glare – an introduction.* Preliminary Thesis, Faculty of Aerospace Engineering, Delft University of Technology, 1990.

[10] M.F.H.C. Boertien, *Strength of Glare after exposure to moisture.* Master Thesis, Faculty of Aerospace Engineering, Delft University of Technology, 1996.

[11] Military Handbook MIL-HDBK-17-1E.

[12] R. Brenninkmeijer, *The effect of moisture ingress on the bonded repair effectiveness*, Master Thesis, Faculty of Aerospace Engineering, Delft University of Technology, 2000.

[13] B. Borgonje, *The influence of elevated temperature on Glare after exposure to moisture.* Master Thesis, Faculty of Aerospace Engineering, Delft University of Technology, 2000.

[14] M.S. Ypma, *Overview of tests concerning the influence of temperature and environmental exposure on Glare*, Report B2V-00-41, Delft University of Technology, 2001.

[15] E.L. Huijzer, *Durability of Glare – The influence of water and temperature on the low-frequency fatigue behaviour of Glare*, Master Thesis, Faculty of Aerospace Engineering, Delft University of Technology, 1992.

[16] O.T. Kieboom, *Fatigue crack initiation and early crack growth in Glare at different temperatures,* Master Thesis, Faculty of Aerospace Engineering, Delft University of Technology, 2000.

[17] L. Ye, G.D. Lawcock, Y.W. Mai and C.T. Sun, 'Effects of fibre/matrix adhesion on carbon-fibre-reinforced metal laminates, I. Residual strength', *Composites Science and Technology*, Vol. 57, No. 12, 1997, p. 1609-1619.

5

Material design allowables and qualification

M.S. IJpma

Abstract – The material design allowable is a statistically determined minimum value of a material strength property. These values are necessary for the design of aircraft. In the case of Glare, material design values can be obtained through the use of the Metal Volume Fraction. The Metal Volume Fraction (MVF) is the amount of metal present in a Glare sheet. For the different material design allowables there is a (linear) relationship between Glare with a low MVF and a high MVF. The second subject concerns qualification of Glare. Qualification of basic Glare includes qualification of the underlying processes and materials. The qualification of spliced Glare is focussed on the qualification of the details in splice. Using the qualified details any splice layup can be made without having to qualify each splice configuration separately.

5.1 Material design allowables

5.1.1 Definition of material design allowables

In structural design, material design allowables are required in order to size the material thickness according to the structural loads. In the aerospace industry, the definition of a material design value or material design allowable is that it is the statistically determined minimum value of a material strength property [1, 2]. There is a difference between material design allowables (material basis values) and design allowable values. Material basis values are an intrinsic property of a material system, whereas design allowable values are application dependent, and consider and include specific additional considerations that may affect the strength or stiffness of the structure. A material design allowable is a material property such as the compressive yield strength, for example, whereas a design allowable takes into account the geometry of the structure, e.g. the presence of stiffening elements and crack stoppers, which affect residual strength.

In a structure a single, critical part needs a higher reliability than a redundant non-critical part. Therefore, different "confidence levels" are used, as are different values of a single material property of a certain material depending on the "confidence" level, e.g. the tensile yield values given in section 5.1.2.

The statistical basis of "single load carrying path elements" is the A-value, and for a "redundant structure" it is the B-value. The definition of an A-value is that at least 99% of the population of values is expected to be equal to or exceed the A-basis mechanical property allowable with a confidence of 95%. The definition of a B-value is that at least 90% of the population of values is expected to be equal to or exceed the B-basis mechanical property allowable with a confidence of 95%. Other statistically based values that are used are the S-value and the typical value. The S-value is the specified minimum value of a mechanical property value, whereas the typical value is a sample mean. This sample mean is defined as the simple arithmetic mean, which has a statistical significance of 50% reliability with a 50% confidence [3]. In general the B-value is used for composite components, even in single load path elements. The reason is that because of high scatter obtaining A-values is too costly.

5.1.2 Influence of material parameters on material design allowables

Many parameters characterise a material, including thickness, alloy elements, fibre orientation, layup, etc. This large number of parameters makes it necessary for material design allowables to be defined for (ranges of) prescribed parameters.
For example, the thickness of monolithic aluminium influences the material properties because the stress-state gradually changes from "plane stress" to "plane strain" between thin and thick monolithic aluminium, while the microstructure and residual stresses may also be different. The different material thicknesses result in different grain sizes and quenching rates during the manufacturing process, and therefore the value of the material allowable changes within certain thickness ranges. For example, the tensile ultimate strength in L-direction (i.e. rolling direction) of aluminium 2024-T3 for different thickness ranges is given in Table 5.1 [3].

Table 5.1 *Tensile ultimate strength in L-direction of aluminium 2024-T3 for different thickness ranges.*

Thickness range	Tensile ultimate strength
0.20-0.23 mm	441 MPa (S)
0.25-3.25 mm	441 MPa (A) / 448 MPa (B)
3.28-6.32 mm	448 MPa (A) / 455 MPa (B)

The material properties of composite material can (for at least a part of the material properties) be calculated from the properties of the raw materials (fibres and resins) and the layup of the material. The "rule of mixtures" is a commonly used method for this. In the event that new combinations of raw materials are to be used, the properties of these raw materials need to be determined.

In Glare, the material characteristics of monolithic aluminium and composites are combined. Due to the use of a small thickness range of the aluminium sheets (0.3 – 0.6 mm) no thickness effects have to be taken into account, neither for the individual aluminium layers nor for the total Glare laminate. As a result, an approach was developed to determine the Glare material property that is very similar to the rule of mixtures used for composites. This approach is the "Metal Volume Fraction (MVF) approach" and is described in section 5.1.5.

As well as the material dependent parameters, the material properties and thus material design values can be influenced by environmental influences.
Composite materials are especially known for their susceptibility to environmental exposure (e.g. moisture) and temperature. Where material design values can be determined for room temperature and normal humidity, the use of materials at high temperatures and/or highly humid environments can require that reduction factors (or so-called 'knock-down factors') be used.
Similar knockdown factors are needed for the material properties of Glare which are governed by the prepreg, e.g. the blunt notch strength.

5.1.3 Use of material design allowables

In structural design, some material properties are directly used for design and stress justification and therefore need a secure statistical basis, while other material properties do not directly affect the structural strength, e.g. the Young's modulus for a strength critical item.

The material properties that in general require a secure statistical basis, i.e. an A- or B-value, are:

- tensile yield strength and tensile ultimate strength
- compression yield strength
- shear yield
- bearing yield and bearing ultimate strength.

Material properties that are not directly related to strength, but which are necessary in most calculations, can be applied as typical values. Examples of such material properties are:

- tensile and compression elastic modulus
- shear modulus
- Poisson's ratio
- physical properties such as specific weight and coefficient of thermal expansion.

The material properties of commonly used aluminium alloys such as 2024-T3 and 7075-T6 are documented in design handbooks, including the military handbooks MIL HDBK 5 for metals [3] and MIL HDBK 17 [4] for composite materials or the Handbuch Struktur Berechnung (HSB) [5] of the German aerospace industry.

If new materials are about to be used in aircraft structures, it should be remembered that the relationship between the various material properties of different materials might change. The critical material property in a design for material A is not necessarily the same as that for material B. As a consequence, different design procedures may be necessary or material properties need to be added to the list of material properties used for conventional materials. For example, for Glare the relatively large reduction (when compared to aluminium 2024-T3) of the tensile strength caused by the presence of a notch (hole) in the material requires that the blunt notch strength is added to the list of material design values [6].

5.1.4 Determination of material design allowables

A material design allowable needs to have a statistical basis. Such a statistical basis is the result of either a direct computation or an indirect computation. The direct computation is defined as a statistical computation of a one-sided lower tolerance limit for one set of data points. The indirect computation is based on a ratio (to another established property), e.g. a knockdown factor, or on a regression (if the property is a function of a fixed factor). In both cases however a large test programme is needed to generate the minimum guaranteed material properties. The direct computation requires many more test results than the indirect method.

A large test matrix would be necessary to determine Glare's statistically based material design values because of the vast number of combinations that are possible with Glare grades (Glare 1, Glare 2, etc.), layups (2/1, 3/2, etc.) and sheet thicknesses (-0.3 or -0.4). For example, the Glare variants Glare 2-7/6-0.4, Glare 2-5/4-0.4, Glare 3-3/2-0.3 and Glare 4A-3/2-0.4 would all need their own material design values. The numerous combinations would result in a very expensive problem. The designers would need a material data table for each Glare grade and layup and numerous tests would be required.

However, a large reduction in this number of data tables (and in the size of the test matrix) is possible through the application of the Metal Volume Fraction approach (MVF). As mentioned in section 5.1.2, the Metal Volume Fraction approach resembles the 'rule of mixtures'-approach of composite materials. The Metal Volume Fraction approach assumes that the basic material properties of a specific Glare laminate only depend on the laminate family (Glare 2, 3, 4, 5 or 6) and on the metal volume fraction relative to the total thickness of the laminate. This means that once the properties of a specific laminate type are known, the properties can be calculated for any other Glare member of that family (without testing).

5.1.5 The Metal Volume Fraction approach

The Metal Volume Fraction (MVF) is defined as the ratio of the sum of the thicknesses of the individual aluminium layers and the total thickness of the laminate, see Equation 5.1.

$$MVF = \frac{\sum_{1}^{p} t_{al}}{t_{lam}} \qquad (5.1)$$

with: t_{al} = thickness of each separate aluminium layer
t_{lam} = thickness of total laminate
p = number of aluminium layers

In words, the MVF is the relative aluminium contribution in the Glare laminate. For example, the Metal Volume Fraction of Glare 3-4/3-0.3 is calculated as follows:

Thickness of single aluminium layer	0.3 mm
Number of aluminium layers	4
Thickness of single prepreg layer	0.13 mm
Number of prepreg layers	3×2 (0°/90°) = 6

$$MVF = \frac{4 \times 0.3\ mm}{4 \times 0.3\ mm + 6 \times 0.13\ mm} = 0.606$$

A Metal Volume Fraction of 1 means monolithic aluminium (well documented material properties) and a MVF of 0 means pure prepreg (a theoretical situation).

The Metal Volume Fraction approach assumes a linear relationship between the material properties at MVF=1, the so-called 'metal layer contribution', and the average of the tested laminate data. This line is extrapolated to MVF=0. The data point at MVF=0 is the so-called "fibre layer contribution", see Figure 5.1. The value of a material property of a specified Glare having any MVF can be calculated using the value of the aluminium and the value of the fibre layer using Equation 5.2.

$$\begin{aligned} \textit{Laminate property} &= MVF \times \textit{``metal layer contribution''} \\ &\quad + (1 - MVF) \times \textit{``fibre layer contribution''} \end{aligned} \qquad (5.2)$$

For example, the tensile yield strength of Glare 3-4/3-0.3 in L-direction is calculated as follows:

The tensile yield of 2024-T3	369 MPa	"metal layer contribution"
The "tensile yield" of prepreg	192 MPa	"fibre layer contribution"
MVF (see above)	0.606	

The tensile yield of Glare 3-4/3-0.4 thus becomes 0.606 × 369 + (1 - 0.606) × 192 = 301 MPa. If this calculated tensile yield is compared with the experimentally determined tensile yield (314.8 MPa) then the result is within 5% accuracy.

Authorities presently accept the MVF approach for the tensile strength, compression strength, shear strength, bearing strength and blunt notch strength.

For a few arbitrary Glare 3 and Glare 4 layups, the material properties according to the MVF-approach are compared with test data in Table 5.2.

Table 5.2 Some Glare material properties according the MVF approach.

Glare type and layup	L/LT	MVF	t (mm)	MVF approach	Test result	Deviation
Tensile yield strength						
Glare 3-2/1-0.3	L	0.706	0.85	317 MPa	322 MPa	- 2 %
Glare 3-5/4-0.4	L	0.667	3.0	310 MPa	307 MPa	+ 1 %
Compression yield strength						
Glare 4A-3/2-0.3	L	0.545	1.65	317 MPa	317 MPa	0 %
Glare 4A-5/4-0.3	L	0.500	3.00	319 MPa	313 MPa	+ 2 %
Blunt notch strength						
Glare 4B-3/2-0.3	LT	0.545	1.65	589 MPa	576 MPa	+ 2 %
Glare 4B-6/5-0.4	LT	0.561	4.28	583 MPa	574 MPa	+ 2 %

The aluminium material data and the material data of the tested MVF is plotted in graphs, see Figure 5.1. In addition, straight lines are drawn, resulting in typical material design values. These graphs can easily be used during the design of structures to obtain the necessary material properties.

The MVF approach is proven to be valid in a range of $0.45 < MVF < 0.85$. The currently used Glare laminates are within this range. At this moment a linear relation has been proven for the typical values of the material properties mentioned above. It is assumed that by dividing the typical values by a factor 1.1, minimum values are obtained and conservatism is introduced that accounts for possible reductions when B-values for Glare material properties will become available. Most of the currently available test data, 98%, are above these minimum values.

With the application of this MVF-approach, a lack of standardisation (as is the case for composite materials) is avoided. This lack of standardisation is a burden, which tempers the enthusiasm of the airline companies to use composite materials.

For more detailed coverage of this subject see Roebroeks [7, 8], upon which this section is based.

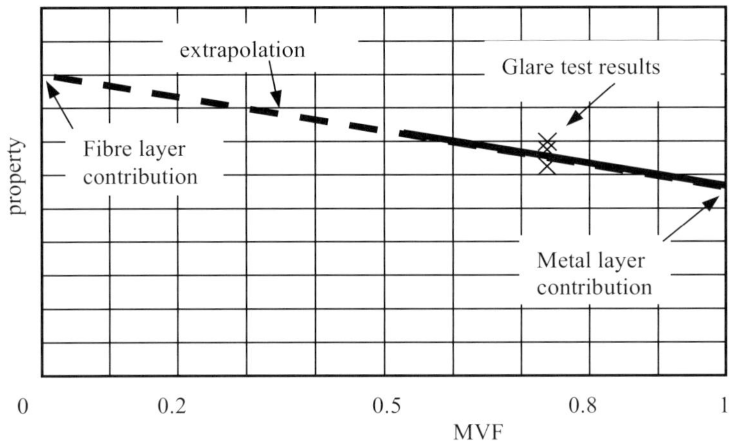

Figure 5.1 *Graphical representation of the Metal Volume Fraction approach.*

5.1.6 Determination of material design allowables for Glare

The MVF-approach is also used to determine the B-values of Glare material design values. For the relevant Glare grades (Glare 2A, Glare 2B, Glare 3, Glare 4A and Glare 4B) two layups are selected, i.e. one with a low MVF and one with a high MVF. The selected material properties of these two layups are then determined. The number of tests is considerably reduced by the application of the MVF approach.

5.2 Qualification

5.2.1 Qualification of the basic Glare material

Variability in composite material property data may result from a number of sources including run-to-run variability in fabrication, batch-to-batch variability of raw materials, testing variability, and variability intrinsic to the material [4].
The manufacturing process and composite build-up require that a qualification programme validates the Glare material properties.

The quality of aluminium is described in international standards such as AMS-QQ-A-250/4 and AMS-QQ-A-250/5 [3]. Aluminium can be purchased according to these standards and the manufacturer is obliged to deliver the material according to the requirements. In case of prepreg materials, the quality of the fibres and matrix needs to be investigated, as do the fibre content and the resin content of the prepreg as well as the (mis)alignment of the fibres in the prepreg [4]. For example, the fibres in the prepreg should not have a misalignment of more than 0.5°.

The manufacturer of Glare material is obliged to deliver Glare according a strict quality control. The quality of the Glare material is guaranteed through a qualification programme. In such a programme all the processes to manufacture Glare are subjected to qualification. These processes include for example the quality assurance of the aluminium sheets, the pretreatment process, the primer application process and the curing process. For all these processes, the quality is validated and shown to be within a certain range. Finally, the quality of the Glare material is tested in order to verify that the quality of the end product, the Glare sheet, will be constant.

The final qualification programme consists of mechanical testing of Glare. The Glare material needs to be manufactured at a qualified plant using different aluminium batches and different prepreg batches ("the proof of the pudding is in the eating"). Consequently, the material properties that are relevant for strength during design are tested at several qualified laboratories, like the National Aerospace Laboratory NLR.

The mechanical test results show small scatter-bands, and prove that the quality of the Glare production is assured by qualifying all the manufacturing processes. The number of tests to be carried out is large, although so far Glare has only been manufactured for a short time. Testing will determine the exact scatter bands within which Glare can be manufactured over the years.

5.2.2 Difference between qualification and material design allowable

In practice, the material data generated during qualification of the material can be used for determination of material design allowables, so long as the required number of test specimens is tested for the necessary statistical basis. The difference between qualification data and data from a material design allowable programme is that the values of the material properties determined during a qualification programme are not important as long as the values are constant or within certain prescribed ranges. The values of the material properties of a material design allowable programme are important as these are used for structural design.

5.2.3 Qualification of spliced Glare

The splice concept offers us the possibility to manufacture Glare sheets that are wider than a single aluminium sheet. Many splice configurations are possible as a result of Glare's layered build-up, e.g. see Figure 5.2. Quality assurance of each splice configuration by performing mechanical tests on each configuration would again require an enormous amount of testing.

A reduction in the test matrix is therefore achieved through the introduction of a qualification philosophy for splices and doubler edges [9]. This philosophy assumes that if the presence of any single detail of a splice or doubler in a laminate does not decrease the strength of the 'basic' laminate, a combination of details in a real splice will also not influence the strength of the 'basic' laminate, provided that the single details do not influence each other.

The single details that are present in a Glare splice are divided in two types, i.e. details due to sheet metal edges and details due to prepreg layer edges.

The details corresponding to the sheet metal edges are the internal and external thickness step, the internal splice doubler and the internal and external splice. Details that can be identified for prepreg layer edges, are the edges of the first, second and third prepreg layers. In Figure 5.3 of both types of splice details an example is given.

From these details, test specimens are derived and the material properties, such as the material design values, of these test specimens are determined. The test results show that in most cases the material design values of the splice details are larger or equal to those of the basic material.

The possible interaction between the splice details is investigated by FEM calculations. The distance necessary to redistribute the stresses due to the presence of a splice detail was determined using such calculations. It was proven that no interaction occurs when a certain minimum distance is taken for the splice.

The purpose (and conclusions) of this qualification philosophy is that any splice configuration can be constructed from the splice details and used without the need for a complete qualification (test) programme.

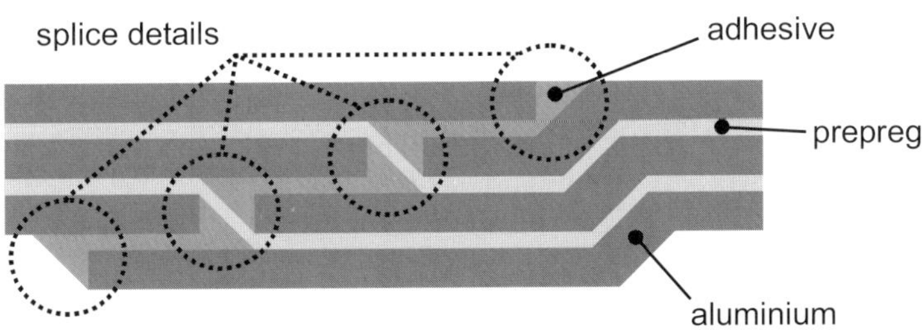

Figure 5.2 *Typical splice layout with splice details.*

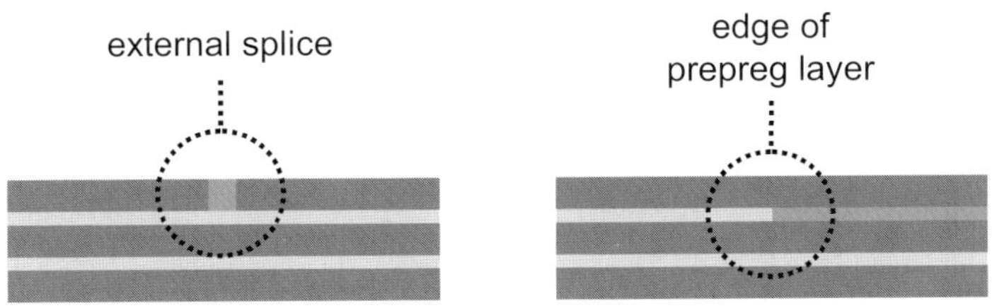

Figure 5.3 *An example of both types of splice details.*

References

[1] JAR 25.613.

[2] ACJ 25.603.

[3] MIL HDBK 5 H.

[4] MIL HDBK 17 E.

[5] Handbuch Struktur Berechnung, HSB.

[6] O.J. Bosker, *Investigation to determine B-value allowables based on normal distributed test data*, TD-R-00-006, SLI, Delft, October 2000.

[7] G.H.J.J. Roebroeks, *The Feasibility of the Metal Volume Fraction Approach for the Calculation of the Glare Blunt Notch Strength*, TD-R-99-005, SLI, Delft, september 2000.

[8] G.H.J.J. Roebroeks, *The Metal Volume Fraction Approach*, TD-R-00-003, SLI, Delft, oktober 2000.

[9] G.H.J.J. Roebroeks, *Proposal for Qualification program for Splices and Doubler Edges in Glare 3 and Glare 4 used in Large Fuselage Skin Panels*, TD-R-97-013, SLI, Delft, December 1997.

6

New aluminium alloys for Glare

A. Auffret and A. Gennai

Abstract – This chapter tries to answer the question: "How do we improve the Glare properties by searching for the most relevant aluminium alloy?" First, a quick overview of the available improvements made in the aluminium industry is given. A brief summarisation of the influence of the aluminium alloy properties upon the residual strength of the laminate follows, and finally the new aluminium alloys most promising for Glare are presented.

6.1 Weight saving and cost-reduction initiatives

The first use of aluminium in aerospace dates back to 1903 when the Wright brothers used an Al-Cu-Mn casting alloy for the engine crankcase. As the design methodologies for aircraft design evolved during the century, so did the drivers for material development. Alloy development criteria have come a long way from simple static strength to today's complex requirements of weight, effective performance, durability, damage tolerance, reliability, manufacturability and maintainability. The intent here is to provide a perspective of the trend that material developers are increasingly being challenged to take. Section 6.2 is based on Lequeu [1].

6.1.1 High formability sheets

Forming of fuselage panels, leading edges or other structural formed parts, can be a very complicated process. It often involves a series of solution heat treatments, quenching and press forming by stretching or drawing. The reason for these complicated processes basically lies in an insufficient forming capability of the alloy used, namely 2024-T3.

Formability is a very complicated science, as those involved in the automotive or canning industries would not deny. This is related to more or less complex strain paths followed by the metal during forming.

Put simply, it can be said that forming of aerospace sheets mainly involves linear strain paths such as biaxial expansion, uniaxial tension and/or plane strain drawing. Formability of an alloy along these various paths can be assessed using the forming limit curve (FLC), as developed in the '50s by the steel automotive industry: the higher the FLC limits, the more formable the alloy under linear paths.

Figure 6.1 compares the FLCs of a high formability grade of 2024 sheets developed by Pechiney, to that of standard 2024. Typically, 20 to 50% improvements in the allowable deformation are obtained with the high formability material over the standard one.

The consequence of these improved formability levels is a reduction in the heat treatment/forming sequences applied to achieve the final shape. Table 6.1 illustrates the typical cost savings achieved. Adoption of one or the other high formability grades depends on the complexity of parts and overall forming elongations.

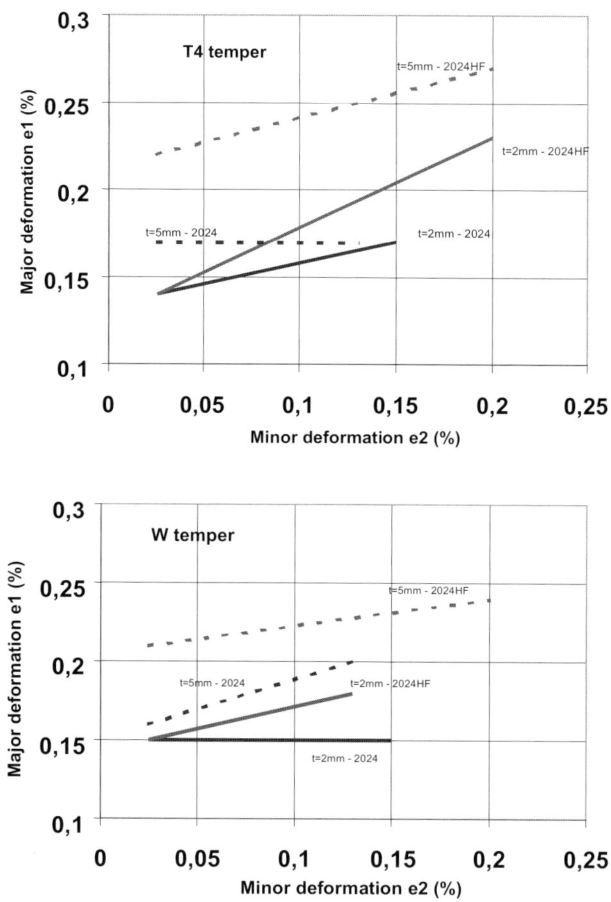

Figure 6.1 *Comparison of the FLCs of high-formability 2024 sheets and standard quality 2024 for two thicknesses and two tempers (T4 and W) (t=thickness).*

Table 6.1 Typical forming sequences for various grades of 2024 sheets.

	2024 Standard	2024 High Formability	
		Grade 1	Grade 2
Initial temper	O/F	O/F	T4
Forming sequence	Solution treatment / quench	Solution treatment / quench	Stretching to final shape
	Stretching to intermediate shape	Stretching to final shape	
	Solution treatment / quench		
	Stretching to final shape		

6.1.2 Age-formable alloys

Age forming is a process developed in the '80s comprising the forming of a part while completing the material ageing treatment. It often necessitates an autoclave, since a vacuum is needed to help the machined part adopt the shape of a matrix. Complex proprietary models are used to account for spring back.
The age forming process has received more and more attention in recent years since it is an efficient way to reduce costs due to:

- the consistency of the final shape, reducing the need for performing shape corrections
- the elimination of expensive mechanical forming

It is recognised that higher ageing temperatures as well as 'reasonable' minimum ageing times are best suited for a good age-forming capability of an alloy. The 7449 alloy for compression dominated structures has been developed in the T7951 and T7651 tempers for this purpose, using standard proprietary two-stage ageing treatments and is available at the customer's request. Such an option has been taken for the age-formed 7449 top wing panels and stringers of the Airbus long range extended capacity A340/500-600 aircraft.

6.1.3 Weldable alloys

One of the most efficient ways to reduce costs (and weight) is to eliminate rivets. Alternative proposals to such classical assembly consist of:

- replacing assembled structures by integrally machined parts
- using welding techniques

Among the many welding techniques available, two show a very promising future in the aerospace industry; namely friction stir welding and laser beam welding.

Friction stir welding was proposed in the mid '90s by TWI in the UK. In short, it can weld virtually any aluminium alloy to any other (identical or different) material! Many types of very attractive applications can be envisaged for an aircraft, but it remains to be proven reliable or to lead to acceptable properties (whether mechanical, damage tolerance or corrosion resistance) in the weld and/or in the heat affected zone.

A dedicated alloy, 6056, has been proposed for weldability purposes. The main welding technique regarded and developed by the manufacturers for fuselage application is laser beam welding. A description of the metallurgical principles underlying this development has been given in [4]. Delivery of 6056 has been proposed as sheets in two versions, clad T6 and bare T78. T78 patented temper has been designed to desensitise the alloy to intergranular corrosion (IGC) while maintaining good levels of static and damage tolerance properties, making it a good candidate for low cost weldable bare sheet and stringers for fuselage panels. In addition, a density of 2.71 g/cm^3 allows for some 2.5 to 3% weight savings when compared to 2024 panels of the same dimensions.
For T6 clad sheets, a specific cladding alloy was designed in order to efficiently protect the core metal against corrosion, taking account the clad/core minimum and maximum potential gap rules.

Typical properties of the two versions of 6056 sheets are shown in Table 6.2 below for a typical thickness range of 1.6 – 3.2 mm.

The improved resistance to IGC of the T78 temper is illustrated in Figure 6.2, which compares the T78 and T6 tempers for sheets of the same thickness. An absence of intergranular corrosion can be seen on the T78 sample, with only a few pits visible. A practical consequence of such outstanding behaviour is that 6056-T78 sheets can be used in the form of bare sheets without impairing the corrosion resistance of the fuselage panel. Cost reduction on the delivered material can be thus achieved.

Table 6.2 *A comparison of typical properties of 6056 and 2024 sheets. R curves are from CCT W760 specimens/ASTM E561. Corrosion is on bare or unclad material.*

Property	6056		Ref 2024
	Bare T78	Clad T6	Clad T3
UTS LT (MPa)	370	380	450
TYS LT (MPa)	340	350	310
Intergranular Corrosion	No	Yes	Yes
R-curve Kc T-L (MPa√m)	125	125	125

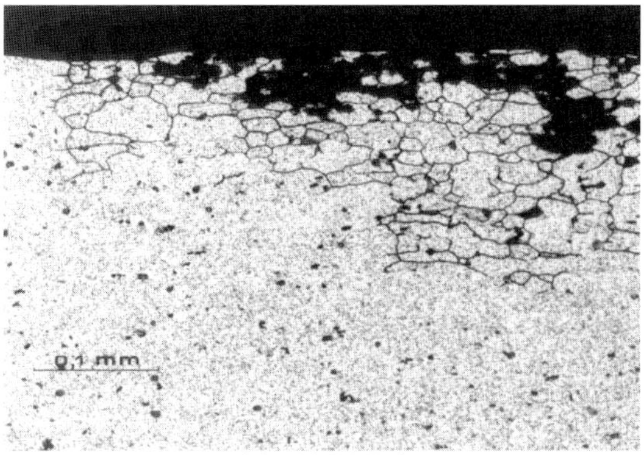

(a) 6056-T6

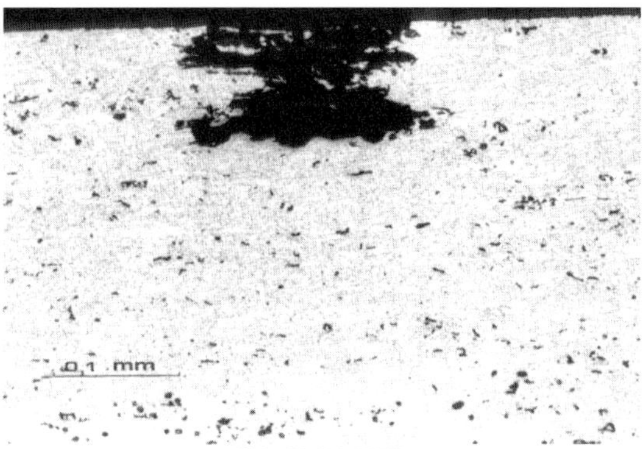

(b) 6056-T78

Figure 6.2 *Typical IGC behaviour of 6056-T6 and 6056-T78.*

The T78 solution has also been developed for fuselage weldable stringers. In this case, the T78 ageing practice is similar to that of the sheets, allowing welding of the stringers to the formed panels in the T4 temper before achieving the final ageing treatment.

6056 solutions have been qualified for use on some single aisle A318 fuselage panels (with both T6 clad and T78 bare sheets) and on Airbus' super-jumbo aircraft, the A380.

6.1.4 Other enablers

Although designers will always desire lighter structures which are safe and long-lasting, this desire is currently being overshadowed by the need for lower-cost structures because of the growing gap between aeroplane cost per seat and yield per passenger. Therefore, the additional major areas of emphasis now are on lean manufacturing, life extension of ageing aircraft for the military and low cost structural concepts to reduce the acquisition and operating costs of future aircraft.

Because manufacturing costs are about 95% of the cost of the airframe, ways to reduce these costs are vigorously attacked. The initial goal is to attain about a 25% reduction in cost of the structure without deviating substantially from conventional aircraft structural design. To accomplish these goals, the following enablers are proposed (section 6.1.5 is based on Liu [5]):

- Superplastically formable sheet-materials offer opportunities to save weight by part consolidation. A new high-strength SPF 7XXX alloy containing Sc has been developed by Alcoa. This emerging material combines the high forming rate of Supral with the strength of 7475-T6. Significantly, straightening costs will be substantially reduced and larger parts can be formed as the new material does not need to be rapidly quenched after solution heat-treatment to develop attractive properties.
- The casting of secondary structures is continuing to displace riveted assemblies. Technologies enabling aluminium castings for primary structures in commercial aircraft either exist or could be developed for cost saving through part consolidation. This approach can move forward rapidly as soon as airframe manufacturers demonstrate their commitment.
- The third generation Al-Li alloy, 2195, is cost effective because its high strength provides superior weight savings. Alloy 2097 offers cost savings because it will eliminate the need of periodic replacement of 2124-T851 bulkheads due to fatigue cracks.
- The Al-Mg-Sc alloy offers opportunities for cost savings because it does not require cladding to provide high corrosion resistance. Currently, costly measures to prevent corrosion are employed to protect alclad 2024 exposed in the belly of the aircraft where cladding is removed by machining. Further cost savings are possible through design improvement using the improved weldability characteristics of Al-Mg-Sc.

6.3 Influence of aluminium alloys on some properties of Glare

The influence of the aluminium layers on the behaviour of Glare is huge, and it would necessitate much more than a few pages to summarise it. The aim here is only to point out two important issues for the designers: the influence of the aluminium alloys on the shear strength and on the residual strength of the laminate (section 6.4 is based on De Vries [3]).

6.3.1 Shear strength

Glare 3 has bi-directional fibres only, so it is highly anisotropic. The fibre orientation has a large effect on the material properties of the material: loading in the fibre direction leads to a higher strength than in the direction perpendicular to the fibre direction. Under off-axis loading, the failure criterion is the aluminium shear strength, see Figure 6.3.

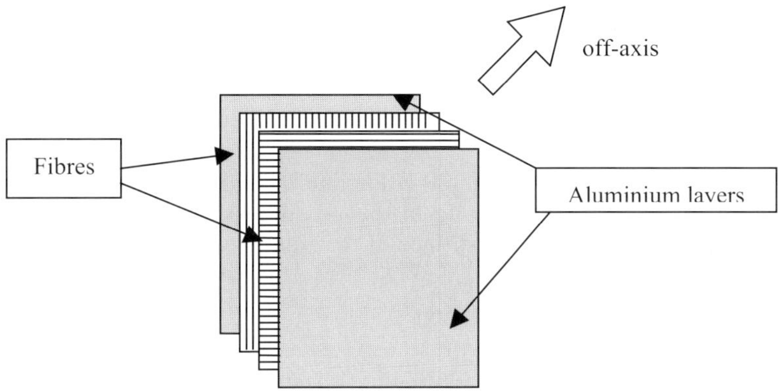

Figure 6.3 *Glare 3 layup.*

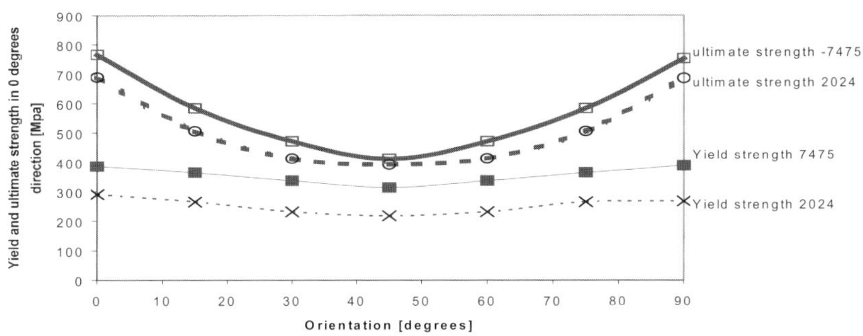

Figure 6.4 *Yield and Ultimate strength of two variants of Glare 3 (with 2024 alloy and 7475 alloy) under off-axis loading* [6].

As we can see in Figure 6.4, the shear strength depends largely on the behaviour of the aluminium alloys. An alloy with higher shear strength can improve the shear property of the laminate up to 15% with respect to the 2024 alloy.

6.3.2 Residual strength

The fracture of Glare with a through-the-thickness crack statically loaded to failure is based on the constituents' properties and the interfaces between them. We can discern three constituents in Glare: the aluminium layers, the fibre layers, and a so-called "resin rich layer".

The **aluminium layers** are responsible for:

- yielding of the materials at high loads
- stable extension before fracture.

The **fibres** are responsible for:

- an extensive strain hardening after yielding
- a high ultimate strength
- a limited ultimate strain.

The "**resin rich layer**", which is at the interface of the aluminium layer and the fibre layer, plays an important role in the possible occurrence of delamination during loading. The quality of the surface of the aluminium sheet can also cause debonding of the layers during loading.

Crack initiation

When a crack occurs at the aluminium sheet, the fibres in the prepreg layer act as a bridge, limiting the crack opening thanks to shear stresses at the interface between the crack-bridging fibre layers and the aluminium sheet, see Figure 6.5. Subsequently, a reduction in stress intensity factor at the tip of the crack occurs.

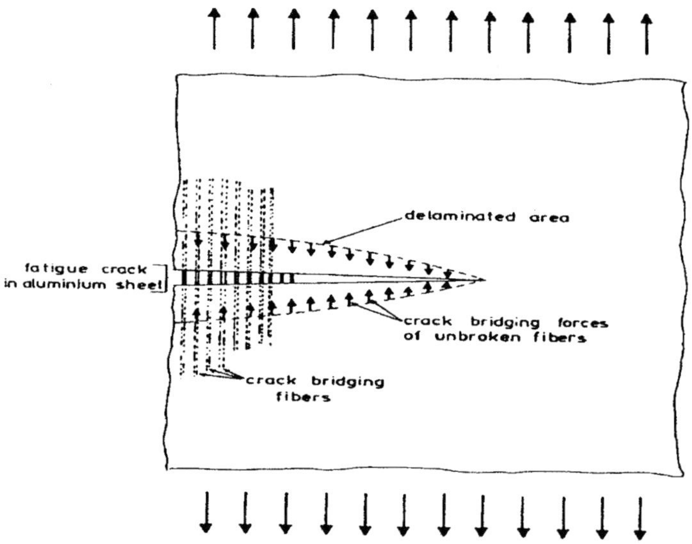

Figure 6.5 Fatigue crack with crack-bridging fibres.

We can schematically point out two effects that contribute to crack initiation, assuming that a cracked sheet is loaded perpendicular to the crack.

- The presence of a crack in a material that is loaded perpendicular to the crack edges causes increased stress intensity in front of the crack tip. Above a given load, the aluminium layer becomes locally plastic, while the fibres remain elastic, see Figure 6.6. This difference of behaviour increases shear stress in the resin-rich layer.
- If the starter crack is not extremely sharp then it will need an initiation location. This causes crack blunting during loading, which reduces the intensity and results in a delayed crack initiation. Since the blunting takes only place in the aluminium layer, additional shear stresses are created in the resin-rich layer at the crack tip.

As a result, both effects raise local shear stresses above the ultimate shear strength of the "resin rich layer" and cause delamination in front of the crack tip. This is illustrated qualitatively in Figure 6.7, where the shear strength distribution of the resin rich layer is plotted in the direction and in front of the crack tip.

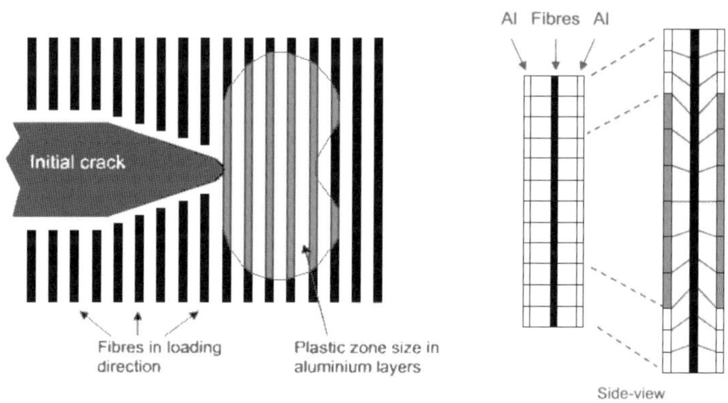

Figure 6.6 *Initial crack and plastic zone in the aluminium layers.*

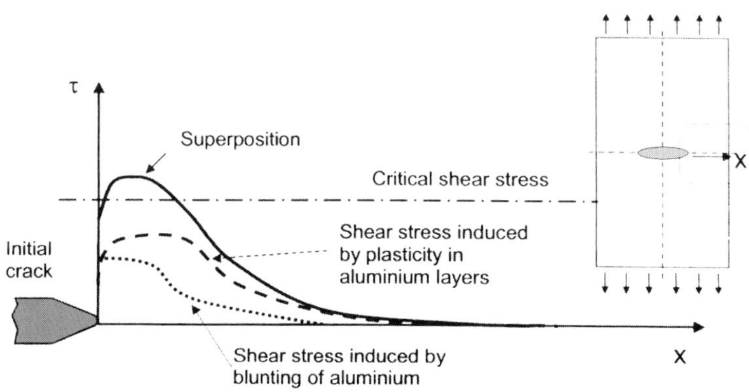

Figure 6.7 *Qualitative distribution of shear stress in front of the crack tip.*

As a result of delamination, fibres close to and in front of the crack tip no longer have to follow the local strain of the aluminium layer at the tip. Moreover, the local deformation can be spread over a much larger area, which decreases the strain in the fibres. Figure 6.8 describes the strain distribution along an aluminium and fibre strip in front of the crack tip in the loading direction. The two dashed areas are equal in size since the elongation of the aluminium layer and the fibres far away from the crack tip is the same.

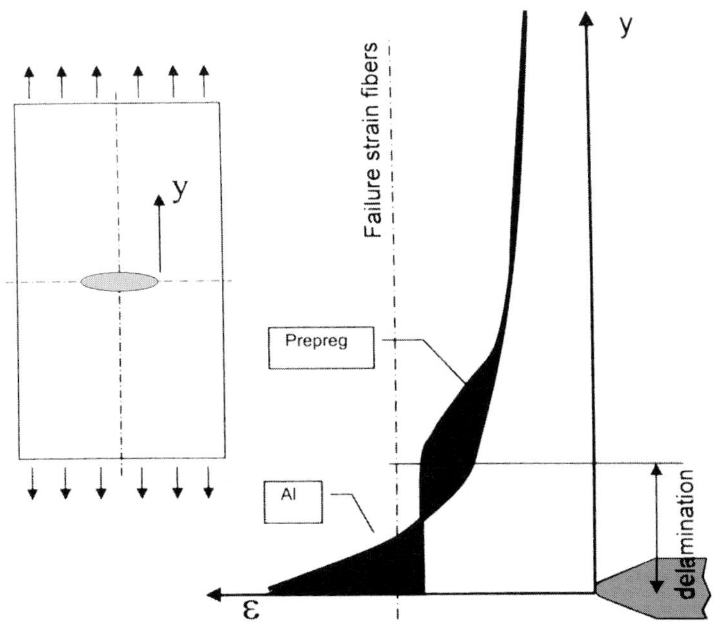

Figure 6.8 *Qualitative distribution of strain along an aluminium fibre strip in front of the crack tip and in the loading direction.*

It must be clear that the occurrence of the delamination is expected to delay the crack initiation. This delamination will occur more frequently for:

- an adhesive layer with a lower shear strength
- an aluminium alloy with a lower yield stress and a higher ultimate stress
- a prepreg layer with higher ultimate strain
- a sharper crack tip, in order to increase the occurrence of blunting delamination

Stable crack extension

The fatigue mechanism changes after the crack has been initiated. The blunting of the crack decreases, inducing a decrease in the shear stresses and therefore a reduction in delamination.

Now the ultimate strain of the fibres, which is much lower of that of aluminium, controls the process of crack extension. This may still be stable thanks to the redistribution of load by the aluminium layers due to their plasticity. In this situation, it is advisable to decrease the aluminium yield stress and to increase the ultimate fibre strain in order to increase the residual strength of the laminate.

Experimental results

Based on these mechanisms, it can be expected that the residual strength is a function of the interface strength, and more importantly a function of the ultimate strain of the fibres and the yield stress of the aluminium layer. Translated to the laminate properties, the residual strength is expected to be a function of the strain hardening, even though the R-curve of the aluminium alloys might be taken into account.

From the statements above, it transpires that the mechanical properties of Glare can be significantly influenced by the characteristics of the aluminium layers, especially for crack initiation and initial crack growth. Tjerk de Vries observed how stress-strain curves and the residual strength of Glare 3 3/2-0.3 laminates were changed by using five different kinds of aluminium sheets. He investigated the high fracture toughness alloys 2024-T3 and 2024-T81, the high strength alloys 7075-T6 and 7475-T761 and the weldable alloy 6013-T6.
The tensile tests pointed out that regardless of the used metal, the characteristic bilinear behaviour of the Glare curve, resulting from the elastic behaviour of the fibres and the elastic-plastic behaviour of the aluminium, was maintained. The fracture of the specimens occurred through fibre failure at a strain close to 4.2 %, i.e. the fibre failure strain, for different stress values independent of the aluminium alloy.

As regards the residual strength of the laminate, it was observed that this is directly related to the strain hardening capacity of the aluminium layers; the higher the strain hardening, the higher the residual strength. The range of variation for the residual strength was between 219 and 240 MPa and it was observed that the stable crack extension process takes place until the load approaches the fracture value.

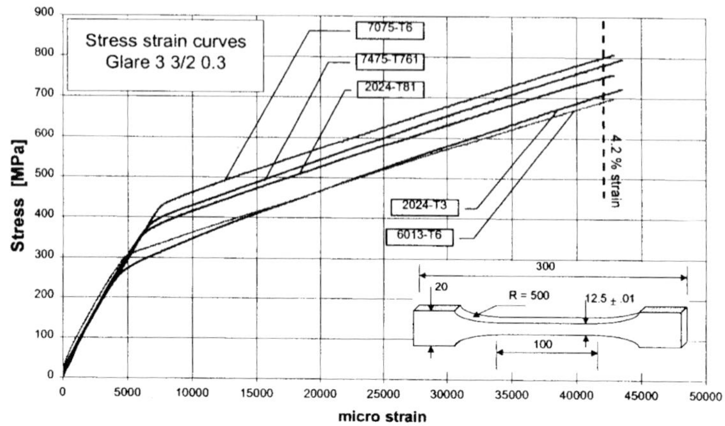

Figure 6.9 *Uniaxial tensile results for specimens with different kinds of aluminium alloys.*

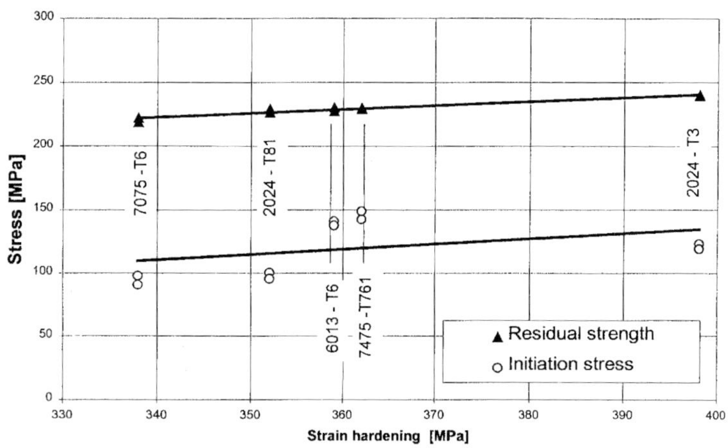

Figure 6.10 *Residual stress and crack initiation stress versus strain hardening.*

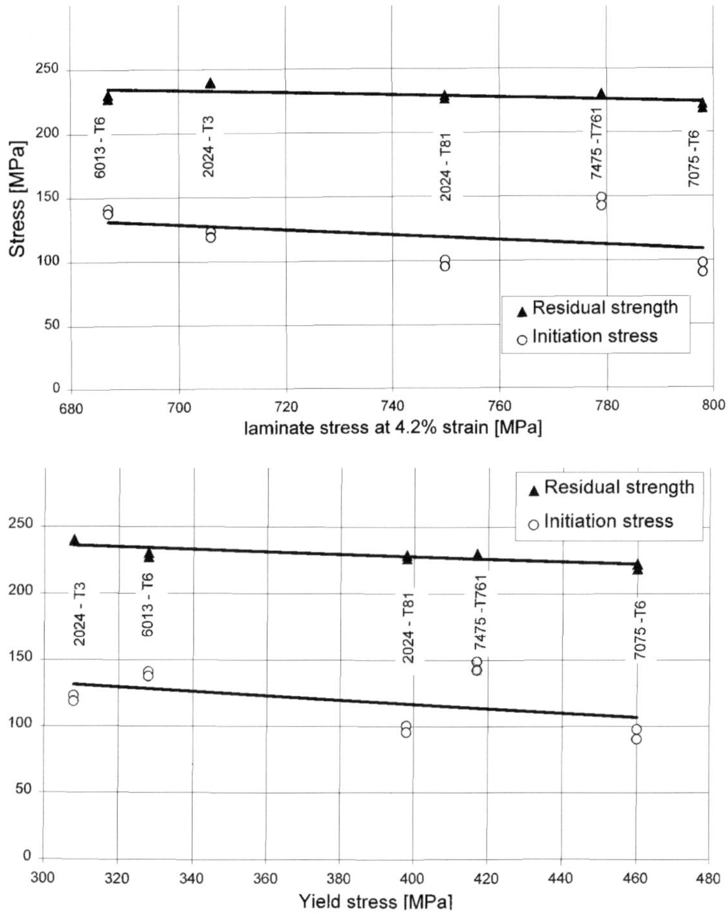

Figure 6.11 *Residual stress and crack initiation versus stress at fibre failure, and versus yield stress.*

It was also observed that the crack extensions in the aluminium sheets are initially longer than in the fibre layers, but after some crack extension and increased load the behaviour of the laminate changes to the opposite. A low yield-stress alloy delays crack initiation by means of the large delamination areas at the crack tip, but a high-strength aluminium does not increase the residual strength of the laminate.

6.5 New aluminium alloys for Glare

The underlying purpose of the programme is to improve the mechanical properties of Glare. The current, commercially available Glare grades consist of layers of glass fibres impregnated in a FM-94 hot curing adhesive and layers of aluminium 2024-

T3. The goal is to improve the mechanical properties of the laminate by using a new kind of aluminium alloy. Six aluminium alloys have been identified that provide significant improvements with respect to Glare. Some of them have never been tested with respect to an aeronautical application. They should provide better mechanical properties, especially crack growth and residual strength, than the 2024-T3 alloy.

6.5.1 Improvement of the static properties: 7xxx alloys

The goal is to deliver sheets with the highest possible strength, good fatigue performance and relatively good short-crack performance. This should improve shear yield strength, blunt notch strength, and crack initiation of the laminate.

Composition and microstructures

To fulfil the requirements of improved static properties, the Al7449 alloy was designed and qualified, and associated with three different tempers corresponding to three different static/corrosion resistance compromises. As shown in Table 6.3, Al7449 is a high Zn version of 7150, with desaturated levels of the other major additional elements Mg and Cu. High strength is achieved largely through precipitation hardening (η'/η phase, containing Mg, Zn and Cu), whereas Zr addition helps keeping an essentially unrecrystallised microstructure.

Table 6.3 Al registered chemistries of 7150, 7449 and 7349 alloys.

Alloy	Si	Fe	Mg	Cu	Zn	Cr	Zr	Ti
Al7150	0.12	0.15	2.0-2.7	1.9-2.5	5.9-6.9	0.04	0.06-0.15	0.06
Al7449	0.12	0.15	1.8-2.7	1.4-2.1	7.5-8.7	0.05	Ti+Zr: 0.25	
Al7349	0.12	0.15	1.8-2.7	1.4-2.1	7.5-8.7	0.10-0.22	Ti+Zr: 0.25	

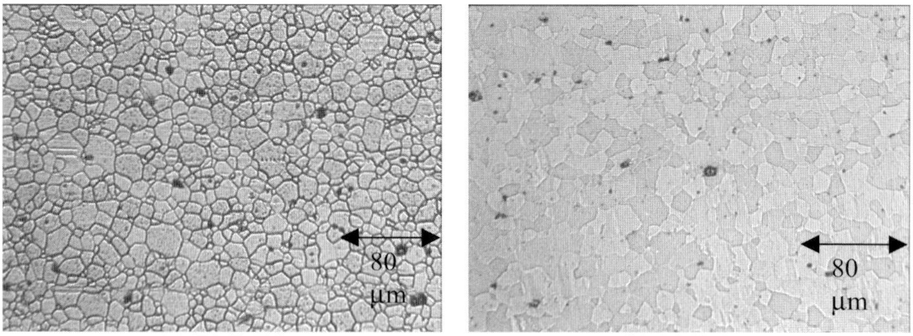

(a) 7449 T6 (b) 7449 T76

Figure 6.12 Microstructures of 7449 T6/T76 alloys.

6.5.2 High damage tolerance alloy: 2524/2024 A sheets

Composition and microstructures

2524 by Alcoa

Following the development of the first Al-Cu-Mg-Mn alloy Duralumin and its derivative 2017, alloy 2024 was developed and has found many successful applications. The importance of corrosion was subsequently demonstrated by the development of alclad 2024-T3, the surface of which corrodes rather mildly and uniformly to protect the high-strength core from localised corrosion attack. It was used as early as 1936 on the DC3 but is now being supplanted by newer alloys. Advances in linear elastic fracture mechanics led to the development of damage tolerance certification rules in the late seventies. Since then, airframers are required to demonstrate that inspection programmes would ensure the early detection of fatigue cracks before they reach a critical size that leads to catastrophic failure. While 2024-T3 has served the need, improvement in damage tolerance is required for new aircraft construction programmes. Alcoa has recently developed 2524-T3, which responds to such a need. It is a 2024 derivative requiring stringent control of alloy purity by limiting Fe and Si. Low Fe and Si minimises the formation of constituents which have been found to exert an adverse effect of fracture toughness, fatigue crack initiation, and fatigue crack growth resistance. In addition, the major alloying elements Cu, Mg and Mn are carefully balanced based on phase diagram considerations to minimise the formation of similarly damaging constituents from these alloying elements.

2024A by Pechiney

In the same way, Pechiney has also developed the similarly composed Al2024A alloys for highly damage-tolerant fuselages. As shown in Table 6.4, 2024A is a dilute, high purity version of the commonly used 2024 alloy. Cu and Mg contents are optimised to levels necessary to achieve the strength without degrading the damage tolerance properties through potential undissolved, intermetallic phases. Low Fe and Si contents also contribute to decrease the size and number density of the associated phases, responsible for lowering overall material damage tolerance.
A conventional thermo-mechanical process is then used to produce fuselage sheets in the T4 temper.

Table 6.4 Relative chemical composition of 2024 and 2024A alloys (in weight %)

Alloy	Si	Fe	Mg	Cu	Mn
Al2024	0.50	0.50	1.2-1.8	3.8-4.9	0.30-0.9
Al2024A	0.15	0.20	1.2-1.5	3.7-4.5	0.15-0.8

(a) 2024

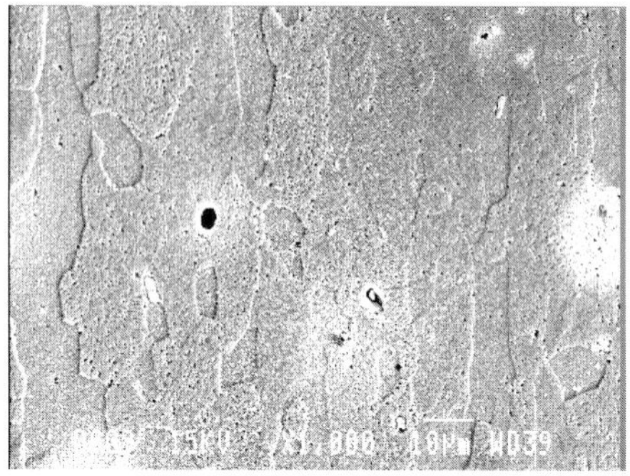

(b) 2024A

Figure 6.13 Fractography of the cross-section of tensile test specimens.

Properties

2524

The carefully controlled alloy composition is then coupled with special thermomechanical processing sequences for alloy production. The resultant microstructure is one that not only provides high toughness, but is also resistant to fatigue crack initiation and propagation, see Table 6.5.

Some specific results of the analysis show that the next alloy, 2524-T3, offers substantial residual strength and cyclic life improvements over 2024 as experimentally verified in the Multiple-Site Damage (MSD) scenario in Figure 6.14.

In addition, these improvements result in an increase of the maintenance interval, which lowers operating costs. Furthermore, the longer critical crack length for 2524 increases the margin of safety, and facilitates easier inspection. The impact of operating stresses on crack growth is calculated for 2524 and 2024 and shown in Figure 6.15 for a longitudinal fuselage skin crack under an intact frame.

2024A

The properties obtained in this way are illustrated in Figure 6.16 and compared with 2024 of the same thickness range. Identical, even slightly better, strength resistance associated with significantly improved damage tolerance is obtained with the new alloy.

Table 6.5 *Typical strength-toughness properties (1.5 mm thickness).*

Property	Test direction	Alclad 2524-T351	Alclad 2024-T351
$F_{t(u)}$ (ksi)	L LT	67 66	67 66
$F_{t(y)}$ (ksi)	L LT	50 45	50 45
Elongation (%)	LT	18	18
Kc (16" M(T)) (ksi√in)	T-L	158	128

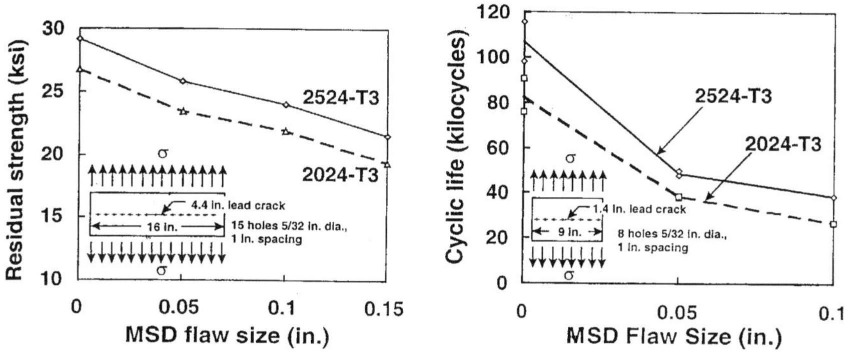

Figure 6.14 *Residual strength and cyclic life capabilities of 2524-T3 and 2024-T3 skin sheet (clad 1.27 mm thickness) in wide, multi-holed panels with central lead crack and varying size MSD crack (two per hole).*

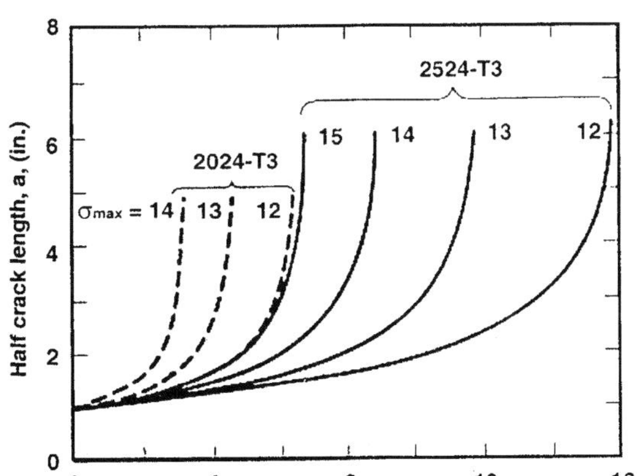

Figure 6.15 *Calculated operating stress and crack growth life for 2524 and 2024.*

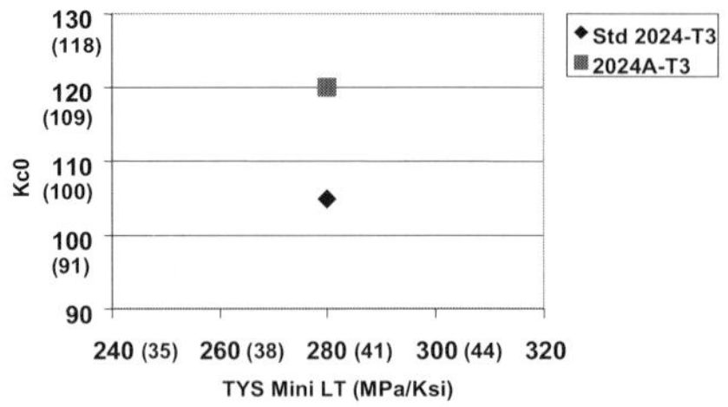

Tensile yield stress versus K_{c0}

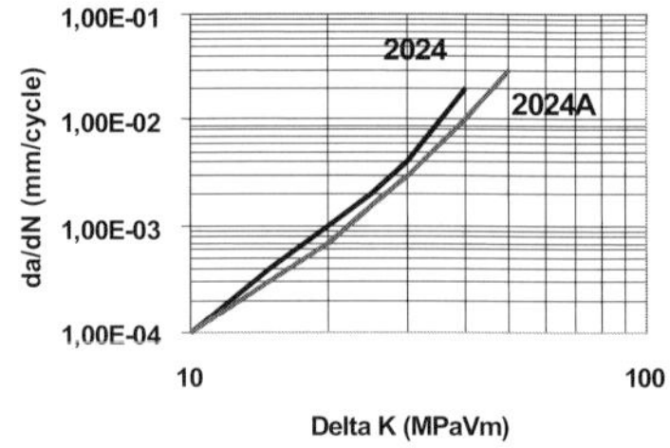

Crack growth rate

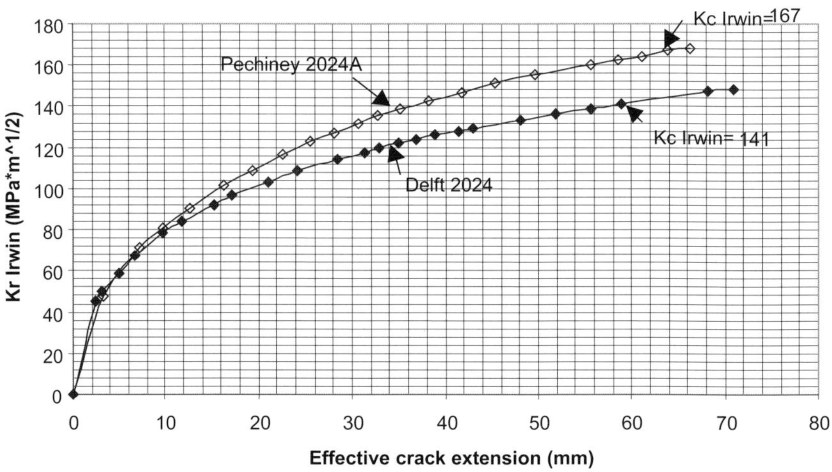

R-curves assessed with the Irwin method

Figure 6.16 Comparison between properties of the 2024 and 2024A alloys.

6.6 Conclusions

This chapter addressed the collaborative relationship between the airframe material, process, design and manufacturing team, which leads to the definition of optimised solutions through improved alloy properties and performances in use. In this way, alloys such as 7449, 2024A, 5186, and qualities such as high-formability and weldable alloys have found industrial applications, leading to weight savings and cost reductions. For Glare, the influence of the strain hardening and the R-curve of the aluminium alloys upon the residual strength of the laminate still has to be investigated, in order to optimise the properties of Glare.

References

[1] Ph. Lequeu, et al., *Int. Journal of Aircraft Engineering and Aerospace Technology*. Volume 73. 2001.

[2] G. Jay, *Comparison of the damage tolerance behavior of two aluminium alloys: 2024 and 2024 A*. TU Delft, 2000.

[3] Tj. de Vries, *Residual Strength Design Tool for Glare*. Internal report. TU Delft, 2000.

[4] R. Dif, et al., *Proceedings of the ICAA6 Conference*. Japan: Toyoashi, July 1998.

[5] J. Liu, et al., *A New Paradigm In The Design of Aluminium Alloys for Aerospace Applications.* ICCA7, 2000.

[6] M. Hagenbeek, *Feasibility of Tsai-Hill criterium for yield and ultimate strength of unnotched material.* Memorandum. TU Delft, 2001.

7

Stress-strain curve

C. van Hengel

Abstract – This section describes several methods that can be used to model the stress-strain curve of Glare.

7.1 Stress-strain

7.1.1 General

A material's stress-strain curve is a deceptively simple graph that in fact means many things to many people. To a structural designer, this curve is the first pointer on how to apply that material in the design of lightweight, robust and safe structures. The material's elastic behaviour in tension, (Young's modulus, yield stress) will determine stiffness and load bearing capability, and the plastic behaviour (Ultimate strength, breaking strain) will give a first indication about brittleness and notch sensitivity. In compression, the shape of the curve (especially near the yield stress) is important with respect to buckling resistance and other stability related behaviour. To a materials scientist the stress-strain curve has additional meaning, because it shows if there is an opportunity for modifying the material's properties by means of cold working such as stretching (which for FMLs, of course, also means modifying the internal stress-state). For manufacturing engineers, the curve is also relevant because it can be used for estimating forming limits in bending and stretching, as well as spring-back behaviour. Finally, as part of quality assurance in production, to the quality inspector the stress-strain curve is a powerful tool to compare "as produced" properties with the specified "as required" values.
Because of this multitude of uses, and the relative simplicity of the test, the stress-strain curve is always one of the first characterisations of a new material to be experimentally determined; clearly a method to predict this curve is then also desirable.

For Glare, various characteristic property values have been determined experimentally for each of the standard Glare types. The results have been collected into the so-called "MVF-method" as described in chapter 5 in this book and by Roebroeks [1]. Considering the characteristic "semi-bilinear" shape of the Glare stress-strain curve, the first and simplest method to calculate a stress-strain curve is by simply basing it directly on the MVF values as shown in Figure 7.1. This approach has the advantage of simplicity and the wealth of experimental data that support it, and may in fact be quite sufficient for many analyses.

For stability calculations however, more information is needed about the precise shape of the curve near the yield point to determine plasticity corrections, and so other methods are needed. Additionally these more detailed methods and the related material models should also preferably be able to deal with one or more of the following aspects:

- off-standard layups, such as a Glare 4 doublers on a Glare 3 skin
- off-axis loading, i.e. in other directions than L or LT, including biaxial loading
- internal stress states due to thermal stress
- effect of plasticity, such as effect on the internal stress state and effect of reversed plasticity (Bauschinger effect)
- out of plane loading (bending)
- effect of temperature.

In short, the ideal material model is able to cover all conceivable Glare combinations (and other imaginable FMLs) while taking into account all relevant mechanical phenomena including thermal stresses, reversed plasticity etc. Furthermore, *constitutive* models are obviously preferred since these require only a limited amount of component data while still being able to cover the full range of laminates. Also, sufficient *validation* of both the models and the input data (especially component values) is always needed to show that the outcome is reliable and accurate.

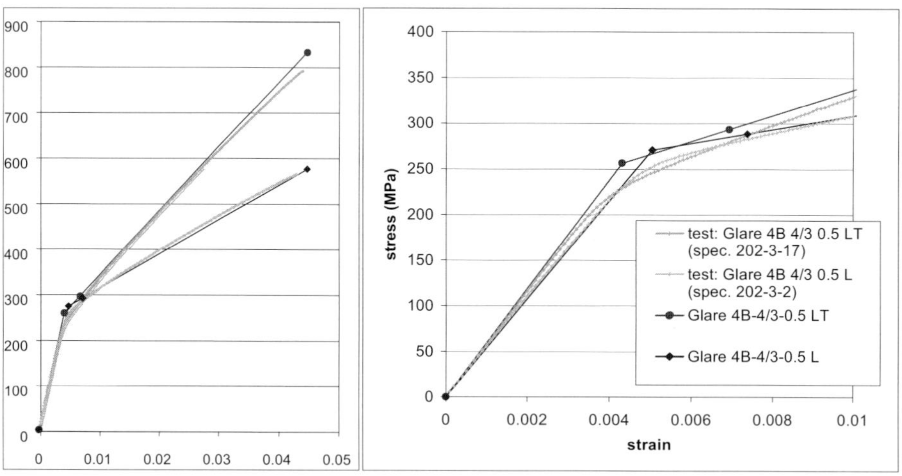

Figure 7.1 *Comparison of measured tensile stress-strain curves with synthetic bilinear curves derived directly from MVF-produced properties (Glare 4B-4/3-0.5, MVF = 64%; full-range curve and magnified curve up to yield shown).*

Obviously in a real world situation simplifications are applied to limit the model's complexity to those cases that are of primary interest. However, a word of caution is called for here in order to avoid over-simplifying such models. For example, for Glare's predecessor, the unidirectional aramid fibre reinforced Arall FML [2], the curing stresses were significantly higher than in Glare, hence the need for a process for their removal (by posture stretching). This in turn drove the need for a model to describe this [3]. With the advent of Glare however [4], temperature stresses were considered marginal because of their greatly reduced value relative to Arall and so they were essentially ignored. Later tests [5] suggest, however, that they do play a role in fatigue initiation, and thus should be included in any comprehensive material model.

In the following section, three unidirectional models are described that meet one or more of the above model requirements. It should be noted that in recent years biaxial models have also become available [6, 7, 8]. These are typically based on Classical Laminate Theory (CLT) with the addition of plasticity for the metal layers. Although such models are clearly believed to be "the way of the future", their treatment is outside the scope of the current book.

7.1.2 Methods for stress-strain curve prediction

In this section, three uniaxial constitutive material models will be described with different application areas. They were developed by (in chronological order) respectively Verolme [9], Hagenbeek [10] and Van Rijn [11].

Following on from his research on the stability of Glare structures, Verolme [9] developed a uniaxial material model, as shown in Figure 7.2, which was later adopted into the standard set of stability calculation methods for Glare [12]. The aluminium component is described by the Young's modulus E, yield stress $S_{0.2}$ and Ramberg Osgood coefficients. After yielding, the aluminium is conservatively assumed to behave ideally plastic. The "prepreg" component (i.e. the unidirectional or crossply glass fibre layer in respectively Glare 2 and Glare 3 / Glare 4) is assumed to behave linearly elastic up to failure. The resulting Glare stress-strain curve then consists of an initial part up to approximately yield that can be described by a Ramberg Osgood formula, and a post-yield linear part that only depends on the stiffness of the prepreg layer.

The main formulae describing this model are as follows (for a complete description see Verolme [9]):

Internal stress in aluminium:
$$\sigma_{i,al} = E_{al} \frac{\Delta\alpha\Delta T E_{pr} t_{pr}}{E_{pr} t_{pr} + E_{al} t_{al}} \qquad (7.1)$$

Internal stress in the prepreg: $$\sigma_{i,pr} = -\frac{\sigma_{i,al}t_{al}}{t_{pr}} \tag{7.2}$$

with: σ_i = internal stress
E = modulus of elasticity
$\Delta\alpha$ = difference in coefficient of thermal expansion (α_{pr} - α_{al})
ΔT = difference in temperature (T_{use} - T_{cure}; -100 K in al cases)

The initial part of the Glare stress-strain curve is then found by calculating the respective metal and prepreg contributions and then adding them according to the rule of mixtures. Taking the compression curve (index c) as an example, this becomes, assuming no influence of the Poisson ratio:

metal part: $$\varepsilon_{Glare} = \frac{\sigma_{al,Glare}}{E_{c,al}} - 0.002\left(\frac{\sigma_{al,Glare}}{\sigma_{cy,al}}\right)^{n_{al}} - \frac{\sigma_{i,al}}{E_{c,al}} \tag{7.3}$$

prepreg part: $$\sigma_{pr,Glare} = \sigma_{i,pr} + E_{c,pr}\varepsilon_{Glare} \tag{7.4}$$

and finally: $$\sigma_{Glare} = \frac{\sigma_{al,Glare}t_{al} + \sigma_{pr,Glare}t_{pr}}{t_{al} + t_{pr}} \tag{7.5}$$

with: ε_{Glare} = strain of the Glare laminate
$\sigma_{al,Glare}$ = stress of the aluminium in the Glare laminate
$\sigma_{ey,al}$ = compressive yield stress of the aluminium
n = Ramberg Osgood shape parameter

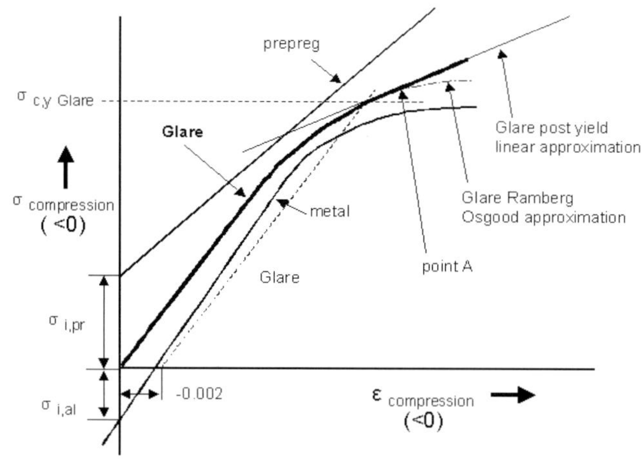

Figure 7.2 Compressive stress strain curve of Glare (schematic).

The compressive yield stress is found by determining the intersection of this last equation and the strain at compressive yield, i.e.:

$$\varepsilon_{0.2,c} = -0.002 + \frac{\sigma_{0.2,c}}{E_{c,Glare}} \tag{7.6}$$

Where the modulus of elasticity of Glare can be approximated by the rule of mixtures:

$$E_{c,Glare} = \frac{E_{c,al} t_{al} + E_{c,pr} t_{pr}}{t_{al} + t_{pr}} \tag{7.7}$$

The Ramberg-Osgood shape parameter n for the Glare can now be determined by also determining the stress at 0.1% offset strain in a similar way, and substituting these values in the formula for n:

$$n_c = \frac{\ln\left(\dfrac{\varepsilon_{0.1,c} - \dfrac{\sigma_{0.1,c}}{E_{Glare}}}{-0.002}\right)}{\ln\left(\dfrac{\sigma_{0.1,c}}{\sigma_{cy}}\right)} \tag{7.8}$$

After metal yielding, the increasing stress of the Glare stress-strain curve can conservatively be assumed to be exclusively due to the glass fibre layers. This minimum E_t is:

$$E_{t,cons} = \frac{E_{pr} \cdot t_{pr}}{t_{Glare}} \tag{7.9}$$

The stress and strain at the point of tangency (point A in Figure 7.3) are then found to be:

$$\sigma_A = \sigma_{0.2,c} \cdot \left\{ \left(\frac{1}{E_{t,coms,1}} - \frac{1}{E_1} \right) \cdot \frac{\sigma_{0.2,c}}{n \cdot 0.002} \right\}^{\frac{1}{n-1}} \tag{7.10}$$

$$\varepsilon_A = \frac{\sigma_A}{E_1} + 0.002 \cdot \left(\frac{\sigma_A}{\sigma_{0.2,c}} \right)^n \tag{7.11}$$

The component values that were used and which, at the time, were the SLC standard values [13], are given in Table 7.1.

As shown in Wimersma [12], the resulting calculated Glare curves fitted the available test data quite accurately for the area of interest, i.e. in compression below yield. As shown in Figure 7.2, Verolmes model also included thermal stresses. Figure 7.3 shows that their inclusion improved the accuracy of the prediction of the Glare yield stress.

However, as already pointed out by Bär [14], the thermal stresses should *not* be used to estimate the effect of temperature changes on Glare properties (especially yield stress). Bär's research included tests under hot and cold conditions, and showed that the effect of temperature on the components is much larger than that of the internal stresses. Incidentally, Bär's material model was similar to that of Verolme but more sophisticated: it included the full biaxial description of the thermal stresses from Classical Laminate Theory, and the aluminium description included strain hardening as well as reversed plasticity.

In an alternative model, aimed at the prediction of non-standard FMLs, Hagenbeek [10] developed a unidirectional model wherein the components are described by quadrilinear behaviour (bilinear tension part plus bilinear compression part). Thermal stresses are included based on Classical Laminate Theory. When using the same set of properties as Verolme, the model gives reasonably accurate predictions in comparison with published Glare properties [11].

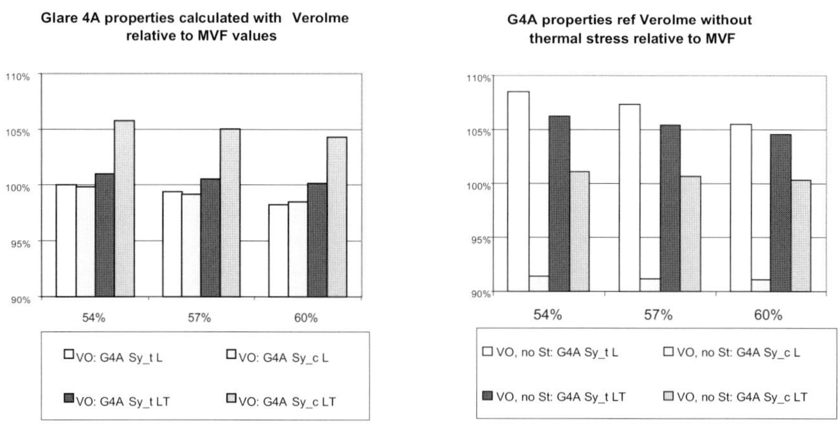

Figure 7.3 *Comparison between tensile and compressive yield stresses in L and LT directions calculated using Verolme's method and the corresponding values according to the MVF method (example for Glare 4A for three MVF values; comparison for typical values). Note that the values including thermal stresses (left) are more accurate than those without (right).*

Table 7.1 Mechanical properties of 2024 T3 and glass/epoxy prepreg.

Material	Direction	E (MPa)	σ_{ty} (MPa)	σ_{cy} (MPa)	α (1/K)	n
2024-T3	L	73700	356.5	-295.8	2.34E-05	10.5
	LT	73700	318.6	-341.3	2.34E-05	10.5
UD Glass	L	53980	-	-	6.15E-06	-
	LT	9412	-	-	2.62E-05	-
CP50 Glass	L, LT	31696	-	-	9.10E-06	-
CP70 Glass	L	39124	-	-	8.32E-06	-
	LT	24268	-	-	1.19E-05	-

Yield stresses: 1.1 times MIL-HDBK-5 values; CP50 Glass = Prepreg of Glare3; CP70 Glass = prepreg of Glare4

The two previous methods started from "a priori" material models and component descriptions, and were then verified by checking their predictions against measured curves. Instead, Van Rijn [11] developed models for Glare 3 and Glare 4 by starting out from a large set of measured Glare stress-strain curves. From these he developed stress-strain curves for the components that, when re-assembled, reproduced the original Glare curves. Parameters in this investigation were Glare type (Glare 3 and Glare 4B), metal thickness (0.3, 0.4 and 0.5 mm), loading direction (tension and compression) and off-axis angle (L, LT and, for tension only, 45 degrees). The van Rijn model does not include thermal stresses, since these are not "visible" at the laminate level (for this reason they also are not included in the MVF method). By way of example, Figure 7.4 shows a comparison of several property values calculated using van Rijn's model with the corresponding values from the MVF method [1].

In general the correlation is quite good, especially when taking into account the fact that the deviation in ultimate strength is due to a difference in definition: for practical reasons, van Rijn's model uses an ultimate strain of 4.2% rather than the more typical value of about 4.5% found in most tests. This good correlation was not a foregone conclusion because the data used by van Rijn are not part of the data used to develop the MVF method. This result indicates that there is a potential for a constitutive model that can replicate data-points from the MVF method with sufficient accuracy.

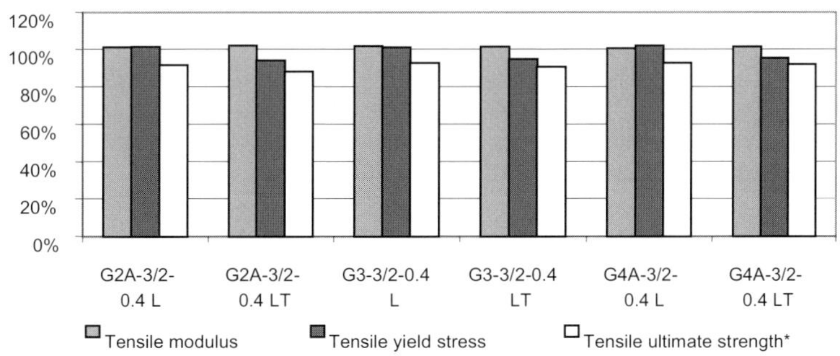

Figure 7.4 Comparison of van Rijn property values and MVF values (index).

References

[1] G.H.J.J. Roebroeks, *The Metal Volume Approach*, Structural Laminates Industries, Technical Report TD-R-00-003 (restricted), Delft, The Netherlands, 2000.

[2] R. Marissen, *Fatigue crack growth in ARALL – A hybrid aluminium*-aramid *composite material*, Faculty of Aerospace Engineering, Report LR-574 and Ph D Thesis, Delft, The Netherlands, Juni 1988.

[3] D.G. van Hengel, *ARALL static properties: An engineering theory*, Faculty of Aerospace Engineering, Masters Thesis, Delft, The Netherlands, November 1980.

[4] G.H.J.J. Roebroeks, *Towards Glare – The development of a fatigue insentsitive and damage tolerant aircraft material*, Faculty of Aerospace Engineering, Ph D Thesis, Delft, The Netherlands, December 1991.

[5] A.U. de Koning, *Verification of the applicability of S-N data of monolithic aluminium for estimation of the fatigue crack initiation life of GLARE coupons*, National Aerospace Labaratory, Report NLR-CR-99161 (restricted), Amsterdam, The Netherlands, April 1999.

[6] M. Kawai et al., *Inelastic behaviour and strength of Fiber-Metal Hybrid Composite: Glare*, Int. J. Mech. Sci. Vol. 40 Nos 2-3, pp. 183-198, 1998.

[7] T.C. Wittenberg and van Baten, *Plastic buckling analysis of flat rectangular FML plates loaded in shear*, in Proceedings of the 32nd International SAMPE Technical Conference (Boston, USA), November 2000.

[8] T. de Jonge, *Plasticity correction for Glare plates loaded in uniaxial compression*, in: "Survey of Taskforce Activities", Fokker Aerostructures, Report B2V-01-13 (restricted), April 2001.

[9] J.L. Verolme, *Prediction of Stress-strain curves of Glare*, Structural Laminates Company, Technical Report TD-R-96-004, Delft, The Nether-lands, May 1996.

[10] M. Hagenbeek, *Estimation Tool for basic Material Properties*, Faculty of Aerospace Engineering, Report B2V-00-29 (restricted), Delft, The Netherlands, March 2000.

[11] J.C.F.N. van Rijn, *A calculation method for the stress-strain curves of Glare 3 and Glare 4B*, National Aerospace Laboratory, Report NLR-CR-2000-172 (restricted), Amsterdam, April 2000.

[12] B. Wimersma, *Stability of Glare Structures – Calculation Methods,* Structural Laminates Company, Technical Report TD-R-97-001, Delft, The Netherlands, February 1997.

[13] Customer data sheet, *Properties of FML Constituents – Prepregs*, Structural Laminates Company, No. 2.200, Delft, The Netherlands, October 1993.

[14] H. Bär, *Verifikation und Ergänzung von Berechnungsmethoden für die statische und dynamische Auslegung von GLARE-strukturen* (in German), Institut für Flugzeugbau der Universität Stuttgart, Diplomarbeit, Stuttgart, Germany, August 1992.

[15] G.H.J.J. Roebroeks, *GLARE®; a structural material for fire resistant aircraft fuselages*, presented October 1996 at Propulsion and Energetics Panel (PEP) 88th Symposium, AGARD conference proceedings 587, September 1997.

8

Shear yield strength

M. Hagenbeek

Abstract – Shear loads in an aircraft fuselage will occur as a result of bending and torsion. Yield values under shear are necessary during the design phase since the material should not deform plastically below the limit load. The Metal Volume Fraction (MVF) method for shear yield strength prediction of Glare laminates was investigated. The maximum deviation of the Iosipescu shear test data from the prediction line is 6.5%. The results indicate a good applicability of the MVF method.

8.1 Predicting the shear yield strength

Shear loads in an aircraft fuselage will occur due to bending and torsion. Yield values under shear are necessary during the design phase since the material should not deform plastically below the limit load. The shear yield strength for a given Glare laminate can be obtained through testing. A prediction method is required to avoid time consuming and costly testing of each separate laminate.

Wu, Slagter and Verolme [1] stated that the Metal Volume Fraction (MVF) method could be used to estimate material properties for Fibre Metal laminates. A linear relationship between the lower and upper bound MVF shear yield strength is assumed. The upper bound is the 100% MVF or (typical) aluminium shear yield strength. The lower bound is the 0% MVF or shear yield strength of the prepreg, which depends on the fibre orientation. The MVF method for shear yield strength prediction of Glare laminates has been investigated. For the validation of the method, test data for Glare 3, Glare 4A and Glare 4B laminates under 0, 45 and 90 degrees relative to the rolling direction were used that were taken from literature [1, 2, 3, 4]. The Iosipescu shear test method used in the literature is described in the next section. The MVF method predictions are given in section 8.3 [5].

8.2 The Iosipescu shear test method

The shear test results that are taken from literature [1, 2, 3, 4], are generated using the Iosipescu in-plane shear test method. This test is described in ASTM D-5379-M-93. It contains a specimen as shown in Figure 8.1, and a test fixture as shown in Figure 8.2.

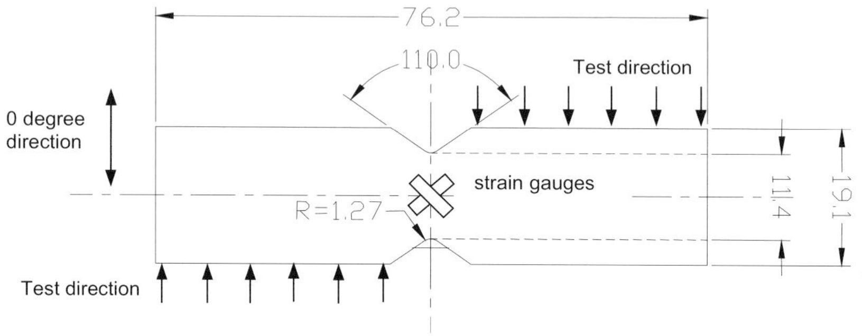

Figure 8.1 Iosipescu specimen with the material orientation and the test direction.

Figure 8.2 Iosipescu test setup.

8.3 The MVF prediction

The MVF method assumes a linear relationship between the lower and upper bound MVF shear yield strength. The upper bound is the 100% MVF or (typical) aluminium shear yield strength. This shear yield strength can be estimated as 0.5 times the tensile yield strength. A typical tensile yield strength of 340 MPa (in the L-direction) and a typical shear yield strength of 170 MPa for aluminium 2024-T3 was used.

The lower bound is the 0% MVF or shear yield strength of the prepreg, which depends on the fibre orientation. Back-calculation of the shear yield strength from the stress percentage in the prepreg for all Glare configurations and layups leads to a mean typical shear yield strength of 15.7 MPa for the 0 and 90 degrees fibre orientation, and 144.5 MPa for the 45 degrees fibre orientation. These calculated shear yield strength values of the prepreg are virtual values, which do not have a physical meaning.

The shear yield strength values as a function of the MVF for Glare 3, Glare 4A and Glare 4B are shown in Figure 8.3, Figure 8.4 and Figure 8.5, respectively The maximum deviation of the test data from the assumed prediction-line is 6.5%. The maximum deviation for each Glare configuration and test direction is listed below in Table 8.1 [5].

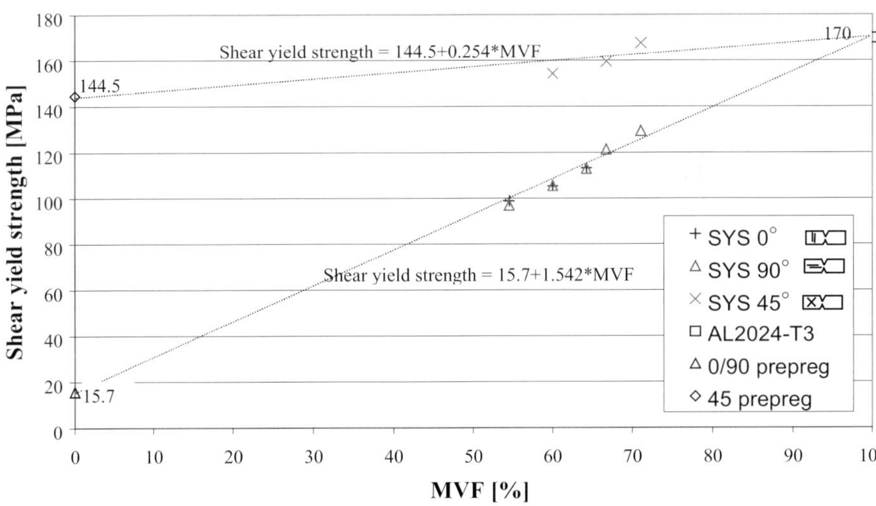

Figure 8.3 Shear yield strength versus MVF for Glare 3 (SYS= Shear Yield Strength).

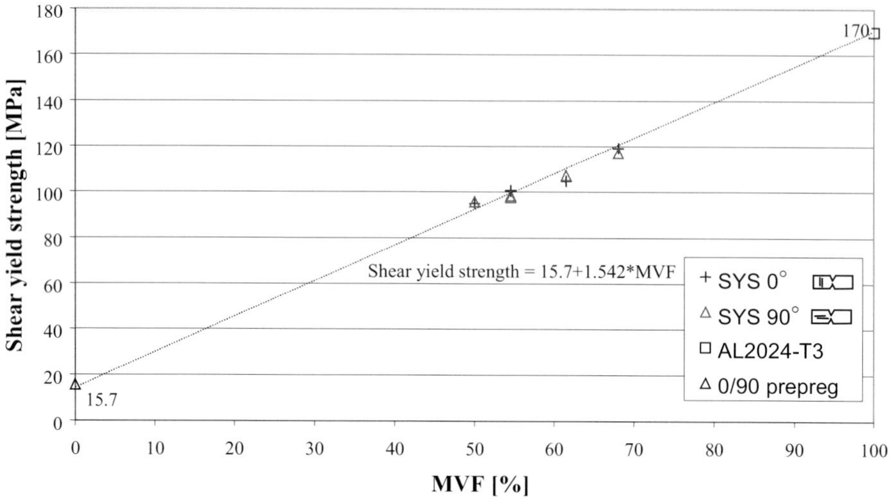

Figure 8.4 Shear yield strength versus MVF for Glare 4A (SYS = Shear Yield Strength).

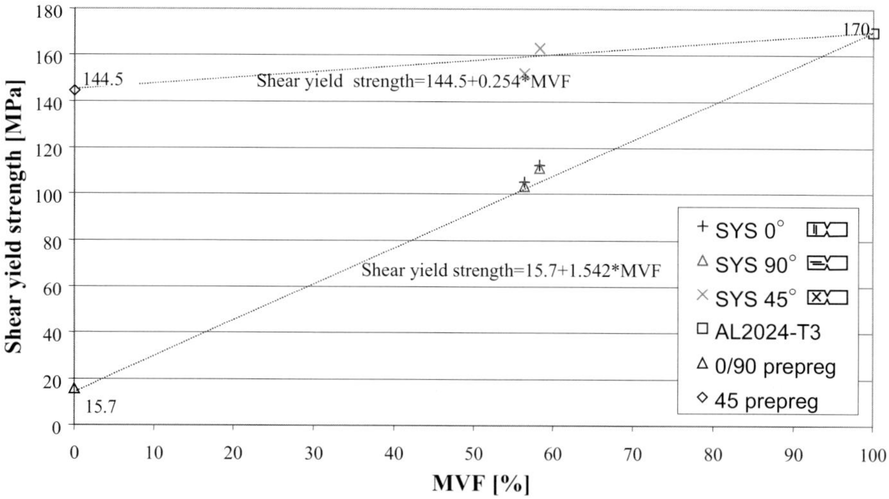

Figure 8.5 Shear yield strength versus MVF for Glare 4B (SYS = Shear Yield Strength).

Table 8.1 Shear yield prediction and test results for Glare 3, Glare 4A, and Glare 4B.

Glare grade	MVF [%]	test direction [degrees]	prediction [MPa]	test [MPa]	max. deviation [%]
Glare 3	60.00	45	159.7	154.3	-3.4
Glare 3	71.02	0/90	125.2	129.4	3.4
Glare 4A	61.54	0/90	110.6	104.8	-5.3
Glare 4B	56.38	45	158.8	152.1	-4.2
Glare 4B	58.26	0/90	105.5	112.4	6.5

8.4 Conclusions

- The results indicate a good applicability of the MVF method.
- The shear yield strength of Glare tested in the 0 and 90 degrees directions is below the value for aluminium 2024-T3, but the strength can be largely improved by using a 45 degrees orientation if necessary.
- The layers with 0 and 90 degrees orientation apparently still have a small contribution to the shear strength.

References

[1] H.F. Wu, W.J. Slagter and J.L. Verolme, *A pilot study of metal fraction approach for fiber metal laminates*, Product Design and Mechanical Division, Alcoa Technical Center, U.S.A., Faculty of Aerospace Engineering, Delft University of Technology, The Netherlands, ATC Report No. 57-92-17, SLC Report No. SL-020-B, December 1992.

[2] A.C. Mattousch, *Shear properties of Glare 3-3/2-0.2 and Glare 3-3/2-0.*3, SLI report TD-R-95-001, 7 March 1995.

[3] P.K. Kuijpers, *The influence of material changes on the shear strength of Glare 3 lamin*ates, report B2v-98-30, Delft University of Technology, November 1998.

[4] P.L. Kuijpers, *Results of shear tests*, report B2v-99-36, Delft University of Technology, October 1999.

[5] M. Hagenbeek, *The MVF method for shear yield strength prediction*, report B2v-99-26, Delft University of Technology, October 1999.

9

Blunt notch strength

O.J. Bosker

Abstract - This chapter describes the work performed to investigate the feasibility of using the Norris failure criterion for blunt notch strength prediction in Glare under arbitrary in-plane loading conditions (especially tension and shear). The described Norris method for Glare blunt notch is based on the experimentally determined blunt notch strengths in L-direction and LT-direction and the blunt notch strength under a 45° fibre off-axis angle.

The Norris failure model was found to be very accurate in predicting multiaxial (biaxial and shear component) blunt notch strengths of Glare based on the Metal Volume Fraction approach. Moreover, a set of formulae is given to calculate the failure strength of an arbitrary laminate.

9.1 Introduction

The blunt notch strength is defined as the strength of a structure containing a hole. This blunt notch strength is an important design parameter because a fuselage structure contains many holes at locations where connections are made (rivets, Hi-Loks, bolts, etc.). Moreover, blunt notches occur in various other forms in an airframe, such as windows, doors and hatches.

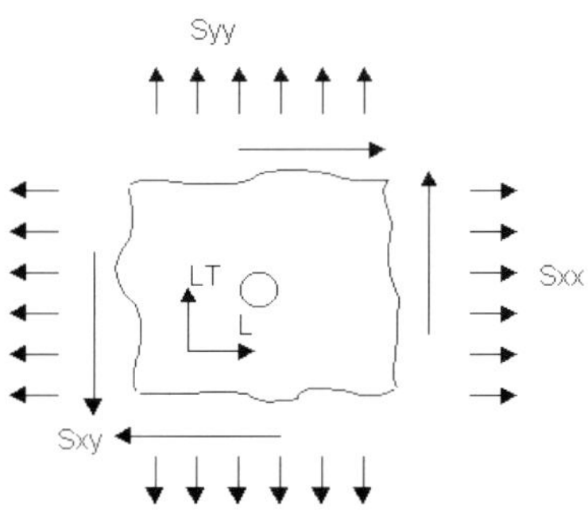

Figure 9.1 Blunt notch under in-plane loading condition.

Combined tension, tension/compression and shear stresses load the fuselage skin as depicted in Figure 9.1. A method to calculate the failure stress levels of the skin with rivet holes is therefore required.

This chapter describes the work performed to investigate the feasibility of using the Norris failure criterion for blunt notch strength prediction in Glare under arbitrary in-plane loading conditions (especially tension and shear). The described Norris method for Glare blunt notch strength is based on the experimentally determined blunt notch strengths in L-direction, LT-direction and under 45° off-axis.

The problem was approached in two steps:

- to check the applicability of the Norris method to predict blunt notch failure
- to extend this method to a Metal Volume Fraction based method.

The blunt notch strength based on the Norris failure criterion for Glare 3, Glare 4A and Glare 4B under arbitrary load combinations will be presented in the form of a number of graphs, wherein for some relevant Glare types[1] the blunt notch strength in L-direction (*Sxx*) is plotted versus the blunt notch strength in LT-direction (*Syy*) for several different shear stresses (*Sxy*).

These graphs are *Sxx-Syy* sections at different levels of *Sxy* of the general failure envelope as depicted in Figure 9.2.
The developed Norris criterion based on the Metal Volume Fraction approach will be validated with the extensive database of experimental blunt notch results. Finally, a set of formulae will be given to calculate the failure strength of an arbitrary laminate.
The presented method currently works for room temperatures only. Note that all strengths in this chapter are net strengths at room temperature and unconditioned.

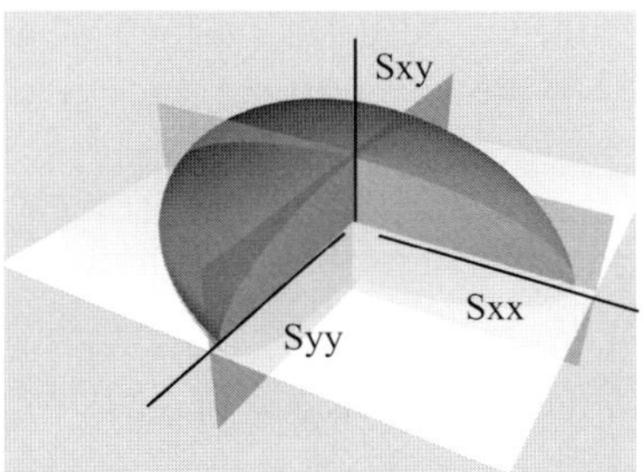

Figure 9.2 Typical Norris blunt notch failure envelope for Glare.

[1] Glare **3** 3/2-0.3, 4/3-0.4 and Glare **4A** 4/3-0.4

9.2 Norris interactive failure criterion

The Norris theory is obtained from the general Hill theory and predicts failure if the following equation is satisfied (*xx* denotes the material L-direction and the *yy* the material LT-direction) [1]:

$$\left(\frac{Sxx}{Sxx_{ult}}\right)^2 - \left(\frac{Sxx}{Sxx_{ult}}\right)\left(\frac{Syy}{Syy_{ult}}\right) + \left(\frac{Syy}{Syy_{ult}}\right)^2 + \left(\frac{Sxy}{Sxy_{ult}}\right)^2 = 1 \qquad (9.1)$$

The theory predicts the multi-axial blunt notch strength by interpolating elliptical curves through three experimentally determined values (Sxx_{ult}, Syy_{ult} and Sxy_{ult}). Sxx_{ult} and Syy_{ult} are the uniaxial blunt notch strengths in respectively L- and LT-direction and can either be determined directly from uniaxial blunt notch tests on a particular laminate or indirectly, i.e. derived from Metal Volume Fraction (MVF) values. The shear blunt notch strength Sxy_{ult} of Glare has not been experimentally determined to date. However, this shear blunt notch strength can be deduced from uniaxial off-axis blunt notch tests. These uniaxial off-axis strengths in a load co-ordinate system can be transformed to stresses in material principal directions using the stress transformation matrix:

$$\begin{bmatrix} Sxx \\ Syy \\ Sxy \end{bmatrix} = \begin{bmatrix} \cos^2(\varphi) & \sin^2(\varphi) & -2\cdot\cos(\varphi)\sin(\varphi) \\ \sin^2(\varphi) & \cos^2(\varphi) & 2\cdot\cos(\varphi)\sin(\varphi) \\ \cos(\varphi)\sin(\varphi) & -\cos(\varphi)\sin(\varphi) & \cos^2(\varphi)-\sin^2(\varphi) \end{bmatrix} \cdot \begin{bmatrix} Sxx' \\ Syy' \\ Sxy' \end{bmatrix} \qquad (9.2)$$

In which *Sxx*', *Syy*' and *Sxy*' are the stresses in load co-ordinate system, and φ is the angle with respect to the material L-direction as depicted in Figure 9.3.

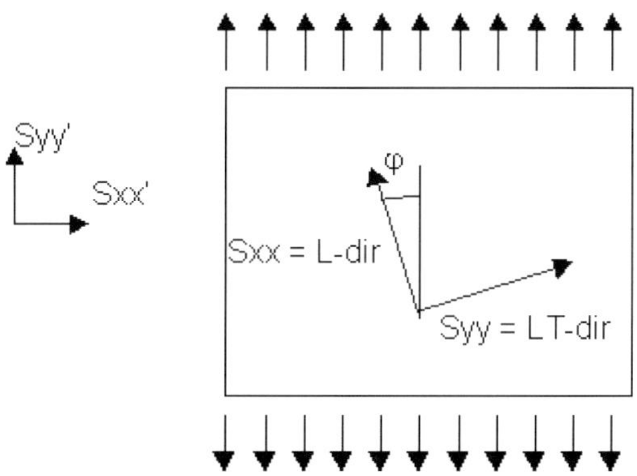

Figure 9.3 Definitions for uniaxial blunt notch experiments.

For the 45° off-axis situation this yields:

$$Sxx = cos^2(45^o) \cdot S_{45} = 0.5 \cdot S_{45}$$
$$Syy = sin^2(45^o) \cdot S_{45} = 0.5 \cdot S_{45}$$
$$Sxy = sin(45^o) \cdot cos(45^o) \cdot S_{45} = 0.5 \cdot S_{45}$$

In which S_{45} is the net strength under a 45° fibre off-axis angle. Sxy_{ult} can now be determined for a specific Glare grade and layout from Equation 9.1:

$$Sxy_{ult} = \sqrt{\frac{S_{45}{}^2}{4 - \left(\frac{S_{45}}{Sxx_{ult}}\right)^2 + \left(\frac{S_{45}}{Sxx_{ult}}\right)\left(\frac{S_{45}}{Syy_{ult}}\right) - \left(\frac{S_{45}}{Syy_{ult}}\right)^2}} \qquad (9.3)$$

Sxx_{ult} and Syy_{ult} are taken as net strengths from Roebroeks [2] and S_{45} is taken as net strength from Bosker [3]. The material properties Sxx_{ult}, Syy_{ult} and S_{45} will be described and calculated in the next section.

9.3 Predicted multiaxial blunt notch strength compared to experimental results

In this chapter, the Norris predicted biaxial blunt notch strengths are compared to experimental results. Within the Glare Technology Development programme an extensive number of blunt notch tests have been carried out. The vast majority of these experiments were uniaxial. To use these (on-axis and off-axis) test results to validate a biaxial blunt notch failure prediction tool, the resulting uniaxial strength values have to be transformed to biaxial stresses in the material direction. This can be done using the transformation matrix Equation 9.2. This calculation has been done for the test results from Bosker [3].

The Norris predicted blunt notch strength can be calculated with actual Sxx_{ult}, Syy_{ult} and Sxy_{ult} values, as will be done in the next section, as well as with calculated MVF-values, as will be done in section 9.4.

In Figure 9.4 the strength in L-direction, Sxx, versus the strength in LT-direction, Syy, is depicted for Glare 3-3/2-0.3 at various levels of shear stress. The Norris predictions from Equation (9.1) are plotted using actual Sxx_{ult}, Syy_{ult} and Sxy_{ult} values (respectively 445.6, 431.4 and 168.9 MPa, see Table 9.1). Experimental values are taken from Bosker [3].

It can be seen that there is a very good correspondence between the predicted values and the experimentally determined values. In the next section this promising result will be extended to a Metal Volume Fraction based method.

Table 9.1 *Experimental uniaxial strength values for Glare 3-3/2-0.3 (in MPa).*

	Off-axis angle (φ)	Test results [3]	Mean test result		
Sxx_{ult}	L-dir.	1: 447.1 2: 444.6 3: 445.2	445.6		
Syy_{ult}	LT-dir.	1: 432.3 2: 427.6 3: 434.4	431.4		
	Off-axis angle (φ)	**Test results [3]**	**Equation 9.3**	**Sxy_{ult} with Equation 9.3**	
				Test	**Mean**
S_{45}	45°	1: 317.1 2: 310.4 3: 316.1	$Sxy_{ult} = \sqrt{\dfrac{S_{45}^{2}}{4 - \left(\dfrac{S_{45}}{Sxx_{ult}}\right)^{2} + \left(\dfrac{S_{45}}{Sxx_{ult}}\right)\left(\dfrac{S_{45}}{Syy_{ult}}\right) - \left(\dfrac{S_{45}}{Syy_{ult}}\right)^{2}}}$	1: 166.4 2: 169.9 3: 170.5	168.9

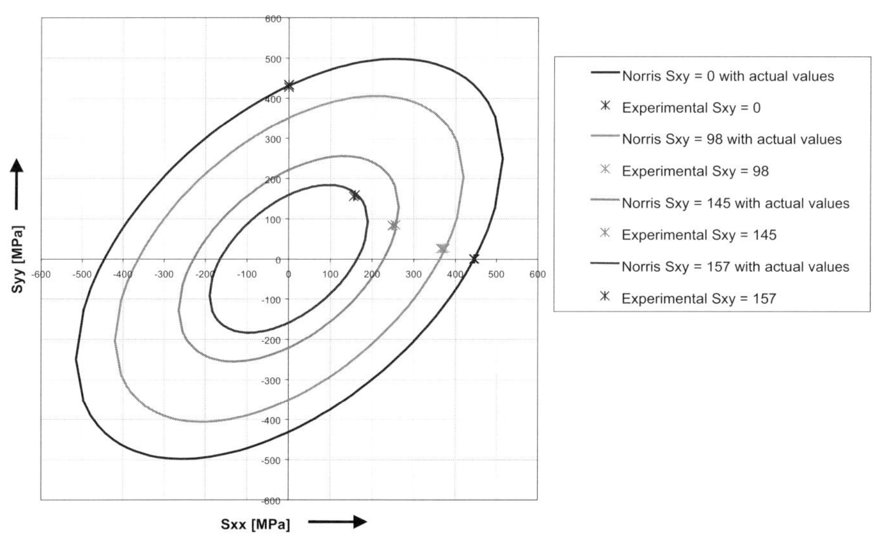

Figure 9.4 *Norris predictions for Glare3-3/2-0.3 compared to experimental data from Bosker* [3], *for Norris actual* Sxx_{ult}, Syy_{ult} *and* Sxy_{ult} *are used.*

9.4 Metal Volume Fraction predicted Sxx_{ult}, Syy_{ult} and Sxy_{ult} values

The necessary material properties, Sxx_{ult}, Syy_{ult} and Sxy_{ult}, used in the Norris Equation 9.1, can either be determined directly from uniaxial blunt notch tests on a particular laminate, as was shown in section 9.3 or indirectly, i.e. derived from Metal Volume Fraction values. The uniaxial blunt notch strengths used in Equation 9.1, will be calculated with the Metal Volume Fraction method [2]:

$$Si_{ult} = MVF \cdot Si_{alu} + (1 - MVF) \cdot Si_{fiber} \tag{9.4}$$

with: i = xx (defined as L-direction), yy (defined as LT-direction)
Si_{alu} = the aluminium blunt notch strength in both material directions
Si_{fibre} = the blunt notch strength of the prepreg layers.

These material properties were derived in [2] and are summarised in Table 9.2.

The shear blunt notch strength of Glare, Sxy_{ult}, can now be calculated from Equation 9.3 and 9.4 and the test values for 45° off-axis blunt notch experiments, S_{45}, from Bosker [3]. The results are listed in Table 9.3 for Glare 3. This has been done for Glare 4A, Glare 4B and Glare 2 as well.

In Figure 9.5, the resulting shear blunt notch strength, Sxy_{ult}, values for Glare 2, Glare 3, Glare 4A and Glare 4B are depicted in one graph. The shear strength of Glare can be expected to be dominated by its aluminium content, since the shear strength of the prepreg is small. This means that the shear strength should only depend on the MVF and not on the Glare grade. Figure 9.5 shows that this assumption holds up reasonably well. The Sxy_{ult} values for all Glare laminates (Glare 2, Glare 3, Glare 4A and Glare 4B) can be calculated with Equation 9.5 (derived by applying least square method on the experimental results.

$$\begin{aligned} Sxy_{ult} &= MVF \cdot Sxy_{alu} + (1 - MVF) \cdot Sxy_{fiber} \\ Sxy_{ult} &= MVF \cdot 248.8 + (1 - MVF) \cdot 51.9 \end{aligned} \tag{9.5}$$

The material properties calculated in this section are summarised in Table 9.4.

With these values and Equations 9.1 and 9.4, the biaxial blunt notch strength for any Glare grade (Glare 2A, 2B, 3, 4A and 4B) and for any layup (2/1, 3/2, etc.) can be calculated using Norris.

Table 9.2 *Raw material contributions to Glare blunt notch strength parameters* [2].

Glare	Layer contribution	L-direction	LT-direction	MVF
Glare 2A	Metal (Si_{alu})	417 MPa	394 MPa	1.0
	Fibre (Si_{fibre})	1193 MPa	0 MPa	0.0
Glare 2B	Metal (Si_{alu})	417 MPa	394 MPa	1.0
	Fibre (Si_{fibre})	0 MPa	1193 MPa	0.0
Glare 3	Metal (Si_{alu})	417 MPa	394 MPa	1.0
	Fibre (Si_{fibre})	597 MPa	597 MPa	0.0
Glare 4A	Metal (Si_{alu})	417 MPa	394 MPa	1.0
	Fibre (Si_{fibre})	795 MPa	398 MPa	0.0
Glare 4B	Metal (Si_{alu})	417 MPa	394 MPa	1.0
	Fibre (Si_{fibre})	398 MPa	795 MPa	0.0

Table 9.3 *Uniaxial strength values (in MPa) for Glare 3 Bosker [3].*

Glare layout	MVF	S_{45} experiment [3]	Sxx_{ult} from (9.5)	Syy_{ult} from (9.5)	Sxy_{ult} from (9.3)
Glare 3 3/2-0.3	0.643	317.1	480.2	466.5	168.6
Glare 3 3/2-0.3	0.643	310.4	480.2	466.5	164.6
Glare 3 3/2-0.3	0.643	316.1	480.2	466.5	168.0
Glare 3 4/3-0.4	0.681	325.4	473.5	458.8	174.0
Glare 3 4/3-0.4	0.681	323.0	473.5	458.8	172.5
Glare 3 4/3-0.4	0.681	323.1	473.5	458.8	172.6
Glare 3 4/3-0.5	0.727	341.4	465.3	449.4	184.5
Glare 3 4/3-0.5	0.727	340.3	465.3	449.4	183.9
Glare 3 4/3-0.5	0.727	348.9	465.3	449.4	189.3
Glare 3 6/5-0.4	0.658	344.3	477.5	463.4	185.4
Glare 3 6/5-0.4	0.658	342.9	477.5	463.4	184.5
Glare 3 6/5-0.4	0.658	340.1	477.5	463.4	182.8
Glare 3 8/7-0.4	0.646	331.7	479.7	465.9	177.5
Glare 3 8/7-0.4	0.646	328.8	479.7	465.9	175.7
Glare 3 8/7-0.5	0.696	345.8	470.8	455.7	186.9
Glare 3 8/7-0.5	0.696	355.3	470.8	455.7	192.9

Table 9.4 *Raw material contributions to Glare blunt notch strength parameters.*

Glare	layer contribution	Sxx_{ult} (MPa) (L-direction)	Syy_{ult} (MPa) (LT-direction)	Sxy_{ult} (MPa)	MVF
Glare 2A	Metal (Si_{alu})	417	394	249	1
	Fibre (Si_{fibre})	1193	0	52	0
Glare 2B	Metal (Si_{alu})	417	394	249	1
	Fibre (Si_{fibre})	0	1193	52	0
Glare 3	Metal (Si_{alu})	417	394	249	1
	Fibre (Si_{fibre})	597	597	52	0
Glare 4A	Metal (Si_{alu})	417	394	249	1
	Fibre (Si_{fibre})	795	398	52	0
Glare 4B	Metal (Si_{alu})	417	394	249	1
	Fibre (Si_{fibre})	398	795	52	0

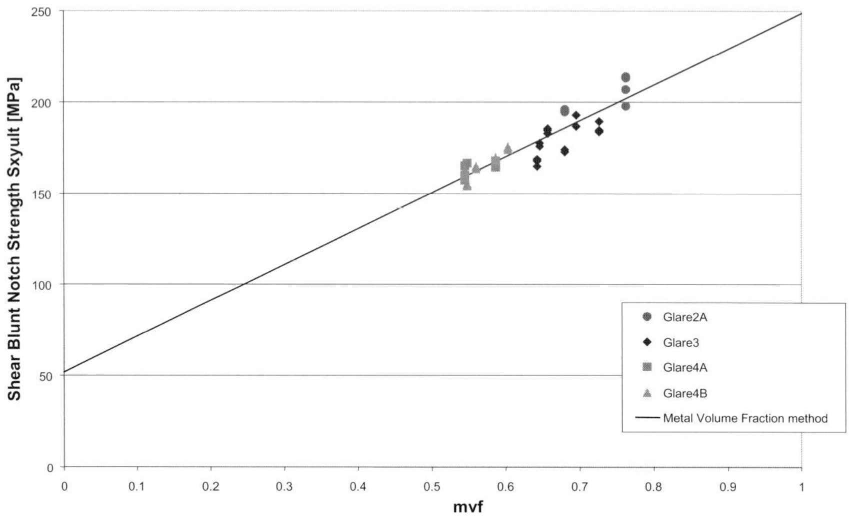

Figure 9.5 *Calculated shear blunt notch strength Sxy_{ult} vs. MVF.*

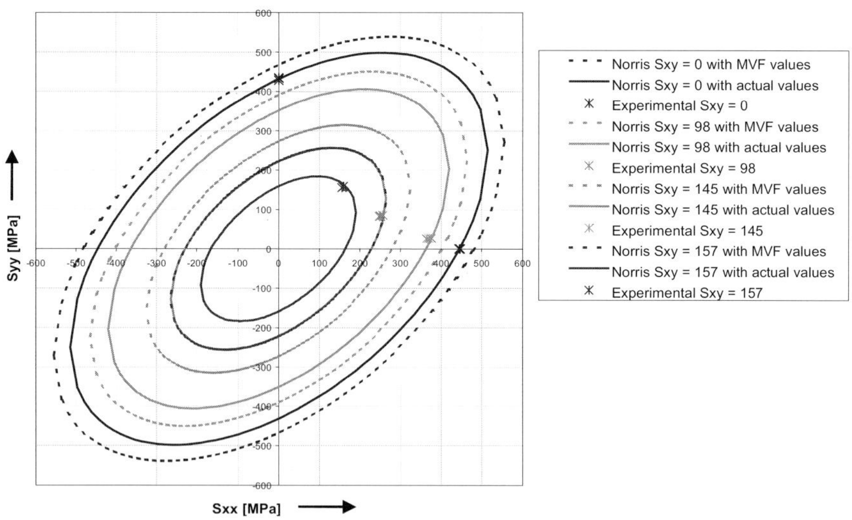

Figure 9.6 *Norris predictions for Glare3-3/2-0.3 compared to experimental data, for Norris both actual Sxx_{ult}, Syy_{ult} and Sxy_{ult} as well as calculated MVF values are used.*

9.5 Validation with experimental results

In Figure 9.6 the strength in L-direction, Sxx, versus the strength in LT-direction, Syy, is depicted for Glare 3-3/2-0.3 at various levels of shear stress. The Norris predictions are plotted with actual Sxx_{ult}, Syy_{ult} and Sxy_{ult} values (respectively 445.6, 431.4 and 168.9 MPa, see section 9.3 as well as with calculated MVF-values from section 9.4 (with Equation 9.4) and Table 9.4 respectively 481.3, 466.5 and 178.5 MPa). Experimental values are taken from Bosker [3].

It can be observed that the Norris failure prediction method is very sensitive towards the used Sxx_{ult}, Syy_{ult} and Sxy_{ult} values, this can also be seen in Figure 9.7 in which Glare 3-4/3-0.4 and Glare 4A-3/2-0.3 are depicted.

The predictions are very accurate, even for high shear stresses, for those laminates where the calculated MVF values Sxx_{ult}, Syy_{ult} and Sxy_{ult} are in close correspondence with the actual values Sxx_{ult}, Syy_{ult} and Sxy_{ult} (as is the case for Glare4A-3/2-0.3 in Figure 9.7 (b)). In general the results are less accurate if calculated MVF values are used. However, the ideal blunt notch prediction tool should preferably be based on the MVF method in order to integrate the tool in a computer programme (like ISSY) and to be able to predict the biaxial blunt notch strength of arbitrary laminates.

To calculate the minimum values from the calculated typical biaxial blunt notch strength, the minimum values for $Sxx_{ult,min}$ and $Syy_{ult,min}$, defined as the typical values (Sxx_{ult} and Syy_{ult}) divided by 1.1 and calculated in Roebroeks [2], can be used. To incorporate the scatter for $Sxy_{ult,min}$ the same reduction factor of 1.1 will be used here as well.
The same results (experimental and Norris) for Glare3-3/2-0.3 are depicted versus the off-axis angle in Figure 9.8. The Norris predicted strength for off-axis loaded specimens ($S_{Norris\ typical}$) was derived from Equations (9.1) and (9.2) and yields:

$$S_{Norris\,typical}(\varphi) = \sqrt{\frac{1}{\dfrac{\cos^4(\varphi)}{Sxx_{ult}^{\;2}} - \dfrac{\cos^2(\varphi)\sin^2(\varphi)}{Sxx_{ult}Syy_{ult}} + \dfrac{\sin^4(\varphi)}{Syy_{ult}^{\;2}} + \dfrac{\cos^2(\varphi)\sin^2(\varphi)}{Sxy_{ult}^{\;2}}}} \tag{9.6}$$

Here it can also be observed that all experimental data lie above the Norris calculated minimum lines ($Sxx_{ult}/1.1$, $Syy_{ult}/1.1$ and $Sxy_{ult}/1.1$).

In Figure 9.9 all experimental results for Glare3 are depicted relative to the typical Norris calculated values ($S_{experimental}/S_{Norris\ typical}$). The calculated minimum Norris value is also depicted in the figure. It can be observed that all experimental values are above the calculated minimum.

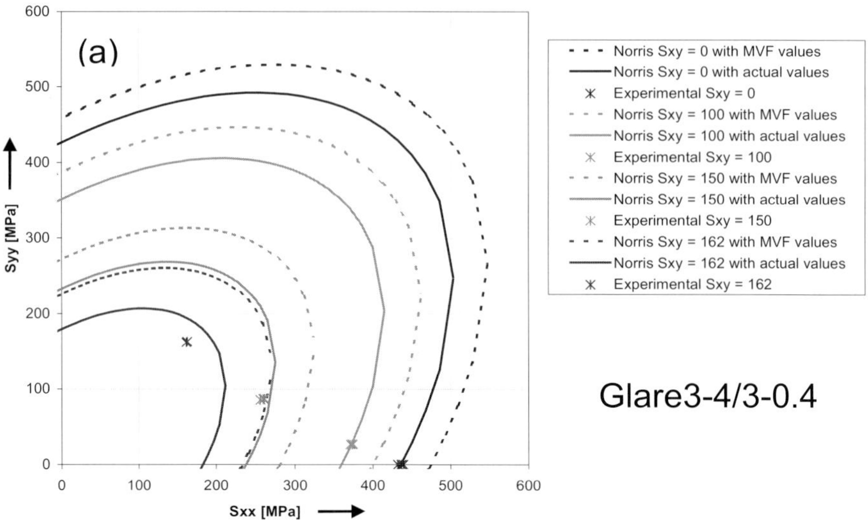

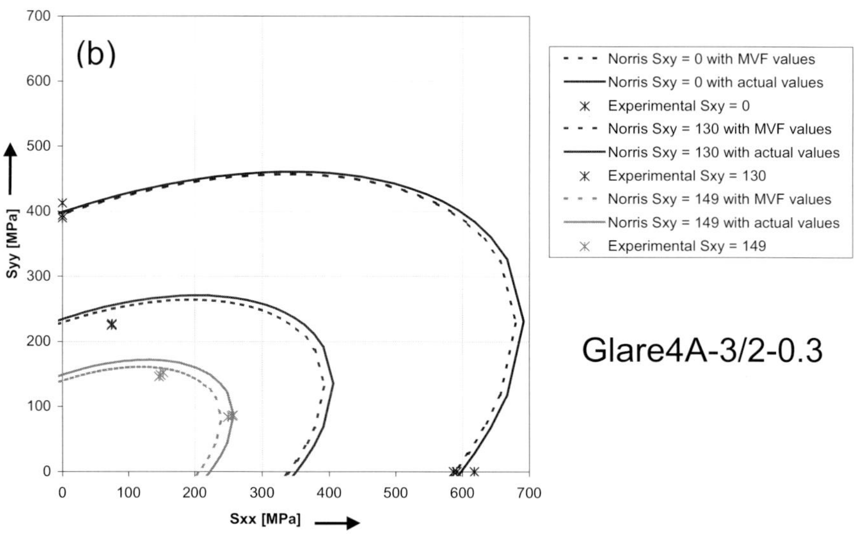

Figure 9.7 *Norris predictions for Glare3-4/3-0.4 (a) and Glare4A-3/2-0.3 (b) compared to experimental data, for Norris both actual Sxx_{ult}, Syy_{ult} and Sxy_{ult} as well as calculated MVF values (first quadrant only).*

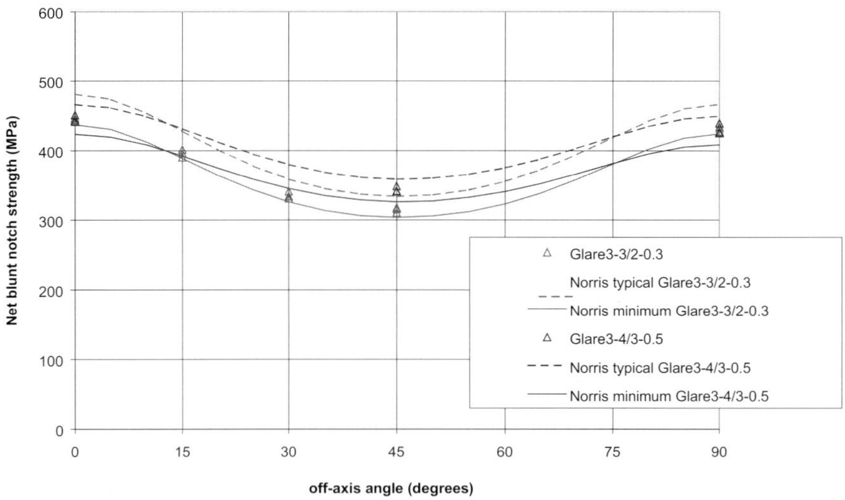

Figure 9.8 *Norris typical and minimum uniaxial strength predictions (with MVF values) for Glare3-3/2-0.3 and Glare3-4/3-0.5 versus the off-axis angle compared to experimental data* [3].

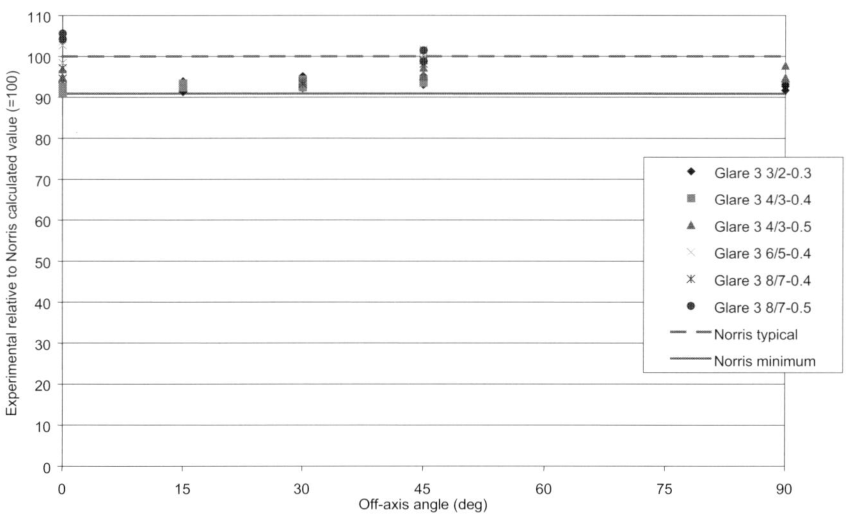

Figure 9.9 *Experimental results relative to the typical Norris calculated values ($S_{experimental}$ /$S_{Norris\ typical}$) versus the off-axis angle for Glare 3.*

As can be seen in Figure 9.10, in which the blunt notch strength for Glare2B (experimental and Norris with Equation 9.6) is depicted versus the off-axis angle, all experimental data for Glare2B lie above the Norris calculated minimum lines ($Sxx_{ult}/1.1$, $Syy_{ult}/1.1$ and $Sxy_{ult}/1.1$). Note also how well the Norris criterion represents the test data for small off-axis angles, even for the highly anisotrope Glare2.
In Figures 9.11 to 13 all experimental results for Glare2B, 4A and 4B are depicted relative to the typical Norris calculated values ($S_{experimental}$ /$S_{Norris\ typical}$), like Figure 9.9. The calculated minimum Norris value is also depicted in the figure. It can again be observed that all experimental values are above the calculated minimum.

9.6 Conclusions and tool description

It has been shown in this chapter that the Norris failure model is very useful to predict typical as well as minimum multiaxial (biaxial+shear component) blunt notch strengths. Even for highly anisotropic materials like Glare 2, the failure loads are very well represented. This Norris model can be successfully combined with the Metal Volume Fraction (MVF) approach and this results in a simple yet powerful tool to predict allowable blunt notch strength.

The governing formula to calculate the typical blunt notch strength is:

$$\left(\frac{Sxx}{Sxx_{ult}}\right)^2 - \left(\frac{Sxx}{Sxx_{ult}}\right)\left(\frac{Syy}{Syy_{ult}}\right) + \left(\frac{Syy}{Syy_{ult}}\right)^2 + \left(\frac{Sxy}{Sxy_{ult}}\right)^2 = 1 \qquad (9.7)$$

In which the Sxx_{ult}, Syy_{ult} and Syy_{ult} are the net blunt notch strength in respectively L- and LT-directions and the shear blunt notch strength as calculated with the MVF approach:

$$Si_{ult} = MVF \cdot Si_{alu} + (1 - MVF) \cdot Si_{fibre} \qquad (9.8)$$

with: i = xx, yy and xy.

The previously unknown blunt notch shear strength, Sxy_{ult}, was successfully calculated from uniaxial 45° off-axis experiments. All material properties, the input for Equations 9.8 and 9.9, are summarised in Table 9.5.

The minimum values can be calculated by using $Sxx_{ult}/1.1$, $Syy_{ult}/1.1$ and $Sxy_{ult}/1.1$ in the above Equation 9.7. All experimental blunt notch results lie above this calculated minimum Norris value.

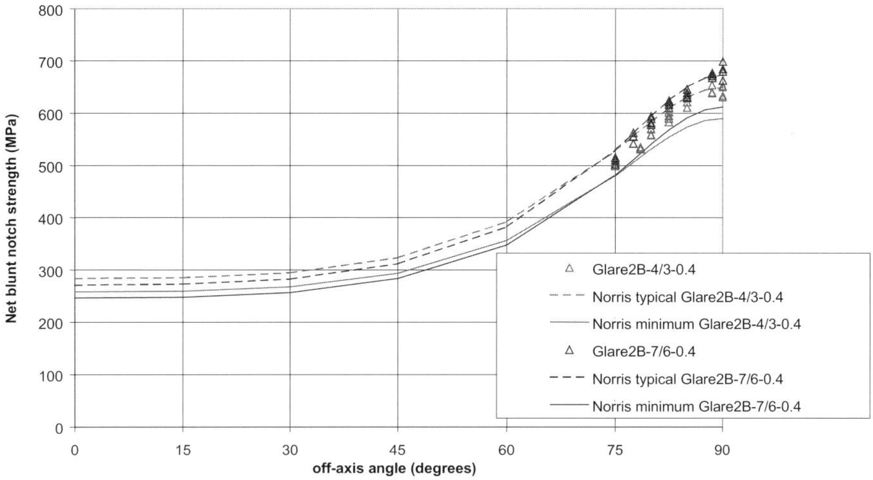

Figure 9.10 *Norris typical and minimum uniaxial strength predictions for Glare2B-4/3-0.4 and Glare 2B-7/6-0.4 versus the off-axis angle compared to experimental data.*

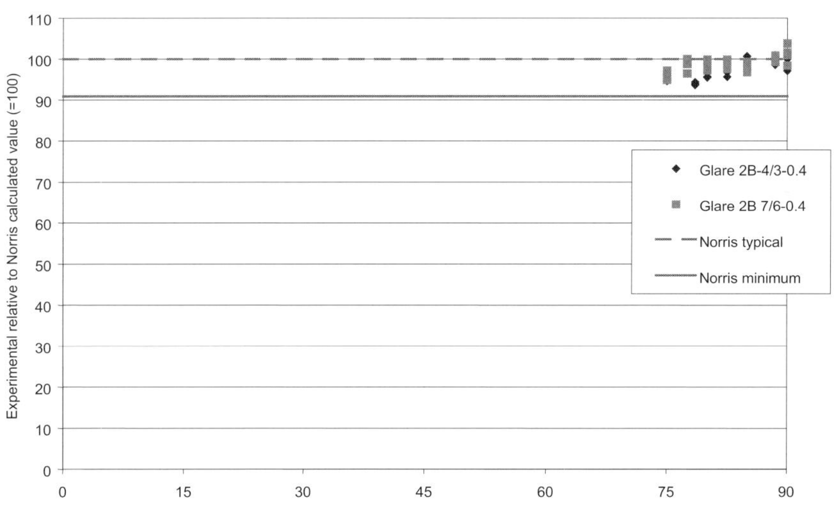

Figure 9.11 *Experimental results relative to the typical Norris calculated values* ($S_{experimental}/S_{Norris\ typical}$) *versus the off-axis angle for Glare 2B.*

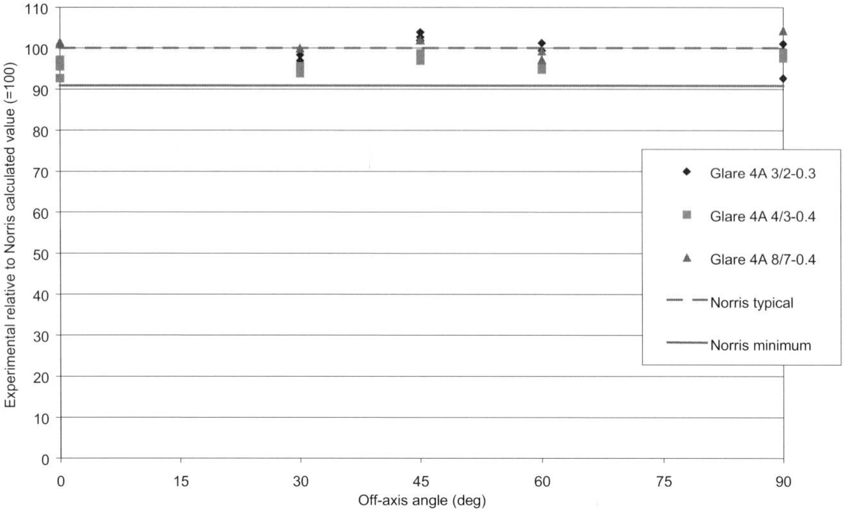

Figure 9.12 *Experimental results relative to the typical Norris calculated values ($S_{experimental}$ /$S_{Norris\ typical}$) versus the off-axis angle for Glare 4A.*

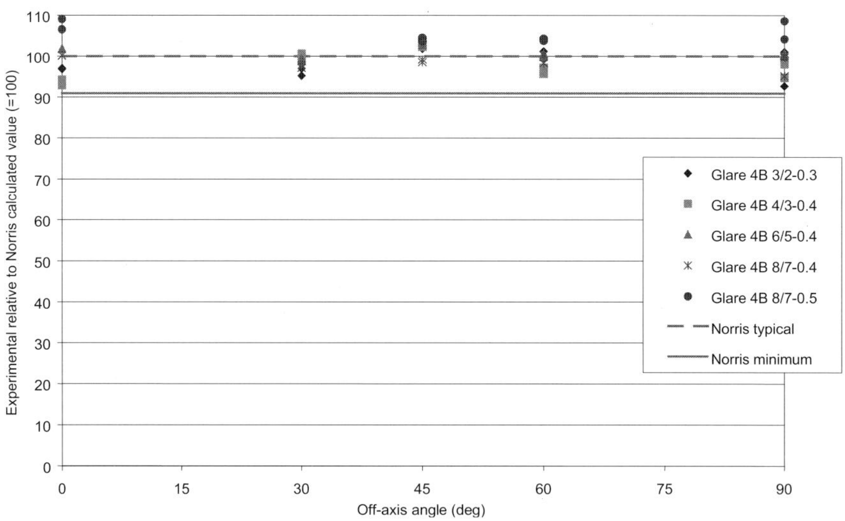

Figure 9.13 *Experimental results relative to the typical Norris calculated values ($S_{experimental}$ /$S_{Norris\ typical}$) versus the off-axis angle for Glare 4B.*

Table 9.5 *Typical Glare net blunt notch strength parameters (in MPa).*

Glare type	Layer contribution	Sxx_{ult} (MPa) (L-direction)	Syy_{ult} (MPa) (LT-direction)	Sxy_{ult} (MPa)	MVF
Glare 2A	Metal (Si_{alu})	417	394	249	1
	Fibre (Si_{fibre})	1193	0	52	0
Glare 2B	Metal (Si_{alu})	417	394	249	1
	Fibre (Si_{fibre})	0	1193	52	0
Glare 3	Metal (Si_{alu})	417	394	249	1
	Fibre (Si_{fibre})	597	597	52	0
Glare 4A	Metal (Si_{alu})	417	394	249	1
	Fibre (Si_{fibre})	795	398	52	0
Glare 4B	Metal (Si_{alu})	417	394	249	1
	Fibre (Si_{fibre})	398	795	52	0

References

[1] C.T. Sun, et al., *Comparative evaluation of failure analysis methods for composite laminates*, U.S. Department of transportation, Federal Aviation Administration, Washington D.C., 1996.

[2] G. Roebroeks, *The Metal Volume Fraction approach*, TD-R-00-003, Structural Laminates Industries, Delft, 2000.

[3] O.J. Bosker, *Glare uni-axial blunt notch test results*, B2V-99-27, Delft University of Technology, Delft, 2000.

10

Stability

T.C. Wittenberg and E.L. Jansen

Abstract – This chapter presents a concise overview of the current design methodologies employed for the static strength of stiffened Glare panels, regarding stability aspects (buckling / post-buckling) under compressive and shearing loads. In the light of the current developments in potential applications of Glare, the focus is on fuselage structures that are representative for Ultra-High Capacity Aircraft (UHCA). A number of experiments on such types of panels have been conducted for shear and compression loading. The results are discussed in this chapter and are compared to predictions from the current design methodologies and from analyses using the Finite Element Method (FEM).

10.1 Introduction

While plate structures possess excellent capabilities in resisting in-plane forces, their performance is generally much poorer when loaded in the flexural direction. Examples of this phenomenon may be found in the relatively poor bending resistance and the buckling sensitivity when subjected to compressive stresses. A possible solution to increase the lateral rigidity of plate structures is to provide them with stiffening elements, such as the stringers and frames that are present in aircraft fuselage structures.

The thin-walled plate segments in a stiffened panel are still prone to buckling phenomena when subjected to axial compressive loads, shearing loads, or combinations thereof. Generally, however, the structure is still able to carry additional loading beyond the point of initial buckling of the individual skin segments, which are bounded by the stringers and frames (Figure 10.1 illustrates local buckling). Structures conforming to this particular feature are said to posses *post-buckling* strength before actual failure of the structure takes place. If, however, the reinforcing elements are too weak to effectively support the skin, they tend to buckle along with it. This phenomenon is called overall buckling and is essentially a failure mode of the structure.

The calculation of the failure load of a stiffened panel generally comprises the following steps, as far as stability aspects are concerned:

1. Determination of the initial buckling load (also referred to as the *critical* load), which is generally done by means of linear (bifurcation) buckling analysis. The deformed shape of the panel at the critical buckling load—the critical mode—is either characterised by wave patterns having nodal lines at the location of the stringers and frames or by buckling waves that extend over the stringers and/or frames. The first type of deformation mode is called local buckling and is thus confined to the individual skin segments, the second type is called overall or global buckling and shows flexural displacements of the stiffening elements. While in the latter case the structure loses its load-carrying capacity (which is equal to failure), the structure generally still possesses strength beyond the critical load when the buckling mode is local. This post-buckling strength is obtained from the stiffening elements, which pick up the (majority) of the additional applied load after initial buckling.
2. Post-buckling stress analysis in case the condition of local buckling is fulfilled. Since the skin segments tend to lose (most of) their load-carrying capability in the post-critical range, the reinforcing elements will pick up the majority of the additional loading.
3. Checking of the post-buckling stresses in the individual panel elements against failure criteria or allowables.

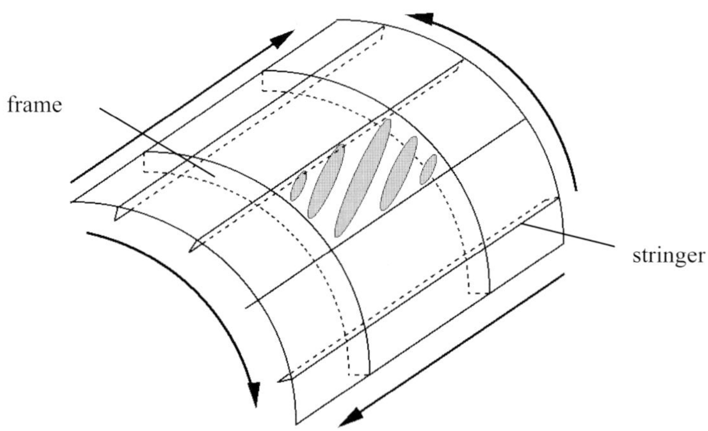

Figure 10.1 Illustration of local buckling in a fuselage panel loaded in pure shear.

This chapter is logically divided into the sub-groups shear loading and compression loading. For each load type, the corresponding design methodology for Glare panels will be concisely outlined. In general, steps 2 and 3 above are combined because of the semi-empirical nature of most practical design methodologies. Therefore, for each respective loading situation the discussion of the analysis method will be divided into one part covering initial buckling, and another dealing with post-buckling and failure of the panel. Subsequently, the extent to which the respective analysis methods have been verified and/or validated with experimental evidence will be discussed. It should be noted that—seen in the light of the current developments in potential applications for Glare material—the discussion in this chapter is focused on panel geometries that are representative for future generation Ultra-High Capacity Aircraft (UHCA).

10.2 Shear loading

The most widely used design and analysis method for stiffened, flat or curved shear panels is the method of *incomplete diagonal tension*, which was presented by Kuhn et al. [1, 2] in the early fifties. This method assumes that the individual skin fields buckle first, where the stiffening elements provide rigid boundary conditions as far as flexural deformations are concerned, i.e. local buckling, see Figure 10.1. Any additional load beyond this point of initial buckling will be mainly carried by the stiffeners, which then serve as compression posts.

The Kuhn method is of a semi-empirical nature; the various formulae are based on a relatively large amount of experimental data on aluminium panels. A modification of the Kuhn method for Glare material is described in Wimersma [3]. The basic features will be discussed below.

10.2.1 Initial buckling

Shear buckling of flat isotropic plates is a well-documented subject, see for example References [4, 5, 6]. In the comprehensive literature the solutions for the shear buckling stress are presented in slightly different ways. In the current methodology for Glare [3] the shear buckling stress is calculated according to the formula

$$\tau_{cr} = 10.92 k_s \frac{\pi^2 D}{b^2 t} \tag{10.1}$$

with: k_s = a non-dimensional buckling coefficient, which depends on the plate edge conditions and aspect ratio

a, b = the length and width, respectively

t = the plate thickness

D = the bending stiffness matrix of the plate.

Since Glare has different values for the bending stiffnesses, the quantity D in Equation 10.1 is tentatively rewritten as $\sqrt{D_{11}D_{22}}$, where the subscripts 11 and 22 refer to the L and LT direction, respectively. Curves for k_s for several boundary conditions of the plate are presented in [7]. Note that it is customary in design practice to consider simply-supported edge conditions of the plate since this gives the highest degree of conservatism.

A physically more correct manner of analysing buckling is to consider Glare as an orthotropic material. Due to the stacking sequence and fibre orientations of the most common Glare configurations, the material can generally be considered to be a *specially orthotropic* laminate. This means that all the coupling terms in the laminate stiffness matrices—referred to as the "ABD-matrix" in classical laminate theory—are identically zero. The governing equation for the buckling load of a flat, specially orthotropic laminate loaded in pure shear is given by

$$D_{11}\frac{\partial^4 w}{\partial x^4}+2(D_{12}+2D_{66})\frac{\partial^4 w}{\partial x^2\partial y^2}+D_{22}\frac{\partial^4 w}{\partial y^4}-2N_{xy}\frac{\partial^2 w}{\partial x\partial y}=0 \tag{10.2}$$

with: D_{ij} = the elements of the plate bending stiffness matrix
w = $w(x, y)$, the out-of-plane displacement of the plate.

The shear-buckling load $N_{xy_{cr}}$ is now found by solving this equation for the out-of-plane displacements w. For simply supported edges, it is assumed that

$$w(x,y)=\sum_{i=1}^{m}\sum_{j=1}^{n}\overline{w}_{ij}\sin\frac{i\pi x}{a}\sin\frac{j\pi y}{b} \tag{10.3}$$

with: a, b = the length and width of the plate, respectively.

Substitution of Equation 10.3 into Equation 10.2 eventually leads to an eigenvalue problem in N_{xy}, the smallest of which is the critical buckling load, $N_{xy_{cr}}$ (note that $\tau_{cr}=N_{xy_{cr}}/t$). In practice, however, it turns out that for most practical Glare configurations the shear buckling stresses according to Equation 10.1 yield very acceptable results compared to the orhotropic plate solution.10.2

10.2.2 Post-buckling and failure

The theory of incomplete diagonal tension for stiffened shear panels is essentially a mixture of two limiting cases. The first is where the skin of the panel is in a state of pure shear stress, which occurs at load levels prior to buckling. The other extreme case is a state of pure diagonal tension where the load bearing capacity in the direction of the compressive stresses is fully exhausted, causing the shear load to be carried fully by tension stresses acting under an angle α (the "angle of diagonal tension") with respect to the stringer direction. The compression forces that are then required for equilibrium are carried by the stiffening elements. With this information in mind, the core of the Kuhn method is based on the assumption that the nominal applied shear stress is split into a pure shear part, τ_S, and a part carried by diagonal tension action of the skin, τ_{DT} see Figure 10.2. Mathematically, this can be represented as

$$\tau = \tau_S + \tau_{DT} \tag{10.4}$$

with

$$\tau_{DT} = k\tau \quad ; \quad \tau_S = (1-k)\tau \tag{10.5}$$

with: k = the "diagonal tension factor".

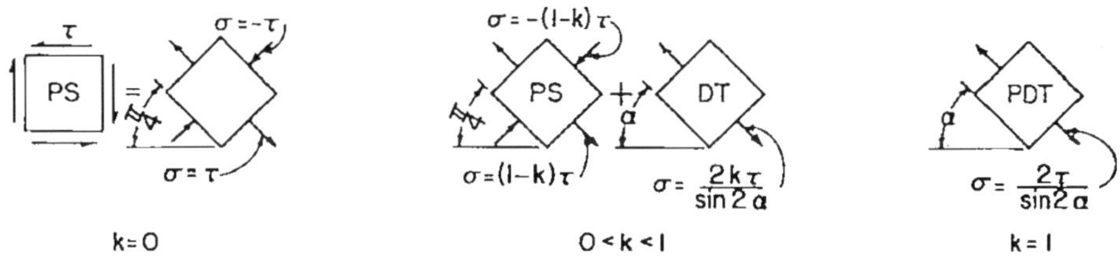

(a) Stress system (PS = pure shear, PDT = pure diagonal tension)

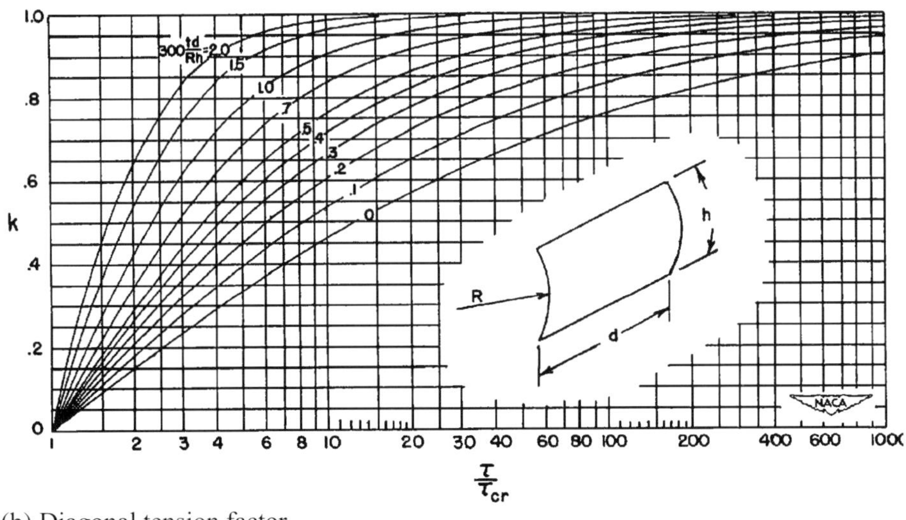

(b) Diagonal tension factor

Figure 10.2 Stress system in the skin and diagonal tension factor as a function of the post-buckling ratio (figures taken from [1]*).*

Empirically derived curves for k are shown in Figure 10.2. It can be seen that k is a function of the loading ratio or "post-buckling ratio" τ/τ_{cr}, where τ_{cr} is the initial buckling stress.

The definition of the nominal shear stress according to Equations 10.4 and 10.5 incorporates the two limiting cases discussed earlier; for $k = 0$ the panel is in a state of pure shear, for $k = 1$ the panel acts in pure diagonal tension. Note that the latter case requires that $\tau/\tau_{cr} \to \infty$, see Figure 10.2. The angle α in Figure 10.2(a) is called the "angle of diagonal tension".

The core of the Kuhn method is described by Equations 10.4 and 10.5 and Figure 10.2. In addition, there are various formulae involved for the calculation of web and stringer stresses, which can be found in Wimersma [3] and will not be repeated here. The modifications of the method for Glare can be found in the directional dependence of the elastic moduli that appear in the various formulae, as Glare is an orthotropic material.

Skin rupture

Failure of the skin is analysed in three different ways. The first check is for general skin failure in shear, for which the Tresca criterion is employed:

$$\tau_{al} = \frac{\sigma_{ult}}{2} \qquad (10.6)$$

with: τ_{al} = the allowable shear stress of the skin
σ_{ult} = the ultimate tensile stress in the L-direction.

The second skin failure criterion is given by the equations:

$$\begin{aligned} \tau_{al} &= C_\tau \sigma_{ult} \quad , \quad \text{with} \\ C_\tau &= 10^{-3} e^{\left(0.6625k^2 - 0.741k + 6.096\right)} \quad , \quad \text{for} \quad 0 \le k \le 0.5 \\ C_\tau &= -0.0316k + 0.3776 \quad , \quad \text{for} \quad 0.5 \le k \le 1 \end{aligned} \qquad (10.7)$$

with: k = the diagonal tension factor.

Equation 10.7 is a close approximation of the curve in figure 19a of Kuhn [1] for a diagonal tension angle of $\alpha = 45°$. The third check for skin failure is the blunt-notch strength of the skin material. Since the angle of diagonal tension is close to 45° in practice, the blunt-notch strength in this particular direction is used. Hence,

$$\sigma_{al} = \sigma_{bn,45^\circ} \qquad (10.8)$$

which is checked against the skin stress in diagonal tension direction, see Figure 10.2(a).

Forced crippling of the stringers
The allowable forced crippling stress of the stringers is calculated according to the equation

$$\sigma_{fc} = 0.051 \frac{\sigma_{0.2,st}}{\sqrt{\frac{\sigma_{0.2,st}}{E_{1,st}} + 0.002}} k^{2/3} \left(\frac{t_{st}}{t_{sk}} \right)^{1/3} \tag{10.9}$$

Forced crippling is essentially a local failure mode of the stringer; the leg attached to the skin is subject to forced deformation by the buckles in the skin.

Column buckling of the stringers
Since the stiffeners are loaded in compression in a post-buckled shear panel, they are prone to failure by column buckling. The column buckling stress of the stringers is calculated with the Euler-Johnson method as used for compression panels, which will be discussed in Section 10.3. This type of failure is only relevant for very slender stringer-skin combinations.

10.2.3 Experimental and Finite Element analysis

The modified Kuhn method for the design of post-buckled Glare shear panels was outlined in the previous sections. However, this method only has an analytical status since it has not been validated with experimental data. Therefore, a programme was initiated which involved the testing of two flat, stiffened Glare panels subjected to pure shear loading. In the light of the current developments in potential applications of Glare material, the dimensions of the test panels were taken to be representative for UHCA-like fuselage structures. Seen against this background, two aspects differ from those of "conventional" aircraft: 1) the application of Glare as skin material, and 2) the large skin thickness associated with the large fuselage diameters that are inherent to UHCA fuselages. Note that the test panels have no curvature (as in the actual fuselage) due to the lack of a suitable test facility. Moreover, it seems more logical to consider the simpler case of flat panels first and then to proceed with curved panels later. Also, from a practical point of view, consideration of flat panels seems to be a good first approximation of the problem as the radii of curvature are—as already mentioned—very large for these types of aircraft.

In addition to the experiments, detailed analysis using the Finite Element Method (FEM) was performed to see whether the test results could be accurately predicted. If so, the computer simulations could replace very costly tests for further verification and development of the design methodology for Glare shear panels.

Design of the shear test panels

The well-known picture frame method was employed for the testing of the shear panels. A special picture frame was designed and fabricated which allows for net square dimensions of 1m × 1m. Both panels were identical except for the applied skin material, which was Glare 3-5/4-0.4 (total thickness 3.0 mm) for one panel (designated S1), and Glare 4-5/4-0.4 (total thickness 3.5 mm) for the other (designated S2). The skin thicknesses, stringer pitch and frame pitch were obtained from a preliminary study on an UHCA conducted at Delft University of Technology using the in-house design and optimisation software ADAS[1]. For sizing of the fuselage structure, the modified Kuhn method described in Wimersma [3] was used by ADAS to calculate the shear allowables of the fuselage panels. A location just above and behind the wing centre section, see Figure 10.3, was selected for the test panels, since the highest shear flows were shown to occur in this particular area. Figure 10.4 shows a photograph of panel S1 after the test and gives a good impression of the configuration of the shear panels.

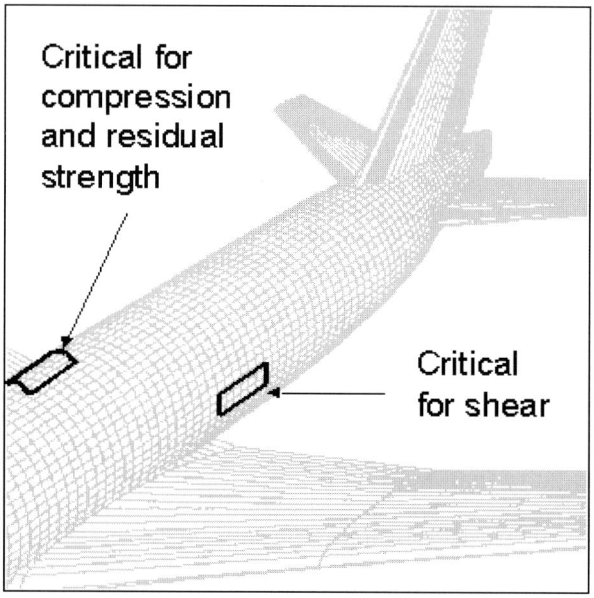

Figure 10.3 ADAS model showing selected location for the test panels.

---

[1] ADAS stands for **A**ircraft **D**esign and **A**nalysis **S**ystem

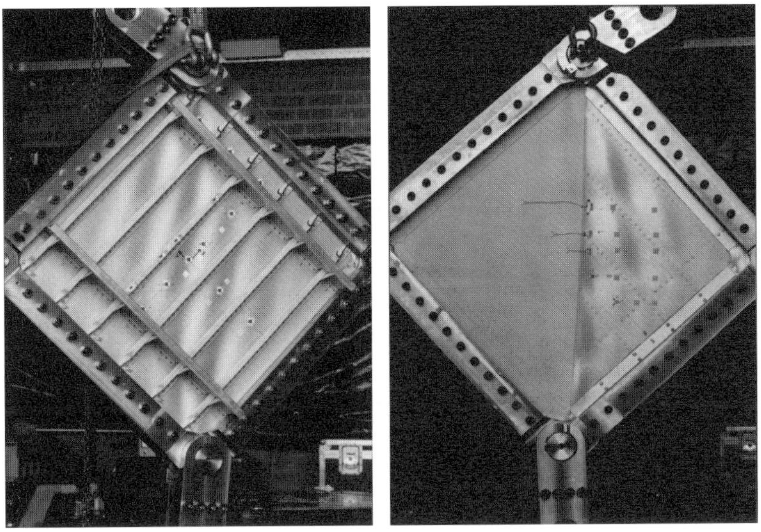

Figure 10.4 *The Glare 3 panel after the test. The buckled shape is clearly visible.*

Analytical Results

Both shear panels were analysed with the modified Kuhn method described in the previous sections [3]. The skin material properties used in the analysis are listed in Table 10.1, which also shows results obtained from the modified Kuhn method. Both panels fail due to skin rupture at the stringer rivet line, as can be seen. For panel S1 the blunt-notch criterion was shown to be critical (i.e. Equation 10.8), while for panel S2 the rupture stress was obtained from Equation 10.7. The calculated initial buckling loads have been corrected for material plasticity. While the buckling load of panel S2 is significantly higher than that of S1, there is only a 10% difference in failure load. In both cases, the panels show some degree of post-buckling behaviour before failure occurs; the post-buckling ratios (calculated as q_{fail}/q_{cr}) are 2.1 and 1.6 for panel S1 and S2, respectively. These relatively low values are due to the high initial buckling loads, which in turn are mainly the result of the large skin thicknesses.

Table 10.1 *Basic material properties and analysis results from the modified Kuhn method.*

Material propeties		Glare 3	Glare 4
E_{11}	[MPa]	59410	52572
E_{22}	[MPa]	59410	58943
ν_{12}	[-]	0.29	0.25
G_{12}	[MPa]	19993	18128
$\sigma_{0.2}$ [1]	[MPa]	269	237
σ_{ult} [2]	[MPa]	614	530
q_{ult} [3]	[N/mm]	336	480
q_{fail} [4]	[N/mm]	700	768
Predicted failure mode		Skin rupture at stringer rivet line	Skin rupture at stringer rivet line

[1] 0.2% yield stress (dependent on loading direction; most conservative value used here)
[2] ultimate tensile strength (dependent on loading direction; most conservative value used here)
[3] critical shear flow (initial buckling), corrected for plasticity
[4] shear flow at failure

FEM and test results

Both test panels were analysed using the STAGS general purpose Finite Element (FE) code [8]. The actual test situation was simulated as closely as possible during creation of the FE model, while at the same time keeping modelling and analysis time to a minimum. The Glare skin was modelled as a laminate with an elasto-plastic representation of the aluminium layers, while the glass fibre layers are assumed to behave linear elastically throughout. Both the skin and the stringers are modelled with 4-node shell elements of the 410 type, while 210 beam elements are used to represent the frames. At the corners, the skin is left out of the model so that a square cut-out is created. This representation of the panel corners is somewhat less stiff compared to the actual situation, but the modelling process is simplified significantly. The picture frame members are represented by beam elements; the kinematics of the picture frame in the FE model is the same as in the actual test. This also holds for the applied loading condition: one corner of the frame is fixed, while a tensile force acting in the direction of the panel diagonal is applied to the opposite corner. As panels S1 and S2 only differ in the applied skin material, the same FE model could be used for both with only minor modifications. For an impression of the FE model see Figures 10.5(a) and 10.6(a).

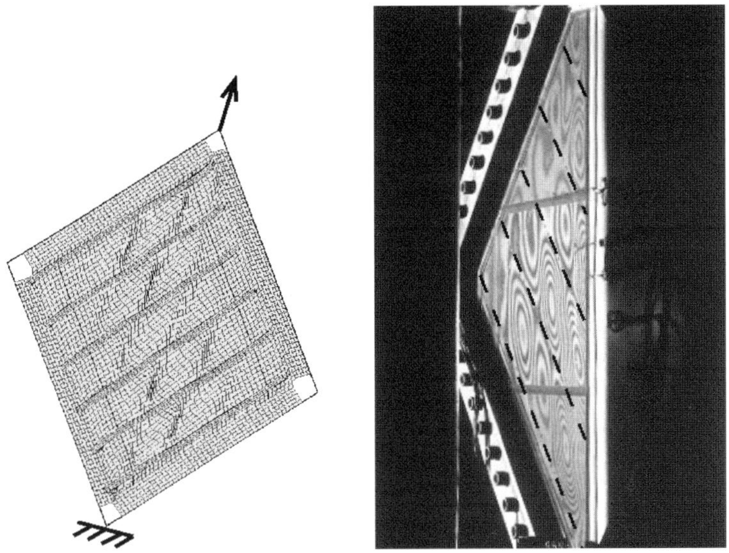

(a) FEM model (b) Moiré fringes (test)

Figure 10.5 *Buckling mode of panel S1 according to FEM and test. Dotted lines in (b) indicate (approximate) stringer locations (top stringer not shown).*

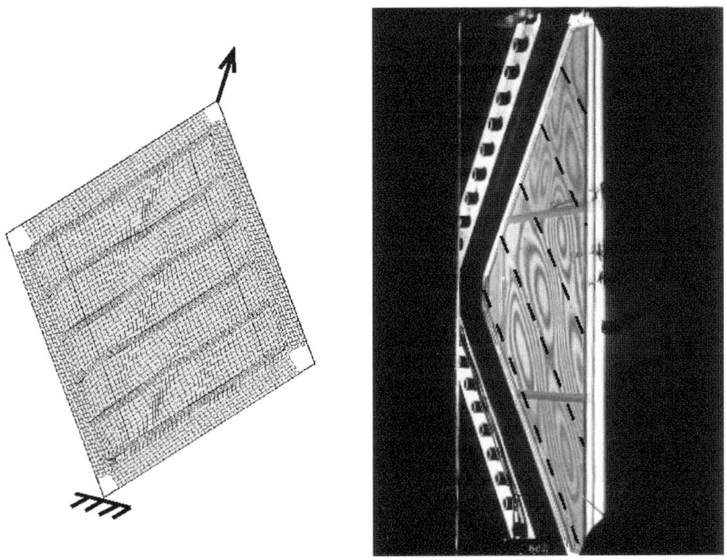

(a) FEM model (b) Moiré fringes (test)

Figure 10.6 *Buckling mode of panel S2 according to FEM and test. Dotted lines in (b) indicate (approximate) stringer locations (top stringer not shown).*

A linear bifurcation analysis was first performed for both panels S1 and S2 to get an indication of the initial buckling loads and the associated modes. These buckling loads were calculated for *elastic* material properties. Then, a combination of these buckling modes was used as an initial imperfection for the non-linear analysis, in which plasticity of the skin was included as explained previously.

The initial buckling loads according to experiments and FEM analyses are presented in Table 10.2, from which it can be concluded that the results agree very well. It should be noted that the tests were stopped at a nearly horizontal slope of the load-deflection curve, which was due to severe plasticity in the skin. In such conditions the capability of the panel to carry additional load is exhausted, and therefore the tests were stopped at these maximum load levels before "physical" failure took place.

Table 10.2 *Initial buckling loads (in terms of applied tensile force).*

Panel	FEM [kN]	Test [kN]
S1	520	510
S2	590	610

Figures 10.5 and 10.6 show the buckling modes of panels S1 and S2, respectively. The FEM and test results show similar deformation patterns. Note that the Moiré fringes only cover one half of the panels, and that the viewpoint is oblique. From the fringes, it can be seen that the buckles tend to "cross" the stringers (this effect is most pronounced for panel S2), or, in other words, that the stringers tend to buckle along with the skin. This phenomenon is also found in the FEM analyses and is referred to as *overall* buckling.

Discussion of design methodology

In the previous section it was determined that the test panels S1 and S2 are prone to overall-type buckling behaviour. This is a violation of a basic assumption in the Kuhn method, which is that the skin should buckle locally, meaning that any additional loads can be carried by the stringers through compression forces. The main reason for this phenomenon to occur is the geometry of the panels—a large skin thickness combined with relatively small stiffener spacing—which causes the initial buckling load of the individual skin segments to be higher than the overall buckling load of the panel.

For comparison of the initial buckling loads obtained from the tests with those calculated from the engineering design method, see Table 10.1, the test results given in Table 10.2 are translated to the applied shear flows. This yields values of $q_{cr}^{test} = 300$ N/mm for panel S1, and $q_{cr}^{test} = 359$ N/mm for panel S2. This is yet another indication of overall buckling behaviour, since the values predicted in Table 10.1 should be conservative in the case of local buckling as they are calculated for simply supported boundary conditions.

10.2.4 Conclusions

It was found that Glare shear panels of UHCA-like dimensions are prone to overall-type buckling behaviour. This means that these panels fall outside the region where the Kuhn design methodology is applicable. Therefore, it is proposed that an engineering procedure be developed to calculate the overall buckling load, including plasticity effects, of stiffened Glare shear panels. By definition, such a procedure will govern a criterion for the applicability of the Kuhn method. Results obtained with the STAGS FE code showed excellent agreement with experimental data, so this method can replace very costly tests during development of a new procedure.

10.3 Axial compression loading

The design of traditional (metal) stiffened panels under compressive loading, operating in the post-buckling region, is usually based on the *Euler-Johnson method* in combination with the *effective width* concept (see for example Niu [9]). In this approach, panel failure is assumed to correspond with that of a thin-walled column consisting of the stringer and an effective width of the buckled skin.

10.3.1 Initial buckling

The initial buckling behaviour of curved panels can be analysed using the governing differential equations. For Glare, the analysis should include the effect of the orthotropy of the material. An analytical expression for the initial buckling load in terms of the geometry, stiffness and wave number parameters, based on Donnell-type equations, may easily be derived for the case of simply-supported boundary conditions, see Verolme [10].

10.3.2 Post-buckling and failure

The failure mode of a thin-walled column depends on its length. For small column lengths, local failure (crippling) occurs, while for large lengths, Euler buckling occurs, see Figure 10.7.

In the intermediate range of lengths, failure is a combination of these two modes of deformation (short column buckling), and in this region the effect of plasticity must therefore be taken into account. An additional complicating effect in the intermediate range is that for thin-walled open sections, such as Z-stringers, buckling can occur in a torsion-bending mode. Because of the complexity of the buckling and failure modes, empirical equations are generally used in the intermediate range. The Johnson's parabola connects the column yield stress σ_{co} with the Euler curve in the column length versus failure stress diagram, see Figure 10.7. In this figure L' is the so-called effective length of the column, which depends on the boundary conditions, and ρ is the radius of gyration of the cross-section,

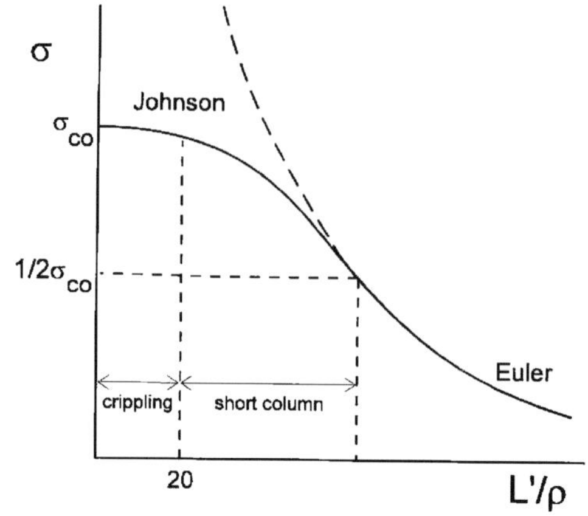

Figure 10.7 *Euler-Johnson prediction method for columns,* [10].

$$\rho = \sqrt{\frac{I}{A}} \tag{10.10}$$

with: I = the moment of inertia
A = the cross-sectional area.

The column yield stress σ_{co}, can be determined from

$$\sigma_{co} = \left(1 - \frac{4\sigma_C}{E}\right)\sigma_C \tag{10.11}$$

with: σ_c = the section crippling stress
E = Young's modulus in the axial direction.

Johnson's modification of the Euler curve can be written as:

$$\frac{\sigma_{EJ}}{\sigma_{co}} = 1 - \frac{\sigma_{co}\left(L'/\rho\right)^2}{4\pi^2 E} \tag{10.12}$$

To estimate panel failure, the panel is considered as a thin-walled column consisting of the stringer and an effective width of the buckled skin. This approach has been extended to include the case of Glare panels [10] and has been implemented in a computer programme [3]. The method involves an iterative procedure. The *effective width of the skin* depends on the buckling strain (stress) and the estimated failure strain. For a given failure strain the crippling stresses for stringer plus effective width of the skin can be calculated using an empirical method, the so-called breakdown method. This gives the crippling load of the structure. Then the Euler-Johnson buckling load is evaluated and a new failure strain can be calculated. The crippling stress of the stringer alone can be used as a starting value of the iteration process. The inter-rivet buckling stress can be used as a cut-off value in the procedure. Verolme [10] showed the applicability of the Euler-Johnson approach for various Glare panels by comparing the results of this method with experimental data.

10.3.3 Experimental and Finite Element analysis

Two curved test were designed and assumed to be representative for UHCA crown panels, see Figure 10.3. The objective of the tests was twofold. The first objective was to verify whether the design approach is applicable to realistic Glare panels with a relatively thick skin, as will be the case in Ultra-High Capacity Aircraft. The second objective was the verification of the finite element models, which had been developed to simulate the behaviour of the panels. If the finite element models are able to predict the observed stability behaviour sufficiently accurately, they may be used to replace expensive tests.

The two panels, a panel with a Glare 3-5/4-0.4 skin (thickness 3.0 mm) and a panel with a Glare 4B-5/4-0.4 skin (thickness 3.5 mm) were tested to failure under axial compressive loading at the static test facility of Daimler Chrysler Aerospace Airbus in Hamburg. The basic material properties of these two Glare types can be found in Table 10.3. For the Glare 4B panel, there are fewer fibres in loading direction than in width direction in the prepreg layers. The panel data can be found in Table 10.3. The material of the stringers, frames and clips is aluminium 7349. The stringer and frame thickness is 3.6 mm, the clip thickness is 2.0 mm.

A schematic drawing of the complete panel is shown in Figure 10.8. The centre section is the representative part. The outer sections have been reinforced at the stringers.

Table 10.3 Test panel configuration.

Panel length	2508 mm
Panel width	1200 mm
Panel curvature	3340 mm
Frame pitch	635 mm
Stringer pitch	200 mm
Number of stringers	5
Number of frames	4

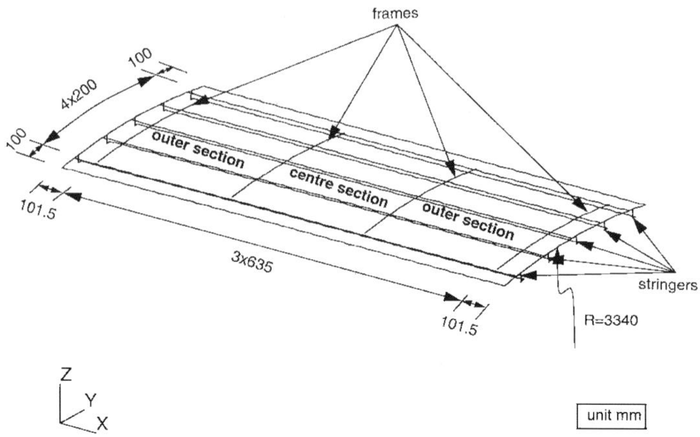

Figure 10.8 *Test panel configuration.*

Finite element analysis
Before the tests, detailed finite element analyses have been performed for these panels using the finite element code ABAQUS. The finite element method gives the possibility to simulate the non-linear behaviour of a panel in the post-buckling region. The finite element discretisation is shown in Figure 10.9.

The frames, the reinforcement components of the outer bays, and the lip of the stringers were modelled with beam elements. The skin, the stringer web, the attached stringer flange and the outstanding stringer flange were modelled with four-node shell elements. Glare was modelled by specification of the properties of the constituents, aluminium 2024 and glass-epoxy, in the layered wall elements of the skin. Complicating effects, such as the post-buckling deformation of the Glare skin, the influence of plasticity in the skin, and the behaviour of the stringer can be taken into account via finite elements. In order to analyse the shifts in buckling pattern in the post-buckling region, an option in the ABAQUS programme was used to stabilise unstable static problems.

The finite element results showed that under increasing loading local buckling first occurs in the bays of the centre section, followed by local buckling in the reinforced outer sections of the panel. The initial buckling mode in the centre section bays consists of three half waves in axial direction, but this mode gives way to a pattern with four half waves, see Figure 10.10. The panels eventually fail in a global (Euler) mode.

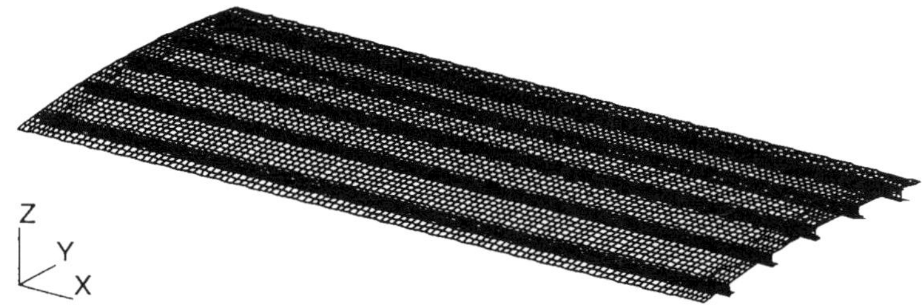

Figure 10.9 *Finite element discretisation of test panel.*

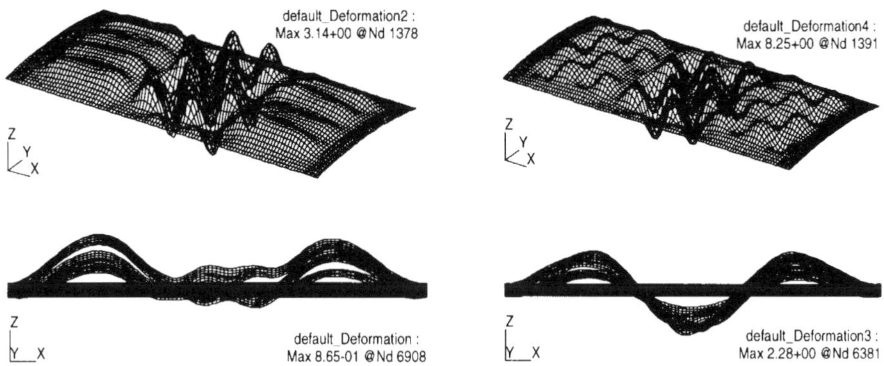

Figure 10.10 *Postbuckling patterns of skin and stringers before (left) and after (right) initial buckling of the outer sections.*

Experimental analysis

In order to capture the phenomena predicted by the finite element analysis, the test panels were equipped with strain gauges at carefully chosen positions in the skin and on the stringers. An optical field method based on photogrammetry in combination with the object grating method was used to measure the out-of-plane deflection of the skin [11]. The panel mounted in the test machine is shown in Figure 10.11.

In order to be able to compare the test results with the finite element calculations, two corrections were made to the FEM results. A correction was added to the total applied load in the finite element calculations to account for additional reinforcement strips at the edges in the experiment. This approximation was also used in the post-buckling region. Moreover, sample thickness measurements revealed that the actual panel thickness was 5% larger than the nominal thickness. This results in a considerable difference in bending stiffness of the Glare skin when compared to that based on the nominal thickness used in the analysis. Therefore, a correction factor was introduced which estimates the increase in bending stiffness due to the larger thickness. Notice also that the finite element modelling of the riveted stringer-skin attachment is an approximation of the actual situation. The support given by the riveted stringer is modelled in the finite element analysis as a line attachment where the toe and heel of the attached flange are free to separate from, or penetrate into the skin.

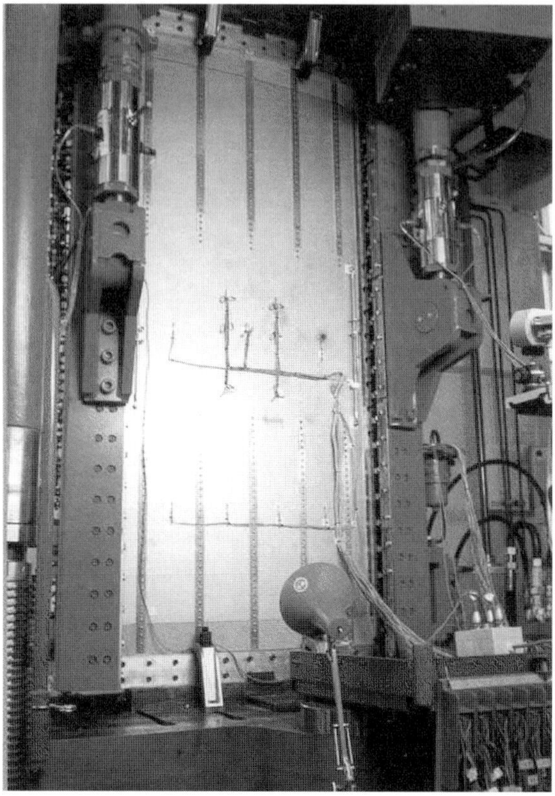

Figure 10.11 Panel mounted in test apparatus.

The initial buckling loads, including a non-linear pre-buckling state analysis, are given in Table 10.4. It should be noted that they are higher than the corresponding values from a linear eigenvalue analysis. The results in Table 10.4 have been corrected for the two effects indicated earlier. The initial buckling load in the test is defined as the load level at which a sufficiently sharp break in the algebraic mean compressive strain can be observed.

The test results show that the initial buckling mode is a local buckling mode in the elastic range. The linear finite element buckling analysis shows that the buckling mode associated with the first (lowest) buckling load consists of three axial half waves in the central section bays. The second buckling load, however, is close to the first one, and the corresponding mode consists of four half-waves. A non-linear finite element analysis shows that the mode with three half-waves gives way to a pattern with four half-waves.

Table 10.4 *Comparison of initial buckling loads (kN). The FEM results and design method results have been corrected to account for the edge reinforcements and for the actual skin thickness.*

Glare type	Test	FEM	Design Method
Glare 4B	725	673	528
Glare 3	575	467	348

The initial buckling prediction of the design method is based on simply supported boundary conditions for the skin. Moreover, in the finite element analysis the effect of a non-linear pre-buckling deformation is taken into account. The result of the design method may therefore be considerably below the finite element result (about 25%, see Table 10.4). The actual boundary conditions at the stringer-skin attachment can be taken into account via more elaborate solution methods, for example the method used by Verolme [10]. This method may be implemented in the present design tool to improve the initial buckling predictions.

In the test, panel failure occurs by buckling of the stringer in an Euler mode. Failure occurred at 1200 kN for the Glare 4 panel, and at 1125 kN for the Glare 3 panel. The finite element post-buckling analysis shows that the mode with four half-waves grows until the panel fails in a global mode. A result of the optical field displacement measurement for the Glare 4B panel is shown in Figure 10.12. This pattern corresponds to a buckling deformation with four axial half-waves in the centre bays.

Table 10.5 gives the failure loads obtained from the FEM analysis and the design method. The finite element predictions of the failure load compare well with the test results. The results have been corrected to account for the edge reinforcements. The prediction of the failure load and mode from the design tool is based on simply-supported boundary conditions for the Euler column, and gives failure load estimates which are about 15% lower than the Finite Element result.

Table 10.5 *Comparison of failure loads (kN). The FEM results and design method results have been corrected to account for the edge reinforcements.*

Glare type	Test	FEM	Design Method
Glare 4B	1200	1217	1045
Glare 3	1125	1142	967

Summarising, it can be concluded that both the initial buckling results and the failure results obtained via the design method are conservative in the present cases. The discrepancies between finite element analysis and test remain to be investigated in more detail. However, the development of a finite element methodology to simulate the behaviour of Glare panels under compressive loading accurately is attainable.

10.3.4 Conclusions

The Euler Johnson method in combination with the effective width concept forms a useful engineering design tool for Glare panels loaded in compression. A further development of a Finite Element methodology to accurately simulate the panel behaviour may reduce the necessity to carry out expensive tests.

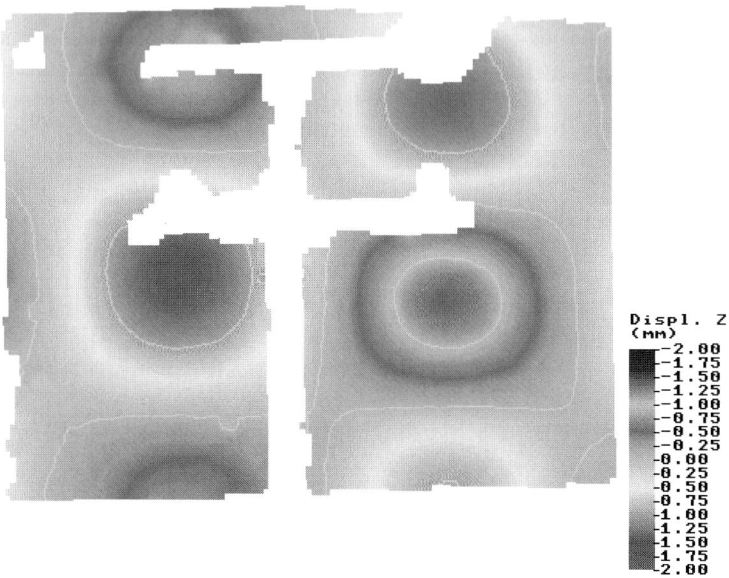

Figure 10.12 *Typical post-buckling deformation of a part of the two centre bays of the Glare 4B panel (the white bands correspond to positions of strain gauge wires).*

References

[1] Kuhn, P., et al., *A Summary of Diagonal Tension. Part I – Methods of Analysis*. NACA TN 2661. National Advisory Committee for Aeronautics, May 1952.

[2] Kuhn, P., et al., *A Summary of Diagonal Tension. Part II – Experimental Verification.* NACA TN 2662. National Advisory Committee for Aeronautics, May 1952.

[3] Van Wimersma, B., *Stability of Glare structures – calculation method.* Report TD-R-97-001. Delft (NL): Structural Laminates Company, February 1997.

[4] Timoshenko, S., *Theory of elastic stability.* 2nd edition. New York (USA): McGraw-Hill, 1961.

[5] Gerard, G., and H. Becker, *Handbook of Structural Stability – Part I, Buckling of Flat Plates.* NACA TN 3781. National Advisory Committee for Aeronautics, July1957.

[6] Galambos, T.V., editor, *Guide to stability design criteria for metal structures.* 5th edition. New York (USA): Wiley, 1998.

[7] Anonymous, *Buckling of flat plates in shear.* ESDU Data Item No. 71005. ESDU International plc, March 1995.

[8] Rankin, C.C., et al., *STAGS Users Manual – Version 3.0.* Lockheed Martin Missiles & Space Co. Inc. December 1997.

[9] Niu, M.C., *Airframe stress analysis and sizing.* Hongkong: Conmilit Press, 1997.

[10] Verolme, K., *The Development of a Design Tool for Fiber Metal Laminate Compression Panels*. PhD thesis. Delft (NL): Delft University of Technology, 1995.

[11] http://www.gom.com description of the ARAMIS optical measurement system.

11

Fatigue

R.C. Alderliesten

Abstract – Fatigue is an important design aspect, especially for the large number of geometrical notches in aircraft structures. The amount of flight cycles needed for a fatigue crack to grow from a detectable length to the critical length determines the maintenance inspection intervals, whereas the initiation life determines the threshold and start of frequent inspections. The longer the fatigue life of a cracked structure, the larger the inspection intervals can be, which reduces maintenance costs. Glare has excellent fatigue crack growth characteristics due to its laminated composition. In this chapter the fatigue mechanisms in Glare will be highlighted.

11.1 Fatigue initiation

Since fatigue initiation affects mainly the aluminium layers in Glare, the fatigue initiation process is analogous to that of monolithic aluminium. A similarity approach can be used: a similar stress state in the aluminium layers in Glare and in the monolithic material will create the same results, i.e. the same initiation life. The fatigue initiation behaviour of the aluminium layers in Glare can therefore be compared to the behaviour of monolithic aluminium if the stress conditions at the notch edge are equivalent for both materials. In other words, the fatigue initiation in Glare is treated in the same way as for monolithic aluminium, using the actual stresses in the aluminium layers at the critical location instead of the applied stresses. Therefore, the actual stresses in the aluminium layers need to be determined as a function of the applied stress for each type of Glare.

The actual stresses in the aluminium layers in Glare consist of stresses due to the curing process and stresses due to the external loading, see Figure 11.1. Internal stresses will occur during the curing process as a result of the differing coefficients of thermal expansion between the layers. This stress system depends on the layup and type of Glare. For each Glare laminate, the residual stresses in the aluminium layers can be calculated as a result of the curing process. The curing stress is indicated by the horizontal line in Figure 11.1 and can be calculated with Equation 11.1.

$$S_{Al_{cure}} = \left[t_{Al_{tot}} E_{Al}\left(\alpha_{Al} - \alpha_{Al}\right) + t_{f,0_{tot}} E_{f,0}\left(\alpha_{f,0} - \alpha_{Al}\right) + t_{f,90_{tot}} E_{f,90}\left(\alpha_{f,90} - \alpha_{Al}\right)\right] * \frac{E_{Al}\left(T_R - T_{cure}\right)}{E_{lam}\left(t_{Al_{tot}} + t_{f,0_{tot}} + t_{f,90_{tot}}\right)} \quad (11.1)$$

with: E_{Al} = Young's modulus aluminium
$E_{f,0}$ = Young's modulus prepreg in fibre direction
$E_{f,90}$ = Young's modulus prepreg perpendicular to fibres
E_{lam} = Young's modulus laminate
$S_{Al\ cure}$ = stress in aluminium layers due to curing
$t_{Al\ tot}$ = total thickness of all aluminium layers
$t_{f,0\ tot}$ = total thickness of all prepreg layers in fibre direction
$t_{f,90\ tot}$ = total thickness of all prepreg layers perpendicular to fibres
T_{cure} = curing temperature
TR = room temperature
α_{Al} = coefficient of thermal expansion aluminium
$\alpha_{f,0}$ = coefficient of thermal expansion prepreg in fibre direction
$\alpha_{f,90}$ = coefficient of thermal expansion perpendicular to fibres.

As well as the curing stresses, the actual stresses in the aluminium layers are affected by the different stiffnesses of the various Glare constituents. This means that the aluminium layers attract more stress than the fibre layers due to their relatively higher stiffness. The stresses in the aluminium layers will therefore be higher than the applied stresses. These stresses can be calculated with Equation 11.2.

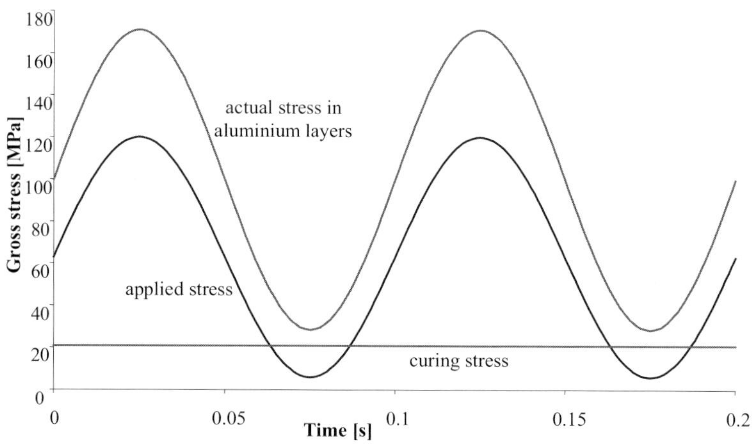

Figure 11.1 *The stress cycle in the aluminium layers in Glare 3-3/2-0.3 L at Room Temperature is a superposition of the stress cycle induced by the applied stress and the curing stress* [1].

$$S_{Al} = \frac{E_{Al}}{E_{lam}} S_{lam} = \frac{E_{Al}\left(t_{Al_{tot}} + t_{f,0_{tot}} + t_{f,90_{tot}}\right)}{E_{Al} t_{Al_{tot}} + E_{f,0} t_{f,0_{tot}} + E_{f,90} t_{f,90_{tot}}} S_{lam} \qquad (11.2)$$

with: S_{Al} = actual stress in aluminium layers
S_{lam} = applied stress on laminate.

This approach of calculating the stresses in the aluminium layers is shown schematically in Figure 11.2.

By superposition of the curing stresses and the stresses induced by the applied fatigue load the actual stresses in the aluminium layers of Glare is obtained, see Figure 11.1. The curing stress is constant in time, whereas the stresses induced by the applied fatigue load vary.

With the actual stresses in the aluminium layers known, crack initiation can be determined in the same way as for monolithic aluminium. The S-N data of monolithic aluminium, i.e. the number of cycles to failure, are recalculated to S-N_i data, the number of cycles to crack initiation. Instead of the applied stress, the actual stress in the aluminium is therefore used to calculate the number of cycles to initiation.

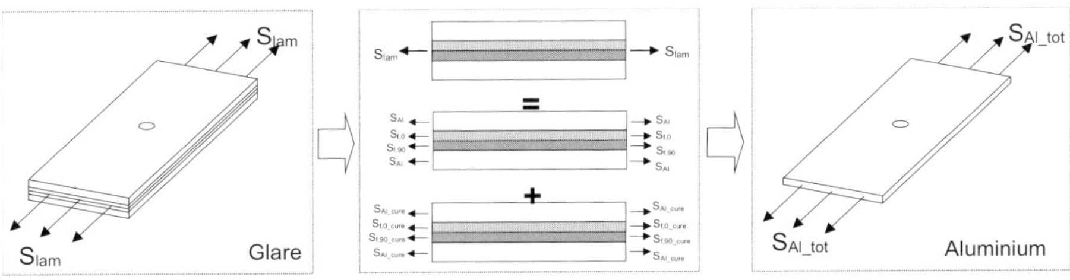

Figure 11.2 Approach of calculating the stresses in the aluminium layers.

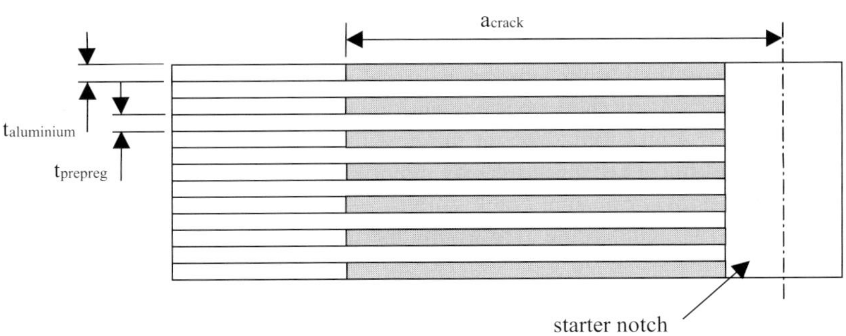

Figure 11.3 Through-the-thickness crack distribution in a through-crack Glare sheet.

11.2 Fatigue crack growth of through cracks

The most important assumption of fatigue crack growth in Glare with cracks through the thickness is that the crack is assumed to be equal in all the aluminium layers, see Figure 11.3. Tests showed that this is realistic for cracks in centre cracked (CCT) specimens starting from a notch and under uniaxial loading. When bending is present in the laminate, as is the case for joints, cracks initiate and propagate in Glare first as part-through cracks. This will be discussed in section 11.3. For through cracks, we also assume that the fibre layers remain intact for this section, which is realistic if glass fibres are applied.

11.2.1 Fatigue crack growth mechanism

During the crack propagation phase a significant part of the fatigue stresses are transferred over the cracks in the aluminium layers by the intact fibre layers, see Figure 11.4. The load is transferred from the aluminium layers through the adhesive layer to the fibre layers, inducing cyclic shear stresses in the adhesive. These cyclic shear stresses cause delamination growth between the aluminium and fibre layers.

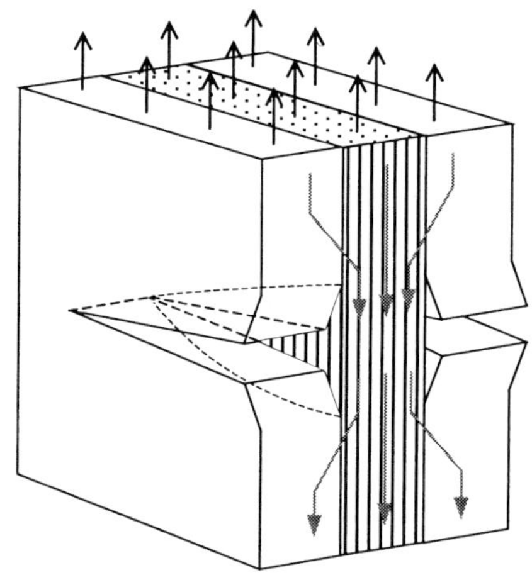

Figure 11.4 Crack bridging of the fibres and delamination of the layers.

Although the stresses in the aluminium layers are higher than the stresses applied to the laminate, as discussed in the previous section, the crack propagation phase is characterised by lower stress intensities at the crack tip compared to the monolithic aluminium configuration. This is caused by the fibre-bridging mechanism. Many tests showed approximately constant crack growth rates in Glare and consequently a constant stress intensity factor in the aluminium layers, see Figure 11.5.

According to Marissen [2] the crack opening during the crack propagation phase occurs due to adhesive shear deformation and delamination between the layers. The load transfer from the aluminium layers to the fibre layers induces cyclic shear stresses in the adhesive layers and shear deformation as a result. Due to this shear deformation the crack opening is constrained, but no longer zero, which induces a stress intensity at the crack tip in the aluminium layers.

The cyclic shear stresses in the adhesive layers are directly related to the fibre bridging stresses. These cyclic shear stresses cause the delamination between the aluminium and fibre layers, which takes place in the fibre/epoxy layers between the resin-rich and fibre-rich layer, to continue to grow, meaning that the length over which the fibres are elongated increases, resulting in lower fibre stresses. In other words, there is a balanced mechanism between delamination growth in the prepreg layers and the bridging stress.
Since the bridging stress affects the stress intensity at the crack tip, one can state that the crack growth and the delamination growth form a balanced mechanism in which both processes are continuously influencing each other during the fatigue life.

Despite the delamination growth, the fibre layers remain intact, and a significant part of the load is transferred over the crack through the fibres, restricting the opening of the crack. This results in a reduction of the stresses in the net section of the aluminium layers and thus a reduction of the stress intensity at the crack tip.

A characteristic difference between monolithic aluminium and Glare is the fatigue life. That of monolithic aluminium consists mainly of the crack initiation phase, since crack propagation is very fast and covers only a few percent of the fatigue life. The fatigue life of Glare, on the other hand, consists for the most part of the crack propagation phase.

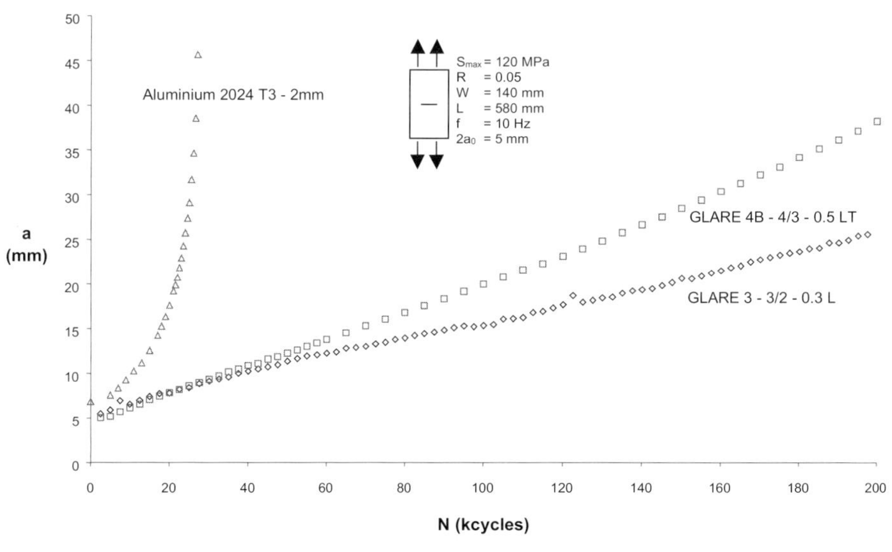

Figure 11.5 *Crack growth curve of aluminium 2024-T3 and Glare 3-3/2-0.3L and Glare 4B-4/3-0.5LT for constant amplitude fatigue loading.*

In a centre-cracked specimen of monolithic aluminium with a constant amplitude loading at the edge, the stress intensity increases with the crack length. However, in Glare the stresses are continuously redistributed over the intact fibres and the net section, resulting in stress intensities at the crack tip that remain approximately constant for the major part of the crack propagation phase. Reduction of the net section and an increase in crack length do not have the same effect on the behaviours of Glare and monolithic aluminium.

Since the fibres restrict the opening of the crack, the fibre stresses are directly related to the level of crack opening. At the beginning of the crack propagation phase the crack opening is very small, which means that the fibres are not yet fully loaded. A certain amount of crack length and Crack Opening Displacement (COD) is needed before the fibre bridging of the crack becomes fully effective.

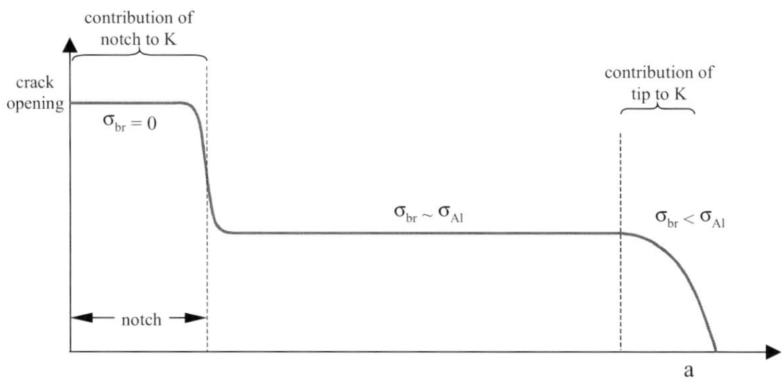

Figure 11.6 *Schematic representation of the crack opening and the bridging stress σ_{br} (σ_{Al} is the undisturbed stress in the aluminium layer).*

Several analytical relationships between the crack opening shape and the bridging stresses are available. Marissen [2] introduced a relation for the contour of the crack opening corresponding with a constant fibre bridging stress along the crack length. Although a decrease of the bridging stresses near the crack tip seems more realistic, the assumption of constant fibre bridging stresses is preferable for modelling purposes and is also something to be expected. If the stress distribution in the fibres is not constant then more load will be transmitted where the fibre stresses are highest. Consequently, the shear stresses between the aluminium and fibre layers are higher and more delamination will occur, which will lower the fibre stress at that location. A more or less constant fibre stress along most of the crack length is therefore a reasonable assumption.

Cox [3] proposed a bridging stress that is constant along the crack length and equivalent to the distant field stresses, except at the crack tip where the bridging stress must decrease to zero. The stress intensity at the crack tip is directly related to the difference between the declining bridging stress and the distant field stress. An additional contribution to the stress intensity factor results from the presence of a starter notch. The crack is initiated at this point and no fibre bridging is possible, see Figure 11.6.

11.2.2 Calculation methodology

The approach for modelling fatigue crack propagation in the through-the-thickness crack configuration is analogous to the stress intensity approach for monolithic aluminium. The crack growth rates are predicted based on the stress intensity factor at the crack tip in the aluminium layers of Glare. The actual stresses in the aluminium layers are calculated from the applied stresses on the laminate. The method used is discussed in the previous section for initiation, and is schematically represented in Figure 11.1.

The characteristics for Glare can be obtained by comparison with the calculated stress intensity factor for monolithic aluminium:

$$\Delta K = F \Delta S \sqrt{\pi a} \qquad (11.3)$$

with: a = half crack length
F = finite width correction factor
ΔK = stress intensity factor range.

The actual stress intensity at the crack tip in the aluminium layers in Glare is determined from the fatigue crack growth tests in Glare using the Paris-Walker relation of the aluminium layers:

$$\frac{da}{dN} = C\left[(1 - R)^{m-1} \Delta K_{eff}\right]^{n} \qquad (11.4)$$

with: C, m, n = Paris-Walker constants
R = stress ratio
ΔK_{eff} = effective stress intensity factor range.

As for Equation (11.3), an expression can be obtained for the stress intensity factor at the crack tip in the aluminium layers of Glare, denoted as ΔK_{eff} in Equation (11.4). Guo and Wu [4, 5] proposed an empirical approach based on a constant crack growth rate in Glare and consequently a constant stress intensity factor, characterised by an equivalent crack length l_0 in equation (11.3) which is a material parameter for the type of Glare:

$$\Delta K_{eff} = \Delta S \sqrt{\pi\, l_0} \qquad (11.5)$$

with: l_0 = equivalent crack length parameter.

Test evidence revealed that this equivalent crack length is not really a material parameter, but depends on the starter notch size and the finite width effect. However, these effects can be incorporated and an empirical relation was obtained for the equivalent crack length for one type of Glare, see Figure 11.7, which can be used to predict crack growth rates, as a function of the applied stress level, for design purposes [6].

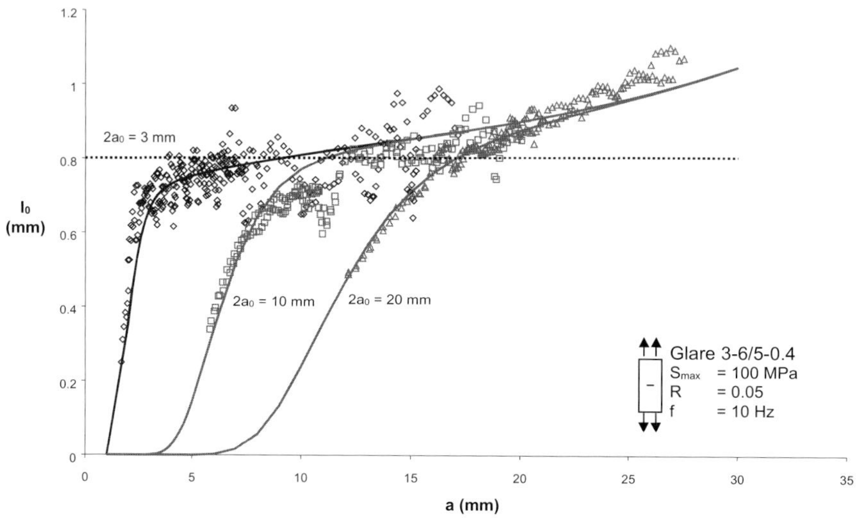

Figure 11.7 Determination of l_0 for Glare 3-6/5-0.4 [7].

11.3 Fatigue crack growth of part-through cracks

11.3.1 Introduction

Instead of fatigue cracks through the thickness, most fatigue cracks in aircraft structures initiate from either scratches or notches and start as surface cracks. For this reason part-through cracks are far more realistic for structures. Part-through cracks can be divided into configurations with surface cracks, i.e. a crack through the outer aluminium sheet, with all the other layers in the laminate intact, and situations in which more layers are cracked, see Figure 11.8 (a).

The crack growth characteristics of part-through cracks with multiple cracked aluminium layers and surface cracks are comparable to the characteristics discussed for through cracks in the previous section. In these cases, the intact aluminium layers also bridge the cracks. The crack growth rates are approximately constant during the crack propagation phase, but for surface cracks the crack growth rate is significantly lower than for through cracks. The bridging mechanism in a surface crack configuration is different from that of the through crack configuration, due to the intact aluminium layers bridging the crack. The intact part of the thickness of the laminate will carry more load, reducing the stresses in the net section of the cracked surface layer. The stress intensity factor at the crack tip will be significantly lower than for a through crack configuration with the same crack length [8, 9].

The starter notch effect

For monolithic aluminium the notch will not have an additional effect once the crack length is long enough. However, for Glare there is an effect for both through cracks and part-through cracks, since the fibres in the notch are cut or broken and no fibre bridging occurs across the notch. Along the fatigue crack in Glare the fibres are intact and reduce the crack growth rate. Since the fibre layers are intact for surface cracks, no additional influence of the starter notch is to be expected on the crack growth behaviour.

Secondary bending

Compared to the through crack configuration, surface cracks and other part-through cracks have an additional factor, i.e. secondary bending. The neutral line of the laminate will shift due to the asymmetry created by the broken surface sheet, which induces secondary bending. This effect is relatively small for a surface crack configuration, but for part-through cracks this may be important.

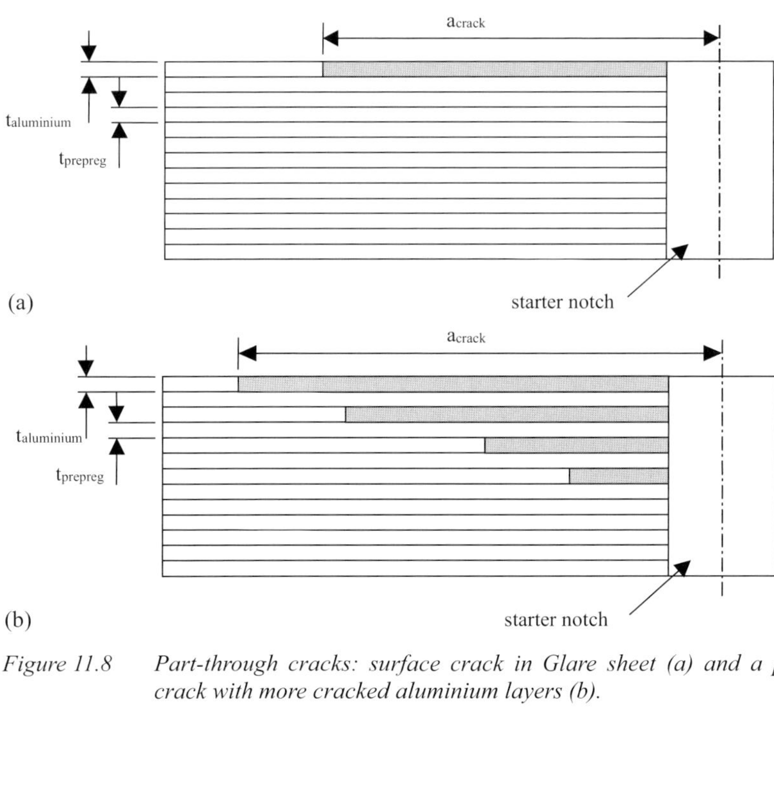

Figure 11.8 Part-through cracks: surface crack in Glare sheet (a) and a part-through crack with more cracked aluminium layers (b).

11.3.2 Calculation methodology

The method discussed in the previous section is available to calculate the crack propagation in through cracks. The method for calculation of crack propagation in surface cracks will be discussed in this section. The calculation method for part-through cracks and surface cracks is based on the following philosophy. Since any part-through crack configuration is between the boundaries of through cracks, i.e. the highest possible crack growth rate, and surface cracks, i.e. the lowest possible crack growth rate, the method to calculate crack propagation in part-through cracks consists of an 'intermediate' solution. This means that a combination of the stress intensity factor definitions for through cracks and surface cracks is used to define the stress intensity factor.

The approach of crack propagation in surface cracks is the same as the method for through cracks, discussed in the previous section. The actual stress in the aluminium layers is calculated from the applied stress. The crack growth rate is predicted based on the stress intensity factor at the crack tip in the surface aluminium layer, itself based on an empirically determined equivalent crack length.

Another approach for determination of the stress intensity at the tip of a surface crack configuration is proposed by De Koning [10]. He proposed to treat the stress system that is opening the crack as a local system near the crack tip, which directly causes the stress intensity at the crack tip and thus the crack growth rate. The load system, see Figure 11.9, is assumed to be independent of the crack length. The effect of secondary bending is ignored and assumed to be covered by the local stress system.

The length over which the stress system acts, δ, depends on the stress S, the thickness of the surface layer and material parameters specific to a Glare type.

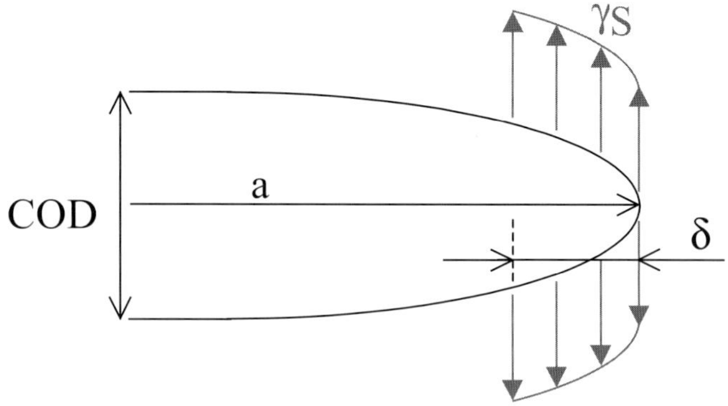

Figure 11.9 Stress system at the crack tip acting on the crack flanks.

The crack in the second layer initiates at the moment the crack in the surface layer has reached a certain length, and its growth is affected by the crack length in the surface layer. The stress intensity relation is then a combination of the through crack approach and the local stress system.

11.4 Variable amplitude loading

11.4.1 Introduction

In many cases a variable amplitude loading sequence is more realistic, causing different fatigue behaviour than deducted from the constant amplitude loading approach discussed in the previous sections. The influence of the successive overloads and low loads on the fatigue behaviour of Glare is different from the characteristics of Glare under constant amplitude loading.

The effect of overloads on the crack growth in Glare can be represented by an example of the simplest variable amplitude loading case, which is the constant amplitude loading with one peak load, see Figure 11.10. After applying a peak load, three different phases can be distinguished; the crack retardation phase, return to original crack growth rate and the phase with the original crack growth rate.

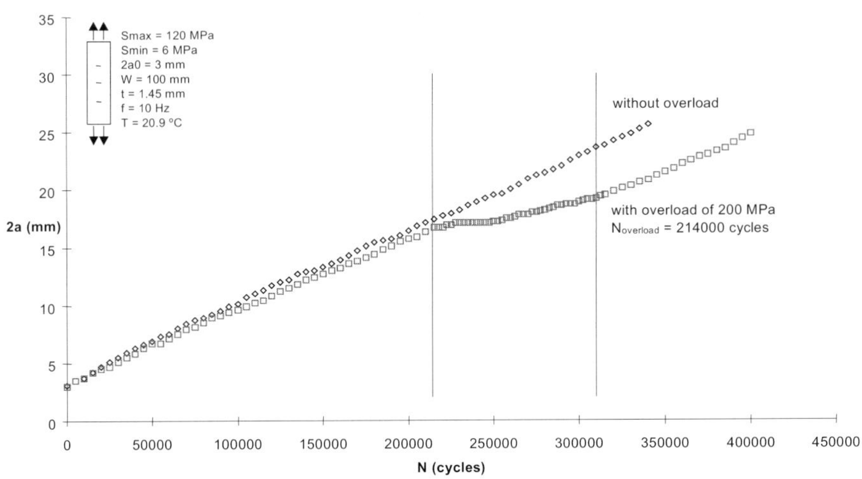

Figure 11.10 *Crack growth curve for through crack in Glare3-3/2-0.3 L with and without a peak load of 200 MPa applied* [11].

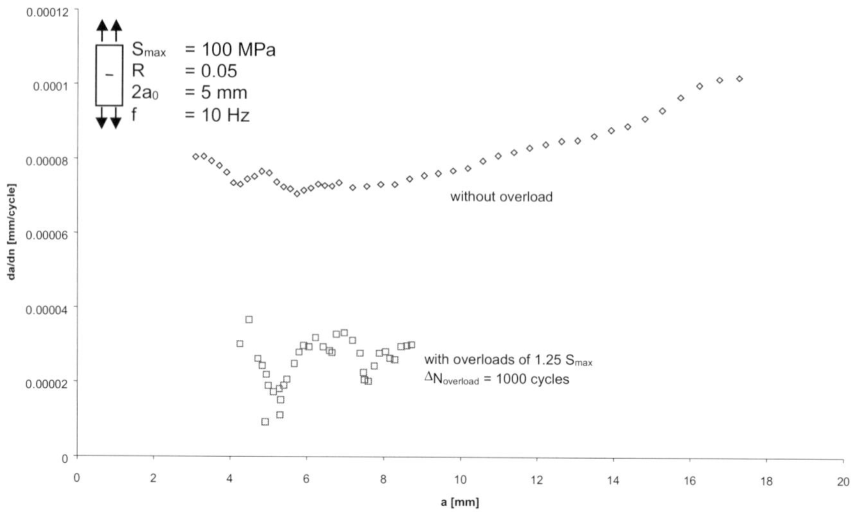

Figure 11.11 *Crack growth rate da/dn versus the crack length a for Glare 4B-4/3-0.5 LT with and without overloads of 1.25 S_{max} at $\Delta N_{overload}$ = 1000 cycles* [7].

The effect of repeatedly applied peak loads during constant amplitude loading, depends on the number of cycles between the peak loads. If the next peak load is applied before the crack growth rate has returned to its original level, the overall crack growth rate will be significantly lower than without applied peak loads. An example is given in Figure 11.11, where the crack growth rate of a constant amplitude loading case is compared with the crack growth rates of a case with a peak load every thousand cycles. Research [11] revealed that the delamination growth is not affected by overloads, thus the crack growth retardation is mainly due to plastic deformation in the aluminium layers. This effect occurs in a more complex manner in a variable amplitude loading spectrum due to the successive high and low loads. Since the phenomenon is governed by the plasticity in the aluminium layers, the crack retardation effect in Glare is similar to that in monolithic aluminium.

11.4.2 General approach

The current approach for modelling crack propagation due to variable amplitude loading in Glare does not yet incorporate the effect of crack retardation as discussed above. A very straightforward but invalid approach is to calculate the crack extension for each cycle with the method discussed for the constant amplitude loading and add the crack extensions to the initial crack length [12]. To include the effect of the present crack length, the crack extension is calculated for blocks of cycles and added to the initial crack length, after which the crack extension is calculated for the following fatigue load block. Due to the absence of crack growth retardation effects, the method will overestimate crack propagation for high loads. However, as result of fibre bridging which reduces the stress intensity factor, the overestimation in prediction of crack propagation in Glare is smaller than for monolithic aluminium.

11.5 Effect of moisture and temperature

Two aspects are especially important for the durability characteristics of Glare with respect to fatigue crack propagation:

- moisture absorption
- temperature

Glare's composition makes it hard to describe the mechanism, due to the individual characteristics of the components with respect to temperature and moisture and their interaction. In the following sections the effects of moisture and temperature will be discussed qualitatively.

11.5.1 Effects of moisture

An aircraft structure will be exposed to moisture in service conditions, therefore the influence of moisture on the fatigue properties of Glare was investigated. The presence of moisture was expected to have a detrimental effect on the fatigue crack growth behaviour of Glare. Put simply, the effect of moisture on the material can be divided in three main effects:

- corrosion and more rapid crack growth rates in aluminium
- decrease of adhesive interface strength between fibre layer and aluminium
- plasticising of the matrix.

The moisture absorption by the matrix results in a decrease of the adhesive interface strength, because it affects the chemical bond between the adhesive and the oxide layer of the aluminium and of the oxide layer itself. The glass transition temperature of the epoxy is also reduced, see Figure 11.12, resulting in a degradation of the mechanical properties and an increase of shear deformation and therefore less crack opening restraint. As a result of the plasticising of the matrix, the glass transition will occur at lower temperatures than in a situation without moisture.

The moisture absorption is dependent on the frequency of the fatigue load. At very low frequencies the effect of moisture absorption can be significant, for frequencies of 10 Hz, the effect is relatively small.

11.5.2 Effects of temperature

The temperature affects the internal stress system in Glare. The curing stresses present in the laminate will change because of the different coefficients of thermal expansion. This can be quantitatively explained using Equation (11.1), with T_R substituted by the ambient temperature. The stresses in Glare due to curing will decrease with increasing temperature.

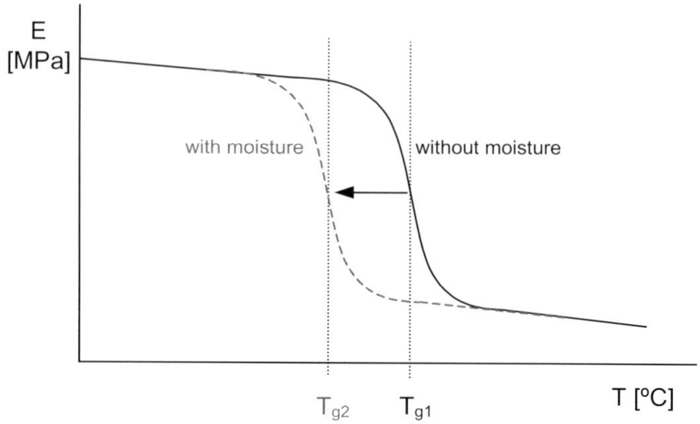

Figure 11.12 *Location of the glass transition temperature for the situation with and without moisture.*

Although the curing stresses decrease due to elevated temperatures, higher temperatures negatively affect the fatigue crack growth behaviour of Glare. This is mainly due to two factors. With increasing temperatures the moisture absorption will also increase, affecting the crack growth behaviour as discussed above. The elevated temperatures cause voids and microscopic defects to open and attract more fluid. Another factor is the softening and weakening of the matrix due to the increase of temperatures. This causes the restraining of the crack opening to decrease, inducing higher stress intensities at the crack tip in the aluminium layers.

The effect of exposure to elevated temperatures on the fatigue behaviour of Glare is illustrated by Figure 11.13. Under such exposure, the crack growth rate increases as a result of the weakened matrix and decreased effective fibre bridging. Alternating fatigue testing and 1000 hours of exposure to elevated temperatures results in an increase of crack growth rate after each exposure period. However, the difference between the crack growth rates of the specimen that is exposed once for 3000 hours and the specimen alternately exposed and fatigue tested is relatively small.

The effect of fatigue testing at elevated temperatures is illustrated in Figure 11.14, where the crack growth curves of 'as cured' Glare 3-3/2-0.3 L are given for three different temperatures and compared with a monolithic aluminium crack growth curve at room temperature. As discussed before, the increase of the temperature negatively affects the fatigue crack growth behaviour of Glare.

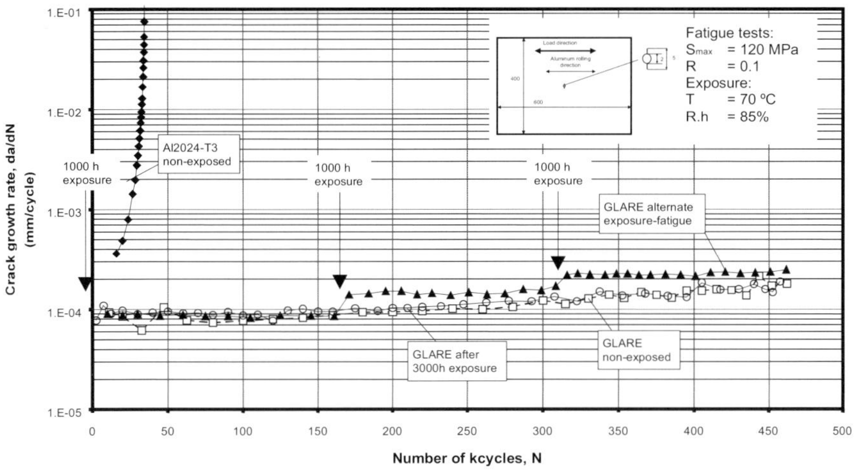

Figure 11.13 *Crack growth rate curves for Glare 3-5/4-0.4L CCT fatigue specimen [13].*

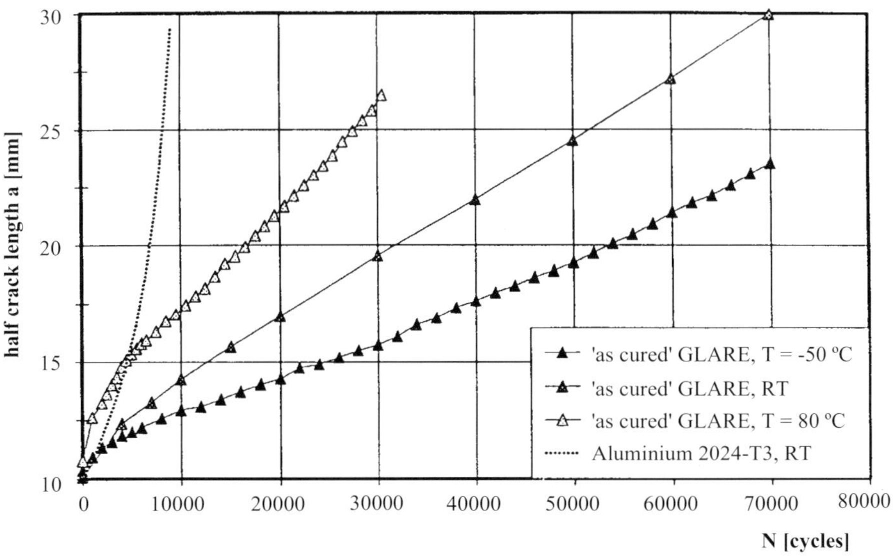

Figure 11.14 *Crack growth curves for 'as cured' Glare 3-3/2-0.3 L tested at three different temperatures with S_{max} = 140 MPa, R = 0.1, $2a_0$ = 20 mm,* [14].

References

[1] O.T. Kieboom, *Fatigue Crack Initiation and Early Crack Growth in Glare at different temperatures.* Master Thesis, Structures and Materials Laboratory, Faculty of Aerospace Engineering, Delft University of Technology, November 2000.

[2] R. Marissen, *Fatigue Crack Growth in ARALL, A hybrid Aluminium-Aramid Composite Material, crack growth mechanisms and quantitative predictions of the crack growth rate*, PhD thesis, June 1988, Report LR-574, Faculty of Aerospace Engineering, University of Technology Delft.

[3] B.N. Cox, *Life prediction for bridged fatigue cracks, Life prediction methodology for Titanium Matrix Composites,* ASTM STP 1253, W.S. Johnson, J.M. Larsen and B.N. Cox, Eds. American Society for testing and Materials, 1996, pp. 552-572

[4] Y.J. Guo and X.R. Wu, *A Phenomenological Model for Predicting Crack Growth in Fiber Reinforced Metal Laminates under Constant Amplitude Loading*, Institute of Aeronautical Materials, Beijing, 100095, China

[5] Y.J. Guo, and X.R. Wu, *A Model for Predicting Fatigue Crack Growth Rates in Fiber Reinforced Metal Laminates (FRMLs)*, Institute of Aeronautical Materials, Beijing, 100095, China

[6] R.C. Alderliesten, *An empirical crack growth model for Fibre Metal Laminates.* Preliminary Thesis, Structures and Materials Laboratory, Faculty of Aerospace Engineering, Delft University of Technology, September 1998.

[7] R.C. Alderliesten, *Development of an empirical fatigue crack growth prediction model for the Fibre Metal Laminate Glare.* Master Thesis, Structures and Materials Laboratory, Faculty of Aerospace Engineering, Delft University of Technology, 14 December 1999.

[8] K.A.M. Gonesh, *Development of a fatigue crack growth model for surface cracks in GLARE.* Master Thesis, Structures and Materials Laboratory, Faculty of Aerospace Engineering, Delft University of Technology, August 2000.

[9] K.A.M. Gonesh, *Crack growth prediction of surface cracks in Glare.* Preliminary Thesis, Structures and Materials Laboratory, Faculty of Aerospace Engineering, Delft University of Technology, March 1999.

[10] A.U. de Koning and L. Schra, *Fatigue crack growth of part through the thickness cracks in Glare 3 and Glare 4B coupons, -Final report GTO subproject 2.4.2 -.* Report NLR-CR-2000-078, Structures and Materials Division, National Aerospace Laboratory NLR, 17 February 2000.

[11] H.J.M. Woerden, *Variable Amplitude Fatigue of Glare® 3, Influence of Peak Loads and Flight Spectrum Development.* Preliminary Thesis, Structures and

Materials Laboratory, Faculty of Aerospace Engineering, Delft University of Technology, May 1998.

[12] H.J.M. Woerden, *Fuselage Spectrum Fatigue Loading on Fiber Metal Laminates.* Master Thesis, Structures and Materials Laboratory, Faculty of Aerospace Engineering, Delft University of Technology, 30 October 1998.

[13] W. van der Hoeven, *Final report Glare Durability program, Results of tests carried out at NLR*, NLR-CR-2000-237, National Aerospace Laboratory NLR, 2000

[14] H. Bär, *Verifikation und Ergänzung von Berechnungsmethoden für die statische und dynamische Auslegung von Glare-Strukturen.* Diplomarbeit, Institut für Flugzeugbau der Universität Stuttgart, August 1992.

12

Fatigue of riveted joints

J.J. Homan, R.P.G. Müller, F. Pellenkoft, J.J.M. de Rijck

Abstract – Mechanically fastened joints have proven to be weak links in an aircraft structure. Therefore, these joints require high quality design and manufacturing processes. This chapter focuses on the joint design optimisation. Glare provides the means to improve the joint design using advanced material design. The enhanced fatigue properties will demonstrate some design advantages. The following key characteristics need to be highlighted: load transfer, secondary bending, influence of biaxiality and curvature and influences of fastener installation. Secondary bending induced by the inherent eccentricities of joints introduces a non-linear bending stress in the joints. A fast and simple means to calculate the stresses in joints is provided in this chapter and it is shown by tests that the neutral line model for this purpose is accurate. A complex three-dimensional stress system exists around rivets, which is affected by the installation procedures. Improved riveting installation procedures, which control interference fits, can delay crack initiation by a factor of ten. The joint fatigue behaviour for Glare is described in three parts, i.e. crack initiation, crack growth and residual strength. Fatigue crack initiation can be predicted using the Fokker severity concept adapted for Glare. Stress levels in equally thick aluminium and Glare sheets show higher stresses for the aluminium layers in Glare. Fibre-bridging and bridging by the intact aluminium layers stop the crack created by the earlier crack initiation in Glare. Crack growth is significantly slower in Glare than in aluminium joints.

Finally two failure modes, i.e. fastener failure and net section failure, dominate the residual strength behaviour of Glare joints. The first failure mode in Fibre Metal Laminates is a somewhat different failure behaviour than for monolithic aluminium, because of the build up of FML. This allows for delamination of the layers during the installation process. Countersunk rivets are most likely to cause delamination in the outer layers, since the contact area is smaller and thus provides less support for the rivet. The fastener will therefore be more easily pulled through the sheet. The net section failure mode phenomenon in Glare is characterised by intact fibres bridging the crack. Experiments and closed form solutions have shown that also rivet pull through can occur. The residual strength calculation of joints with fatigue cracks is based on the remaining net section strength of the aluminium. Residual strength is up to 1.5 times higher, dependent on the metal volume fraction for Glare, than for aluminium joints.

The higher fatigue crack growth resistance makes lower thickness or higher stresses in Glare possible. Many tests indicated that a reduction of 30% in thickness still leads to higher fatigue lives compared to aluminium 2024 joints.

12.1 Introduction

Aircraft structures have many mechanically fastened joints. These joints are introduced because of manufacturing requirements and to limit the dimensions of the sheet size as well as due to logistic requirements. Mechanically fastened joints are highly loaded due to local pin loading and load path eccentricity. In pressurised cabins they are therefore prone to fatigue cracking due to the pressurisation cycle during each flight. A dangerous type of damage is Multiple Site Damage (MSD) failure of small cracks in rivet holes linking up to a lead crack as occurred in the well-known Aloha Airlines Boeing 737 accident in 1988. The drive for increased design lives in combination with higher design stress levels requires improved joint designs and/or materials to prevent fatigue-induced failures. This chapter focuses on the joint design optimisation, using Glare. It demonstrates the enhanced fatigue properties of Glare joints. This chapter first treats important joint characteristics in section 12.2, which control the joint fatigue properties that are discussed in section 12.3. A summary will be given in section 12.4.

12.2 Joint characteristics

12.2.1 Load transfer

Load transfer in mechanically fastened joints occurs due to the different fastener rows transferring the stresses from one of the mating sheets to the other. For joints with two fastener rows, rivet loads are equally distributed over the rows. For three row fastened joints, as commonly applied in aircraft industry, the load transfer per row depends on the stiffness of sheets and fasteners. Closed form solutions based on deflection compatibility allow calculation of load transfers in the joint. It is necessary to calculate the global load transfer stresses at the outer rows. Local load transfer in the joint depends on the tolerance of the fastener in the hole. This phenomenon will be separately discussed in section 12.2.

A closed form solution has been developed to calculate the load transfer stresses for global load transfer. For Glare the rule of mixtures is applied to calculate the mechanical properties of the sheet. An example for a three-row riveted joint in aluminium and Glare is given below. Using the simplification that the load transmission by friction is neglected, the closed form solutions then become similar to the Jarfall model [1]. The interference effects are accounted for by the fastener flexibility. Two other assumptions are made to allow for a simple displacement compatibility method:

1. Displacements due to fastener flexibility are assumed to be linearly related to the fastener-bearing load. This will only be valid for low loads on the joint, e.g. fatigue loads.
2. A uniaxial model is assumed. It consists of elements with stiffness equal to the stiffness of the sheet, interconnected by the fasteners with certain flexibility.

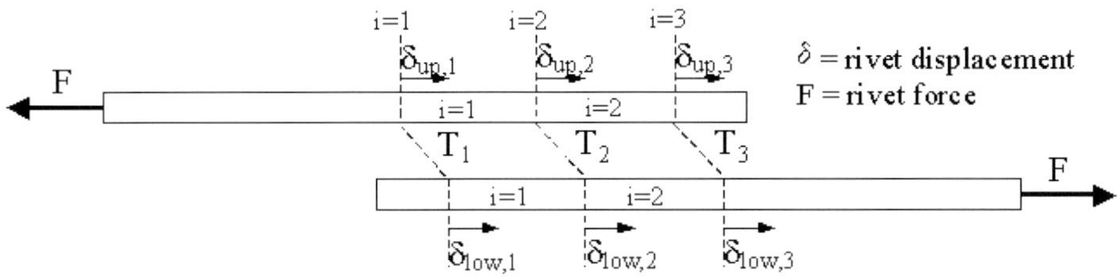

Figure 12.1 *Simple model of a riveted lap joint, used for displacement compatibility calculations.*

The model is shown in Figure 12.1, and is drawn for a lap joint with three rivet rows. The method described below is a more generalised method and is set up for n rows of fasteners.

The bearing load transmitted by the fasteners: T_i (i = 1 to n) equilibrium:

$$F = \sum_{i=1}^{n} T_i \qquad (12.1)$$

Bypass loads in the upper sheet and lower sheet:

$$F_{up,i} = F - \sum_{j=1}^{j=i} T_i \qquad (i = 1 \text{ to } i = n - 1) \qquad (12.2)$$

$$F_{low,i} = \sum_{j=1}^{j=i} T_i \qquad (i = 1 \text{ to } i = n - 1) \qquad (12.3)$$

Displacements in the upper and the lower sheet: $\delta_{up,i}$, $\delta_{low,i}$ (i = 1 to n). Hooke's law in the upper sheet and lower sheet yields:

$$\delta_{up,i} - \delta_{up,i-1} = f_{sheet,up} \cdot F_i \qquad (12.4)$$

$$\delta_{low,i} - \delta_{low,i-1} = f_{sheet,low} \cdot F_i \qquad (12.5)$$

with: f_{sheet} = the sheet flexibility.

Assuming equally spaced fastener rows, the sheet flexibility can be written as:

$$f_{sheet} = \frac{s_{row}}{E \cdot t \cdot s_{rivet}} \qquad (12.6)$$

with: s_{rivet} = fastener pitch in a row
s_{row} = distance between rows.

The displacements due to fastener flexibility are related as follows:

$$\delta_{low,i} - \delta_{up,i} = f_{rivet} \cdot T_i \qquad (i = 1 \text{ to } n) \tag{12.7}$$

Without any loss of generality it can be assumed that:

$$\delta_{up,1} = 0 \tag{12.8}$$

We now have 3 equations from which T_i, $\delta_{up,i}$ and $\delta_{low,i}$ ($i = 1$ to n) can be solved. It turns out that $T_1 = T_3$ for equal thickness of a lap joint with three fastener rows. The following ratio is found for T_2:

$$\frac{T_2}{T_1} = \frac{f_{rivet}}{f_{rivet} + f_{sheet}} \tag{12.9}$$

If the $f_{rivet} = 0$ then $T_2 = 0$. In other words, if the fasteners would not rotate at all, the two outer rows would each transmit half of the load in the specimen. However, if there is some fastener flexibility, T_2 becomes more positive, although according to the ratio T_2/T_1 T_2 will always be smaller than T_1.

12.2.2 Secondary bending

Knowing the load transfer, and the fastener bending moments resulting from the fastener shear stress, the bending moments in the sheets can be derived. These bending moments lead to out-of-plane deformation of the sheets, the so-called secondary bending. The effect of secondary bending in many cases doubles the membrane stresses due to in plane loading and therefore plays an important role in the fatigue initiation and crack propagation life of joints. Aspects that influence secondary bending in general are:

- The geometry of the joint, e.g. fastener row pitch, number of fastener rows, sheet thickness etc.
- The mechanical properties of the skin material.
- Structure stiffening

For Glare a further discrimination is required due to the long crack growth periods compared to aluminium joints, which will be discussed in section 12.3.2. In these cases, cracking of individual layers should be corrected for by means of stiffness response changes associated with continued cracking.

An example of the stress distribution created by combined tension and secondary bending in a Glare joint is shown in Figure 12.2.

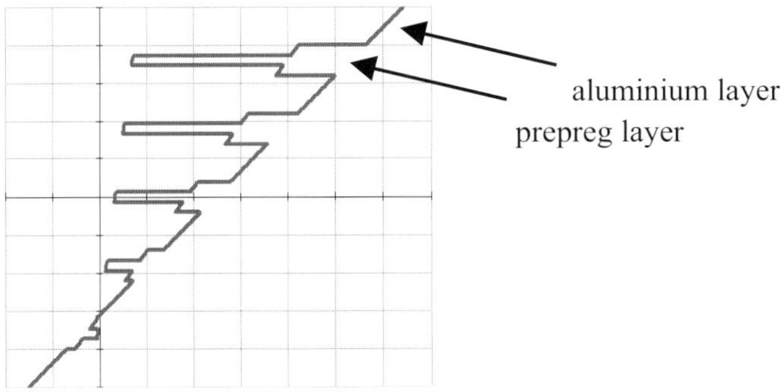

Figure 12.2 *Theoretical stress distribution in Glare 4-6/5-0.5 under combined tension and secondary bending conditions* [2].

The bending in a joint can be simulated by modelling its neutral axis: the "Schijve" neutral line model [3,4,5] for a two-row riveted joint, which was extended to joints with three or more rivet rows by Müller [6]. De Rijck and Fawaz [7] demonstrated that this model also applies for more complex joints with built-up doublers and longitudinal stiffeners. A neutral line represents the load path eccentricity in the joint, see Figure 12.3. Closed form solutions can be used to describe the load transfer over the different rows as well as the secondary bending. The load transfer initially calculated using Jarfall's theorem for the displacement criteria gives the neutral axis step for each row. Secondary bending calculation is based on the bending moment causing deformation in the mating sheets. This secondary bending behaviour can be described by a second order differential equation.

The deflection of the neutral line in the loaded condition (load P) can be described with the following differential equations:

$$M_x = P \cdot w = E_i I_i \cdot \frac{d^2w}{dx^2} \Rightarrow \frac{d^2w}{dx^2} - \alpha_i^2 \cdot w = 0 \qquad (12.10)$$

with: $\alpha_i^2 = \dfrac{P}{E_i I_i}$

which have the following solutions:

$$w_i(x) = A_i \cdot \sinh(\alpha_i \cdot x_i) + B_i \cdot \cosh(\alpha_i \cdot x_i) \qquad (12.11)$$

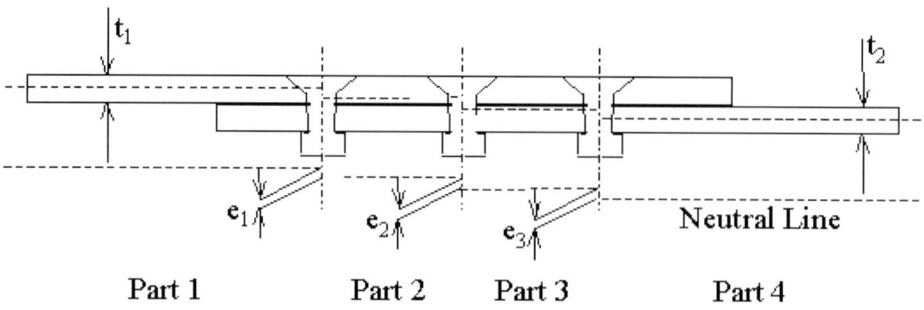

Figure 12.3 *Schematic representation of a three row fastened lap joint.*

For Glare, the flexural rigidity EI is calculated with the tensile modulus of elasticity of the laminate: E_{Glare}. This is only accurate for symmetrical laminates and joints where both sheets are the same materials. But for a first approximation this E_{Glare} will provide a sufficient base for calculations. To take the influence of non-symmetrical laminates into account the modulus of elasticity will be made up out of the independent layers (using flexure formulas for composite beams, [8]), this will allow sheets made of different materials.

The boundary conditions for the different parts are derived from the equal displacements and slopes of the deflection curve at both sides of the fastener rows and the clamping conditions at the end. With the solutions of the boundary value problem, the bending stress in the sheet thickness direction can be obtained with:

$$\sigma_{bending} = \frac{M_x}{I_i} \cdot z = \frac{P \cdot w_i(x)}{I_i} \cdot z = \frac{d^2 w_i(x)}{dx^2} \cdot E_i \cdot z \qquad (12.12)$$

For the model, the following assumptions are made [6]:

1. Fasteners are infinitely stiff.
2. The eccentricities remain the same regardless of the applied load.
3. The analysis is based on the elastic beam theory. Transverse shear stiffness is therefore not included.
4. Stresses in lateral direction are excluded.
5. Simple clamping conditions at the end of the lap joints specimen are used. It depends on the clamping devices and the specimen length, whether clamping conditions can be assumed to represent hinged or a fixed clamping. Schijve already noted a very small effect of the clamping conditions at the end of the sheet.
6. The secondary bending is treated as a 1-D problem; i.e. there is no variation of the bending in width direction.

For a joint with more than two rivet rows, the model can be extended by using the load transmissions γ through the different rivet rows to calculate the eccentricities [6], see Equation 12.13:

$$e_i = \frac{T_i}{P} \cdot e_{tot} = \gamma_i \cdot e_{tot} \qquad (12.13)$$

For the calculation of the stresses in Glare materials, the different elastic properties of the layers must be considered. When assuming a linear strain through the thickness, the bending stress of Equation 12.13 can be written as:

$$\sigma_{bending} = w'' \cdot z \cdot E_{layer} \qquad (12.14)$$

With this method of calculating the bending stresses on the basis of the elastic bending, the internal stresses caused by curing of the laminate or possible thermal stresses can also be incorporated.

The fibre layup of Glare has some influence on the bending stiffness. This influence is the largest for thinner Glare laminates where the influence of the different layers on the flexural rigidity is the largest. However, the influence is then 10% for realistic bending factors or bending ratios k_B ,i.e. ≤ 2.5 [2]. This bending factor is defined as the ratio of the bending stress and the applied stress and is frequently used to describe the severity of the secondary bending:

$$k_B = \frac{\sigma_{bending}}{\sigma_{applied}} \qquad (12.15)$$

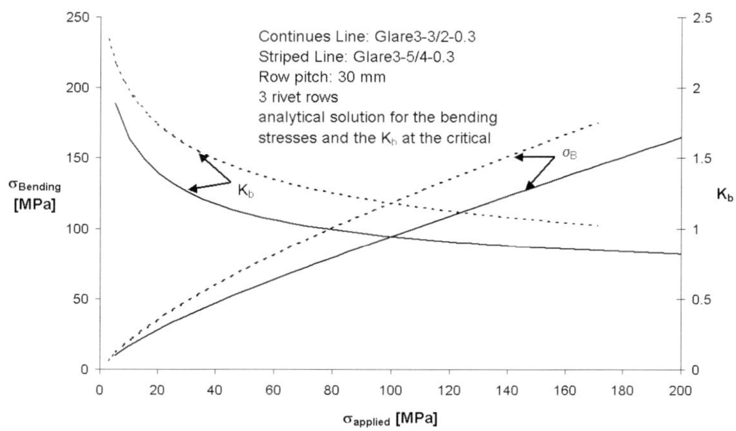

Figure 12.4 *Secondary bending stress and the ratio k_B at the critical rivet location as function of the applied stress, The results are obtained from the neutral line model.*

Calculations show that for an increase in thickness an increase in bending stiffness can be observed in monolithic aluminium joints. For comparable Glare joints with a similar sheet thickness the bending factor k_B results in a lower peak stress at the most critical fastener row. This can be explained by the already mentioned lower flexural rigidity of Glare. Figure 12.4 shows the increase in K_b for increased thickness for the Glare3 grade.

12.2.3 Effect of applied fasteners and fastener installation process

The fastener choice for a joint is in general driven by the required static strength and by the production costs of the joint. For joints in fuselages of aircraft this will often result in the use of solid aluminium rivets for longitudinal joints and high strength fasteners with a titanium pin (such as Hi-Loks and Lockbolts) in circumferential joints.
The choice of the fastener type determines the fatigue life to a large extend. The following aspects are important:

- Clamping forces.
- Fastener flexibility
- Fastener fit
- Installation method

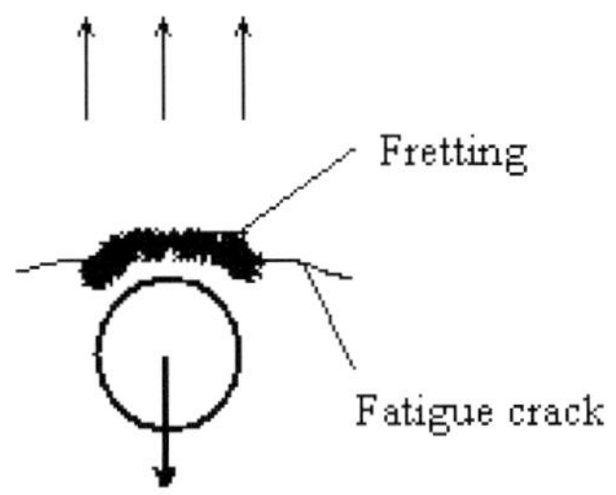

Figure 12.5 Fretting near a rivet hole.

Longitudinal joints

As mentioned above, for longitudinal joints usually solid aluminium rivets are applied. Solid rivets are deformable shank fasteners, meaning that their shank is deformed in the grip area during normal installation processes. This deformation can be described as follows:
During squeezing, radial expansion of the rivet in combination with axial compression results in a complex 3D residual stress state. The intensity of the squeeze force determines this residual stress state. Calculations and experiments demonstrated that at higher squeeze forces circumferential residual compression stresses are present at the hole perimeter. These residual stresses not only lower the stress amplitude and mean stress, but in case of continuous interference during

loading also have a reduced rivet tilting and a twofold increased bearing area, resulting in longer crack initiation periods, see Figure 12.6. Increasing residual stresses also lead to a shift of the initiation location from the edge to a location away from the hole, see Figure 12.5 and 12.6.

As mentioned in section 12.2.1, the rivet installation determines local load transfer, i.e. the bearing pressure at the hole edge. The squeeze force determines the hole filling, and at higher squeeze forces a level of interference can be achieved that even can be maintained throughout the loading cycle. The optimum squeeze forces can be verified by simple flexibility tests revealing higher stiffness and together with that smaller rivet tilting lower hole perimeter stresses. Despite the two times lower rivet flexibility with interference fit, the load transfer shedding to the outer rows is of benign effect to the reduced rivet tilting, i.e. enhanced crack initiation periods can be obtained. For Glare this occurs with every aluminium layer and through the thickness cracking is postponed due to this lateral crack bridging.

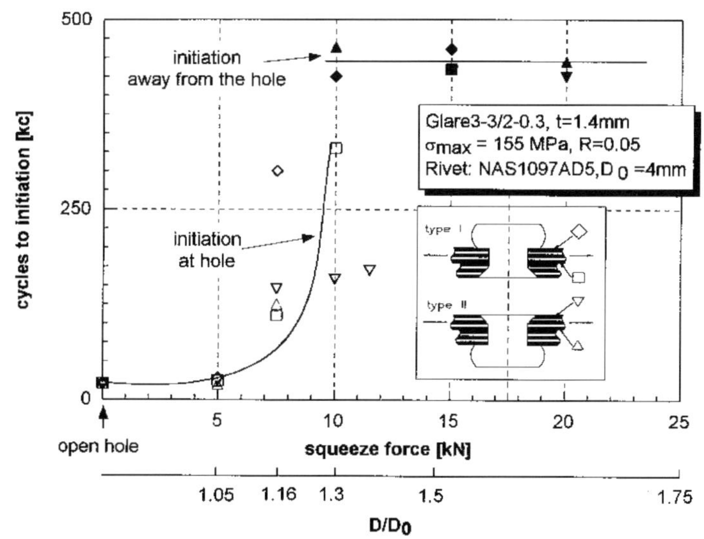

Figure 12.6 *Crack initiation fatigue lives for specimens with a single unloaded rivet. Effect of the rivet squeeze force.*

Circumferential joints

High strength fasteners used in circumferential joints such as Hi-Loks and Lockbolts are non-deformable shank fasteners. Non-deformable shank fasteners are those fasteners whose shank does not deform in the grip area during normal installation processes. These fasteners can be installed either with a clearance fit or with interference fit. Interference fit causes the hole to expand resulting in favourable stresses around the hole just as with solid rivets. Since these fasteners introduce

much more clamping forces than solid rivets, for example, the effect of pin loading on the fatigue life will be less. Because of the large clamping forces, initiation is mainly driven by fretting, see Figure 12.5.
Initial fatigue tests [9] showed, contrary to expectations, less good fatigue endurance for interference joints than for clearance joints in Glare probably due to created damage like delamination.
Fretting can significantly decrease the fatigue life of joints. The amount of fretting depends on the clamping forces. Ongoing research shows interfaying sealant can prevent fretting in Glare. Further, sealant has some load carrying capabilities, which results in higher initiation lives.

12.2.4 Biaxial stresses in fuselage joints

As shown in the pervious section two common types of joints can be distinguished in aircraft fuselages, i.e. longitudinal lap-splice and circumferential butt joints. An aircraft pressure cabin has two principal stresses due to pressurisation, i.e. these stresses cause a biaxial stress state. Figure 12.7 shows these two principal stresses. As can be seen the ratio between both stresses (i.e. the biaxiality ratio) is 0.5 for an unstiffened skin.

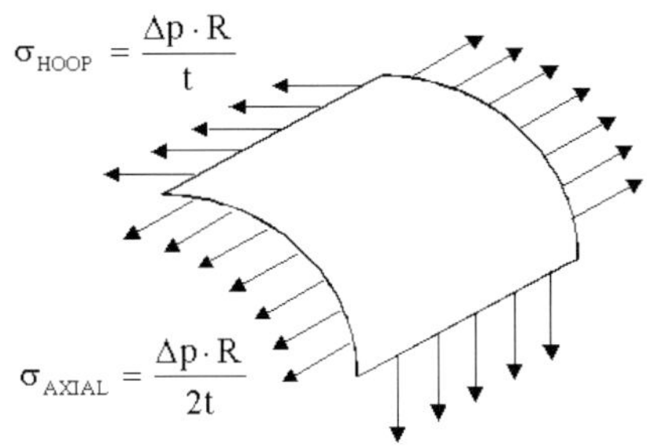

Figure 12.7 Stresses in skin material for unstiffened fuselage.

The longitudinal lap-splice joints are subjected to the highest loads, i.e. hoop stresses. This implies that these joints have to transfer the highest load. Mechanical fasteners will in most cases transfer the loads in these joints from one fuselage panel to another.

A sheet with a row of holes subjected to a uniaxial stress field has a higher stress concentration at the edges of the holes than a sheet with a row of holes subjected to a biaxial stress field [10]. The theoretical K_t for an open, circular hole in isotropic material subjected to pure tension is 3.0, whereas for a biaxiality ratio of 0.5 the K_t changes to a lower value of 2.5. For joints, a biaxial stress has a negligible effect on the secondary bending of the joint. Tests done by Müller [6] revealed only an average 5% reduction of the stresses around the rivets for biaxiality ratio of 0.5 compared to a uniaxial stress field.

12.2.5 Influence of curvature on fatigue of joints

Apart from biaxiality of the stress field, also curvature is typical for pressurised aircraft fuselages. Monolithic aluminium and Fibre Metal Laminates show a different behaviour with respect to curvature and pressurisation. Monolithic aluminium panels with a fatigue crack showed strong bulge-out effects under pressurisation, whereas Fibre Metal Laminates (Arall in this case) did not reveal bulge-out effects at all [11]. In the FML the fibres remained intact and this restrains the displacements at the crack flanks.

Also Müller [6] extensively determined the difference between the deformation in joints of flat coupons and in a realistic situation in the fuselage. Tests were performed at the reduced scale barrel test setup from the Faculty of Aerospace Engineering as well as the full-scale Airbus A340 barrel test setup in Hamburg [6]. The comparison revealed frames and stiffeners to have a high influence on secondary bending. Away from the frames, stresses were similar to flat uniaxially-loaded panels.

Riveted lap-splice joints in a cylinder with frames

The total deformation in a cylinder with frames is different from an unstiffened cylinder. At the frame location the fuselage shell will be prevented to expand outwards, see Figure 12.8.

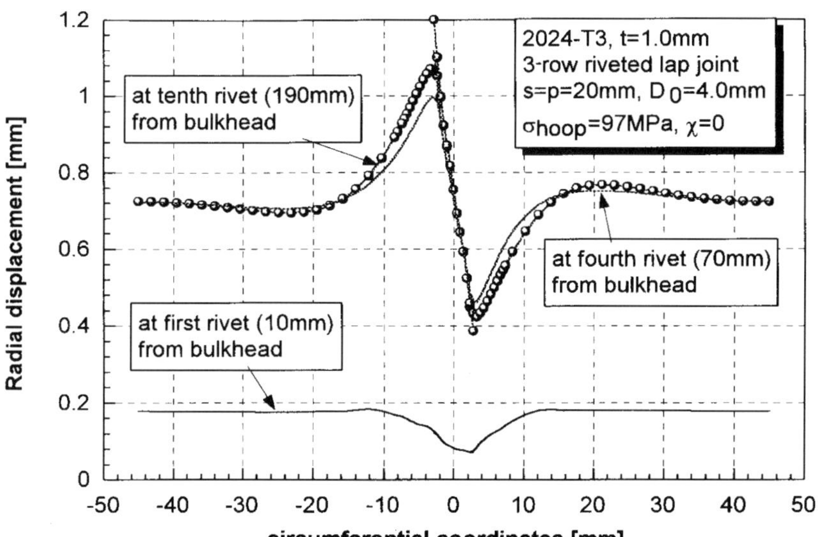

Figure 12.8 Radial displacements and rotation of the splice area at different locations away from rigid frames, according to finite element calculations by Müller [6].

The maximum displacement will now occur between the frames, also know as the 'pillowing effect'. This is more pronounced for large fuselages and stiff frames with a stiff frame skin connection. For smaller fuselage structures the lap-splice joint is subjected to a more uniform hoop stress and secondary bending. At the position of the frames the secondary bending is restrained, only between frames the secondary bending can fully develop. Three phenomena occur with joints in stiffened structures compared to flat panels:

1. Pillowing of the joints reduces the maximum hoop stress near the frames, but increases the bending stress in longitudinal direction at the frame.
2. Secondary bending restraint at the frames. Secondary bending is not influenced by biaxiality and bending away from the frames as is similar to the results found for coupon testing.
3. Sheet deformation due to curvature and pillowing change the initial stress state as well as the stress intensity factor related to cracking. For Glare, barrel test results are similar to test results of flat panels.

Concluding: the biaxial loading has an influence on the fatigue crack initiation and only a minor influence on the fatigue crack growth. The radius of a fuselage has no influence on the measured secondary bending [6], and since secondary bending is a major factor in fatigue crack growth it appears that the radius does not have much influence on the fatigue life of joints.

12.2.6 Durability

As can be seen in chapter 28 of this book, exposure to moisture and aggressive environments can affect the matrix as well as the aluminium layers. With Glare riveted joints it appeared that a moderate increase in crack growth rate occurred for testing in salt water [6]. The delaminations were somewhat larger. The reduction in fatigue properties was only small and will not be included in the joint fatigue modelling methods described in section 12.3 but can be accounted for using a small knockdown factor.

12.3 Joint fatigue modelling

12.3.1 Fatigue crack initiation

A calculation method for predicting the fatigue life of riveted joints was developed by Fokker Aerostructures several years ago [12]. This method is restricted to sheets of monolithic aluminium. To make the method useful for FMLs like Glare, the method needs to be extended. Figure 12.2 shows the stress distribution through the thickness indicating the highest stresses occur at the mating surface, resulting in sub-surface crack initiation and the cracks to emanate at the mating surface and grow through the thickness, the so-called elliptical crack fronts as described by Fawaz [13]. With Glare this through-the-thickness crack growth requires re-initiation at all aluminium layers separately.

Basic calculation method

The calculation method for riveted lap joints is based on what is called the "stress similarity concept". This stress similarity concept is based on the assumption that the stresses at the hole perimeter determine the fatigue initiation life of the joint, regardless of the stress distribution, etc. This implies that if two different joints have the same stress level at the hole perimeter, they will have identical fatigue lives. Using this assumption, the fatigue initiation life of any arbitrary joint can be calculated using the fatigue life data of another arbitrary (reference) joint.
The principle of the method can be divided into four parts:

1. For the actual joint, the peak stresses at the hole perimeter are calculated from the load cycle.
2. The calculated peak stress is applied on the reference joint:
   $N_{actual} = N_{ref}$ if $\sigma_{min,peak,actual} = \sigma_{min,peak,ref}$ and
   $\sigma_{max,peak,actual} = \sigma_{max,peak,ref}$
3. From the peak stresses in the reference lap joint, the remote stresses (load cycle) for the reference joint is calculated backwards.
4. The fatigue life of the reference joint, which is identical to the fatigue life of the actual joint, can be read from the fatigue life data of the reference joint. If necessary, stresses should be corrected beforehand if the stress ratio does not equal the stress ratio for which fatigue life data of the reference joint are available.

The peak stresses are defined as:

$$\sigma_{peak} = K_t \cdot \sigma_{net} = K_t \cdot \frac{S_p}{S_p - D} \cdot \sigma_{remote} \qquad (12.16)$$

The stress concentration factor at the hole perimeter can be defined as:

$$K_t = (1-\gamma)K_{t,hole} + \gamma K_{t,pin} + k_b K_{t,bending} \qquad (12.17)$$

with: γ = load transfer at the fastener to be analysed ($0 \le \gamma \le 1$).

$K_{t,hole}$ = theoretical stress concentration factor of a hole in a finite width sheet.

$K_{t,pin}$ = theoretical stress concentration factor of a pin loaded hole in a finite width sheet.

$K_{t,bending}$ = theoretical stress concentration factor of a hole in a finite width sheet, loaded in bending.

k_b = ratio of the bending stress due to secondary bending and the remote gross stress, according to [14]. For more complex structural details, k_b can be established using FE calculations.

Since k_b depends on the remote stress level, part III of the calculation method is in fact an iterative process. Note that the K_t is related to net section stresses. K_t values can be obtained from different sources such as [15].

The calculation method takes geometry into account. However there are some other significant parameters. These parameters should be accounted for by either including them in the reference joint or by developing correction factors.
The presence of countersinks affects the stress distribution near the initiation site and therefore the stress concentration factor. In general, the presence of a countersink decreases the fatigue life.

Extension for Glare

The complexity of the crack initiation mechanism in Glare is related to the subsequent initiation in each individual layer. This complex reduction in stiffness versus increased eccentricity forms the basis to model the peak stresses in Glare and the related crack growth [2]. The extension for Glare of the method described above relies on the following assumptions:

- Peak stresses can be calculated for each individual aluminium layer.
- The fatigue initiation behaviour of an individual aluminium layer is the same as for monolithic aluminium.
- The same K_t factors as for monolithic aluminium (viz. open hole, pin loading, bending) can be used.
- Reference joints for monolithic aluminium can be used.

With these assumptions only part I of the calculation method needs to be modified. These modifications are based upon the following boundary conditions for the peak stress calculation:

- Bending factor depends on layup and layer to be analysed.
- Difference in stiffness included.

The equation for the calculation of the peak stress in the facing aluminium layer can now be written as:

$$\sigma_{peak,al} = \frac{S_p}{S_p - D} \cdot \left[(1-\gamma)K_{t,hole} + \gamma K_{t,pin} + k_b K_{t,bending}\right] \cdot \sigma_{remote} \cdot \frac{E_{al}}{E_{FML}} \qquad (12.18)$$

As compared to the equation for monolithic aluminium sheets, one term is added to the above equation, viz. a correction for the difference in stiffness in the aluminium layer and the complete laminate. This correction term is defined as:

$$\frac{E_{al}}{E_{FML}} = \frac{E_{al}}{\int_{layer\ i} E_i t_i} \qquad (12.19)$$

Further, the values of factor k_b (ratio of bending stress and remote stress) and factor γ need to be adapted to the FML characteristics (different Young's modulus). The k_b is explained in section 12.2.2.
All other factors remain unchanged. However, it should be noted that the internal stress levels in the FML due to curing should be accounted for. Adding the internal (tensile) stresses in the aluminium layers to the remote stress can do this. After doing this, the equation for the peak stress can be written as:

$$\sigma_{peak,al} = \frac{S_p}{S_p - D} \cdot \left[(1-\gamma)K_{t,hole} + \gamma K_{t,pin} + k_b K_{t,bending}\right] \cdot \left(\sigma_{remote} \cdot \frac{E_{al}}{E_{FML}} + \sigma_{al,curing}\right) \qquad (12.20)$$

Peak stresses can be calculated for each required aluminium layer by adjusting the ratio k_b, i.e. choosing the appropriate bending stress level.
It is obvious that as a result of larger bending stresses, the aluminium layers in the mating surfaces will initiate first. Crack initiation in the subsequent layers will be affected by crack growth in the first layer(s). The layers that are already cracked decrease the net section, and thus increase the net stress. The factor k_b will also be affected by the layers that are cracked.

The fatigue initiation life, which is calculated using the method already described, is the life at which a crack is initiated. In monolithic material it is common for this crack size to equal the detectable crack size and so a crack growth analysis will be performed using this initial crack size. For Glare, however, the transition from initiation to crack propagation should be carefully defined. Since the aim is to develop a prediction method that includes crack initiation, crack growth and residual strength, the crack size after initiation can be chosen such that an optimum can be achieved in method development.

One of the assumptions for the extension of the method is that reference curves for joints in monolithic aluminium are possible to use. On first impressions, this is a requirement that is easy to fulfil since that initiation in Glare takes place in the aluminium layers. However, it should be noted that a correct choice of the reference joint is important with respect to fastener choice and fastening process in the reference joint (see also section 12.2.).

12.3.2 Fatigue crack growth

The stress distribution through the thickness, see Figure 12.2, and the phenomenon of independent crack growth of the aluminium layers in Glare will lead to a crack distribution through the thickness, see Figure 12.9, i.e. a part-through crack. The calculated stress distribution will be the assumption for further crack growth calculation.

If the stress in the aluminium layer on the visible or outer side of the joint is lower than the crack initiation threshold stress, this layer will not experience crack growth, which has a positive effect on the residual strength of the joint. The crack growth in the aluminium layers in which cracks will occur can be modelled with the crack growth models described in chapter 11, so the effects such as influence of internal stresses on crack closure are also taken into account. As the tip of the crack in the aluminium layer at the mating surface of the joint will be ahead of the cracks in the underlying layers, the crack will behave as a surface crack.

When one or more layers experience crack growth, as in Figure 12.9, the net section between the fastener bore holes will decrease, which will lead to an increase of the net stresses. As mentioned before, the reduction of the net section results in a decreasing stiffness, and since the layer aluminium at the faying surface will show crack growth the eccentricity increases. The increase in flexibility of the joint due to crack growth compensates for the larger eccentricity jump.

Although crack initiation at the faying surfaces in the joint occurs faster than in monolithic joints, the crack propagation properties are much better. The presence of the fibres in Glare provides crack bridging capability and therefore slow crack growth behaviour. Through-the-thickness penetration of cracks is also slowed down. The presence of the fibres requires that each individual layer must initiate separately. Due to the lower bending stresses in the subsequent layers, initiation occurs much later than in the faying surface layers, resulting in a significantly lower through-the-thickness crack growth rate than in monolithic aluminium.

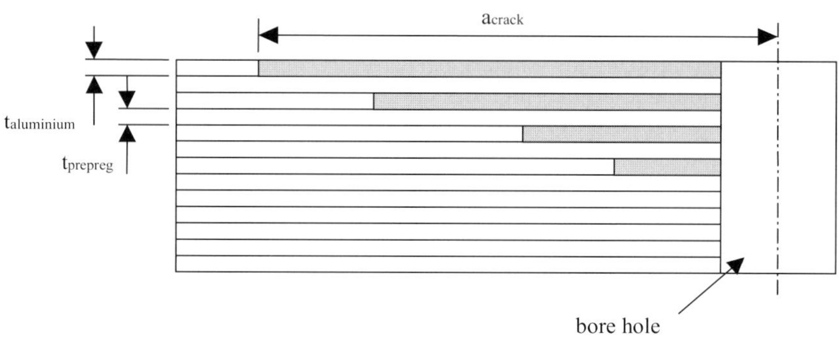

Figure 12.9 Through the thickness crack distribution in a Glare joint.

12.3.3 Residual strength; rivet pull through and rivet shear

Residual strength in fatigued joints is defined as the load a structure can carry with damage, in this case fatigue crack growth due to cyclic loading of the aircraft fuselage. The residual strength depends on two constituents of the structure, the fasteners and sheet material. Although only two constituents are used, three different failure modes exist: Rivet Pull Through (RPT), Rivet Shear Failure and Sheet Net Section Failure. Before going deeper into the failure of the sheet material, the failure of the fasteners can be just as important. For certain sheet thickness over fastener diameter ratios (t/D), the failure is characterised by the fastener type, see Figure 12.10.

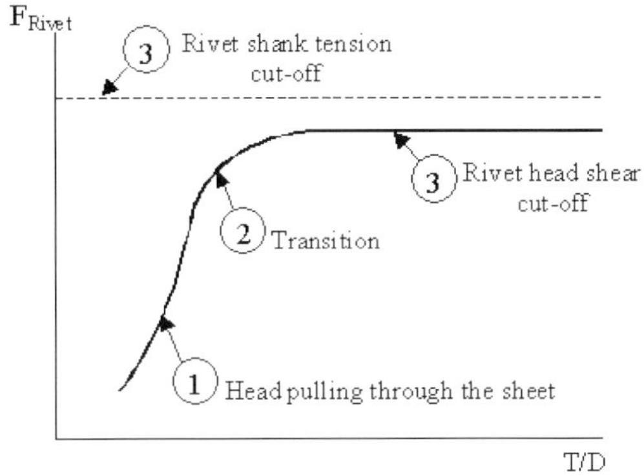

Figure 12.10 *Schematic diagram of rivet strength ratio versus sheet thickness/rivet diameter.*

Figure 12.10 shows a static joint analysis in Glare by Slagter [16]. This diagram clearly demonstrates the ability to describe typical ultimate load carrying capabilities of riveted joints in similar fashion in Glare compared to aluminium. In the design development phase preventive designing can be performed to avoid rivet failure or RPT:

1. Failure mode is the rivet head is pulled through the sheet.
2. Transition area in which the deformation or partial failure of the rivet heads allows for premature rivet pull through. The geometrical layout of certain non-symmetrical joints such as single lap-splice joints and single butt joints inhibit the tilting of the mechanical fasteners used for joining the two sheets. The tilting of these fasteners is induced by the basic eccentricity inherent to multi layer joints. This results in the fasteners being partially loaded in tension and not just in shear. An experimental study proved that rivet pull-through occurs more frequently in FML then in monolithic materials. This even happens for commonly used sheet thickness / rivet diameter ratios.
3. Failure mode is either by rivet shear of the head or tensile failure in the shank. Rivet shear failure occurs when the maximum shear stress of the rivet is exceeded, the shear strength depends on the rivet material and the cross-sectional area of the rivet.

Fibre Metal Laminates have somewhat different failure behaviour than monolithic aluminium material. The build up of the FML provides less support with respect to a perpendicular loading on the layers. This allows for delamination of the layers as early as during the installation process. Especially countersunk rivets can cause delamination in the outer layers since the contact area is smaller and thus less support for the rivet, see Figure 12.11. The fastener will more easily be pulled through the sheet.

Delamination is of less importance for straight shank fastener holes, this is because there is no difference in contact behaviour for each layer, e.g. the contact area is for all layers similar [17]. In reality the outer metal layer in the countersunk hole do not carry the full load, these outer layers do not contribute to the total thickness of the laminate. In [17] a theoretical model is developed based on Slagter's model [16] to calculate a theoretical strength value for this problem.

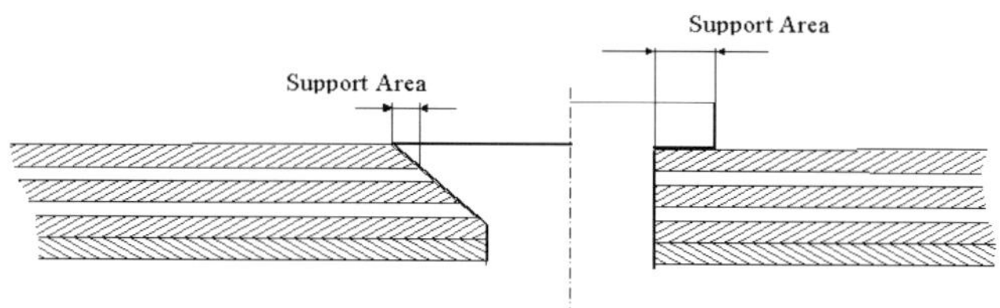

Figure 12.11 Difference in support area for countersunk and protruding head rivets.

12.3.4 Residual strength - Net section failure

The weakest area in a joint is the area with the smallest cross-sectional area; this is due to the holes and the stress concentrations at those holes. Two methods are available to calculate the strength of fatigued joints. The first method proposed by Müller is based on research done by Vermeeren [18], the second method is based on the Metal Volume Fraction (MVF) and on research by Müller.

The residual strength calculation of joints with fatigue cracks is based on the remaining net section strength of the aluminium. In the case of a FML the residual strength can be calculated using the remaining intact aluminium layers in combination with the un-cracked glass fibres based on the blunt notch strength of the un-cracked joint.

Calculation of the static strength of un-cracked lap-splice joints

Without any presence of cracks at the edges of the fastener holes, the fastener holes can be regarded as blunt notches. Calculation of the blunt notch strength will be dealt with in two ways:

I) Blunt notch approach:
This approach uses a blunt notch K_{bn} factor as suggested in [6]. The empirical value of K_{bn} according to Vermeeren is approximately constant for hole type notches (excluding sharp notches). The problem with this method is that K_{bn} has been validated for the grade Glare 3.

$$\sigma_{bn} = K_{bn} \cdot \frac{\sigma_U \cdot A_{net}}{W \cdot t} \tag{12.21}$$

with: σ_{bn} = blunt notch strength based on the gross section
K_{bn} = blunt notch factor
σ_U = tensile strength of the sheet material
A_{net} = net section area
W = width of sheet
t = thickness of sheet

The laminate failure strain is equal to 4.7%; the stress in the aluminium layers is for 4.7% strain approximately 420 MPa. At the point of fibre failure, the stresses in the aluminium layers increase dramatically and will cause immediate failure of the sheet. Since the fibre failure is the critical phenomenon, the strength of a Glare 3 sheet can be calculated using the rule of mixtures.

$$\sigma_{U,Glare} = \frac{\sum t_{al}}{t_{tot}} \cdot \sigma_{al(\varepsilon=4.7\%)} + \frac{\sum t_{pL}}{t_{tot}} \cdot \sigma_{U,pL} \tag{12.22}$$

with: $\sigma_{al(e=4.7\%)}$ = stress in aluminium sheets at 4.7% strain (420 MPa)
$\sigma_{U,pL}$ = ultimate tensile strength of the prepreg layers in loading direction (2640 MPa)
t_{al} = thickness of the aluminium sheets
t_{tot} = total thickness of sheet
t_{pL} = thickness of the prepreg layers with fibres in loading direction. Fibres perpendicular to the loading direction do not contribute to the laminate strength

It should be noted that Vermeeren used central notched specimens, and Müller's tests were performed using lap-splice joints. So in the latter, filled holes and load transfer influences the notch effects. Future investigations will shed more light on the influence of load transfer and filled holes on the results of these calculations.

II) MVF approach:
This method is based on the amount of aluminium material in the Fibre Metal Laminate. If the MVF is equal to 1, it is easy to understand that the blunt notch strength is equal to the blunt notch strength of monolithic aluminium. With decreasing MVF, adding prepreg layers the blunt notch strength of the fibre metal laminate increases.

$$MVF = \frac{t_{Al}}{t_{tot}} \tag{12.23}$$

$$\sigma_{BN\,Glare} = MVF \cdot \sigma_{BN\,Al} + (1 - MVF) \cdot \sigma_{BN\,Prepreg} \tag{12.24}$$

with: $\sigma_{BN\ Al}$ = blunt notch strength Al2024-T3 (LT-direction) (428 MPa)
σ_{BN} Prepreg = blunt notch strength of a single UD-prepreg layer (1162 MPa)
σ_{BN} Glare = net blunt notch strength of the sheet

These methods provide the means to calculate the blunt notch strength of un-cracked notches, for cracked notches the calculation becomes more complicated. As mentioned above, the influence of filled holes and load transfer are not taken into account and it has been shown by Müller that the method works for Glare 3 grades. Ongoing investigations into other Fibre Metal Laminates will help to improve the method.

Calculation of the static strength of cracked lap-splice joints

Fibre failure determines the residual strength of a fatigue cracked lap-splice joint in many cases. The engineering approach suggested in [6] provides a way to use the cracked area of the aluminium layers as an indication of the residual strength of the lap-splice joint. The residual strength of a cracked lap-splice joint can be determined using

$$W \cdot t_{Glare} \cdot \sigma_{res,cracked} = W \cdot t_{Glare} \cdot \sigma_{res,uncracked} - A_{cracked} \cdot \sigma_{Al(\varepsilon=4.7\%)} \tag{12.25}$$

$$A_{cracked} = m \cdot (2a_{average}) \cdot (3t_{Al}) \tag{12.26}$$

$$\sigma_{res,cracked} = \sigma_{res,uncracked} - \left(\frac{2a_{average}}{s}\right) \cdot \left(\frac{3t_{Al}}{t_{lam}}\right) \cdot \sigma_{Al(\varepsilon=4.7\%)} \tag{12.27}$$

Equation 12.27 can be rewritten in a form that incorporates the MVF and the cracked area of aluminium layers.

$$\sigma_{res,cracked} = \sigma_{BN\ Glare}(gross) - MVF \cdot \frac{A_{Al_{cracked}}}{A_{Al_{uncracked}}} \cdot \sigma_{BN\ Al} \tag{12.28}$$

where:

$$A_{Al_{uncracked}} = m \cdot s \cdot t_{Al} \tag{12.29}$$

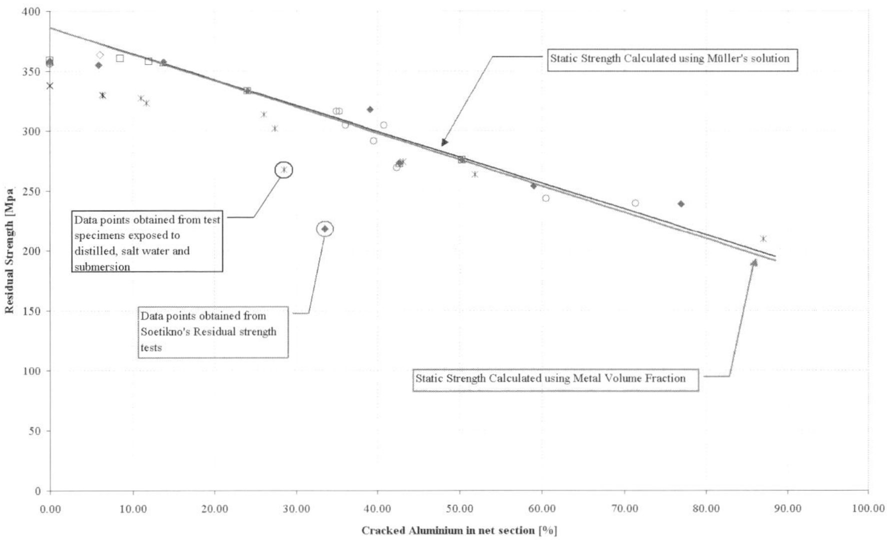

Figure 12.12 *Residual strength prediction and test results, all results are taken from* [8], *except for Soetikno's residual strength tests* [19].

12.3.5 Additional factors

The fatigue models for initiation, crack growth and residual strength described above only approach reality. Many aspects are not (yet) included in the models. Some of these aspects are:

- Fastener type and installation.
- Biaxiality.
- Complex local structural details.
- Durability aspects
- Relation coupon scale testing vs. full-scale testing.

Some aspects can be implicitly included in the models (for example by selecting an appropriate reference S-N curve in the initiation model). Other aspects, e.g. durability, can be accounted for by using a correction factor or a knockdown factor. Such factors still need to be established.

12.4 Joint design in Glare

The foregoing analysis has demonstrated insight into the complexity of mechanically fastened joints. These key characteristics have been described in this chapter for fatigue prediction. As a reference, aluminium joints have been considered.

It appeared that the crack initiation could be well described with the adapted conventional models for monolithic aluminium. For crack growth a new model was given which includes fibre bridging. For residual strength the blunt notch characteristics and pragmatically derived aluminium stresses helped to describe failure loads in joints.
The baseline fatigue properties of Glare joints can be well described and good mathematical insight has been achieved based on analysis and numerous experiments. Not only do the models describe the behaviour of joints well, they also confirm the large difference between Glare and aluminium. It can even be shown that joints in Glare will have the ability to withstand ultimate load after an aircraft's economic life.

References

[1] L. Jarfall, *Shear loaded fastener installations*, SAAB-SCANIA, report KHR-3360, 15 April 1983.

[2] F.J. Pellenkoft, *Secondary bending in Glare*, Master Thesis, Delft University of Technology, Faculty of Aerospace Engineering, 2000.

[3] J. Schijve, *Calculation on the moment distribution in riveted lap joints with bonded doublers*, NLR, Report S.159

[4] J. Schijve, *The influence of the dimensions of riveted lap joints and strap joints on the secondary bending*, NLR, Report NLR TR 68026 LI, 1968.

[5] J. Schijve and A. Hartman, *The effect of secondary bending on the fatigue strength of 2024-T3 Alclad riveted joints*, NLR, Report NLR TR 69116 U, 1969.

[6] R.P.G. Müller, *An Experimental and Analytical Investigation on the Fatigue Behaviour of Fuselage Riveted Lap Joints*, Ph.D. Thesis, Delft University of Technology, Faculty of Aerospace Engineering, 1996.

[7] J.J.M. Rijck and S.A. Fawaz, *A simplified approach for stress analysis of mechanically fastened joints*, Fourth Joint DoD/FAA/NASA Conference on Aging Aircraft, St Louis, May 2000.

[8] J.M. Gere and S.P. Timoshenko, *Mechanics of materials*, Chapman & Hall, UK, 1991.

[9] S.J.A. Vissers, *Detail design of a fuselage crown panel with a butt joint for the A3XX*, Master Thesis, Delft University of Technology, Faculty of Aerospace Engineering, 1998.

[10] J. Schijve and A. Vlot, *Damage and Fatigue Crack Growth of Aircraft Materials and Structures*, Delft University of Technology, Faculty of Aerospace Engineering, 1996.

[11] D. Chen, *Bulging of fatigue cracks in a pressurised aircraft fuselage*, Ph.D.Thesis, Delft University of Technology, Faculty of Aerospace Engineering, 1991.

[12] J.J. Homan and A.A. Jongebreur, *Calculation method for predicting the fatigue life of riveted joints*, Proceedings of International Conference on Aircraft Fatigue 1993, pp 175-190, Stockholm, 1993.

[13] S.A. Fawaz, *Fatigue crack growth in riveted joints*, Ph.D. Thesis, Delft University of Technology, Faculty of Aerospace Engineering, September 1997.

[14] J. Schijve, *Some elementary calculations on secondary bending in simple lap joints*, NLR, Report TR 72036U, 1972.

[15] R.E. Peterson, *Stress concentration factors; charts and relations useful in making strength calculations for machine parts and structural elements*, New York, Wiley, 1974

[16] W.J. Slagter, *Static strength of riveted joints in fibre metal laminates*, Ph.D. Thesis, Delft University of Technology, Faculty of Aerospace Engineering, 1994.

[17] R van Rooyen, *Behaviour of Fasteners under Tensile Loads*, TZ-Thesis, Delft University of Technology, Faculty of Aerospace Engineering, 2000.

[18] C.A.J.R. Vermeeren, *The Blunt Notch Behaviour of Metal Laminates: ARALL and Glare*, report LR-617, Aerospace Engineering, Delft University of Technology, January 1990.

[19] T.P. Soetikno, *Residual Strength of the Fatigued 3 Row Riveted Glare3 Longitudinal Joint*, Master Thesis, Aerospace Engineering, Delft University of Technology, May 1992

13

Residual strength

T.J. de Vries

Abstract – The residual strength of fatigue-cracked Glare specimens is virtually independent of the relative initial crack length. The number of cut fibres and the amount of delamination determine the residual strength instead.
To obtain fracture, the energy available for crack growth (energy release rate G) must be larger than the crack resistance R of the material. With this knowledge a K_R-curve can be determined experimentally. The K_R-curve is better for Glare 2 than for Alclad-2524-T3 as regards specific weight. From experimental data it follows that for Glare the K_R-curve is a function of: the Metal Volume Fraction, the properties of the metal and fibre layers, the interfaces between these layers, the rolling direction in comparison with load direction and the quantity of fibres in de load direction. A prediction model for the residual strength of Glare was developed, but it can only be used for of Glare types that are based on the 2024-T3 alloy and loaded in either L-T or T-L direction. Additional research is needed to extend the model.
To fulfil the two-bay crack criterion after FOD (necessary for newly designed aircraft), the residual strength of the Glare skin can be improved. This can be done by adding extra "crack stopping" fibre layers in the laminate. A test on a skin panel showed that this idea works and the crack tended to flap to the centre of the stiffener bay.

13.1 Introduction

The residual strength of a material is defined as the remaining static strength of the material in the presence of any damage that can occur during the service life of an aircraft. This is an imprecise definition since the term damage is not properly defined and can vary between a smooth dent and a sharp crack. For design verification it is conservative to represent damages in Glare as sharp cracks. However, a distinction must be made between two different kinds of damage in the Glare laminate:

1. *Cracks that occur in one or more aluminium layers while the fibres in the prepreg layers that bridge the crack are still intact.*
   These cracks are mainly caused by fatigue loading and originate from rivet holes, for example. In this case one or more of the aluminium layers may be broken but the fibres remain intact. They bridge the crack and reduce the stress intensity at the crack tip. Other causes of "part-through cracks" in Glare are scratches, corrosion damage or lightning strikes.

2. *Cracks that occur simultaneously in the fibre prepreg layers and aluminium layers.*
   These cracks occur due to severe Discrete Source Damage (DSD) when an object penetrates the structure. This damage creates "through-the-thickness cracks" with broken fibre and aluminium layers. Examples of sources that create DSD are engine non-containment, a bird-strike and runway debris.

The residual strength of a fatigue crack in Glare is superior to the residual strength of a fatigue crack in aluminium or a through-the-thickness crack in Glare. Section 13.2 describes the fracture mechanism of a through-the-thickness crack and a part-through crack respectively in Glare sheets, which are loaded up to fracture. This chapter describes the residual strength characteristics of Glare with through-the-thickness cracks. Section 13.3 discusses the use of the K_R-curve to describe the residual strength of Glare laminates and presents a general calculation method. Section 13.4 uses this method to predict the residual strength of a curved fuselage panel including back-up structure that contains a two-bay crack.

13.2 Fracture mechanism

This section discusses the fracture mechanism of Glare materials with through-the-thickness cracks and part-through cracks respectively.

13.2.1 Through-the-thickness cracks

Glare material with a through-the-thickness crack that is loaded to fracture exhibits a capability for slow stable crack growth prior to rapid failure [1]. This rapid failure or unstable crack growth occurs when the specimen is loaded under force control. The fracture behaviour is similar to the response of monolithic aluminium sheet material. Figure 13.1 shows a typical curve of the gross stress versus the physical crack extension for a Glare laminate, obtained with a uniaxially loaded Centre Cracked Tension (CCT) specimen. The crack extension is the extension measured in the visible outer aluminium layers.

Figure 13.2 shows some images of the crack tip in a uniaxially loaded Glare 2 specimen. The initial saw cut in this specimen was made with a jeweller's saw and the specimen was first fatigue loaded to sharpen the starter crack.

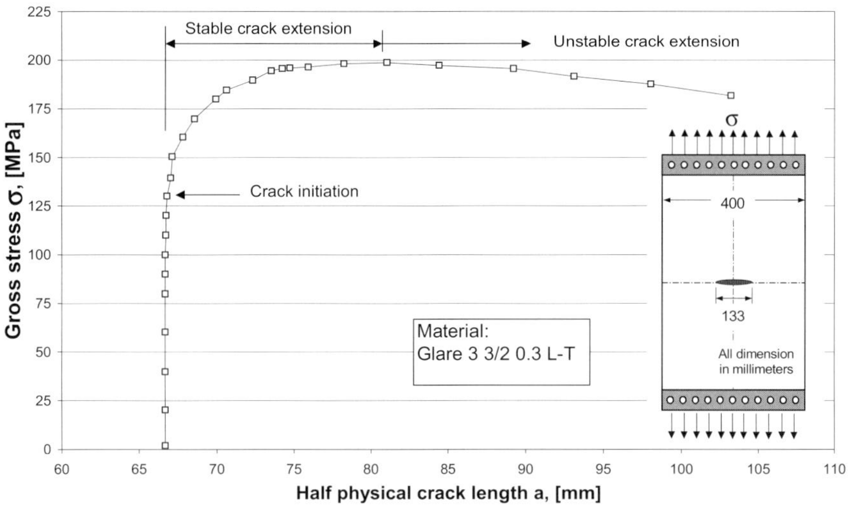

Figure 13.1 *Typical crack extension versus applied load curve for Glare.*

Fatigue pre-cracking

The crack extension created during fatigue loading is visible in Figure 13.2 (a). During fatigue cycling, the fibres stay intact and bridge the crack in the aluminium layers. Consequently at the start of the residual strength test, the crack length within the prepreg layers differs from the crack length in the aluminium layers. As a result of the fatigue pre-loading, debonding occurs between the fibre prepreg and the aluminium layers above and below the fatigue crack in the aluminium layers where the fibres are still intact. This kind of debonding is called "fatigue" delamination.

Static loading

During static loading, the fatigue crack in the aluminium layers opens, see Figure 13.2 (b). At a certain load, the fatigue crack in the aluminium layer starts to propagate. At approximately 90% of the residual strength of the specimen, the first fibre bundle in the visible prepreg layer fails at the tip of the fatigue crack, illustrated in Figure 13.2 (c) where the broken bundle appears brighter compared to the surrounding and intact prepreg layer. During further load increase, random failure of fibre bundles is observed between the tip of the saw cut and the tip of the fatigue crack in the aluminium layer.

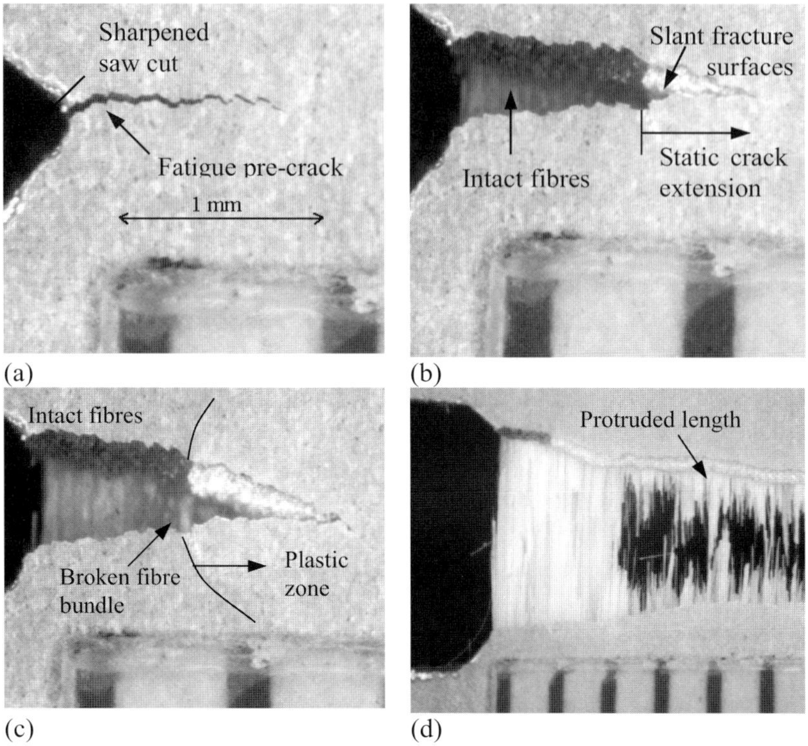

Figure 13.2 *Crack growth sequence for a Glare 2 CCT-specimen* [1].

Extensive research illustrated that the crack extensions in the prepreg layers and the aluminium layers become approximately equal during the remaining fracture process. However, this is not visible in Figure 13.2 due to the slanted crack faces of the outer aluminium layer. During the loading process, "static" delamination is created in front of the crack tip due to high strain gradients and plasticity in the aluminium layer. This reduces the strain in the glass fibres, which have an ultimate strain of only 4.5%, and consequently delays fibre failure. The glass fibre layer controls further crack extension up to fracture. After fibre failure, the adjacent aluminium layer is unable to take over the high loads and will consequently fracture.

13.2.2 Part-through cracks

Compared to the extensive stable crack growth in statically loaded specimens with through-the-thickness cracks, specimens with part-through cracks do not show any stable crack growth at all when loaded up to failure.

D'Allesandro et al [2] and Holleman [3] observed differences between the fracture mechanism of statically loaded fatigue cracks in Glare 2 and Glare 3. This is illustrated in Figure 13.3, which shows the starter crack and the development of the crack for both materials. During loading the crack opened slowly in both materials, effectively restrained by the intact fibres. In the Glare 2 material, static delamination occurred between the prepreg and the aluminium layer at loads close to fracture. This delamination grew in fibre direction and resulted in larger crack opening of the aluminium layers. Final fracture was initiated as fibre failure near the fatigue crack tip while the fibres in the wake of the fatigue crack remained intact.

In contrast to the behaviour observed for Glare 2, fracture in Glare 3 was initiated in the fibres near the hole at the centre of the fatigue crack, see Figure 13.3. This results in a larger crack opening and subsequent fibre failure from the centre towards the tips of the fatigue crack. After all the bridging fibres failed, an almost sudden fracture occurred.

The residual strength increase of a fatigue crack compared to a saw-cut of the same length is significant. This is illustrated in Figure 13.4 for Glare 3-3/2-0.3. The residual strength of the fatigue-cracked specimens is nearly independent from the relative initial crack length. Thus, the amount of fibres cut and the amount of delamination determine the residual strength.

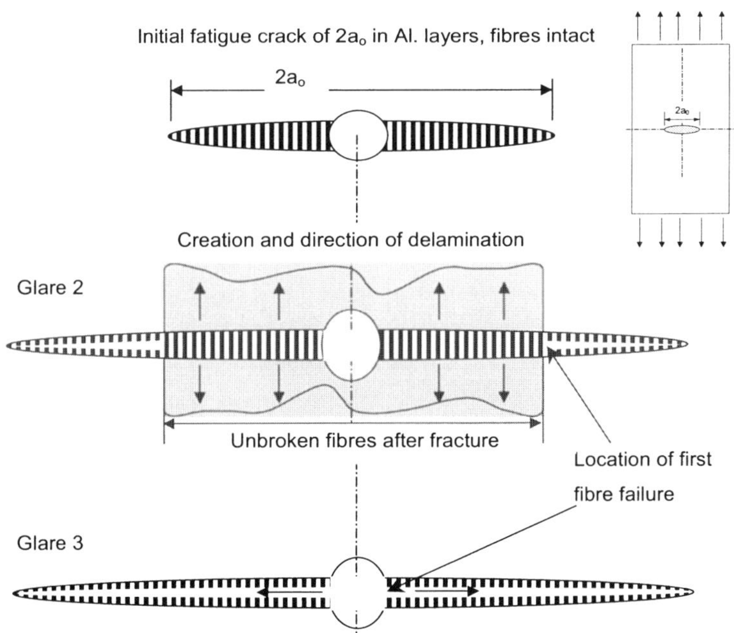

Figure 13.3 *Schematic overview of fracture process within Glare 2 and Glare 3, originating from a fatigue starter crack.*

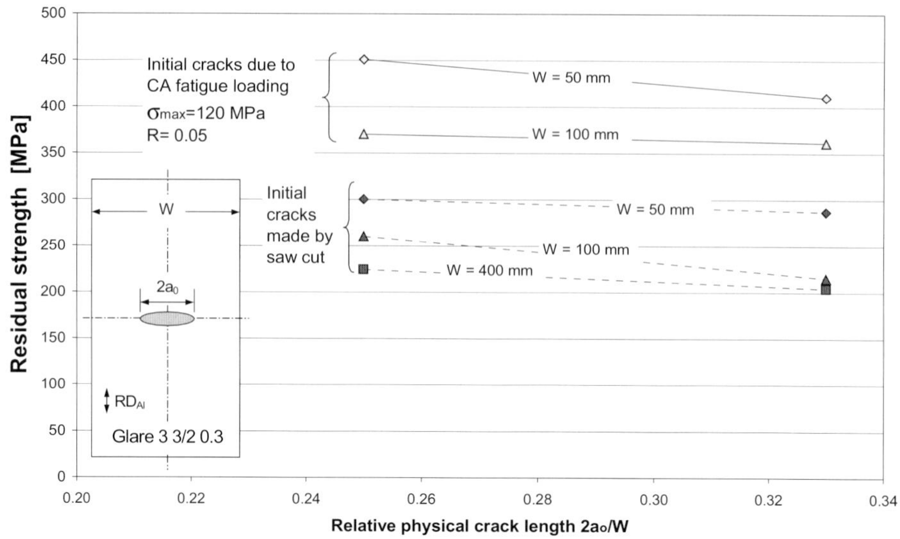

Figure 13.4 *Residual strength as a function of relative crack length and initial crack creation for Glare 3-3/2-0.2.*

13.3 R-curve as a material parameter

To successfully predict the fracture behaviour and residual strength of materials, a fracture criterion independent of the width of the specimen and the length of the crack is required. This section discusses the applicability of the R-curve or energy release rate as a characterising residual strength material parameter for Glare laminates with through-the-thickness cracks.

The R-curve concept is based on an energy balance:

To obtain fracture, the energy available for crack growth (energy release rate G) must be larger than the crack resistance R of the material.

In general, stress intensity units are used instead of energy release rate units. The relation between them in case of plane stress is:

$$G = \frac{{K_G}^2}{E} \quad \text{and} \quad R = \frac{{K_R}^2}{E} \qquad (13.1)$$

with: E = Young's modulus of the material in loading direction.

This approach is able to cope both with stable crack extension and a limited amount of plasticity. The plasticity is reflected in the use of a corrected crack length to calculate the energy release rate: the effective crack length, a_{eff}. There are generally two approaches to calculate an effective crack length. The first approach calculates the plastic zone size in front of the crack tip as a function of the applied load and the physical crack length, a_{phys},. An example of such a method is the Irwin plastic zone correction. The second method uses a global approach and compares the opening of the real physical crack with the opening of a crack in an elastic material. By making these two equal, an effective crack length is obtained. This approach is based on the compliance method. The shape of the K_R-curve can be determined experimentally from the K_G-values and the related crack extension during slow stable crack extension.

Several researchers investigated whether the K_R-curve concept applied to metallic sheets is also applicable for Glare material with through-the-thickness cracks. This is discussed in section 13.3.1. Prediction of the residual strength of fatigued specimens using the K_R-curve approach is not possible since there is no stable (physical) crack extension.

13.3.1 K_R-curve approach for Glare

Section 13.2 illustrated the dependency of the laminate residual strength properties on the fibre and aluminium material characteristics and the bonding strength of the prepreg with the aluminium layers. The metal layer is responsible for the stable crack extension, yielding of the material and the redistribution of the stress away from the crack tip. The fibre layers are responsible for the large strain hardening of the material and final fracture. Without aluminium layers or in a case of bad bonding quality, the fibre layer in front of a crack would split as illustrated in the pre-treatment research and show no stable crack extension but fracture suddenly. Because the Glare laminates illustrate residual strength behaviour that is more comparable with the behaviour of aluminium than for composites, the development of a calculation method is based on methods that are applicable for aluminium.

Vermeeren [4] Bresser [5] Sutton [6] and De Vries et al. [1,7,8] investigated the applicability of the K_R-curve concept for Glare. They all concluded that the K_R-curve could function as a material parameter but within certain limitations of the specimen width and the initial crack length. Small values of W and large values of $2a_o$ should be avoided. Figure 13.5 contain K_R -curves obtained with a plastic zone correction based on the *compliance method* for Glare 3. The curves were obtained in different laboratories at the Universities of Pisa (Italy) and Delft and at McDonald Douglas. As a reference, the K_R -curves obtained with the *Irwin plastic zone correction* for the same data are also included in these charts as lines. The curves obtained with the *compliance method*, especially for Glare 3, are identical for different W and a_0.

For both laminates, the initial part of the K_R-curve obtained with compliance correction lies above the curve obtained with Irwin correction. This is due to the fact that the Irwin method calculates a plastic zone and thus an effective crack extension from the moment that the applied load is larger than zero. In contrast, effective crack extension according to the compliance method starts to increase when physical crack extension takes place. It is important to notice that the K_R-curve shapes are identical for the different methods. For a certain a_{phys}, a_{eff} is significantly different depending on whether it is obtained with the compliance- or Irwin plastic zone correction. Consequently, the critical K_R-value differs significantly, depending on the method used.

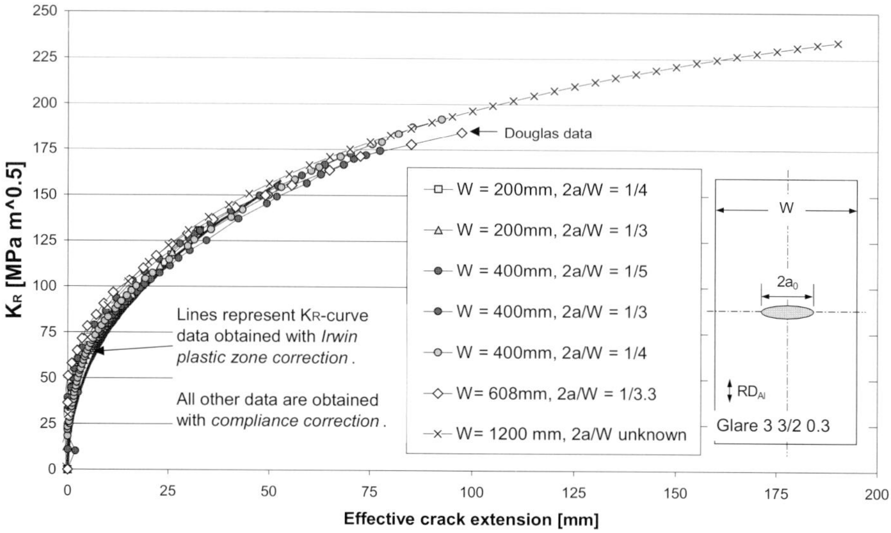

Figure 13.5 *K_R curves for Glare 3-3/2-0.3 for different W and a_0, based on compliance plastic zone correction* [8].

13.3.2 Glare laminates versus current aluminium alloys

The aluminium alloys currently used in fuselage structures are mainly Alclad 2024-T3 and the new alloy, Alclad 2524-T351. This latter alloy is applied in the Boeing 777. Although it contains the same static properties as 2024-T3, its fracture toughness is significantly improved.

In Figure 13.6, the K_R-curve fits of Glare 2A and Glare 4A in L-T direction are compared with the curves of the aluminium alloys Alclad 2024-T3 and Alclad 2524-T351 in L-T direction. The residual strength of Glare 2 in L-T direction is superior. The K_R-curve of Glare 4A is above Alclad 2024-T3 but below Alclad 2524-T351. The K_R-curves for the Glare 3 laminates are not illustrated since these are comparable with the curve for Alclad 2024-T3.

An important reason to use Glare instead of aluminium is the lower specific weight of the fibre prepreg compared to monolithic aluminium. Therefore, to compare Glare with other materials, it makes sense to divide the maximum allowable stress by the specific weight. The specific K_R-curves based on these specific weights and compared with the specific weight of 2024-T3 are plotted in Figure 13.7. The superiority of the Glare 2A laminate, tested in L-T direction remains the same. However, the K_R-curves for the Glare 4A laminates are now comparable with the one for Alclad 2524-T351 and the Glare 3 curves are significantly above the curves for Alclad 2024-T3.

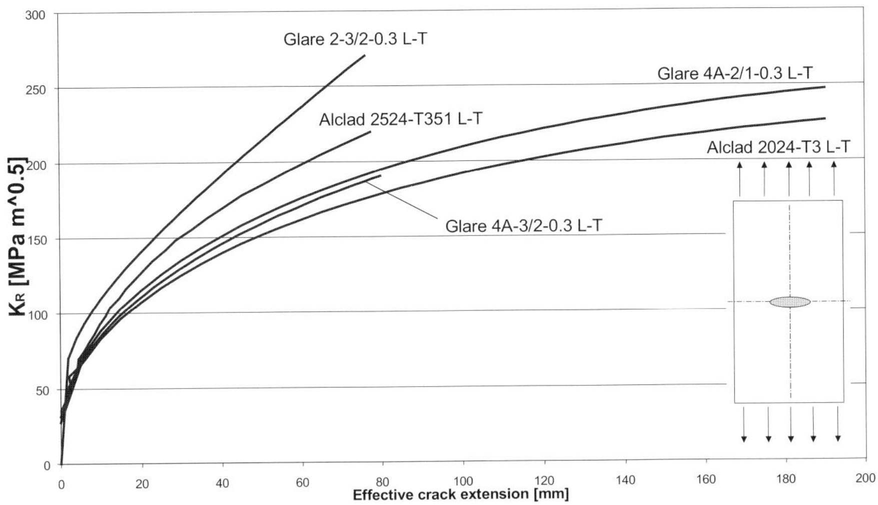

Figure 13.6 *K_R-curves obtained with compliance correction for different Glare laminates and aluminium alloys in L-T direction* [8].

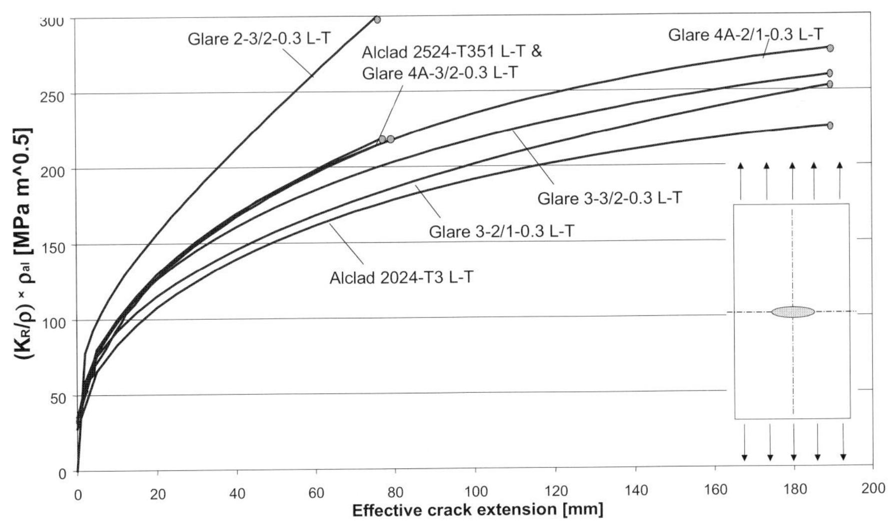

Figure 13.7 *Indexed K_R-curves obtained with compliance correction for different Glare laminates and aluminium alloys in L-T direction.*

13.3.3 Effect of constituents

An extensive investigation has been carried out [8] to determine the influence of the constituent properties on the residual strength (Aluminium characteristics, fibre prepreg characteristics and bonding quality). A distinction is made between the crack initiation process and the crack growth / fracture process.

Crack initiation
The crack initiation stress of a laminate that is statically loaded increases when an aluminium alloy with a lower yield stress is used. The use of a stiffer fibre prepreg advances crack initiation.

Crack growth / fracture
The residual strength of a laminate is directly related to the strain hardening capacity of the laminate and thereby the aluminium layers in relation to the prepreg; a larger strain hardening increases the residual strength. This is illustrated in Figure 13.8. An increase of the failure strain and strength of fibres also increases the residual strength of a Glare laminate.

Static delaminations are expected to release the fibre layers from following the large strain concentrations of the aluminium layers at the crack tip. Because the ultimate strain of the fibres is significantly less than for aluminium, failure of the fibres is delayed by this kind of delamination. As a result, the aluminium layers control crack initiation and initial crack growth but the fibre layers will dominate larger crack extensions and fracture.

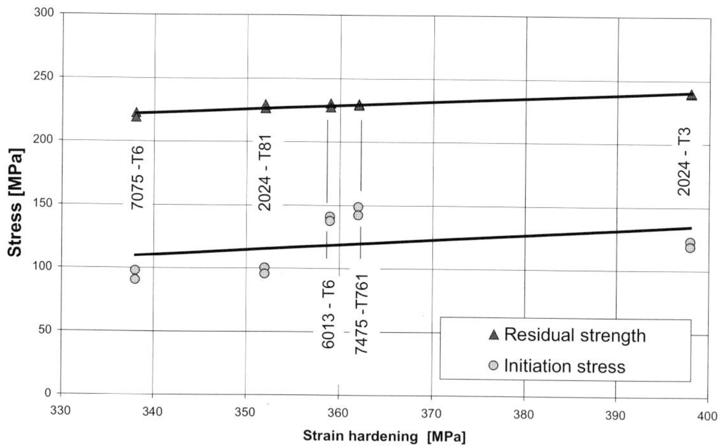

Figure 13.8 *Residual strength results (L-T) and crack initiation stress results versus strain hardening in L-direction.*

Semi-destructive- and non-destructive inspections on the specimens that were kept under a constant load confirmed the creation of static delamination before dynamic delamination. The etching method showed that static delamination advanced ahead of the crack tip. Fracture of the fibre layers occurred in bundles, randomly but in close proximity of the crack tip. It was also observed that the crack extensions in the aluminium layers are initially longer than in the fibre layers. However, after some crack extension and increased load, the cracks in the prepreg layers also become longer.

In order to investigate the effect of delamination on the residual strength, the bond strength between the prepreg and aluminium layer was influenced by applying different pre-treatments to the aluminium layers. A ranking of the bond strength was based on both Lap-shear and Roller peel tests, which gave the same ranking. Fatigue tests illustrated that a better bond strength results in lower crack growth rates and lesser fatigue delamination. On the contrary, the residual strength of a laminate increases with lower bond strength.
As expected, no influence of the laminate thickness on the residual strength is observed. A slight influence of the aluminium layer thickness is seen as all the laminates show a relation with the MVF. However, for the investigated aluminium layer thicknesses between 0.3 and 0.5 mm, this influence is small.

Based on the experimental results obtained during this study, it can be concluded that the residual strength of a Fibre Metal Laminate, or more generally, its K_R-curve, is a function of:

- the amount of aluminium in the laminate, represented by the MVF
- the material properties of the metal layers
- the rolling direction of the metal sheets compared to the loading direction
- the amount of fibres in the laminate in loading direction
- the material properties of the fibre layers
- the interfaces between the different layers.

The fibres perpendicular to the loading direction are expected not to add any strength in this direction.

13.3.4 Calculation method

As described, the K_R or crack growth resistance curve of a specific Glare laminate is able to describe the laminate residual strength properties. This curve is a material "parameter" since it is independent of the geometry. However, every Glare variant has its own K_R curve that needs to be determined experimentally. Therefore, it would be easier if the K_R curve of a specific laminate could be based upon the constituent properties. De Vries [8] investigated whether the use of the Fibre Volume Fraction representing the amount of fibres in the loading direction, FVF_{LD}, and the MVF can be used to describe the residual strength of a Glare laminate with through-the-thickness cracks.

Figure 13.9 contains K_R curves based on the compliance correction for a wide variety of Glare laminates in the L-T direction, all based on the same aluminium alloy 2024-T3. It is expected that the residual strength ranking for these laminates is based only on the MVF and the FVF_{LD} because the laminates have the same fibre- and pre-treated aluminium layers and are all tested in the rolling direction. Figure 13.9 also contains a reference curve for 2024-T3 L-T and a table with the MVF and FVF_{LD} values for the different laminates.

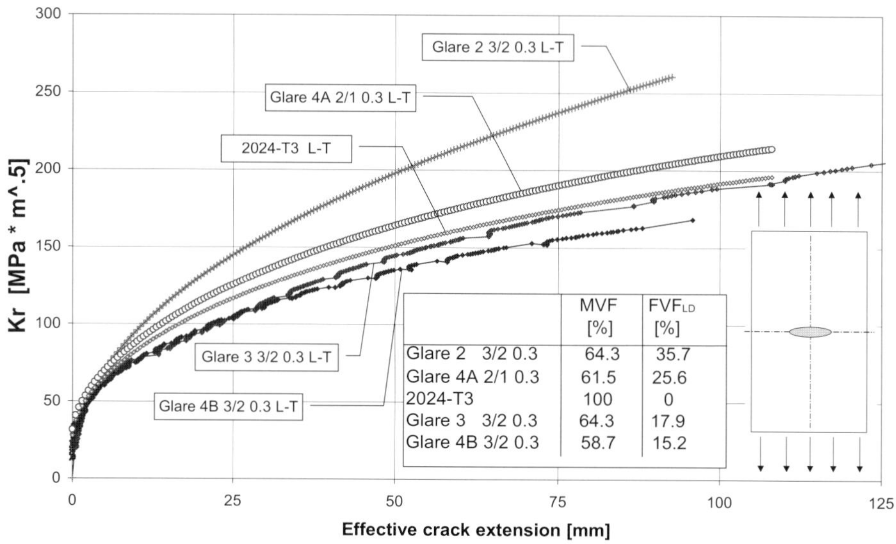

Figure 13.9 *K_R curves in L-T direction for different Glare variants and 2024-T3, based on compliance plasticity correction.*

To obtain a relation between the MVF, the FVF_{LD} and the K_R curve, experimental residual strength values for several laminates were obtained from 800 mm wide CCT panels with 2a/W = 0.25. To combine the MVF and the FVF_{LD}, a parameter T is introduced to represent the effective load-carrying fraction in the laminate.

$$T = \alpha \cdot FVF_{LD} + MVF \qquad (13.2)$$

where α is a constant that represents the contribution of the fibre layers to the residual strength in comparison to the contribution of the metal layers. α depends on the properties of the aluminium alloy, the properties of the fibres and the aluminium rolling direction. The value of α is determined empirically, such that the Glare and aluminium residual strength data for a specific test direction are represented by a second order equation ($\sigma_{res} = a \cdot T^2 + b \cdot T + c$). For Glare based on 2024-T3, we obtain values of α = 1.84 and 2.1 in L-T and T-L direction respectively. The results are illustrated in Figure 13.10.

Table 13.1 Prediction tool.

General formulae:	$T = \alpha\, FVF_{LD} + MVF$				
	$M(T) = (a\,T^2 + b\,T + c)/(a+b+c)$ and				
	$K_{R\text{-}2024\text{-}T3} = k\,\Delta a_{eff}^{\,n} + l\,\Delta a_{eff} + m$				
Material:	α	A	b	c	Validated range T
Glare L-T (2024-T3)	1.84	164	-125	171	0.82 - 1.3
Glare T-L (2024-T3)	2.1	307	-561	441	1.02-1.18
K_R curve fit	k	L	m	n	Source
2024-T3 L-T (1.6mm)	16.12	-1.06	27.53	0.612	DERA
2024-T3 T-L (1.6mm)	18.08	-0.51	21.49	0.520	DERA

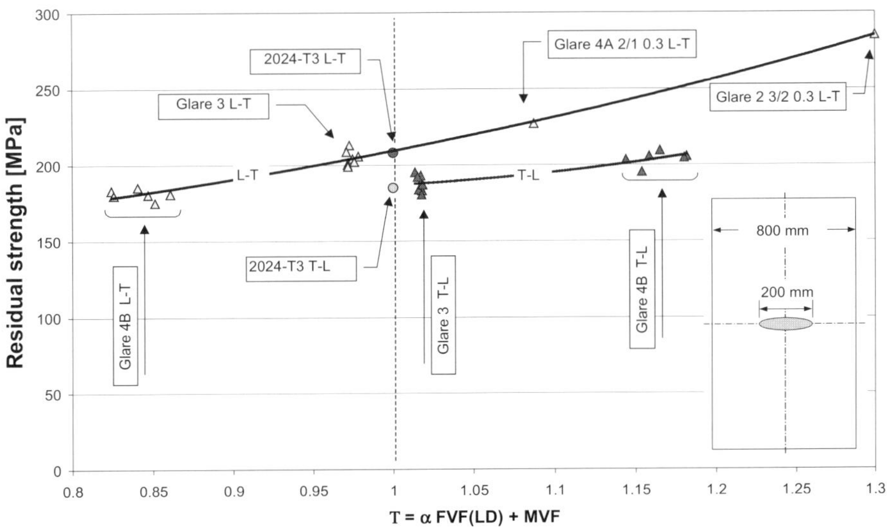

Figure 13.10 Residual strength as a function of T.

The obtained curve fit constants a, b and c are presented in Table 13.1. Within the prediction tool, a factor is necessary that relates the K_R curve of a specific Glare laminate to the residual strength of the used aluminium alloy (T=1). Therefore, a new variable M(T) is introduced:

$$M(T) = \left(\frac{a \cdot T^2 + b \cdot T + c}{a + b + c} \right) \qquad (13.3)$$

This factor is used to multiply the standard K_R curve of the aluminium alloy in the laminate with the same rolling direction.

$$K_R(T) = M(T) \cdot K_{R-Al-alloy} \tag{13.4}$$

To be precise, a K_R curve of a wide aluminium sheet with the same thickness as the single aluminium layers should be used for this purpose. However, a large test programme that was carried out on 2024-T3 CCT specimens of thicknesses from 0.8 to 2.0 mm illustrated that the difference in K_R curves was within the scatter [9]. More importantly, the shape of the curves was identical, which means that use of a different but identically shaped K_R curve is adjusted by the factor α.

The K_R curves obtained for different laminates are plotted as lines in Figure 13.11 together with the experimental K_R curves (dots). It is clear that this approach can be used to predict the K_R curves for different laminates.

The prediction model is summarised in Figure 13.12. It is based on the assumption that α is known for the aluminium constituent. When a new aluminium alloy is introduced, CCT residual strength tests with a comparable geometry must be carried out for different layups, covering a large range of possible T-values. With this data a new α is derived that is related to the new aluminium alloy.

The derived prediction model is only validated for Glare variants that are based on 2024-T3 alloys and loaded under tension in either the L-T or T-L directions. Additional research is necessary to extend this prediction model for different loading conditions.

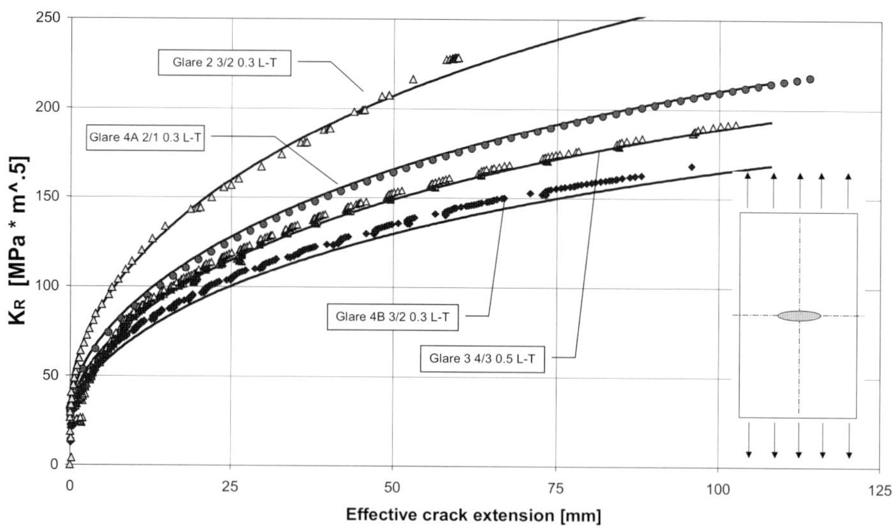

Figure 13.11 Relationship between residual strength and T for L-T direction.

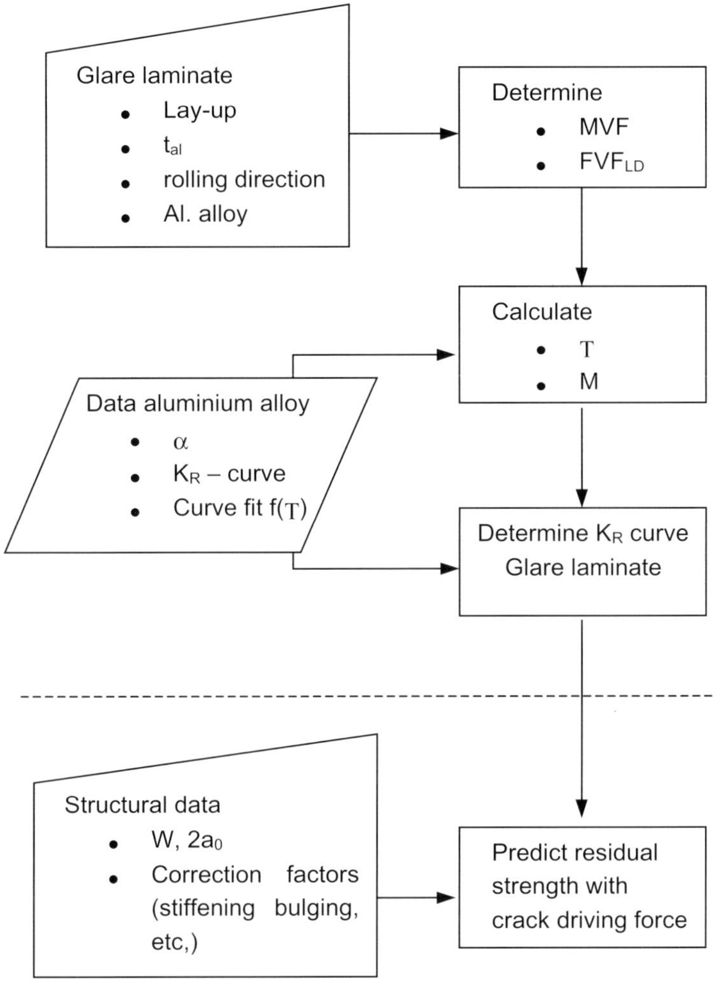

Figure 13.12 Residual strength prediction model.

13.4 Internal crack stopper bands

There is a push within the civil aircraft industry to introduce the so-called two-bay crack criterion as a requirement for newly designed aircraft. Airbus Industries requires that the A380 fuselage will be capable of resisting a two-bay crack while Boeing claims that the 777 already fulfils the two-bay crack requirement.

If this crack is caused by FOD, the two-bay crack requirement can become a critical design parameter for the application of Glare as fuselage skin material. For this kind of damage the fibres in the wake of the crack are broken, which reduces the residual strength of the material significantly compared with a fatigue crack where the fibres remain intact. As a consequence, the weight benefits due to the excellent fatigue behaviour might not be used to their full extent. This is especially the case for cracks in longitudinal direction where the frames are spaced with a large internal distance and are often flexibly connected to the skin, see Figure 13.13. Consequently, the crack-arresting capabilities of the frames in case of a two-bay crack are limited and the residual strength is largely dependent on the residual strength capabilities of the skin material [10].

There are several possible solutions to improve the residual strength in the circumferential direction. The research presented in this section was confined to the local addition of extra fibre layers to the Glare skin material that function as crack stoppers, see Figure 13.14. The baseline fuselage material investigated was Glare 3-3/2-0.4 with a thickness t of 1.7 mm and the goal was to increase the residual strength above the residual strength of 2524-T3.

Figure 13.13 Illustration of a two-bay longitudinal crack in a fuselage.

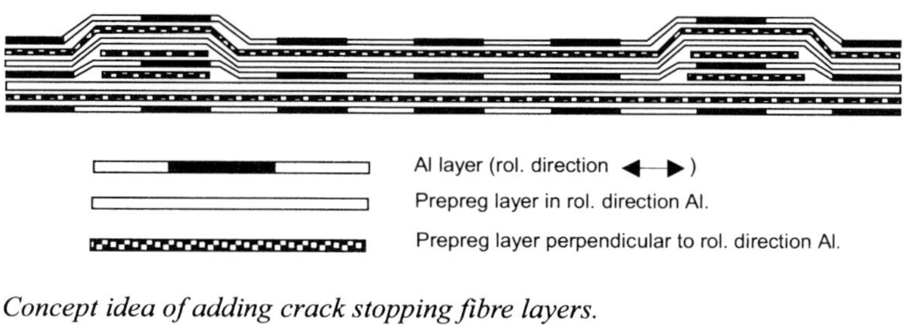

Figure 13.14 Concept idea of adding crack stopping fibre layers.

This prediction is carried out with the residual strength calculation method described in the previous section. It assumes that the residual strength of a crack stopper panel can be predicted as if the whole sheet is made of this crack stopper material

In this way, the number of extra fibre layers necessary to obtain a certain residual strength can be estimated. This is illustrated in Figure 13.15. The residual strength for a 2524-T3 T-L panel with W = 800 mm and $2a_0/W = 0.25$ is 232 MPa. The figure shows how large the combination of FVF_{LD} and MVF should be for a Glare material in the T-L direction to outperform 2524-T3 in the T-L direction. It was found that T must be larger than 1.2. Table 13.2 contains the T-values for a Glare 3-3/2-0.4 laminate with a different number of additional fibre layers. An addition of 4 extra fibre layers is sufficient. This means that the thickness increases from 1.7 mm between the stiffening elements to 2.2 mm below the frames and the stiffeners need to be joggled.

Table 13.2 *FVF_{LD} and MVF for different Glare laminates.*

Material	FVF_{LD} [%]	MVF [%]	$T = \alpha \cdot FVF_{LD} + MVF^*$
Glare 3-3/2-0.4 T-L	14.7	70.6	0.99
Glare 3 + 2 extra fibre layers	25.6	61.5	1.11
Glare 3 + 3 extra fibre layers	30.1	57.8	1.16
Glare 3 + 4 extra fibre layers	34.1	54.5	1.21

* $\alpha = 1.94$ in T-L direction.

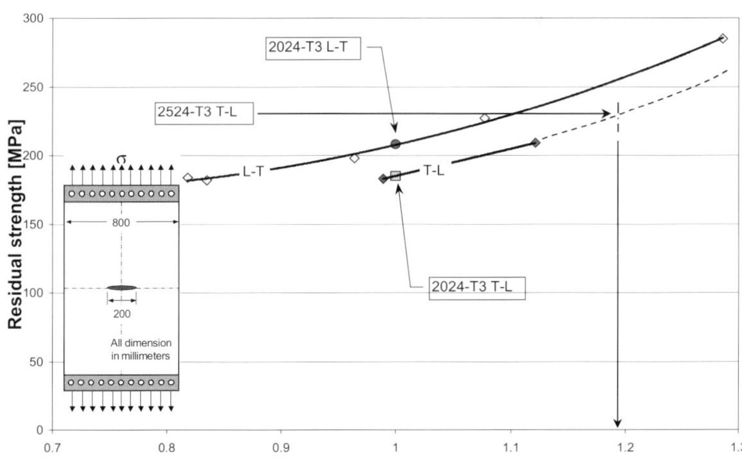

Figure 13.15 *Residual strength properties of Glare laminates versus 2524-T3 in T-L direction.*

Based on this prediction a full-scale test on an A380 like fuselage section was carried out at the IMA facility in Dresden, Germany. This panel was curved, contained stringers, frames and lap joints and was loaded in both longitudinal and circumferential directions and by internal pressure. The panel contained 175 mm wide crack stopper bands below frames consisting of 4 extra S2-glass fibre prepregs. To improve the rivet pull-through strength, extra 0.4 mm thick and 40 mm wide 2024-T3 strips were included in the crack stoppers [11], which resulted in a small decrease of the residual strength based on gross stresses but also a small increase of the maximum load. The crack stopper configuration is illustrated in Figure 13.16.

The testing of this panel was divided into two parts. The panel contained an initial through-the-thickness crack of 200 mm. During the first part, the panel was fatigue-loaded with 1-Δp flight pressurisation cycles, which resulted in a maximum circumferential skin stress of 90 MPa. The crack growth was monitored as a function of the number of flight cycles. Since the fatigue crack growth in the Glare material tends to be arrested during fatigue loading due to the fibre bridging, the cracks were extended several times with a saw. This included cutting of the fibres. When the two-bay crack in longitudinal direction has reached a length of two full frame bays, all the other damage is repaired and the residual strength for this particular crack is tested. This is the second part of the test.
When the crack was sawed to 500 mm in this panel, it became unstable during the additional fatigue loading process. However, an operator was able to stop the test before rapid crack growth occurred. At that time, the crack was more or less 100 mm in front of the crack stopper band and the total crack length was 940 mm.

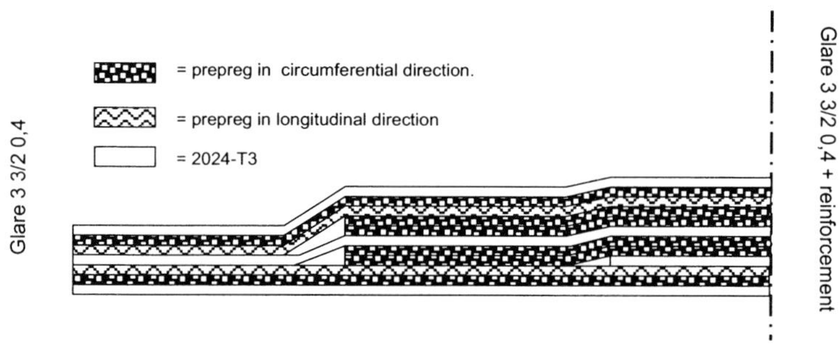

Figure 13.16 Layup crack stopper bands in XB-6.1 panel.

Residual strength testing

It was decided not to extend the crack by sawing directly, but to start the residual strength test and to see how the crack would behave. During loading, the crack started to grow clearly below the 1*Δp situation (1*Δp = 605 mbar). When 1*Δp was reached, the crack had extended up to the crack stoppers and was stopped.

Additionally, the panel with a two-bay crack of 1100 mm was fatigue loaded with 500 additional pressurisation cycles. During this fatigue loading, stable crack growth in the order of 30 mm took place in the aluminium layers. Both cracks showed the tendency to flap to the centre of the stiffener bay, see Figure 13.17.

The panel was subsequently tested for residual strength. At 1.6*Δp, the crack flapped at both sides up to next stiffener, remained there and allowed stable decompression. This is illustrated in Figure 13.19. A summary of the test procedure and results is given in Figure 13.18. Figure 13.18 also illustrates the residual strength results for a comparable 2524-T3 panel and a Glare 3-3/2-0.4 laminate with crack stopper bands assembled from two extra fibre layers. It is clear that the Glare 3-3/2-0.4 with four extra fibre layers behaves better than 2524-T3 as predicted.

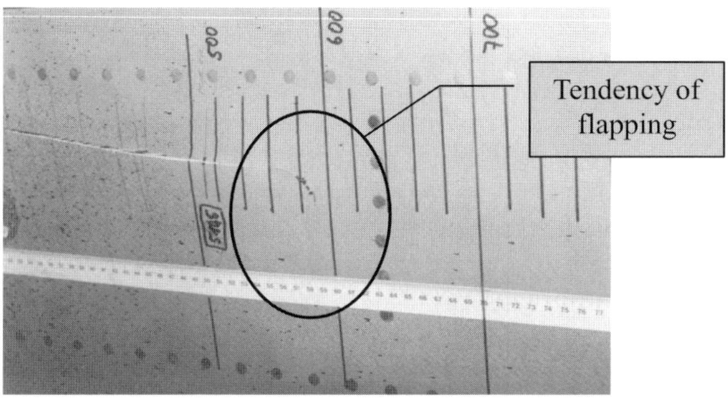

Figure 13.17 Illustration of flapping tendency during fatigue loading.

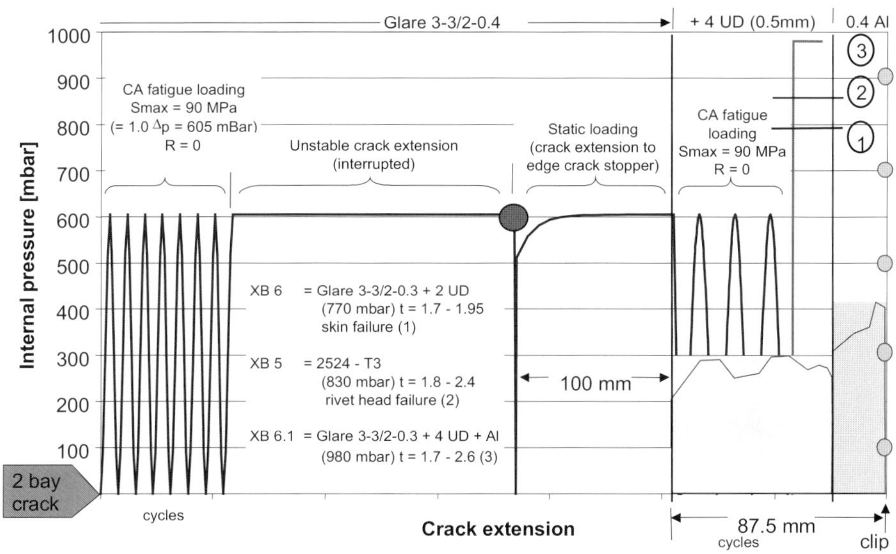

Figure 13.18 Test procedure and results.

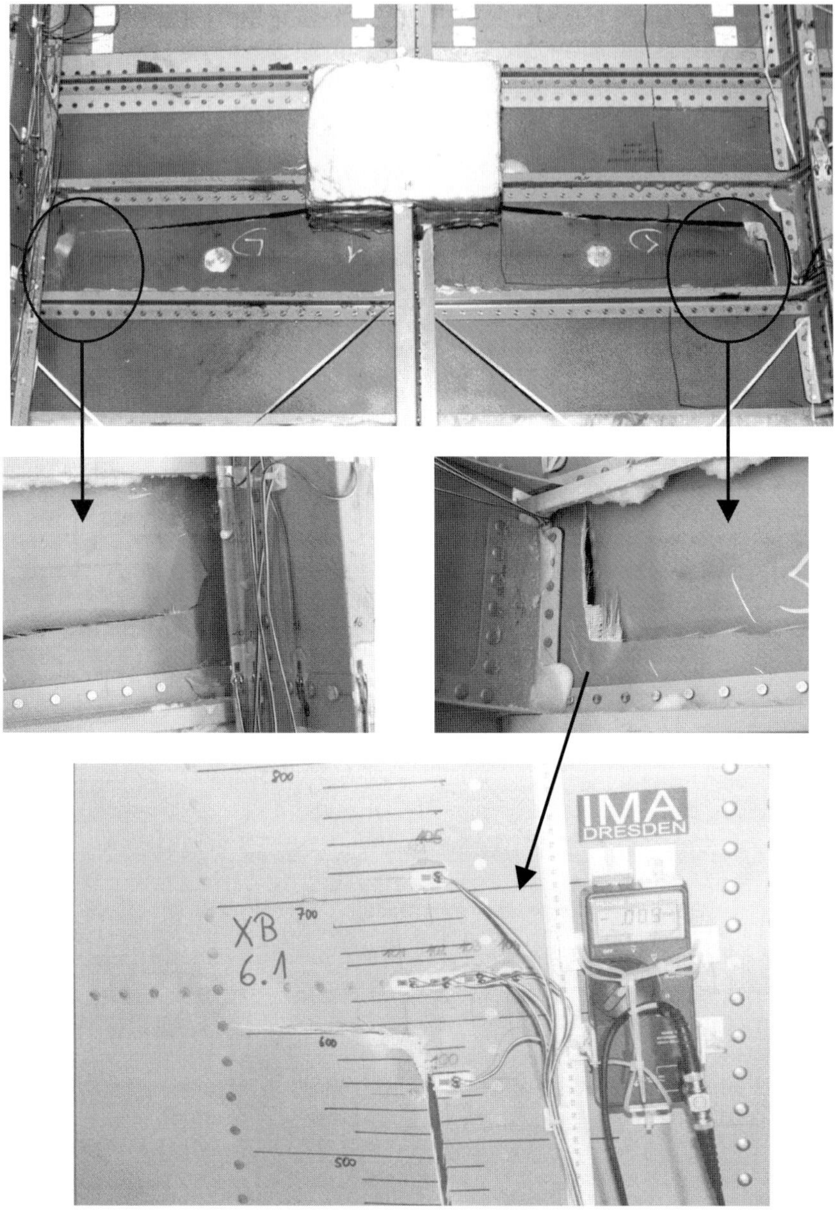

Figure 13.19 Illustration of final failure.

References

[1] T.J. de Vries and M. Pacchione, *The application of the R-curve concept for the prediction of the residual strength of fiber metal laminates,* Proceedings of the 1995 USAF Structural Integrity Program Conference, WL-TR-96-4093, United States, 1996.

[2] C. D'Allesando et al., *Evaluation of residual strength of Glare by experimental measurement of the J-integral*, Journal of Engineering Fracture Mechanics, 1994.

[3] E. Holleman and T.J. de Vries, *Residual strength of some Glare grades as a function of specimen geometry (BE2040 subtask 3.1-e)*, Memorandum M-717, Faculty of Aerospace Engineering, Delft University of Technology, The Netherlands, 1995.

[4] C.A.J.R. Vermeeren, *The Residual Strength of Fibre Metal Laminates*, PhD thesis, Faculty of Aerospace Engineering, Delft University of Technology, The Netherlands, 1995.

[5] D. Bresser, *Survey on the Methods for Determination of the Residual Strength of Fibre Metal Laminates*, TN-0047/95, Deutsche Aerospace Airbus, Germany, 1995.

[6] G.R.Sutton, *The damage tolerance of fibre reinforced aluminium alloy laminates*, DRA/SMC/CR962001, Defence Research Agency, United Kingdom, 1996.

[7] T.J. de Vries, *Various parameters concerning the residual strength of aircraft materials,* Master Thesis, Faculty of Aerospace Engineering, Delft University of Technology, The Netherlands, 1994.

[8] T.J. de Vries, *Blunt and sharp notch behaviour of Glare laminates*, The Netherlands: Delft University press, 2001.

[9] T.J. de Vries, *Plane stress fracture toughness test results of 2024-T3 clad sheet, 2024 T351 and 7475-T7351 plate aluminum alloys*, B2V-96-04, Restricted, Delft university of Technology, 1996.

[10] T.J. de Vries and L. de Kraker, *Crack stoppers in Glare laminates*, B2V-00-38, Delft University of Technology, Delft, 2000.

[11] T.A.J. Meesters, *Rivet Pull Through Strength*, B2V-99-37, Delft University of Technology, Delft, 1999.

14

Damage tolerance aspects

Th. Beumler

Abstract – The spirit of Glare is its crack bridging mechanism, which provides superior damage tolerance properties. Depending on the specific property, Glare shows either monolithic metal or composite behaviour. This challenges the definition of strength justification and certification procedures. Airworthiness regulations have to be interpreted for Glare in order to guarantee the same level of safety as obtained for aircraft structures made of other materials and to take at the same time benefit of its particular properties.

14.1 Continued airworthiness

14.1.1 Accidental damages

The technical concept of continued airworthiness of aircraft structures in case of fatigue or accidental damages is contained in the regulations FAR/JAR 25.0571, ACJ 25.0571 and ACJ 25.0603. The latter is focusing on the specific properties of composites. FAR/JAR 25.1529 specifies the administrative off set. The general idea is simple: to make sure that damage will be detected before its size becomes critical and leads to the loss of the aircraft, and repair the damage and restore the original strength. Consequently, the inspection becomes an integrated part of the damage tolerance concept during the aircraft life. However, some structures may be susceptible to damages but not be inspectable at all. In this case, FAR/JAR 25.571 paragraph (c) requires the safe life concept, which demands high safety and scatter factors on strength, and guarantees that the applied loads (ultimate) can be transferred with non-detectable damage sizes.

For monolithic aluminium the accidental damage event is easy to handle with the damage tolerance concept, since this type of damage is always visible. In any practical case a deformation takes place, resulting in a dent, a scratch or even a penetration, e.g. as result of a lightning strike. An exception is the delamination of bonded metallic structures after local overheating, for instance due to ground equipment being operated too close to the aircraft structure. This damage is not visible up to 300 °C, when a discoloration of the paint system can be observed [1].

Non-destructive test equipment is required to identify a possible delamination. Once the damage size is recorded, analytical methods provide reliable information concerning the residual strength of the structure. Conservative inspection intervals can be determined if some further flights have to be carried out before repair. Limit load capability must be justified with any damage size.

In carbon fibre structures mechanical impacts do not always lead to visually detectable damage. Therefore a different strength justification procedure is a common procedure. Non-visible defects (usually delaminations) are considered while sizing the component. With both analytical methods and tests, no growth of the damage is demonstrated due to fatigue and static loads.

Concerning detectability, Glare behaves similar to monolithic aluminium in the event of a mechanical impact. Dents, scratches and lightning strike damage are visually detectable [2] and can be covered by the damage tolerance approach according to FAR 25.571. As a worst-case assumption, the damaged location can be considered a completely lost load path, i.e. a penetration with all fibres broken. The residual strength can be calculated with the R-curve approach [3], similar to monolithic aluminium.

A fundamental difference exists between the damage extension and the residual strength behaviour of monolithic aluminium and Glare. Due to the crack bridging mechanism, the damage size extension is much slower in Glare than in monolithic aluminium, especially in case of part-through damage [4,5]. The residual strength of Glare can decrease significantly in the presence of part-through damage, but it starts from a high level. For example, Hagenbeek investigated the residual strength of a 320 mm wide Glare 3 panel with a 0.65 mm deep and 158 mm long scratch, i.e. one aluminium layer and one glass fibre layer completely cracked [4]. The residual strength dropped from 690 MPa (tensile strength of the material, i.e. no design criterion) to 403 MPa. However, despite the large reduction the latter value is still higher than the blunt notch strength (i.e. 381 MPa, gross stress), which is the typical design criterion for all riveted locations. The remaining intact fibres in Glare are able to carry significant loads, occasionally leading to allowable damage sizes beyond common sizes in the aircraft industry.

As an illustration, a stress check can be performed on the Glare demonstrator panel, which is installed in the forward upper fuselage section in A310 A/C 484, see Figure 14.1 and Figure 14.2. The basic panel is made from Glare 3-3/2-0.4. The following hoop stress situation can be analysed for the undisturbed far field:

- maximum fatigue stress at internal pressure (Δp) : 85 MPa
- limit load stress ($1.15\Delta p$) : 97 MPa
- ultimate load stress (2 x pressure release valve pressure) : 182 MPa

Material properties:

- tensile ultimate strength (typical divided by 1.1 [6]) : 594 MPa
- blunt notch stress (gross): 363 MPa

Figure 14.1 *A310 aircraft 484 with Glare demonstrator panel, first flight, 4th of October, 1999.*

Figure 14.2 *Glare panel installed in aircraft 484.*

The reserve factor for the undamaged panel is 594/182 = 3.2 and the ultimate stress is approximately a factor 2 lower than the blunt notch stress. Large-scale damage in the longitudinal fuselage direction can therefore be accepted, in fact the entire panel could be scratched 0.4 mm deep. From a residual strength point of view, the aircraft may continue operation without a repair in the discussed case, but the responsible stress design engineer's engineering judgement would not provide him a warm comfortable feeling. Time dependent strength degradations like corrosion and delaminations along the scratch have to be considered for a safe flight continuation. Additionally, load conditions other than the tension mode may become critical with increased damage sizes. Consequently, the stress design engineer will define a much lower tolerable damage size than the material can afford in tension, balancing the detectable damage size with the acceptable damage size.

However, the structural reserve of a damaged Glare panel for tension load cases offers the customer a comfortable time window to define and perform a repair, without the necessity to ground the aircraft immediately. High flexibility is offered to both operator and repair shop [7].

14.1.2 Fatigue damages

The fatigue Structural Significant Items (SSIs) in a pressurised fuselage skin panel are holes, e.g. bore holes and cut-outs. Cut-outs can be designed with the focus on visual inspectability. The considered edge crack is a single entity, usually covered by a directed inspection.

Much more complicated is the damage tolerance justification for riveted joints, which are, for economic and aerodynamic reasons, often designed as single shear joints. The jointed sheets experience the highest bending stresses at their mating surfaces, in other words on the visually not inspectable side.

Since the Aloha accident with a Boeing 737 in 1988, the rulemaking organisations tried to further specify the damage tolerance regulations in order to avoid catastrophic failures due to cracks, which are not detectable by non-directed inspections. The idea is to distinguish between single and multiple load path structures on the one hand, and the level of the inspectability on the other hand. If a multiple load path structure is investigated, in which a fatigue crack can be detected by non-directed inspections, the fatigue approach is adequate to determine the threshold inspection [8]. If either a single load path structure or a multiple load path structure, in which fatigue cracks are undetectable by non-directed inspections, is investigated, a crack propagation calculation is recommended for the threshold determination. The crack propagation method takes manufacturing flaws in all bore holes into account, taking poor manufacturing quality into account. Occasionally, the so-called rough flaw must be considered, e.g. in a longitudinal joint once per frame bay. Under discussion are a manufacturing flaw size of 0.005 inch and a rough flaw of 0.05 inch.

The number of flights achieved when a crack becomes detectable (N_{DET}), divided by a scatter factor, equals the inspection threshold (N_{TH}). The detected crack length must be below the critical crack length (at limit load). Commonly applied scatter factors for aluminium structure are between 2 and 3 based on full-scale test results, see Figure 14.3.

The number of flights achieved when the cumulated cracks become critical (limit load, N_{CRIT}), divided by a scatter factor, equals the inspection threshold, see Figure 14.4.

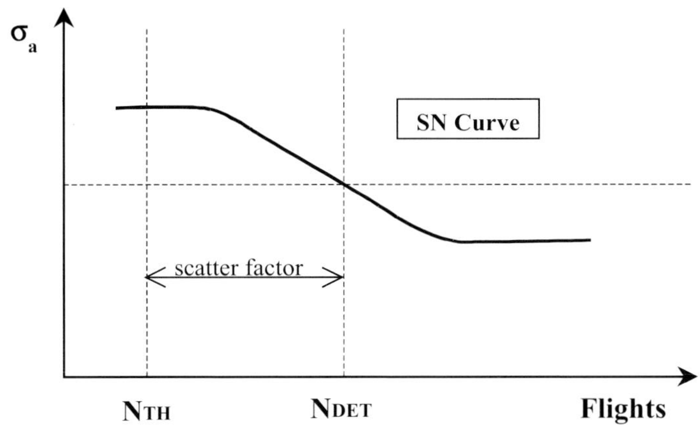

Figure 14.3 *Fatigue concept, recommended for a multiple load path structure if readily detectable.*

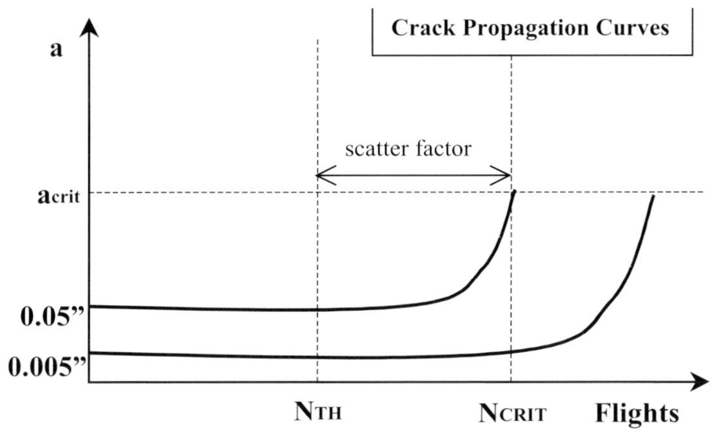

Figure 14.4 *Initial flaw concept, recommended for single load path structure and multiple load path structure if not readily detectable.*

In the past at least, the fatigue concept did not really consider a multiple site damage situation but focused on a single crack without taking adjacent damage into account. If so, the true safety margin of the determined threshold inspection remains unclear since the residual strength capability of the structure at the time of the detection of the single crack is not considered. The applied scatter factor has to account for this circumstance. This is possible if the crack propagation in aluminium is considered to be fast and the scatter factor is high. In that case the flights to crack initiation and the flights to detectable crack length are relatively close to each other, and the determined threshold inspection is not significantly influenced by inclusion of the crack propagation part.

The initial flaw concept considers adjacent damages for the residual strength prediction and therefore seems to provide a more accurate prediction of the level of safety for pressurised fuselage joints.

Circumferential and longitudinal skin splices are Principle Structural Elements according to ACJ 25.0571, whose failure could result in catastrophic failure of the aircraft. It is stated: “Damage at multiple sites due to prior fatigue exposure must be included (in the damage tolerance evaluation) where the design is such that this type of damage can be expected to occur.”

A lost skin panel [9] made the aircraft community aware of the fact that the structural aluminium joints in pressurised fuselages are without any doubt susceptible to multiple site damage. Furthermore it has a single load path character, since the frames did not prevent the significant failure. However, the recommended crack pattern (0.005 / 0.05 inch) for an initial flaw calculation is arbitrary. The method cannot account for the influence of hole filling on crack initiation due to different squeeze forces for aluminium fasteners or the fitting of bolts. Since the hole-filling dependent factors on the crack initiation life are significant (factor >3 [10]), the advantage of the arbitrary crack lengths can be discussed. A promising alternative is the statistical evaluation of damage accumulations at all bore holes along a rivet row [11]. If damage accumulates to 1 at a particular location, the point of crack initiation (N_i) is reached (Miner rule). It follows a crack propagation calculation for each individual location, when its damage has been accumulated to 1. The author recommends this method, since it combines the benefit of smart design and high manufacturing quality (fatigue approach) with the accuracy of an MSD calculation for the residual strength and inspection threshold prediction.

A riveted Glare joint behaves differently compared to an aluminium joint. Due to the internal stress system and lower stiffness of the prepreg layers, its aluminium layers tend to crack earlier than in a monolithic structure when the same external load is applied [12]. The crack propagation rates in Glare are orders of magnitudes lower than in monolithic aluminium [5,13]. Figure 14.5 illustrates the behaviour of a Glare joint. Slow crack propagation occurs after initiation (N_i), accompanied by a gradual loss of residual strength. At a certain time (N_{UL}) the ultimate load capability will be reached (F_{UL}). The associated crack length is the critical crack length $a_{c,\,UL}$. A similar relationship exists for the limit load (N_{LL}, F_{LL}).

As mentioned above, fatigue cracks always initiate on the mating surfaces of single shear joints. Due the slow crack growth, the cracks in riveted Glare joints remain undetected for a long period of time. Figure 14.6(a) presents a test example in a Glare 2-7/6-0.4 butt strap. Four of the seven aluminium layers are cracked, all fibres are intact.

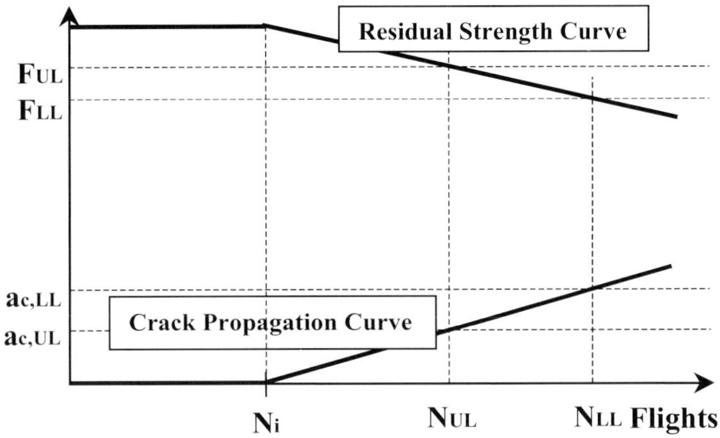

Figure 14.5 *Schematic crack initiation, crack propagation and residual strength behaviour of a Glare structure with crack. N_i = point of crack initiation.*

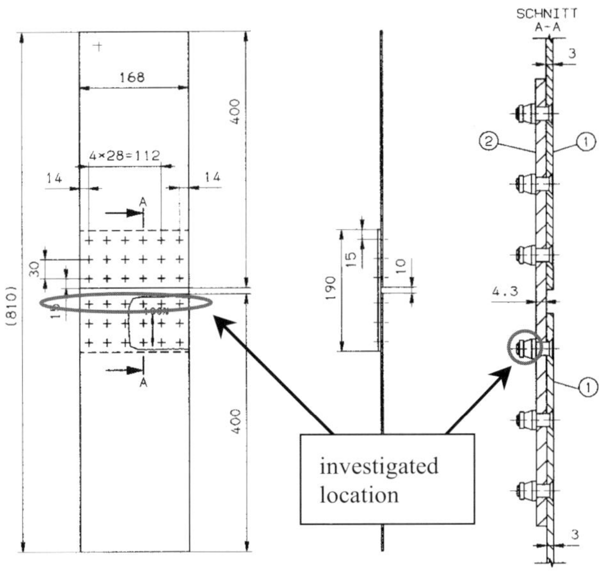

Figure 14.6(a) Circumferential joint specimen with Glare 2-7/6-0.4 butt strap.

The crack illustrated on the left side of the bore hole in Figure 14.6(b) is the lead crack. The average crack length is the total of all cracks divided by the number of possible fatigue critical locations: in this situation $a_{avj} = \Sigma a_{ij}/12$. For aluminium layer 1 (j=1): $a_{av1} = 3.7$ mm.

Figure 14.6(c) compiles some details of the crack situation in layer 1. The detectable crack length is estimated as 7 mm (NDT). One crack, 8.3% of the total number of cracks, exceeds the inspectable crack size, the inspectability of the specimen is therefore obtained after approximately 180000 cycles. The average crack length is included in Figure 14.6(c) as well. It is an indicator of the residual strength of the joint. After 180000 cycles, 9.2% of the total aluminium in the Glare butt strap is cracked (all layers). The blunt notch strength of the butt strap for this situation can be calculated with Müllers rule [10]:

$$\sigma_{res.cack.} = \sigma_{BN} - R * MVF * \sigma_{al(\varepsilon=4.7\%)} = 625 - 0.092 * 0.651 * 420 = 599\ MPa \qquad (14.1)$$

with:
- σ_{BN} = blunt notch stress (net stress) [6]
- R = ratio of damaged aluminium to total amount of aluminium
- MVF = metal volume fraction
- $\sigma_{al(\varepsilon=4.7\%)}$ = stress in aluminium layers at laminate failure strain at room temperature

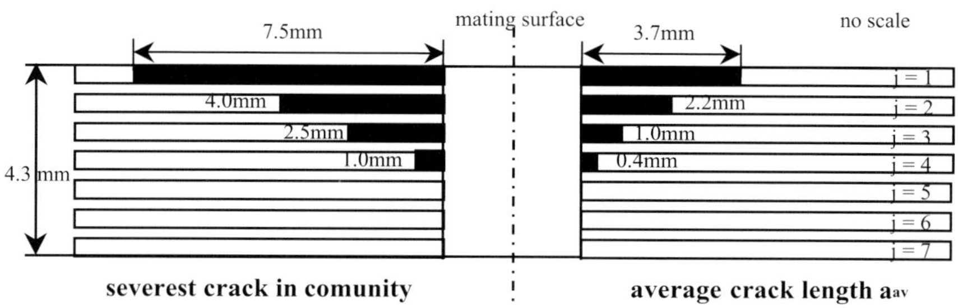

Figure 14.6(b) Crack sizes in Glare-2-7/6-0.4 butt strap after 180000 cycles.

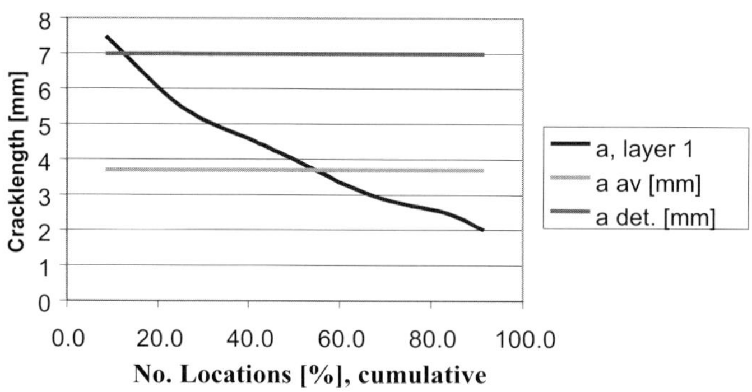

Figure 14.6(c) Crack lengths distribution butt strap, layer 1.

The determined residual strength of the butt strap, including cracks, still exceeds the ultimate load of aircraft fuselage circumferential joints. Therefore the design probably would not require inspection.

14.2 Engineering factors

Appropriate factors to account for material scatter, load spectrum scatter and environmental influences have to be applied on test results for either the verification of the design service goal and the determination of the inspection threshold. The aircraft manufacturers look back on a long history of experience with aluminium structures. So-called test-to-structure factors have been developed for coupons and components, which have to be applied on stress or life in order to obtain a full-scale result. They are used during pre-design for a quick interpretation of test results.

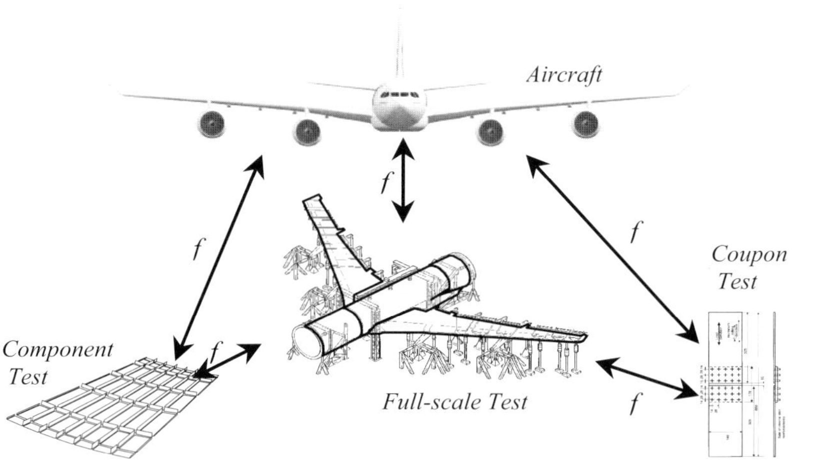

Figure 14.7 *f = engineering factors; including scatter factors, required for the interpretation of test results provided by specimens of different sizes and at different times of the project.*

Fatigue results for small aluminium specimens, i.e. coupons, are related to final failure, and the crack propagation part of the test is considered to be negligible. Consequently, the applied factors, which link the coupon with the full-scale result, are dependent on the crack initiation behaviour of the metal, not on the crack propagation behaviour. Factors of 3 to 5 on the life of a constant amplitude coupon test result are commonly used in order to match the full-scale result, see Figure 14.7.

The factors, which have to be considered between the full-scale result and the operational aircraft, will be discussed in the following sub-chapters. In general, the factor provides a certain safety standard, e.g. Airbus' target is to achieve a probability of survival for the monolithic aluminium structure of about 95%, i.e.

only 5% of the aircraft will be subject of fatigue damages before reaching the Design Service Goal [2]. This level of safety should also be kept for Glare structures.

The full-scale test result is valid for the applied mission mix only, i.e. the anticipated usage of the aircraft. The aircraft will be certified for the applied load spectra, there is no factor accounting for load variations. If necessary for future aircraft derivatives, an assessment for range or weight can be performed analytically. However, a conservatism can be integrated in the full-scale test procedure by truncating the peak fatigue loads, in order to avoid the positive effects of plastification at bore holes.

Corrosion is a time and user dependent item, which is covered by dedicated corrosion inspection tasks. Therefore it is not considered in the full-scale test evaluation as well.

Since the dimensions of both aircraft and full-scale specimen are similar, the engineering factor f between both is mainly dedicated to the material scatter and the environmental influences, which may decrease the material properties.

14.2.1 Material scatter

The schematic diagram presented in section 14.1.2, see Figure 14.5, does not contain all information required for an inspection threshold determination which is justified with a statistical damage distribution. According to the example presented for the Glare 2 butt strap, see Figure 14.6, two crack propagation curves should be included in the diagram. One curve represents the lead crack, which determines the number of flights until a crack becomes inspectable. The other curve represents the average crack length for the determination of the load capability. An important variable is the Design Service Goal. Its relation to the point of crack initiation, the ultimate load capability and the limit load capability defines whether the Glare structure has to be inspected for fatigue or not. Some case studies will be discussed.

The slope of the average crack propagation curve (a_{av}) determines the number of flights at which the limit load capability is reached (N_{LL}), e.g. for a riveted joint. N_{LL} is the absolute minimum tolerable strength for the aircraft structure. In order to account for material scatter, N_{LL} (full-scale result) needs to be divided by an appropriate scatter factor for a conservative aircraft inspection threshold determination. Note that N_{LL} is a calculated value, since the lead crack is not detectable at N_{LL} in example 1, see Figure 14.8(a). In contrast to aluminium structures, the scatter factor is dependent on crack propagation, not on crack initiation. The determined inspection threshold in example 1 is lower than the Design Service Goal (DSG), but the DSG is reached before the ultimate load capability is passed (N_{UL}). The low slope of the Glare crack propagation curve provides a high number of flights between ultimate load capability and limit load capability, and the safety margin at N_{UL} is high. No directed fatigue inspection is required since the DSG will be reached before N_{UL}.

The DSG in example 2, see Figure 14.8(b), is moved between the ultimate strength capability and the limit load capability. The safety is similar to that in example 1, a directed inspection is required at N_{UL}.

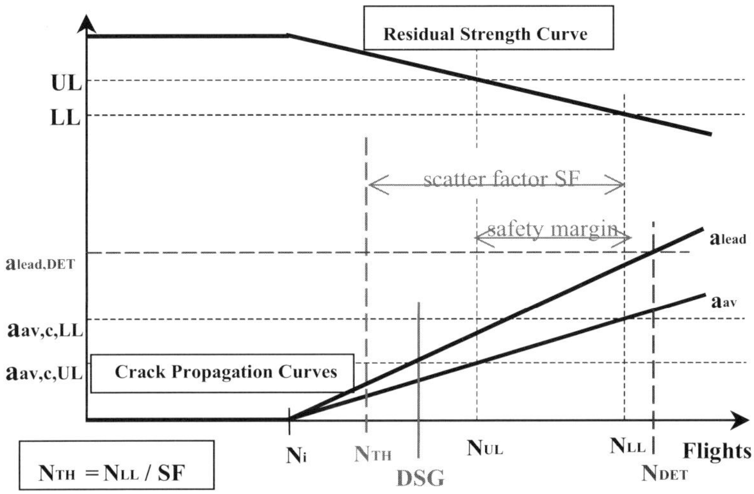

Figure 14.8(a) Inspection determination, example 1, $N_{TH} < DSG < N_{UL}$.

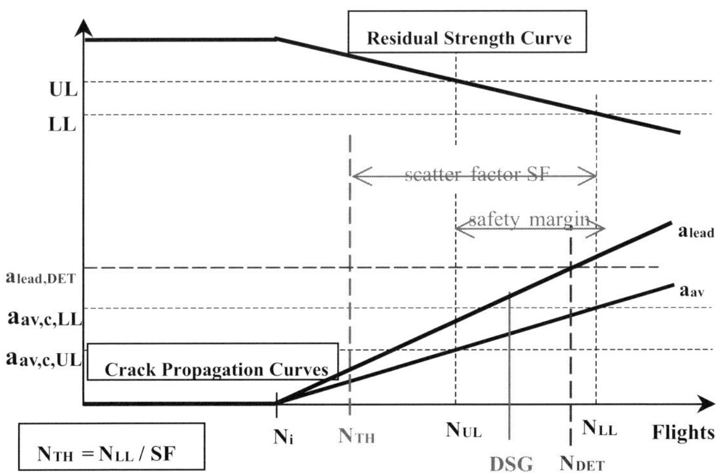

Figure 14.8(b) Inspection determination, example 2, $N_{TH} < N_{UL} < DSG < N_{LL}$.

Note that the lead crack is inspectable before the limit load capability is reached. But it has no influence on the threshold determination at all, as long as pure analysis is acceptable.
The lead crack propagation curve (a_{av}) is valuable for the full-scale test evaluation. A tear-down inspection at the end of the test will provide both the slope of the lead crack and the multiple site damage distribution. The latter will define the slope of the average crack length curve and finally the values N_{UL} and N_{LL}.

Example 3, see Figure 14.8(c), illustrates a non-acceptable situation. The limit load capability is passed before the design service goal is reached. This situation requires corrective actions, i.e. either a reduction of the design service goal to at least N_{LL} or a structural modification at N_{SMP} (structural modification point). In the latter case it is recommended to start inspections at N_{TH}. Common scatter factors on full-scale results are 2 for N_{SMP} and 3 for N_{TH} where aluminium structures are concerned. For Glare these factors need to be determined.

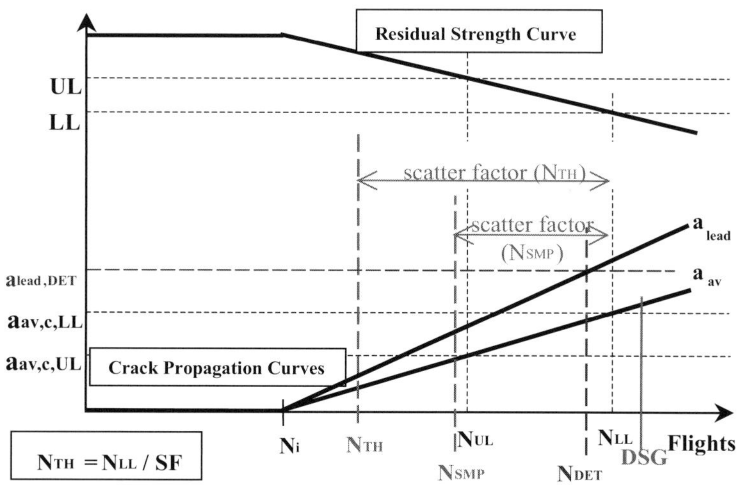

Figure 14.8(c) Inspection determination, example 3, N_{LL} < DSG.

14.2.2 Environmental influences

JAR 25.603 requires one to “take into account the effects of environmental conditions, such as temperature and humidity, expected in service.” The Glare laminate contains two materials with different coefficients of thermal expansion, i.e. aluminium and glass fibres. Since the material will be cured at 120 °C and operated in a temperature range between approximately –50 °C and +70 °C, temperature related effects have to be taken into account. The fibre matrix is epoxy based and susceptible to moisture absorption. However, the moisture absorption is limited to the sheet edges and bore holes, due to the barrier function of the outer aluminium sheets. Time consuming cycling tests have been conducted by Graafmans [14] in order to investigate the delamination behaviour of Glare edges. It was concluded,

that the plasticising effect of the resin locally decreases the shear stresses between aluminium and glass fibres and therefore environment-related delamination has been observed.
Kieboom [12] investigated the crack initiation behaviour of Glare bore holes at different temperatures and found similar results for 70 °C and 20 °C. The crack initiation life increased significantly at –55 °C, similar to the effect on monolithic aluminium. Consequently, the temperature effect does not need to be considered for crack initiation in Glare. The crack initiation life of the aluminium layers in Glare can be predicted from the monolithic aluminium curves, if the stress levels in these sheets are calculated accurately. Since the moisture level in the resin has no significant influence on the stresses in the aluminium, no ageing influence is expected on crack initiation. Verification tests are under evaluation.

A significant influence of temperature on crack propagation rates has been identified by Bär [15]. The crack propagation rates increase at higher temperatures and decrease at lower temperatures, see Figure 14.9. It has to be mentioned that a high stress intensity was necessary to obtain crack growth. However, an investigation is required concerning the influence of the flight mission related temperatures on crack propagation. Comparative crack propagation calculations have been performed for a bore hole, using a short range mission profile [16]. In the first case the entire load spectra have been calculated with room temperature crack propagation data. For comparison, the same spectra have been divided into three temperature sections, see Figure 14.10.

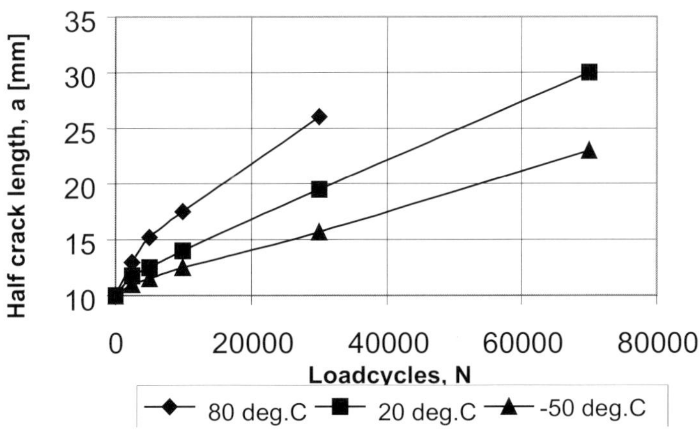

Figure 14.9 *Measured crack propagation data at different temperatures* [15], *CCT specimen, Glare 3-3/20.3, initial saw cut length 2ai = 20 mm, maximum stress 140 MPa, R=0.1, constant amplitude.*

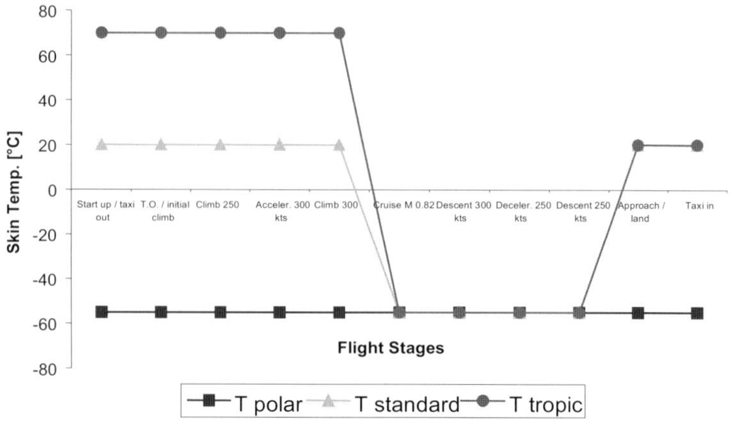

Figure 14.10 Simplified temperature profiles for flight missions [16].

The tropic (skin) temperature profile has been selected as worst case for any flight. All ground and flight stages before maximum fuselage differential pressure have been considered at 80 °C, cruise and descent at –50 °C, landing and after flight ground cases at room temperature. The calculation at constant room temperature predicted higher crack propagation rates than the calculation with the temperature mix. It can be concluded that the low crack propagation at low temperatures overcompensates the high crack propagation at high temperatures. Since the portion of loads at low temperatures is smallest for short-range aircraft, long-range missions are also covered by this investigation. The author recommends repetition of the analytical investigation for any aircraft due for certification. If a full-scale test is conducted at room temperature, no special factor to cover temperature effects has to be considered for the Glare parts.

The residual strength properties of Glare joints decrease after ageing [17]. This behaviour should be accounted for during the residual strength tests of full-scale certification articles, i.e. higher loads have to be applied in order to simulate the degradation effect on an unaged structure.

References

[1] W. 'tHart and R. Diks, *Laboratory tests to simulate APU heating on GLARE3-4/3-.4*, NLR, Report NLR-CR-99338, 1999.

[2] *Glare Application For Primary A3XX Airbus Structure,* DaimlerChrysler Aerospace Airbus Document, EMF-723/99, Issue 3, 2000.

[3] T.J. de Vries, *The residual strength of Glare laminates with foreign object damage*, Delft University of Technology, *EADS Technology Seminar* 2001.

[4] M. Hagenbeek, *Fatigue and residual strength effect of scratches*, Delft University of Technology, Report B2V-00-14, 2000.

[5] A. de Koning and L. Schra, *Fatigue crack growth of part through the thickness cracks in GLARE3 and GLARE4B coupons*, NLR, Report NLR-CR-2000.078, 2000.

[6] G.H.J.J. Roebroeks, *The Metal Volume Fraction Approach*, SLI Report TD-R-00-003, 2000

[7] *A310 NSN484 GLARE Panel Repair Approval Sheets*, DaimlerChrysler Aerospace Airbus,EMR, 1999.

[8] H.J. Schmidt, *Current and Future Regulations, Design Goals and Industry Standards*, DaimlerChrysler Aerospace Airbus; EMF-1/99, 1999.

[9] Aircraft Accident Report: *Aloha Airlines, Flight 243, Boeing 737-200, N73711, near Maui, Hawaii, April 28, 1988, NTSB/AAR-89/03*. Washington DC: U.S. National Transportation Safety Board, 1988

[10] R. Müller, *An Experimental and Analytical Investigation on the Fatigue Behaviour of Fuselage Riveted Lap Joints*, Delft University, PhD-Thesis, 1995.

[11] H. Trey, *Analysis of Aircraft Structure Regarding Widespread Fatigue Damage*, EADS-EMF, EADS Technology Seminar 2001

[12] O. Kieboom, *Fatigue Crack Initiation and Early Crack Growth in Glare at Different Temperatures*, Master Thesis, Delft University of Technology, 2000.

[13] J.Homan, R. Alderliesten, *Prediction Methods for Fatigue Crack Propagation in unstiffened GLARE Through Cracks*, Delft University of Technology, Report B2V-99-40, 1999

[14] G. Graafmans, *Thermal Behaviour of Fibre Metal Laminates*, Delft University of Technology, Master thesis, 1995

[15] H. Bär, *Verifikation und Ergänzung von Berechnungsmethoden für die statische und dynamische Auslegung von GLARE Strukturen*, Universität Stuttgart, Diplomarbeit, 1992.

[16] Th. Beumler, *Crack Propagation Justifications for Aircraft Fuselage Riveted Joints under Consideration of Operational Temperatures*, Delft University, Report B2V-01-07, 2001.

[17] B. Borgonje, *The influence of elevated temperature on Glare after exposure to moisture*, Delft University of Technology, Master thesis, 2000.

C. Design Aspects

15

Fuselage barrel design and design for manufacturing

G.P. Wit

Abstract – To assess the potential of Glare as a skin material for ultra large aircraft, several Glare skin panels were designed and manufactured for the Megaliner Barrel Programme. This programme is being executed as a joint effort between Airbus, Fokker Aerostructures and the Dutch laboratory for Aeronautical Engineering. This chapter gives an overview of the Glare development process for the Megaliner Barrel and focuses on requirement definition and the iterative sizing and design process. Some attention is also given to the complications that occurred due to a simultaneous development of the Glare shells, the Fabrication Technology for Glare components and the tools and data to design using Glare. The article starts with a short description of the structural layout of the Megaliner Barrel.

15.1 Overview of the Megaliner Barrel

The Megaliner Barrel represents a typical fuselage section of an ultra large aircraft with two passenger decks. It consists of 3 full-scale barrel sections (A, B, D) with a geometrically constant cross-section. The sections are joined together with butt-straps and stringer couplings. Each of the sections A, B and D consist of several skin panels with z-shaped stringers attached. The application of various materials for the skin panels is indicated in Figure 15.1. The back-up structure attached to the skin panels includes clips, frames, frame support structure, floor beams and the floor support structure. Door and window cut-outs are included in the structure.

The fatigue and damage-tolerance behaviour of the complete structure will be tested by simulating in-service loads. The applied loads include internal pressure, shear loads in 2 directions (Sz, Sy), force (Sx) in direction of flight (DOF), torsion (Mx) and bending moments in 2 directions (My, Mz): see Figure 15.2. Another important result from this test will be the validation of the Finite Element Model (FE-model) by correlating the strain gauge measurement from the barrel test to the predicted internal load distribution in the various structural parts of the fuselage.

The majority of the butt-straps at frame C64 and C72 are Glare parts: Glare is used from the lower floor beams up because of its excellent resistance to fatigue loads. The stringer couplings are z-shaped parts that transfer the stringer loads at the sections butt joints. At C64 the stringer couplings are all machined aluminium parts. At C72 some of the stringer couplings in the top panel are Glare parts.

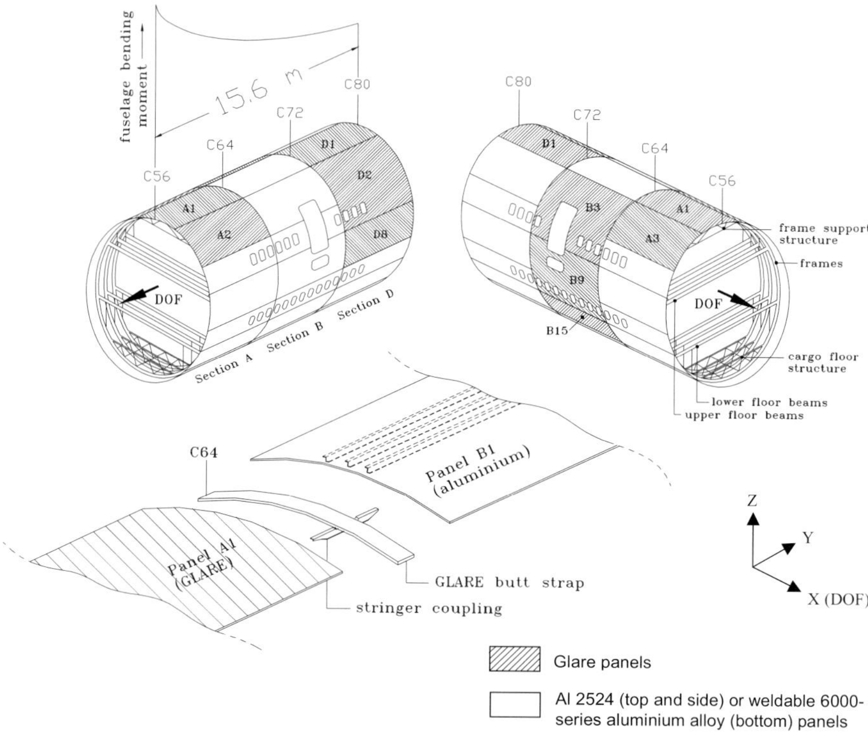

Glare panel definition

Four basic Glare types (for Glare type definitions, see Kok [1]) have been applied, best suited to the loading condition of the part. Butt-straps are all made of Glare 2B (in some cases with local reinforcements). The Glare type distribution for the panels is as follows:

A1, A2, A3, D1:	Glare 4A top panels with Glare 2A stringers (bonded)
B3:	Glare 3 upper door surrounding panel with bonded Glare doublers and aluminium stringers (some bonded, some riveted)
B9:	Glare 3 lower door surrounding panel with riveted Glare doublers and aluminium stringers (riveted)
B15:	Glare 3 bottom panel with aluminium stringers (bonded)
D2:	Glare 3 upper side panel with Glare 2A and aluminium stringers (bonded)
D8:	Glare 3 lower side panel with aluminium stringers (bonded)

Figure 15.1 Layout of the Megaliner Barrel.

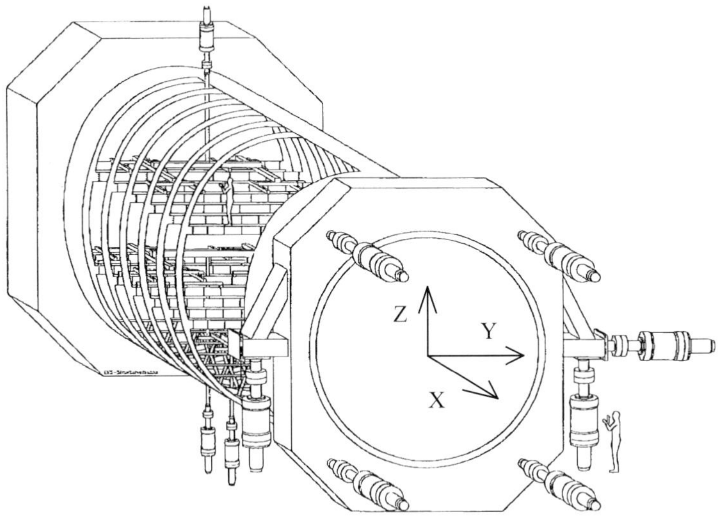

Figure 15.2 *Test setup for the Megaliner Barrel (ref. Airbus test department).*

15.2 The development process for the Glare shells in the Megaliner Barrel

Attempting to introduce a new material in the aircraft industry is a time consuming and costly process. It requires a major investment in research to assess the behaviour of the material under its predicted use. Even when the potential of the material is confirmed for all relevant material characteristics, the process of qualifying the material must result in the formal proof and acceptance for its application. Not only must the material characteristics be generated and accepted, but also the analytical methods (substantiated by sufficient test results) to analyse the strength and damage tolerance behaviour of the components. These aspects are further referred to as the "Material Development" Programme of Glare.

Glare also has its limitations and opportunities with respect to manufacturing. Because of its composite build-up, several manufacturing concepts can be proposed to create a final Glare product. Whether the proposed concept is feasible and cost effective for industrial application needs to be verified by production trials. The investigations related to manufacturing have been defined in the Glare "Fabrication Technology Development" Programme.

The material and fabrication-technology developments of Glare have reached their final stages. Major steps forward in both areas were achieved in the same period that the Glare shells for the barrel were developed. Many of these advances were made because the Glare components in the barrel are typical for all structural details that may occur in ultra large aircraft. The information how to design and manufacture all these parts was obviously required to complete the barrel project successfully. The process that has been followed to develop the Glare components in the barrel is schematised in see Figure 15.3.

The interaction with the Glare Material Development and Fabrication Technology Development Programmes is explained in section 15.2.1. In section 15.2.2, the four phases are further described. Finally in section 15.2.3, the first phase (establish requirements) is discussed in more detail.

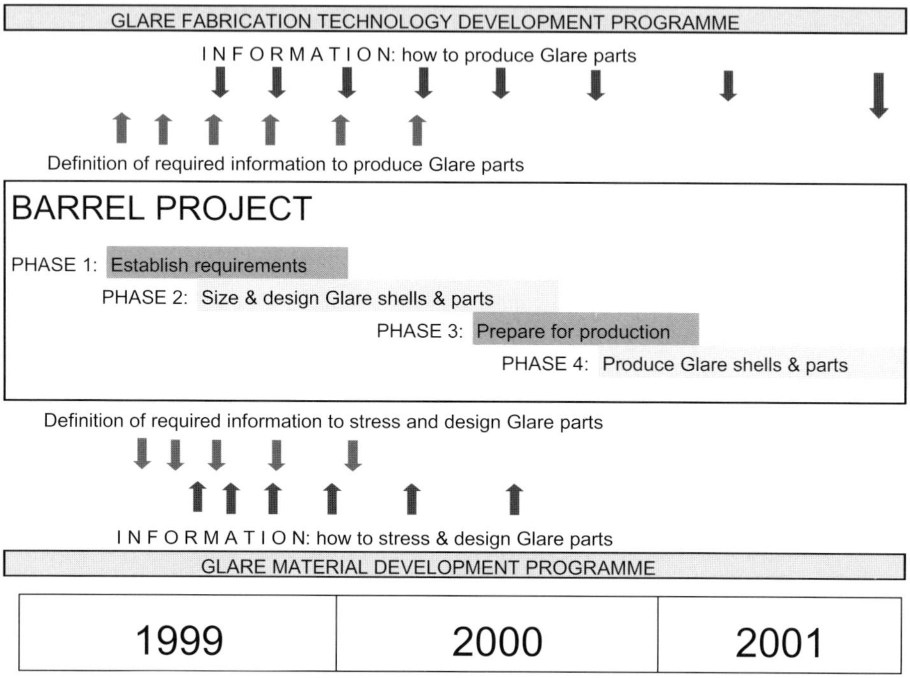

Figure 15.3 *Phases in the development process for the Glare shells in the Megaliner Barrel.*

15.2.1 Interaction between the Megaliner Barrel Programme and the Material Development / Fabrication Technology Programmes

At the start of the Megaliner Barrel Programme, a lot of the information required to develop the Glare products was not available. This included information related to the sizing and design phase and information related to the fabrication of the parts. Adequate planning of tasks and resources was required to assure that all parts of information became available in the right phase of the programme.

Examples are given in the list below of information that was essential to the Megaliner Barrel Programme and generated in the Glare Material and Fabrication Technology Programmes.

- *allowable stress levels at door corner cut-outs*
  See section 15.2.3 (stress requirements) and Brügemann [2].
- *allowable stress levels at edges of bonded doublers*
  Some verification (fatigue) tests have been performed to assure that fatigue cracks at the edges of bonded doublers will not occur, see Figure 15.4.

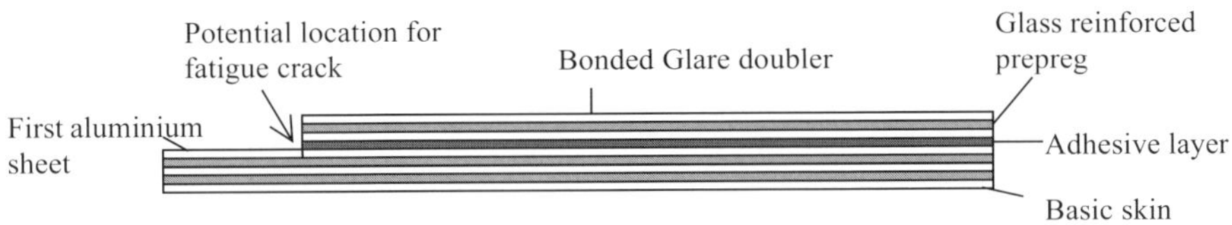

Figure 15.4 Fatigue cracking at edges of bonded doublers.

- *allowables for different types of fasteners*
  Tables for static fastener allowables have been generated for several types of fasteners and Glare laminates, based on coupon tests.
- *stress analysis methods and tools, suitable to size Glare structure*
  The basis for the Glare stress tools are the methods already developed for aluminium structure, adapted to Glare. Nevertheless, some special methods had to be developed for Glare. This process is still on-going.
- *manufacturing concept for the Glare stringers and stringer couplings*
  Up until the production of the Glare stringers for the barrel, manufacturing tests were performed to ensure that the final Glare stringer fulfils the requirements. For more details about Glare stringer manufacturing, see [3].
- *manufacturing principles for drop-offs of interlaminar doublers*
  Drop-offs of interlaminar doublers must follow certain rules to avoid quality problems at the drop-off.
- *optimum splicing concept*
  Several concepts to splice Glare laminates were evaluated in Roebroeks [4] and one of these was selected.
- *achievable tolerances for Glare parts (laminate thickness and sheet location)*
  Due to limited production experience, the tolerances that can be achieved in full-scale Glare panels had to be determined with production trials at the end of the barrel design phase.
- *inspection procedure to ultrasonically inspect Glare panels and stringers*
  Ultrasonic inspection of composites is a well known process in the aircraft industry and procedures have been established at many companies. However, the aluminium sheets in the Glare laminate result in specific ultrasonic characteristics of Glare, requiring adaptations of the existing procedures.

- *C-scan machine*
  A completely new and unique C-scan machine was designed and manufactured to scan the large Glare panels for the barrel.

15.2.2 Phasing of the Glare Megaliner Barrel Programme

In this section, the four phases in the development process are briefly discussed, see Figure 15.3.

Phase 1: Establish requirements

To start the sizing and design process, a set of requirements for the final product must be defined. Also, the means of compliance that may be used to show that the product complies with the requirements should be defined (e.g. testing or the use of certain analytical methods). In the ideal situation, all requirements are frozen before the start of the design phase. In this case however, requirement definition progressed with the design. The main reason for this is that Glare has some unique characteristics compared to monolithic aluminium, for which new requirements or means of compliance must be defined. Requirements that have been used in the Megaliner Barrel Programme are discussed in section 15.2.3.

Phase 2: Size and design Glare shells

The iterative sizing and design process used in the Megaliner Barrel Programme is extensively described in section 15.3.

Phase 3: Prepare for production

Production preparation included:

- tooling design and manufacturing: this includes moulds for the bonding process, tools for the layup process, frames and tools for the pre-treatment process of the thin aluminium sheets, tools for C-scanning, assembly and painting.
- generation of production control documents and instructions from the Engineering drawings.
- quality concept for Glare production. This includes procedures for the ultrasonic C-scans of the material.
-

Phase 4: Produce Glare shells

The following Glare parts and assemblies were produced, see Figure 15.5:

- skins including interlaminar doublers and longitudinal splices
- stringers
- stringer couplings
- butt-straps
- doublers for the door surrounding panels
- bonded assemblies: bonding of stringers and doublers to the Glare skin.

Figure 15.5 *Examples of Glare products: panel, butt strap, stringer and stringer coupling.*

15.2.3 Requirements for the Glare panels in the Megaliner Barrel Programme

In this section, the requirements used for the Glare panels in the Megaliner Barrel Programme are listed. This list is not intended to be complete, but highlights some requirements typical for Glare or requirements that have had a major impact on the design.

Stress requirements

Stress requirements are divided into three categories.

1. *Static strength*
   Requirements for ultimate and limit load conditions were specified, based on the FAR 25 requirements and company policies. A typical company policy requirement is a set of minimum reserve factors for all different failure criteria[1].
   Compared to monolithic aluminium, Glare has two unique characteristics with respect to static strength:

[1] The reserve factor is defined as the allowable stress level of the material/structure divided by the actual stress level in the construction. Reserve factors must always be larger than 1.0. "Target" values exceeding this minimum are usually defined to reduce the risk of expensive re-designs because of small increases in loads. Such a list was also defined for the barrel design.

1.1. The blunt notch behaviour of the laminate for different loading conditions (combined tension and shear loads) [5]. For this failure mode, a conservative approach had to be used since an analytical tool to predict laminate failure under arbitrary loading conditions had not yet been finalised.

1.2. The influence of temperature and humidity on the allowable stress levels in the laminates. The influence of this ageing behaviour was not taken into account in the barrel for two reasons:

1.2.1. The final "knock down" factors had not yet been established because of a lack of sufficient test data.

1.2.2. The barrel would not be tested at elevated temperatures or in a humid environment. The application of knock down factors may have resulted in too conservative a design which would not give valuable test results (the limits of the material are not tested).

2. *Damage tolerance*

Damage tolerance requirements are also based on the FAR 25 requirements and company policies.

- A typical example of a company policy requirement that is not part of the FAR requirements is the 2-bay crack criterion[2].
- A unique requirement for Glare was defined to exploit the excellent damage tolerance behaviour of the material: the Glare structure must be able to withstand ultimate load with damage occurring from cyclic loads simulating a complete life of the aircraft (taking the proper scatter factors for fatigue and crack growth into account). When this requirement can be met, operators may eliminate expensive inspection programmes that are intended to find and repair this fatigue damage.
- Cut-off stress levels for fatigue load cases were defined. Two examples that had significant impact on the final design are:
  1. Allowable stress level in the door corner for fatigue conditions. The allowable stress was substantiated by component tests on door corner specimens [2]. This criterion determined the maximum thickness in the door corners and was a weight driving criterion for door surrounding panels.
  2. Allowable stress levels for aluminium stringers under fatigue conditions. The combination of a Glare skin with aluminium stringers has the disadvantage that the relatively stiff stringer attracts more loads than the skin. Even when taking the Poisson ratio effect for the skin into account, this results roughly in a 5% higher fatigue stress level in the (fatigue critical) stringer compared to the Glare skin (which can sustain much higher fatigue stress levels). In areas where fatigue loads

[2] This criterion requires that the structure is able to withstand limit load with a cracked skin and a broken central stiffener (this can be a frame or a stringer). The crack extends in both skin bays next to the central broken stiffener.

are dominant, the average stress level in the skin + aluminium stringer must be reduced to a level such that it can eliminate all weight savings. For that reason, this criterion determines the location where Glare stringers must be applied to maintain approximately 15% weight savings.

3. *Deformations*
   Deformation requirements were defined for the door surrounding panels to ensure that the door would remain in its proper place under all (ultimate) load conditions and that the air seal would remain airtight under all (limit) load conditions.

Design requirements

There were many design requirements. For the barrel, these requirements resulted in a set of design principles according to which the detail design (drawings) were made. Three examples of design requirements affecting Glare are:

1. *The need for single clad aluminium sheets*
   A clad layer is required on the outside of the aircraft skin for protection against corrosion. Since the contact area of the aluminium sheets with the glass prepreg must be bare material, single-side clad material must be applied for the outer aluminium sheet. Not many manufacturers of thin aluminium sheets are willing to supply single-side clad material.

2. *Geometrical limitations due to standardisation*
   Parts such as stringers, frames and clips are often standardised to keep the parts count and associated cost low. An example of such standardisation is the fixed height of stringers. This height was based on aluminium extrusions and a design principle for a frame – stringer intersection for aluminium parts. Glare stringers have a slightly larger bend radius and subsequently less space to install fasteners in the upright flange. 4 mm more space would have given some more design freedom, see Figure 15.6.

3. *Tolerance requirements*
   - for the location of interlaminar doublers and splices:
     The location of thickness steps in the laminate determines the detail design of the stringers and sub-structure such as clips and frames that are directly attached to the skin. The tolerance in the location of the thickness steps must be limited to avoid interference of the sub-structure during the assembly process.
   - for the thickness of laminates:
     For the upper door-surround panel B3, special tooling was manufactured to keep the tolerances of the bonded doubler package within the required range.

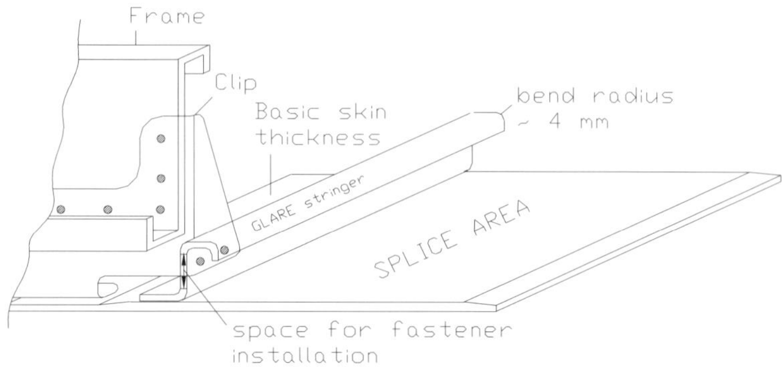

Figure 15.6 *Skin / stringer / frame attachment by clip.*

Production requirements

Production requirements are often defined to assure that the production process of the parts is as simple as possible, as a simple production process will result in a cost-effective product. This kind of "cost optimisation requirements" is further discussed in section 15.4. Some other production requirements are "hard" requirements, defined to assure that the product can be made (without incurring exceptional costs). Two examples of the last category are:

1. *No shimming during assembly*
   When mismatches occur during assembly processes, shimming is sometimes used to fill the gaps that exceed allowable values. Production processes that have large tolerances often require shimming. In the highly automated plants of Airbus such as the shell assembly plant in Nordenham, shimming disrupts the process and is therefore not allowed.

2. *Maximum dimensions of Glare panels and parts*
   Due to limitations of production resources, e.g. autoclave size, size of pre-treatment baths and riveting machines, there are maximum dimensions of Glare panels and parts.

15.3 The iterative sizing and design process

Based on the stress requirements, a sizing tool was developed to optimise the (weight of) the Glare structure in the barrel. After the iterative sizing process, design principles were made and finally the 3D models and production drawings were prepared. The process is shown in Figure 15.7.

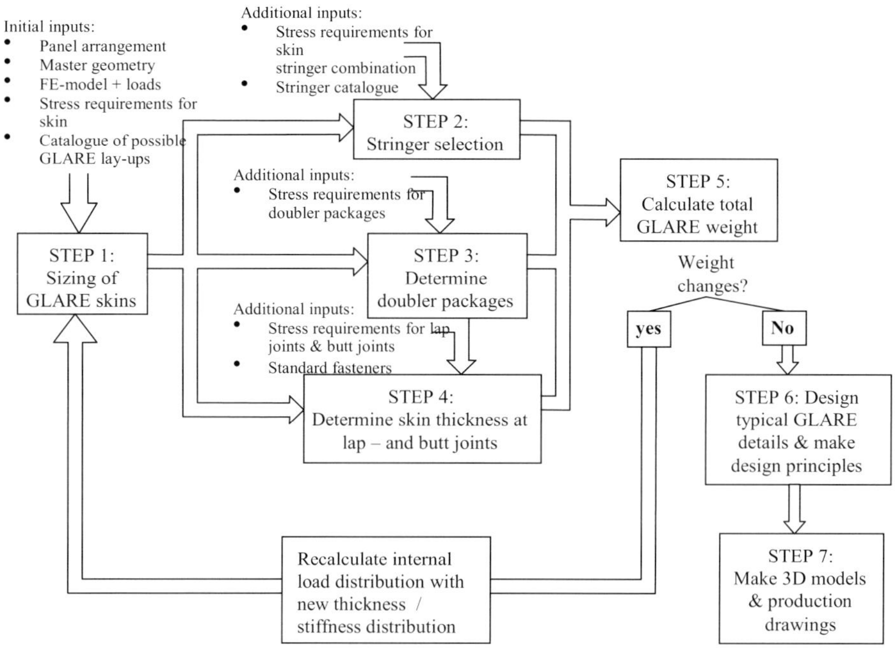

Figure 15.7 The iterative sizing and design process for Glare shells.

Step 1: Sizing of the basic skin thickness and selection of a Glare type and layup

Input:

- Panel arrangement
  The panel arrangement and material distribution as described in section 15.1 was chosen to obtain a maximum of information for both Glare and the 2524 aluminium alloy from the test. All main structural items such as door-surround and window panels were tested for both Glare and 2524. A disadvantage of the asymmetric material distribution is that the stiffness distribution is also asymmetric, shifting loads from the Glare panels to the stiffer aluminium structure.
- Master geometry
  The master geometry consists of the outer loft, stringer pitches and frame pitches. These parameters were frozen during the preliminary design phase on the basis of marketing requirements and parametric weight studies.

- FE-model + loads
  The external loads (static and fatigue) were applied to the FE-model. An initial thickness distribution for the structure was used to calculate the first internal load distribution.
- Stress requirements for skins
  This topic has been briefly discussed in section 15.2.3. Approximately 20 stress criteria were used to size the Glare skin panels. In some cases, preliminary material allowables have been used.
- Catalogue of Glare layups
  This catalogue consists of the Glare laminate layups that can be used. Availability of aluminium sheets and standardisation were the criteria by which to compose the list.

Output:

- Selected Glare type per panel (Glare 4A, 4B or 3) based on minimum weight criterion: for each Glare type, the minimum achievable skin thickness and corresponding weight was calculated. This resulted in a selected layup per panel including locations of interlaminar doublers. An example of such a panel description is given in Figure 15.8.

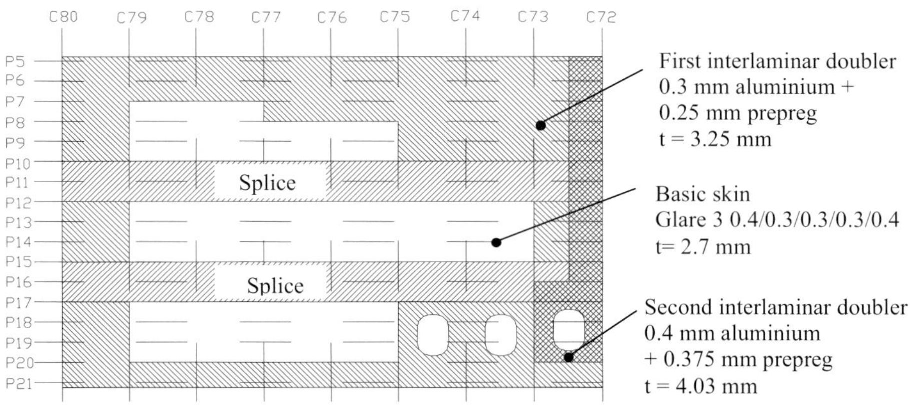

Stringers:

- All stringers are bonded except p5
- Stringer p5 to p10 are Glare
- Stringers p 11 to p 21 are aluminium

Figure 15.8 Glare type and layup for panel D2 (lower side panel).

Step 2: Selection of stringers and fastening process (bonding or riveting)

Input:

- Stringer catalogue
  Given the skin layup, an aluminium or Glare stringer was selected from a list of standardised stringers (z-shape).
- Additional stress criteria for the stringer / skin combination
  The main stress criteria for the skin / stringer combination were overall panel buckling under compression loads (Euler-Johnson), fatigue cut-off stresses for the stringer and residual strength for a cracked skin with a broken central frame (2-bay crack criterion). The lightest stringer that just fulfils all the stress criteria was selected. In the top panels (A1, A2, A3, D1 and upper part of D2) Glare stringers had to be applied to allow for the high fatigue stress levels in these panels[3].

Output:

- Selected stringers and fastening process (bonding or riveting) for all panels. Wherever possible, the bonding process was selected to attach the stringer to the Glare skin[4]. For some stringers running across doublers near the door cut-out, the bonding process could not be applied.

Step 3: Determining doubler packages at door corners

Input:

- Stress requirements for doubler packages
  The sizing stress criterion for the door corner thickness and corresponding doubler package around the door was the allowable stress level for fatigue load cases. Component fatigue tests were performed to validate the allowable. An important result from these fatigue tests was that titanium doublers are not required in case of a Glare doubler package as opposed to the design solution for the corresponding aluminium door surrounding panels. This is a major cost reduction factor for Glare compared to aluminium. Other criteria used for the door doubler package were:
  - local static failure under compression loads (based on tests)
  - tensile failure at ultimate load and yield at limit load
  - allowable fatigue stress level at edges of bonded doublers, see Figure 15.4
  - deformations of the door surrounding panels: these deformations may not lead to air leakage at limit load or dislocation of the door at ultimate load.

---

[3] The application of Glare stringers has been kept to a minimum as the manufacturing process for these parts is still under development

[4] The bonding of stringers is a relatively inexpensive process and the behaviour under compressive loads is slightly better than for riveted stringers. The expensive mould for the bonding process was already required for the Glare skins and therefore no additional investments were required to bond the stringers to the skin.

Output:

- Definition of Glare doublers around the door cut-out. This includes:
  - the layups of each doubler (Glare type and layup)
  - the orientation of the fibres
- fastening process of the doubler to the skin. To test both the bonded concept and the riveted concept, the lower door surrounding panel (B9) was riveted to the skin and the upper door surrounding panel (B3) was bonded to the skin, see Figure 15.9.

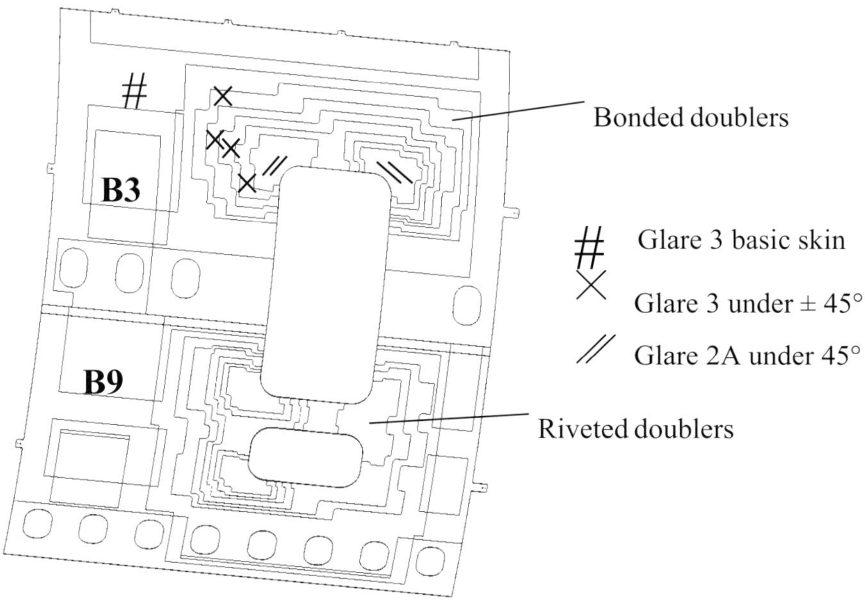

Figure 15.9 Glare panels B3 and B9.

Step 4: Determining skin thickness at lap and butt joints

Input:

- Layup of the basic skin at the joint locations
- list of standard fasteners to be used
- allowables for the applicable fasteners in Glare laminates
- stress requirements for joint sizing are:
  - the joints should be able to sustain ultimate load after a complete fatigue life (including an appropriate scatter factor). This requirement resulted in cut-off stresses for fatigue loads, based on coupon tests of joints.
  - the joints must not yield at limit load or fail at ultimate load.

Output:

- Skin thicknesses at lap joints and butt joints
- selected fastener for each joint (type and diameter).

Step 5: Calculation of weight of Glare panels and repetition of steps 1 to 4 when not converged

Given the output from steps 1 to 4, the minimum weight of each Glare panel was calculated. The difference compared to the previous sizing loop was a measure for the convergence of the process. In the case that process had not converged to the optimum solution, the FE-model was re-run using the new thickness and stiffness distribution from steps 1 to 4. This resulted in a slightly different internal load distribution and steps 1 to 4 had to be repeated.

Step 6: Design of typical Glare details and definition of design principles

Input:

- Panel layout definition from the sizing process (steps 1 to 4)
- design principles of the sub-structure (clips, frames etc.)
- additional design requirements for door doubler packages (e.g. reparability of the doubler package: space to install two rows of fasteners).

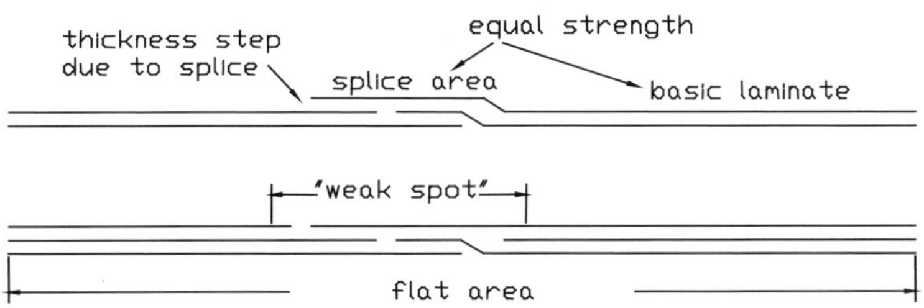

Figure 15.10 *"Weak spot" in a flat spliced Glare laminate.*

Output:

After convergence of steps 1-5, the Glare details could be designed. These included

- the exact location of splices and interlaminar doublers.
- a detailed layer by layer composition of the laminate. Special attention had to be paid to the requirement of a flat area under bonded doublers. If the doublers around the door cut-out were bonded then the mating surfaces had to be completely flat (to assure a good quality of the bond layer). If a splice was located underneath one of the doubler packages then the interruption of the aluminium sheets would create a "weak spot" in the laminate, which had to be located in a low stress area, see Figure 15.10.
- exact contours of the doubler packages in the door surrounding panels.
- concepts for the Glare stringer couplings and butt-straps. At the intersection of lap-joints and butt-joints, a geometrically complicated connection must be designed. The available space to install fasteners has been the limiting factor.

Especially for Glare stringers and stringer couplings, the available space to install fasteners is minimal due to the minimum achievable bend radius of approximately 4 mm (compared to 2 mm for extrusions or milled parts).

Per panel and butt-strap, a design principle was made that included all the information from steps 1 to 4 and the Glare design details. This information was the basis for the production drawings.

Step 7: Make 3D models and production drawings

Input:
Design principles for each panel and butt-strap from step 6.

Output:
Complete product definition dossier. This includes:
- Geometrical information of the products
- bill of material
- reference to the applicable production processes
- reference to the applicable inspection procedures.

15.4 Design for manufacturing for Glare

The term "design for manufacturing" is used to express the importance of taking the limitations and opportunities with respect to production into consideration during the design phase. For Glare this was a major issue during the development process as the cost of the final Glare components is largely influenced by the design and selected manufacturing concept.

The following design guidelines for Glare can be used:
- Preferably use Glare 3 instead of Glare 4.
  Glare 3 contains less glass prepreg, which is one of the more expensive components of a Glare panel. If the weight of a Glare 4A or 4B panel is only 1 or 2% less then Glare 3 is preferable.
- Apply the bonding process as much as possible.
  Bonding of stringers is a relatively inexpensive process and the behaviour under compressive loads is slightly better than for riveted stringers. The expensive mould for the bonding process is already required for the Glare skins and therefore no additional investments are required to bond the stringers to the skin.
- Minimise the number of thickness steps in the laminate.

Thickness steps in the basic skin often require thickness steps in the sub-structure (e.g. stringer joggles). For this reason, Glare 3 improved[5] should only be used if a significant weight saving can be achieved.

- Avoid the application of Glare stringers and stringer couplings.
  Glare stringer and stringer couplings are complicated parts, built-up of several break-formed Glare profiles and relatively expensive to manufacture. Glare stringers should only be used when significant weight savings can be achieved (see also section 15.2.3; stress requirements; damage tolerance).
- Minimise the number of unique aluminium sheets.
  To minimise the parts count, the large number of aluminium sheets should contain as many identical sheets as possible.
- Minimise the number of splices.
  The introduction of a splice not only creates additional weight, it also increases manufacturing costs (parts count and layup effort increase).
- Avoid the application of single side clad aluminium sheets.
  Aluminium sheets with a single-side clad layer are significantly more expensive than bare material and increase the cost of logistical control.
- Avoid the use of titanium doublers in door surrounding panels.
  Titanium is much more difficult to machine or drill than Glare or aluminium. Especially when applied in a package with softer material, special tools must be used to assure a good quality of the holes.

Given an optimal design, the selected manufacturing concept to produce the Glare parts also largely influences the final product cost. Examples of cost reduction in manufacturing are:

- Make the Glare product in the mould instead of using flat sheets.
  In the past, curved Glare panels were manufactured from flat plates by a rolling process. Performing the layup process in the curved mould eliminates a production step. An additional advantage is that the mould can also be used to bond the stringers in a second shot. In the near future the feasibility of a one shot curing process will be investigated (simultaneous curing of the Glare skin and bonded stringers).
- Allow overlaps in prepreg.
  The accuracy with which prepreg must be positioned on the mould is a parameter for the layup time and cost. If prepregs may overlap, layup time can be considerably reduced and the chance of panel rejection at the ultrasonic quality check is less. Overlapping results in small thickness steps in the final laminate of approximately 0.1 mm.
- Try to avoid hard tooling for bonding.

---

[5] Glare 3 improved has the same basic layup as Glare 3 with additional bands of glass fibres, located underneath each frame to act as built-in crack stoppers. This results in thickness steps in the skin at each frame location, requiring joggling of all stringers at each frame location.

The hard tooling concept as developed for the door surrounding panels in the barrel results in excellent tolerance bands for the thickness, but is time consuming during layup of the doublers. If it can be shown that a soft tooling concept can also be used, this is to be preferred.

- Minimise the C-scan inspection effort.
  The inspection of large Glare panels using ultrasonic techniques is a time consuming process. When SPC (Statistical Process Control) can be used to eliminate part of the inspection effort, this can save a substantial part of this cost.

References

[1] J. de Kok, *Manufacturing schedule for Glare sheet*, FAE/EMP/99-12.002. Fokker Aerostructures, Papendrecht, Netherlands, 2001.

[2] V.P. Brügemann, J.L.C.G de Kanter, *Fatigue allowables for the door corners, B2V-00-54*. Delft, Netherlands: Delft University of Technology, 2000.

[3] G. Roebroeks and E. Kroon, *Glare stringer manufacturing concept*. Delft, Netherlands: Delft University of Technology, 1998.

[4] G. Roebroeks, *Optimum splice geometry for large fuselage*. Delft, Netherlands: Delft University of Technology, 2000.

[5] T.J. de Vries: *Blunt and sharp notch behaviour of Glare laminates*. Delft, Netherlands: Delft University of Technology, 2001.

16

Cut-outs; door surrounding

J.L.C.G. de Kanter

Abstract – Cut-outs are highly fatigue sensitive due to the large stress concentrations. In aircraft fuselages these cut-outs are quite large in the case of the windows and doors. Much effort is usually needed to limit the stresses and displacements around these cut-outs. Displacements are mainly controlled by edge members and stresses are controlled by doubler packages. The stress level may be increased through the application of Glare in the doubler packages, due to the improved fatigue behaviour compared to conventional aluminium. Glare also presents the possibility of tailoring the material to the load, i.e. fibres aligned with the load. For the door corners this may result in the use of Glare with a 45 degrees orientation. FE analysis defined the total doubler package and a test programme was run to confirm the behaviour of the material and to predict the crack behaviour of the Glare door corner.

16.1 Introduction

Large cut-outs exist in aircraft fuselages for doors, see Figure 16.1, and windows. These cut-outs are highly fatigue sensitive due to the large stress concentrations and in the case of doors also due to in-service accidental damage. Crack initiation and crack growth must be limited for reasons of safety and (repair) cost. This fatigue control is carried out by keeping stress levels below carefully defined maximum values. The use of Glare presents us with the possibility of increasing the stress level compared to conventional aluminium alloys, reducing the weight of the structure.

Reduction of stresses around cut-outs is realised by means of doublers and reinforcing members along the edges. The doublers are built-up packages attached to the basic skin. The edge reinforcements constitute of edge frames plus intercostals, stringers and a door or window pan. In the case of a door, the pan is used for sealing and protection against accidental damage only. Window-pans are more structural, load-carrying members. The edge reinforcements compensate for the removed material and redistribute the loads.

Loading of the windows is dominated by pressure and shear and is critical in shear buckling and fatigue. The door loading is dependent on the type of door used, i.e. plugged or non-plugged and stressed versus non-stressed. Passenger doors (and windows) are generally non-stressed and plugged. Such a door is only loaded by pressurisation.
Typical of the design of doors and windows is that their location is frequently modified, not only in the design process but for the different aircraft types as well, e.g. freighter versions and as a result of specific carrier wishes. This requires a flexible design process.

The subject of this chapter will be the door corner doubler design, but many aspects will be the same for the window corners, in particular for the doubler packages surrounding the cut-out area. The chapter remains at a very global level, to be able to show the general actions taken to come to a reliable door corner and to show the typical aspects of the application of Glare.

The structure of the chapter is as follows: first an overview is given of the door structure, consisting of sills, frames, intercostals and doublers. This overview is extended with typical issues concerning the application within large aircraft and the application of alternative materials such as titanium and Glare. The third section describes the numerical modelling of the fuselage and door corner in particular. Tests were performed, which are discussed in section four. Both design and test come together in section five, which validates the design and material. Finally the conclusions will be given.

Figure 16.1 *Fuselage door.*

16.2 Door surrounding, effect of fuselage size and materials

The door surrounding consists of edge reinforcements, see Figure 16.2, and doublers. Doublers are built-up packages attached to the basic skin. Optimal doublers are located on both sides of the skin, but due to aerodynamics and aesthetics in fuselage design, most doublers are only located on the inside instead. This shifts the neutral axis and creates additional stress concentrations due to secondary bending. An exception to this is repair, where repair patches are attached to the outside.
Different production techniques are used for the doublers. Chemical milling is one of the possibilities to create optimal doubler patterns, though at the cost of bad crack stopping capabilities due to integral structure characteristics, i.e. once a crack has initiated in the material, there is no inherent crack stopping capability. Composed packages, which are bonded or riveted, are better. The material boundaries create natural crack boundaries, preventing further crack growth. Bonding is preferred since rivets present crack-initiation sites. In the case of Glare the use of bonded doubler packages is most logical. The laminated character of Glare also enables the doublers to be inter-laminarily installed, i.e. between the layers of the basic skin laminate.

The edge frames are reinforced frames, which are stabilised by intercostals attached to the surrounding frame (auxiliary frame). Stabilisation is necessary to prevent torsion of the edge frames due to loading imposed by the door stops. Generally the area between the edge and consequent frame is closed to create a torsion-box, with the intercostals as ribs between the frames.
The sills are compensation for the material removed by the cut-out in the axial direction. Like the frames, the sills are generally made as box beams. This provides torsional stiffness, limiting additional stresses due to the presence of the cut-out. The sills are not loaded by door stops or latches.

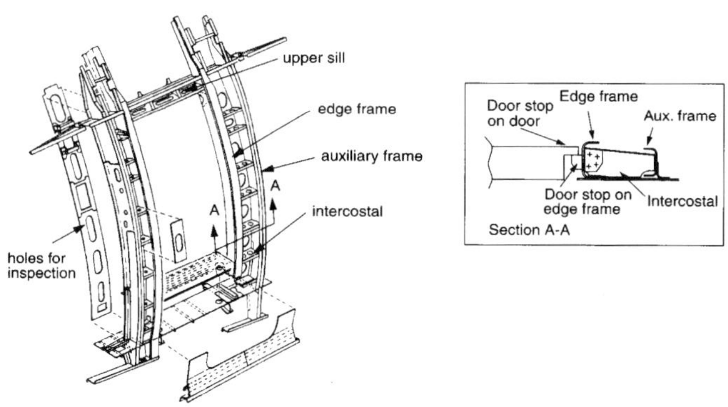

Figure 16.2 Door surrounding structure [1].

The door treated as an example in this chapter is one between the wing and the tail of the aircraft, which is the highest loaded part of the fuselage. For large aircraft the size of the door itself is not different from existing doors, though the surrounding and loading are. As the skin thickness does not grow in the same rate as the radius of the fuselage, the fuselage becomes more flexible and more prone to eccentricities and changes in stiffness. For example, to reach the same design stress for the same internal pressure, the thickness t must be proportional to the radius of the fuselage R. However, the displacement of a cylindrical pressure vessel is proportional to R^2/t. Therefore the displacement increases with the fuselage radius. Moreover, the lower stiffness of Glare compared to monolithic aluminium causes more flexibility in the fuselage. The displacements are as essential as the stresses in the door corner, as the displacements must be limited to prevent door movement or possible relief from the door stops. As the door must remain equally stiff, the local door surrounding becomes more stressed due to the size of the aircraft.

The displacement and stress levels are design allowables, the doubler thickness is a design variable and a boundary condition can be set to this variable as the maximum doubler thickness. This boundary condition can make it necessary to introduce a different doubler material, such as titanium with its high stiffness and strength per volume, although the application of a different material is not advantageous with regard to production and maintenance costs.

Focusing on the doublers and the use of Glare, the first thing to observe is the stress location and orientation at the corner. The highest stress occurs approximately in the middle of the corner at an angle of 45 degrees. The basic Glare skin material has the lowest strength and stiffness characteristics in that direction, making it quite unsuitable for the corner. However, because doublers have to be added, there is also an advantage in the anisotropy of the material, as the fibres can also be oriented along the line of highest stress, i.e. under 45 degrees. This adds substantial stiffness in that direction and thus reduces strain and stress. It also includes the crack bridging capacity since the fibres will be located perpendicular to any possible crack. Nevertheless, it may be possible that stresses remain too high locally and that the application of a stiffer material such as titanium is required. It is important to note that the use of doublers is very local and that they have very little influence when displacements need to be controlled, due to their local character. As the stresses have a very steep gradient at cut-outs, they are very local and well suited to being relieved by doublers.

16.3 Numerical modelling and analysis of fuselage and door surrounding

The door corner design is performed in roughly three steps of Finite Element (FE) model refinements. The first is a global model of the fuselage, the second a coarse model with enough detail for doubler packages and the third is a refinement of the door corner for accurate fatigue stress analysis. These three steps will be discussed here. Note that the design of the door surrounding is limited here to the skin and doublers, which are made of Glare.

16.3.1 Global fuselage model

The global fuselage model is used to investigate the different load cases and the response of the fuselage to them. As the model represents the whole fuselage, restrictions are set on the element sizes. The result is that the model has elements with the size of stringer and frame pitches, and some very rough elements around cut-outs and other stress concentrations. For the door corner itself this results in a coarse mesh with triangular shell elements, which is considered inadequate for realistic stress calculations at the edge. However, due to experience with previous designs and the current goal, the model is acceptable. In the case of the door corner the model yields a definition of the severest load cases for the door corner, a fatigue load case and some limit and ultimate load cases.

It is questionable whether a linear analysis suffices for the fuselage, or that a non-linear analysis is necessary due to the fuselage size and therefore flexibility. The observations made from a non-linear analysis are [2]:

- the difference in overall deformation is negligibly smaller in non-linear analysis compared to linear analysis.
- deformations at maximum displacement locations are significantly smaller (up to 15%) compared to linear analysis.
- the mentioned locations have lower stress levels compared to linear analysis.

The reason for this outcome is the redistribution of stress after deformation and the stiffness increase, such as a more effective membrane function of the skin after deformation. The substantial deviations are found at eccentricities and stress concentrations, such as door cut-outs, which are still very roughly modelled and will not be used for detail. It can be concluded from the comparison of non-linear and linear analysis that a linear calculation is conservative, although acceptable for this purpose.

16.3.2 Coarse model, doubler package design

In general, the doubler packages, see Figure 16.3, are designed over an area of two frame bays next to the cut-out and six stringer pitches above and below the cut-out. The basic skin material in this case is Glare 3 in its original orientation, i.e. fibres in longitudinal and circumferential directions of the fuselage. However, the doubler packages can have any orientation, though a more refined orientation than in discrete steps of 45 degrees is considered to be too complex and impractical.

To limit the stress at the corner, two means can be recognised: 1. adding material at that location and 2. transferring the load from the corner. The load transfer from the corner is basically impossible because the load will always pass the cut-out as closely as possible. Different cases have been studied of doubler layouts, but all show the necessity of the thickest doubler at the corner edge, diminishing over the given area to the basic skin. The doubler geometry will not vary much and is only a matter of optimisation to level all stresses below the allowables, which are set.

Under 0/90 degrees the Glare part will be critical in shear (due to the off-axis strength of Glare), though the application of doublers under 45 degrees raises the strength significantly and makes the corner fatigue critical, as corner packages in monolithic aluminium structures generally are. Glare 2, which has fibres in only one direction, can be used under 45 degrees, and the special Glare variant designated here as 'Glare 3S' can also be used, which has an additional layer of fibres under ±45 degrees.

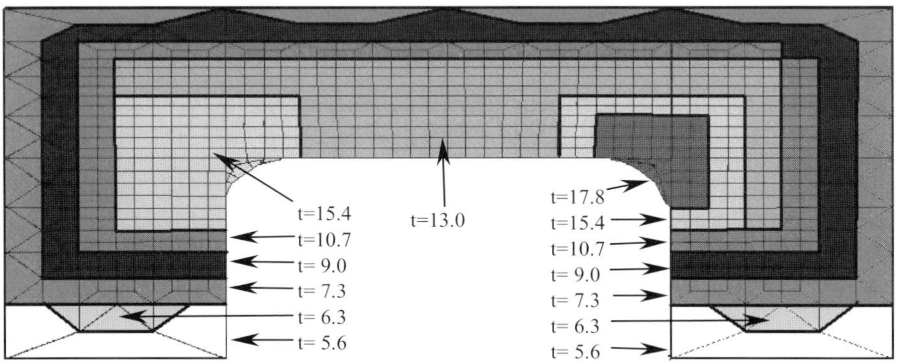

Figure 16.3 Doubler package layout.

16.3.3 Details in doubler design

Details to be considered during the design of the doublers are the eccentricities at the doubler edges, since all doublers are on one side (internal). Secondary bending is caused due to the shift of the neutral axis. The bending can result in a stress increase of 60% close to the thickness step (if the doubler has an equal thickness as the basic material), though it diminishes with increasing force and distance [1]. The result is that a linear analysis will overestimate the stresses and will produce conservative results. For the door corner itself this issue is not significant, because no thickness step is present there, and the pillowing/stretching effect is limited by the thickness of the laminate and the stiffness of the edge members.

A non-linear case study indeed showed some stress relief at the highly stressed areas and a stress increase at the low stressed areas, keeping the strain energy constant. This effect is only interesting for ultimate load cases and not for the fatigue load case that determines the corner doubler geometry.

A second detail issue is the stress gradient at the door corner. In the radial direction this gradient is extremely high and for accurate fatigue stress analysis a large number of elements is required in a FE analysis. A study by Heerschap [1] shows an estimated minimum of twelve elements in radial directions. The stress field retrieved thus enables an accurate prediction of fatigue life.

16.4 Physical Glare door corner testing

The Glare fuselage door surrounding is designed to control a stress of around 180 MPa on the cut-out edge in a case of fatigue loading, for example. This fatigue stress was validated by fatigue tests on coupons. Figure 16.4 gives a representation of the coupon geometry and location at a door cut-out.

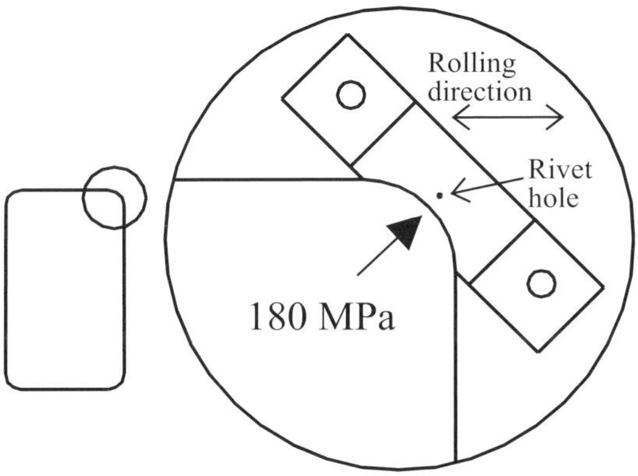

Figure 16.4 Coupon definition.

16.4.1 Test specimen

Coupon geometry and material

The coupon geometry represents a fuselage door cut-out. The coupons are not a complete doubler package, but distinct laminates of Glare 2, Glare 3 and Glare 3S, see Figure 16.5. These separate laminates are used to be able to determine the separate laminates' behaviour in the complete doubler package. For the worst case scenario a rivet hole is drilled in the coupon near the cut-out, see Figure 16.4, which may occur after repair.

Support

A well-defined load introduction was also defined for the specimen. Two options were tested, see Figure 16.6, one simulating fully clamped support, i.e. the edge loaded by five pin-loaded holes, and one with one pin-loaded hole that is free to rotate. The advantage of the single loaded hole is that by redesigning the location of the hole the stress gradient in the middle of the specimen can easily be changed and optimised to achieve a realistic situation. Another advantage, which was proven during testing, was that the 5-pin-support could introduce secondary (out of plane) bending in the specimen, where the single hole would not.

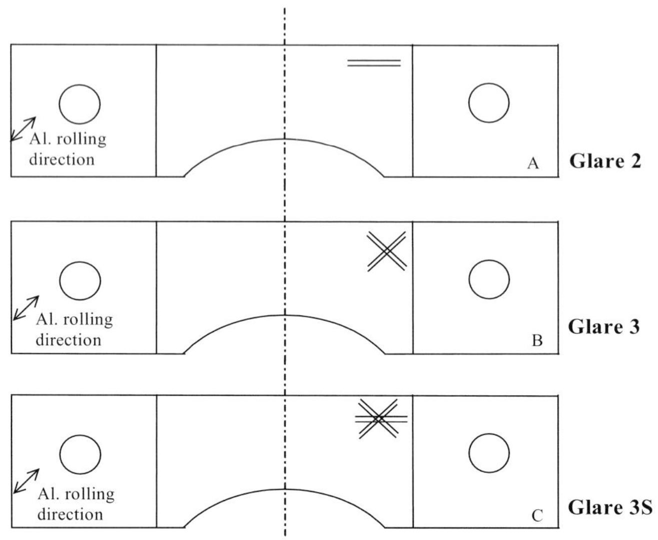

Figure 16.5 *Material and orientation definition of the coupons.*

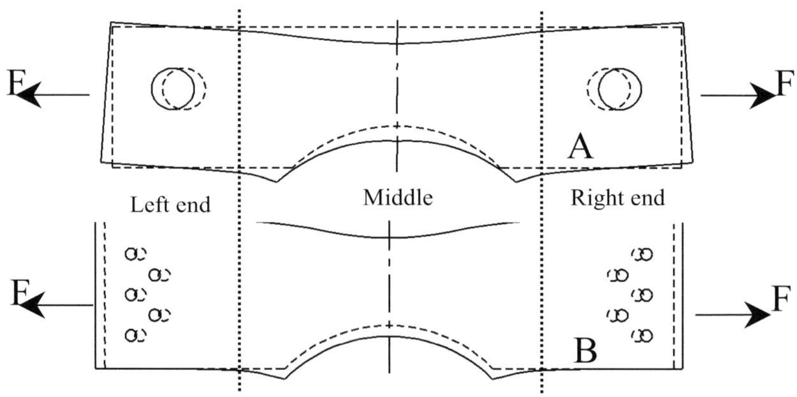

Figure 16.6 *Load introduction definition, rotating and clamped.*

16.4.2 Test

Measurements

The coupons had strain gauges attached along the middle of the specimen and along the corner, to give a total picture of the stresses (and stress gradients) along the cut-out and in the radial direction. Strain gauges were also used to determine the load that was necessary to reach the maximum fatigue stress of 180 MPa at the critical location.

Cracks were measured on both sides of the coupon, one side with a digital camera with an accuracy of 0.01 mm and other side with a microscope with an accuracy of 0.1 mm, see Figure 16.7. The cracks were measured every 5,000 cycles.

Test

The fatigue tests were performed with Constant Amplitude loading (R=0.1) and a maximum number of 180,000 cycles. The maximum load was defined for each specimen after strain measurement, which should cause the defined edge stress of 180 MPa. Different loads for the different laminates and different supports were obtained.

16.4.3 Test results

The main results are the crack growth curves, see Figure 16.8, [3]. Crack initiation occurred between 55 and 65 kilocycles for all materials. The first detectable crack, defined as a=5 mm, occurred at 110 kilocycles for Glare 3 and 3S. The crack in the Glare 2 laminate remained below the detectable length for its whole life of 180 kilocycles.

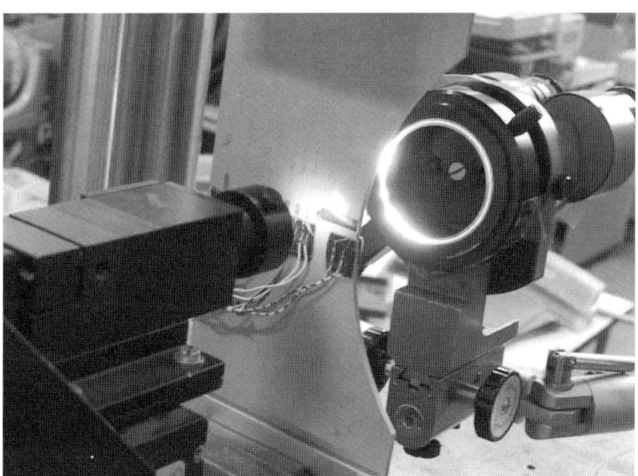

Figure 16.7 Test setup with camera and microscope.

Crack growth was stable for Glare 3S and Glare 2. A constant da/dN was found. The Glare 3 specimen had unstable behaviour with an increasing da/dN after approximately 110 kilocycles. This showed the limited strength of Glare 3 under 45 degrees loading. Glare 3 was the only laminate that failed within the 180 kilocycles (155 kilocycles).
In all cases, the cracks initiated at the rivet hole in the specimen. Not only at the corner side, but also at the straight edge side of this hole. This is the result of the local stress concentration from the rivet hole upon the global stress concentration from the corner cut-out. However, the crack at the straight edge side did not grow significantly and remained below a length of 5 mm. It was clear that the highest loaded area was near the corner radius. Only after failure of the material between the corner and the rivet hole did the crack at the straight edge begin to grow further.
The difference between the clamped and rotating support was small. A significant life reduction was only found for the Glare 3 specimen (40 kilocycles).

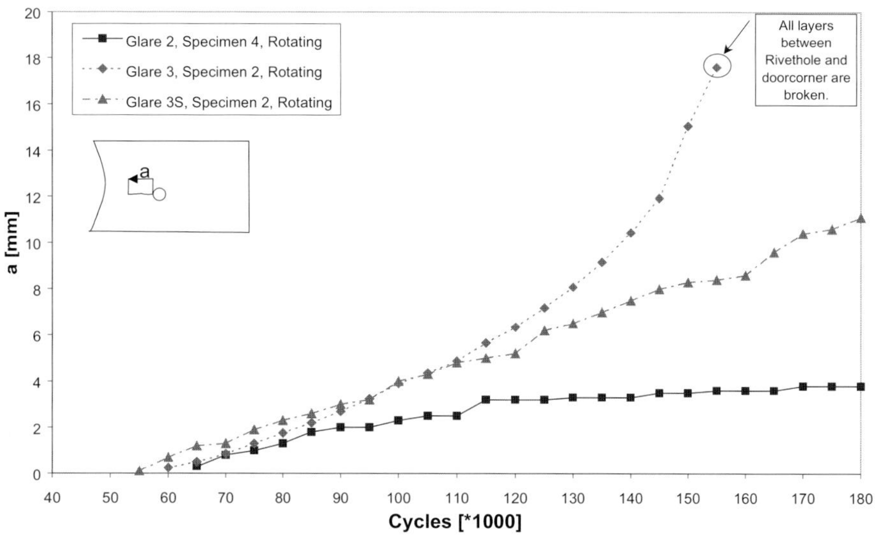

Figure 16.8 *Crack growth curves coupon test.*

16.5 Door corner design verification

The doubler package is designed to hold a stress level of 180MPa, and the tests confirm that this is possible. However, the edge stress is not the only important factor. Once the stress gradient differs too much from reality, the average stress is higher than it should be and a different crack behaviour will be the result. Secondly, crack initiation occurs at the rivet hole in the tested situation, making the stress at that point more important than at the corner edge. This requires a more refined FE analysis of the corner radius to control the stress gradient. Results of the coarse model and the test specimen show reasonable effects, the coupon test specimen

exhibits a somewhat lower gradient, resulting in higher stresses over the examined area as the edge stress is maintained at 180 MPa. Therefore tests give reliable, perhaps conservative, fatigue behaviour results.
As mentioned, for a more accurate prediction it is necessary to refine the corner in the FE fuselage model and to have an accurate model of the coupon test (to be able to redesign the test specimen towards an exact representation of the corner or to be able to translate the coupon test to the real life situation for different door corner radii and laminates).

16.6 Conclusions

Increased size and use of lower stiffness material results in an increased flexibility of the fuselage and impacts the door corner design, which requires equal stiffness to prevent door movement.
The stress level around the cut-out is limited by doubler packages. Through the application of Glare it is possible to increase the stress level or increase the fatigue life of the door corner, as has been demonstrated by tests. The doubler packages are best bonded, although repair with rivets has to be taken into account. As can be seen from the tests, this reduces the allowable stress level or fatigue life considerably, as the crack will initiate and grow from the rivet hole.
The tailoring of Glare towards the load enhances the fatigue life. The Glare 2 and 3S specimen that have fibres in the loading direction have stable crack growth and pass the 180 kilocycles design life, whereas Glare 3 fails at 155 kilocycles.
For a more accurate prediction of the fatigue life it will be necessary to refine the FE-models to describe the stress gradient with sufficient accuracy at the corner. This is especially important because the stress at the rivet hole rather than the stress at the corner edge determines the fatigue life. It would be useful to have a good representation of the coupon specimen and the fuselage corner section to translate the test results into a faster prediction for the real fuselage.

References

[1] M.E. Heerschap, *Design of Cut-outs in Shell Structures*, PhD-thesis, Delft University of Technology, 1997.

[2] A.G. Reis, *Titanium Doublers, Non-linear analysis and new concept for stress control in the door surrounding area*, Master Thesis, Delft University of Technology, 2000.

[3] V.P. Brügemann and J.L.C.G. de Kanter, *Fatigue allowable for the door corner*, B2V-00-54, Delft University of Technology, 2000.

17

Detailed design concepts

O.C. van der Jagt and B.C.L. Out

Abstract - This chapter handles some aspects of the detailed design of aircraft structures in Glare, and the design of splices, internal doublers and riveted joints in particular. In order to apply Glare in very large fuselage panels, a splice concept was developed, which allows a number of longitudinal splices to be cured in the same curing cycle as the basic material. Through the introduction of this splicing concept, the width of a panel is no longer limited to the maximum width of the aluminium sheet. In addition, internal local reinforcements (doublers) can be integrated into the panel during layup. A brief review of the development from splice concept to splice design is given. The use of doublers is discussed. The chapter concludes with a discussion on the design of riveted joints in Glare.

17.1 Introduction

The design process of a structure in general can roughly be divided in three steps, see Figure 17.1. First the conditions a structure is exposed to must be determined. Based on this list of requirements, a global or conceptual design is made, which provides the initial specifications of a structure, and in which design problems can be identified. In some cases several design concepts are developed simultaneously, and at some point in the design process, a trade-off is made. The next step is the detailed design, in which the previously mentioned design problems are solved, and the results are used to make a new global design incorporating the design details. This is an iterative optimisation process and is repeated until the specifications of the design meet the requirements. Eventually a final design is drawn up. Of course this is a tremendous simplification. In reality the design process is much more complex, especially for an aircraft structure.

Since the detailed design is strongly dependent on the specific structure, it is impossible to cover all detailed design problems that a designer may encounter when designing an aircraft structure in Glare. Therefore three detailed design problems typical for Glare are addressed and design directives given.

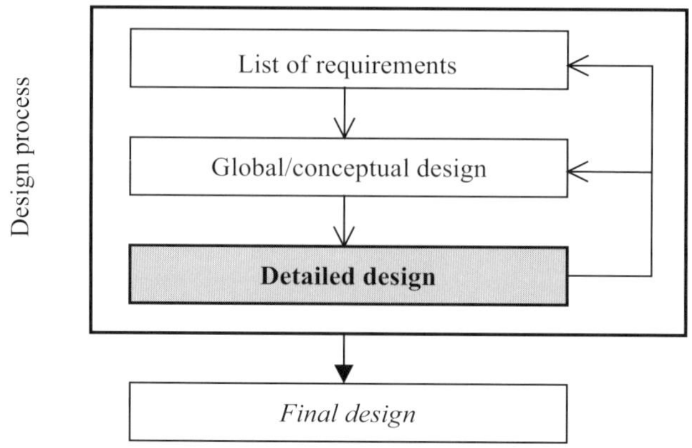

Figure 17.1 Schematic flow chart of a design process.

The first problem is how to produce large fuselage skin Glare panels efficiently. A possible solution is to rivet or bond separate small Glare panels together. However, Glare offers a unique way of creating large Glare panels by means of splices.
The second problem is how to handle local stress concentrations. In general these stress concentrations can be dealt with by locally increasing the thickness of the material (when using this solution care must be taken not to increase the problem. A local reinforcement also increases stiffness locally, and may thus attract *extra* loads). In this case Glare also offers a unique way of increasing the thickness locally by using internal doublers.
The third problem is designing riveted joints (common in almost any aircraft structure) in Glare. An approach is offered for designing a riveted joint.
In section 17.2 the splice concept is introduced and explained. In section 17.3 the internal doubler concept is treated and in section 17.4 riveted joints are considered. Finally, this chapter closes with some concluding remarks.

17.2 Splice concept

17.2.1 Introduction

The dimensions of a Glare fuselage skin panel (consisting of alternating aluminium and prepreg layers) are limited by the width of the aluminium layers. The length of the aluminium layer is not a limiting factor. The available prepreg dimensions are also not limiting. The next critical items limiting the maximum dimensions of a Glare skin panel are the dimensions of the pretreatment facility, the autoclave curing facility and C-scan facility. Another critical item may be the handling and transportation of the panel, depending on the tooling available.
Currently only aluminium sheets with widths up to 1524 mm (60 inch) can be manufactured to the required accuracy with the necessary nominal thicknesses between 0.3 and 0.5 mm. This limitation would imply the necessity of applying

many costly mechanical longitudinal or circumferential joints (at a maximum distance of 1524 mm) in an aircraft fuselage. To avoid this disadvantage, the internal splicing concept has been developed.
The splice concept basically consists of butted aluminium sheets. To assure proper load transfer (the aluminium layers are interrupted and thus not capable of transferring loads) a doubler (extra layer of aluminium) is necessary, see Figure 17.2. In the remaining part of this chapter the word 'splice' also refers to the splice concept as a whole.

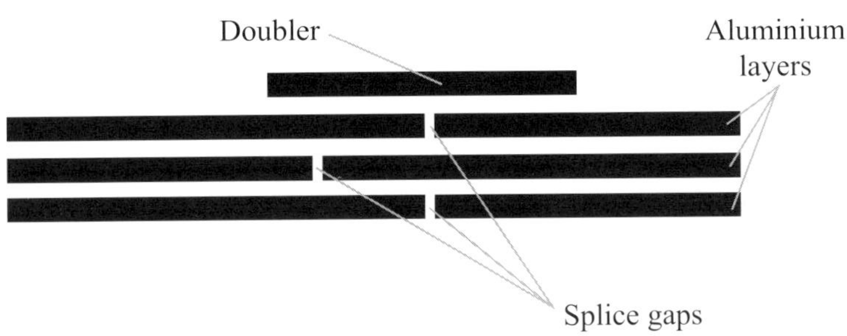

Figure 17.2 Splice concept.

17.2.2 Splice design

The design of a splice will be illustrated by looking at the development of the splice concept in general. In order to prove the applicability of spliced Glare it was necessary to design and test spliced Glare panels extensively [1].

Of course it is necessary to have requirements before a design can be produced. The requirements are given in a qualitative way:

- maintain all the advantages of standard (unspliced) Glare
- minimum interference between the details of the splice with the location of stringers, frames, doublers and mechanical joints
- maximum strength and durability behaviour of the splice
- minimum cost of the production of a spliced panel
- minimum weight
- maximum sheet smoothness at the outside fuselage skin
- maintain operational advantages of a monolithic aluminium skin, e.g. alclad outer surface of the skin and the possibility of paint removal

Glare splice with doubler using prepreg as adhesive

The development of the splice concept into a (optimised) splice design started with the implementation of a local reinforcement (doubler) in order to compensate the strength reduction due to the interrupted (spliced) aluminium layers. The configuration is shown in Figure 17.3. In [1] it was found that the resistance against delamination of the prepreg layers (used as an adhesive between the basic spliced material and the aluminium doubler) was not sufficient. At approximately 30% below the unspliced material tension strength, delamination between laminate and doubler (the prepreg/metal layer interface) occurred. This configuration did not meet the requirements and was therefore considered unacceptable.

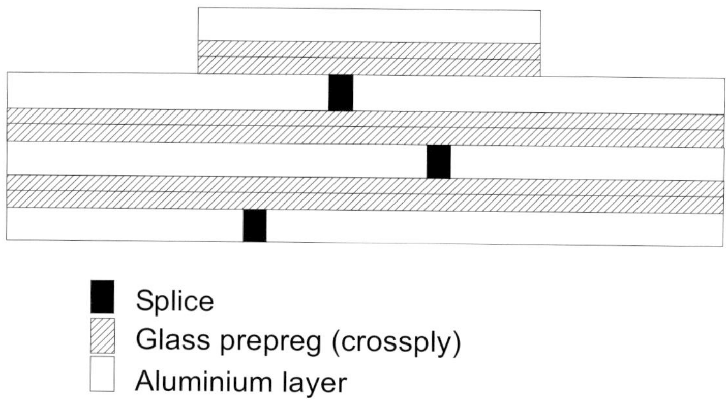

Figure 17.3 Glare splice with doubler using prepreg as adhesive.

Glare splice with doubler using adhesive

In the second configuration a thin Glare doubler is bonded over the splice area, using standard adhesive film, see Figure 17.4. Tests have shown appropriate behaviour on all evaluated properties, meeting all the requirements mentioned.

Doubler configurations

Once it was known that the doublers should be bonded to the spliced Glare with adhesive film (instead of prepreg) a doubler configuration could be designed. Basically there are two options [2]:

1. use of one-sided thick doublers, see Figure 17.5(a)
2. use of two-sided thin doublers, see Figure 17.5(b)

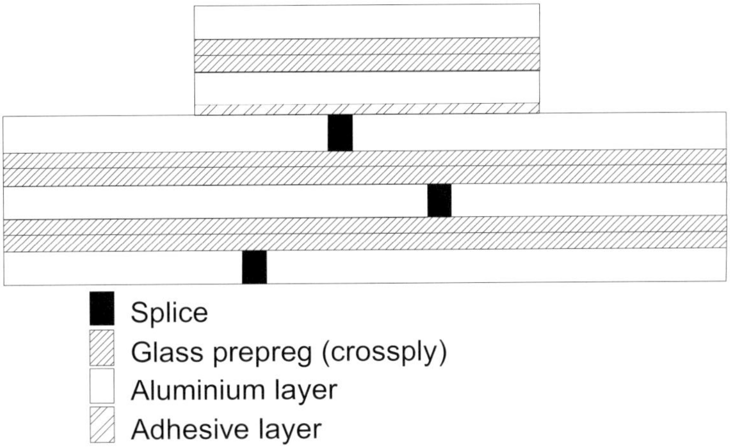

Figure 17.4 *Glare splice with doubler using adhesive.*

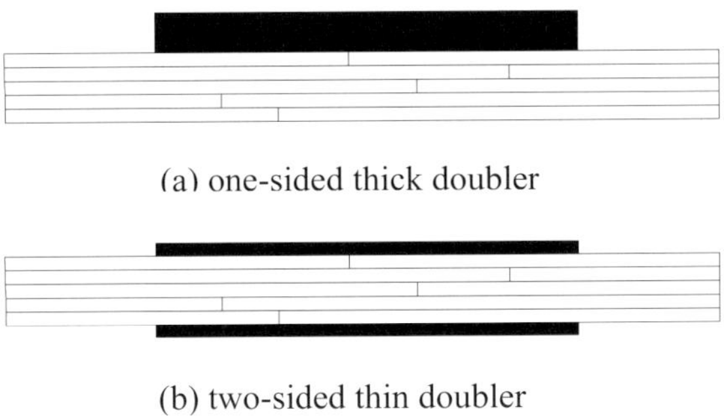

Figure 17.5 *Options for the doublers at the splice area.*

The thin doublers are considered to be superior with respect to structural performance (delamination growth, durability, fatigue crack initiation, panel weight) of the splice area due to the relatively thin thickness step. The only disadvantage of two-sided thin doublers compared to the one-sided thick doubler, is the non-flat outside of the fuselage panel. The thin doubler (approximately 0.3 mm) is visible, although it is much flatter than a conventional lap-joint in an aluminium panel. As such, it is expected that the effect on aerodynamic performance is small, when compared to conventional lap joints.

Self Forming Technique

For both thin and thick doublers an additional autoclave cycle is required to bond the doublers over the splice area in the laminate, thus increasing manufacturing costs. From this point forward a new production method, the Self-Forming Technique

(SFT) [2], was developed to manufacture the required doubler. Figure 17.6 shows a schematic configuration obtained when the SFT is applied to a spliced Glare laminate, including a doubler on the outside and on the inside of the laminate. The drawing clearly shows that the aluminium layers are no longer flat after curing.

The SFT can also be used to construct a ply drop-off, see Figure 17.7. The inner aluminium layer edge gently deforms over the end of the prepreg layer to the thinner adhesive bondline under the pressure used to cure the total package in the autoclave. At that location of load transfer (at the edge of the ending layer) an adhesive is used, because for the aluminium layer thicknesses, delamination does not occur for the ply drop-off configuration obtained with the SFT. Without the SFT a ply drop-off in Glare results in a significant strength reduction because of delamination over the prepreg/metal layer interface initiated at the thickness step as discussed for the first configuration (also see [1]).

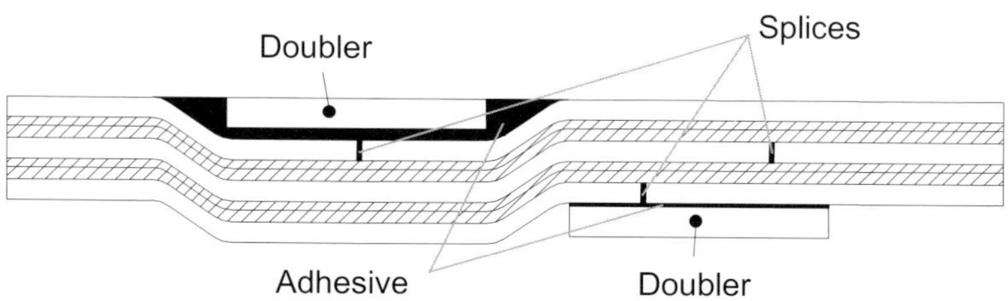

Figure 17.6 *Doubler geometry for spliced Glare using the SFT.*

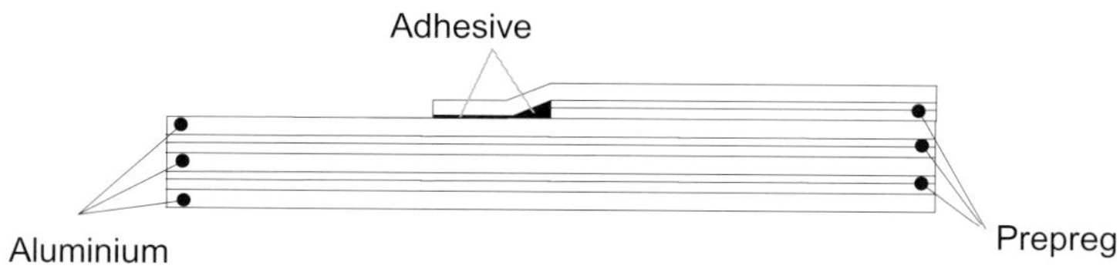

Figure 17.7 *Ply drop-off using the SFT.*

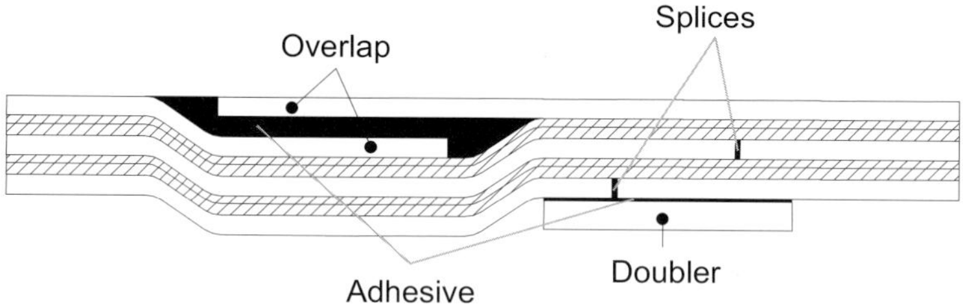

Figure 17.8 *Metal layer overlap using the SFT.*

With the SFT the use of separate doublers is no longer necessary, because the splice including bonded doubler can be formed by a simple overlap of the outer aluminium layers, see Figure 17.8. This creates only one visible line filled with adhesive at the outside of the laminate, where a doubler manufactured according to Figure 17.6 will create two visible lines. Despite all the advantages of configuration in Figure 17.8 over the earlier configurations, some problems still exist:

- The outside of the Glare fuselage skin will probably consist of single side clad aluminium, allowing the aircraft users to have 'bright-skin' (unpainted) aircraft. However, with the configurations shown in Figure 17.7 and Figure 17.8, this implies that the clad layer will also be in the bondline of the outer doubler. The aircraft industry will hesitate to allow the use of clad layers in the bondlines, because this may result in lower bond line strength.
- Inspection of aircraft could require paint removal over the fuselage skin several times during the life of the aircraft. The paint removal techniques (chemicals, water-jet paint stripping, etc.) may damage the adhesive fillets on the outside of the skin.
- For 'bright-skin' aircraft the adhesives are clearly visible (even if uncoloured adhesive is used) since environmental effects and/or polishing operations may make the splice lines clearly visible during the use of the aircraft.
- Polished fuselage skin requires non-pretreated (outside) aluminium sheets of the Glare laminate. However, that part of the aluminium layer that is within the overlap needs to be anodised and primed. This implies that that part of the outer aluminium sheet needs to be protected from the pretreatment fluids; an operation which is much more complex than for single side pretreatment (as used for the standard bright-skin Glare fuselage sheets).

The configuration that solves all the problems of the configuration in Figure 17.8, is shown in Figure 17.9. For the outer surface the clad layers are no longer in the bondline (the doubler between first aluminium layer and prepreg layer is bare material). The adhesive flow-out can be limited to a 0.1-0.5 mm gap between the edges of the spliced aluminium layer. For the inner surface the overlap technique, see Figure 17.8, is used. This reduces the number of aluminium sheets required for the laminate. However, there are some advantages when connecting the inner aluminium layer by a separate doubler according to Figure 17.9. In this way it is possible to increase the number of individual aluminium layers with the same mean dimensions. This is of particular importance when a panel will be spliced at a regular global splice pitch.

So far, the Glare splice technology used in Figure 17.9 appears to have a large advantage when compared to a monolithic aluminium aircraft structure. By applying this concept considerable benefits can be achieved, since much wider Glare panels can be produced than with the first splice concept. However, during production trials with large panels, e.g. the first Glare A310 demonstrator, it turned out to be difficult to achieve the desired local splice gap dimensions and tolerances. During the trials, a gap dimension of 1 mm, with a tolerance of ± 1 mm was required. These tolerances were considered to be an acceptable requirement for this type of splice. During the production trials, local overlaps occurred at several locations, also internally. For double-curved panels in particular, it proved difficult to meet the required tolerances. Although it is believed that the overlaps can be prevented (for instance by developing additional tooling) it is felt that the severe tolerances can only be met at increased costs. For these reasons the splice concept according to Figure 17.10 has been developed. In this configuration all metal layers are spliced by an overlap. The metal layers are bonded with adhesive film.

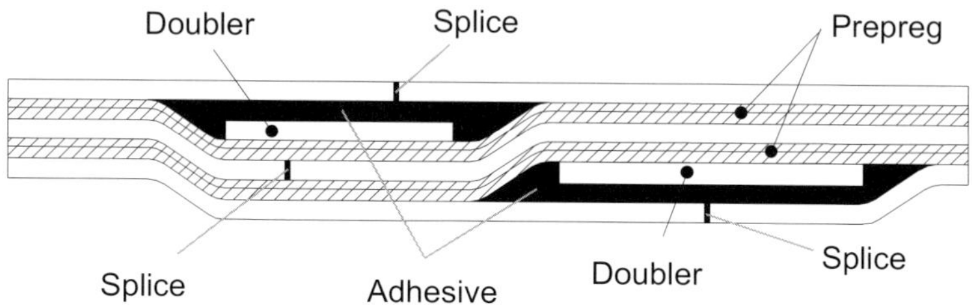

Figure 17.9 Optimised splice design.

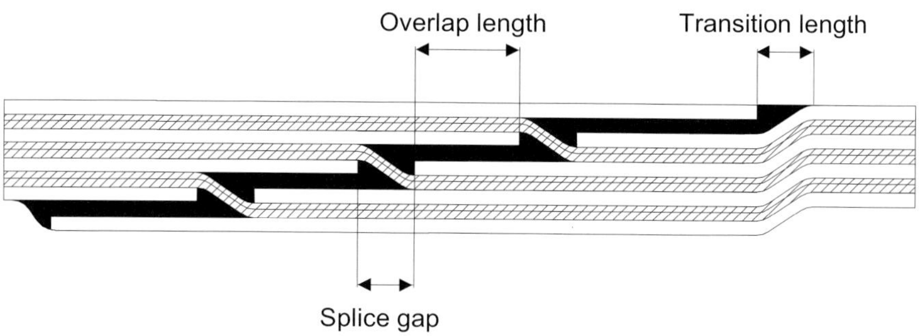

Figure 17.10 Overlap splice.

It is expected that the nominal dimensions could have a tolerance of ± 5 mm. Bonding to the clad surface of the outer metal layer is a disadvantage. However, satisfactory experience to do so is available for several existing aircraft. Another disadvantage could be the behaviour of the adhesive fillet on the outside of the skin during paint stripping. A solution can be to replace the fillet by some foil during the production of the laminate and afterwards remove the foil and subsequently fill the resulting gap with sealant.

In the past some details of the proposed new splice concept were already evaluated. Advantages of this concept are:

- the tolerance problem is reduced, especially for double curved panels
- no splice doublers are required
- the uniformity of the splice geometry is enhanced

For example, the Glare A330 barrel test panel manufactured by SLI and delivered to DCAA in 1997 for evaluation in a barrel test contained the overlap splice technique for one of the metal layers in the laminate.
The very first single curved Glare panel manufactured in Delft also contained the splice overlap for the outer metal sheet, resulting in a smooth outer surface of the laminate.

17.3 Doubler concepts

At points in the structure where extra strength or stiffness is necessary, extra material is needed. Glare offers the possibility of locally adding material layers. Such reinforcement is called a doubler. Up till now three types of doublers (local reinforcements) have been considered for application, see Figure 17.11:

1. Interlaminar doublers used to connect the aluminium layers in a splice according to Figure 17.11(a). It represents the so-called 'old' splice concept. It consists of both an outer doubler as well as an inner doubler, the 'splice doublers'. These were discussed in the previous section.
2. A bonded doubler consisting of a finished laminate bonded to the skin using an additional adhesive layer, see Figure 17.11(b). This doubler type is referred to as 'external doubler'.
3. A ply drop-off where the material steps down from a Glare layup (n+1/n) to layup (n/n-1), see Figure 17.11(c). This doubler is referred to as a 'ply drop-off doubler', or 'ply drop-off' for short. The thickness steps are applied internally and consequently are not visible on the outside of the laminate. Moreover, this concept is manufactured in the same curing cycle as that of the splice, in the same manner as the previously mentioned Self Forming Technique (SFT).

It is believed that at the step of the doubler to the basic skin, fatigue problems may occur in the basic skin, especially in the first aluminium layer, unless the fatigue loads normal to the doubler edge are low. It is also possible that delamination of the prepreg layers near the edge of the doubler will occur at higher static loads of the panel. For these reasons it is advisable to design the first thickness step with a Glare 2/1 laminate, i.e. a Glare 3/2 one as indicated in Figure 17.11 and repeat such a step at a certain distance. Another solution could be to design a stepped doubler edge according to Figure 17.12, although it must be acknowledged that this concept is much more costly than that in Figure 17.11 due to the extra production steps necessary. It may be possible to design the contour of the doubler so as to avoid high stresses due to the abrupt thickness step.
Another aspect of the external doubler is the fact that it is not possible to bond the doubler to the skin at locations where there are thickness steps at the inside skin, e.g. at the thickness steps of the splices, see previous sections. So in general it is only possible to apply this type of doubler on the flat (single curved) area on the inside or outside of the skin.

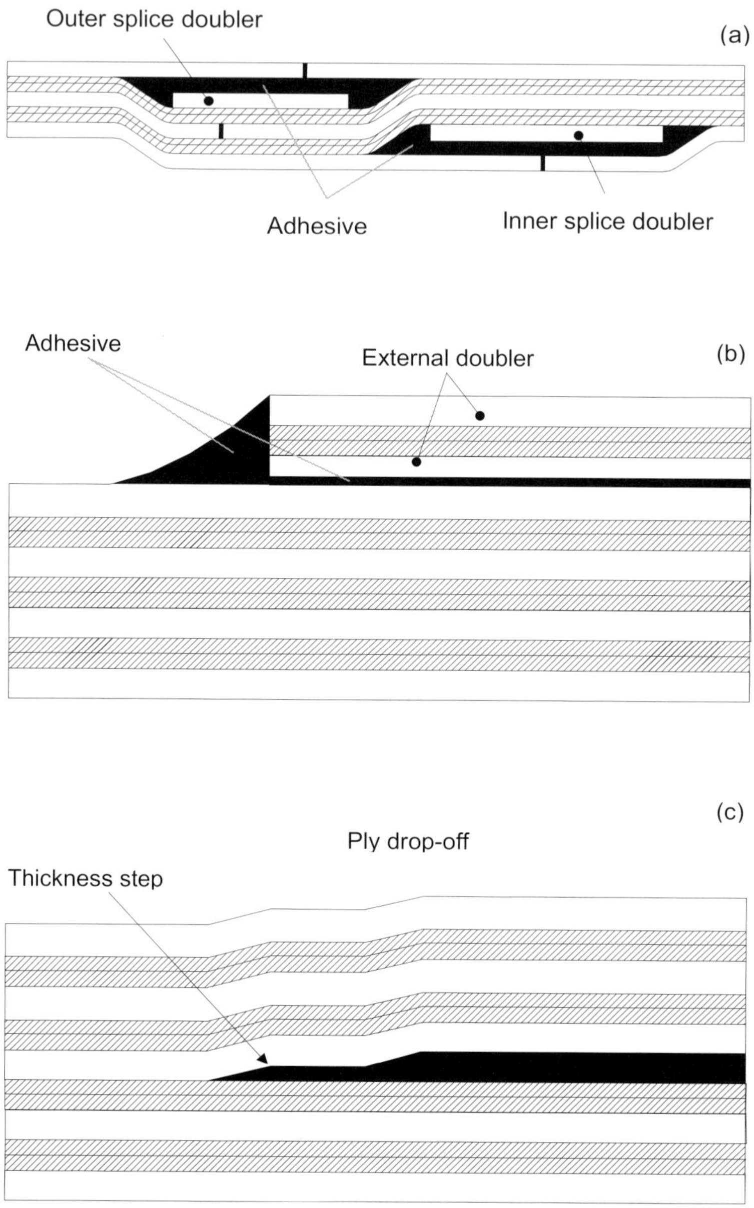

Figure 17.11 *Doublers; (a) interlaminar doubler, (b) bonded doubler, (c) ply drop-off.*

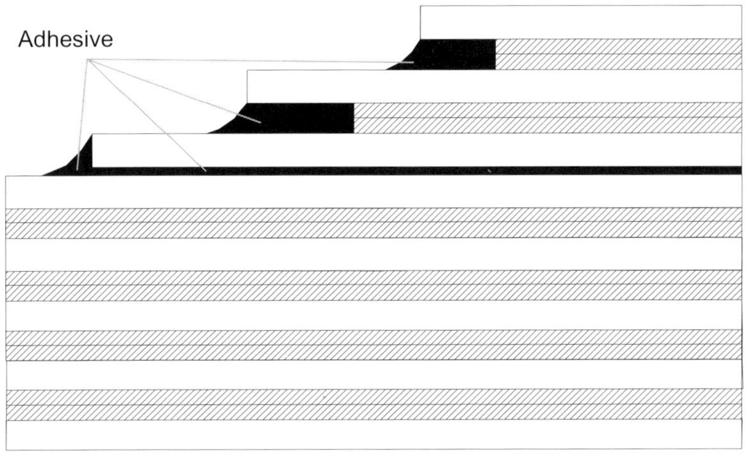

Figure 17.12 Stepped external doubler edge.

17.4 Riveted joints

As mentioned in the introduction, it is impossible to address all detailed design problems a designer may encounter when designing an aircraft structure in Glare. This is also the case for riveted joints. Many configurations are possible, and the final design is dependent on many parameters. The most important parameters are (in no particular order):

- type of joint, e.g. lap joint or but-joint
- materials to be joined, e.g. Glare-Aluminium or Glare-Glare
- type of Glare
- type of rivets
- mechanical loads on the joint
- production requirements
- maintenance requirements

The designer, who is accustomed to designing for aluminium most of the time, must be aware of some distinct differences between aluminium and Glare. Normally, the designer acquires the (multi-dimensional) design space based on calculations or values from a design handbook (yielding constraints), and then determines the design objectives, e.g. minimum weight or minimum costs. This is illustrated in Figure 17.13.

The optimum (in this example) can then be found at the intersection of the two constraints. This simple example illustrates how to come to an 'optimum' design. In reality, for most design problems, there are many more degrees of freedom, more constraints and more complicated goal functions. Besides which, most of the time a desired failure mode is necessary. When the final design is made according to the example in Figure 17.13, the failure mode would be uncontrolled. A designer would deviate a little from the optimum, in order to assure one (desired) failure mode.

The following example is used to illustrate the approach that can be used to design a riveted joint in Glare, see Figure 17.14.

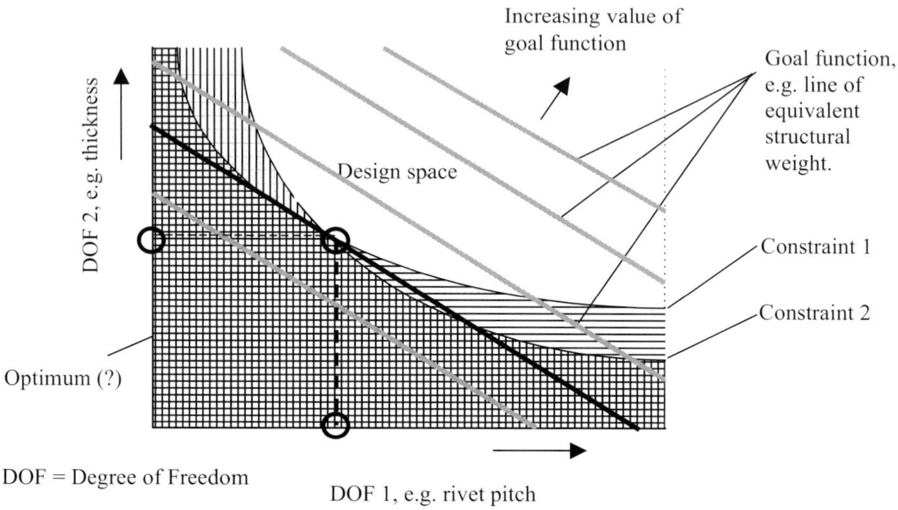

Figure 17.13   Two-dimensional design space with two constraints and one goal function.

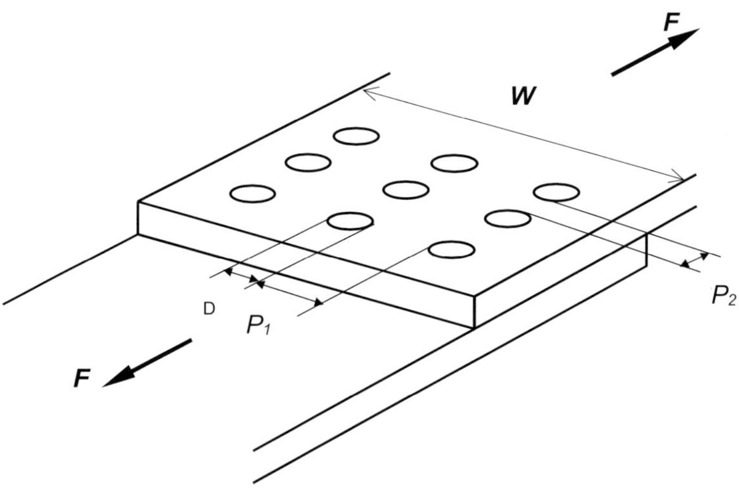

Figure 17.14   Example of riveted lap joint.

The following assumptions are made:

- There are only two degrees of freedom: rivet pitch P_1 perpendicular to load direction and rivet row distance P_2 in load direction. The width of the specimen is fixed and therefor the number of rivets is variable. The number of rows is fixed at three. All other design variables (e.g. type of joint, type of rivets, hole diameter D, material thickness, type of Glare and edge distance) are fixed.
- The load case is one load F (positive or negative) as indicated in Figure 17.14.
- Constraints: The minimum rivet pitch P_1 is determined by the net section (static/blunt notch strength and fatigue strength) and by production requirements. The minimum rivet row distance P_2 is determined the fatigue life (see chapter 12 and [3] for a calculation method on this subject). The maximum rivet pitch P_1 is determined by rivet strength. The maximum rivet row distance P_2 is determined by inter-rivet buckling.
- The goal function is minimum weight. The weight of the joint can be approximated with the following function (excluding the weight of the rivets):

$$M = K + C_1P_1 + C_2P_2 + C_3P_1P_2 \qquad (17.1)$$

with: K = constant (determined by the other (fixed) design variables)
C_1, C_2, C_3 = constants (determined by the other (fixed) design variables)
M = the weight of the joint

The values used in Figure 17.15 are purely intended for this example, and therefore only indicative. The dotted lines inicate the constraints. The design space is within the dotted lines.

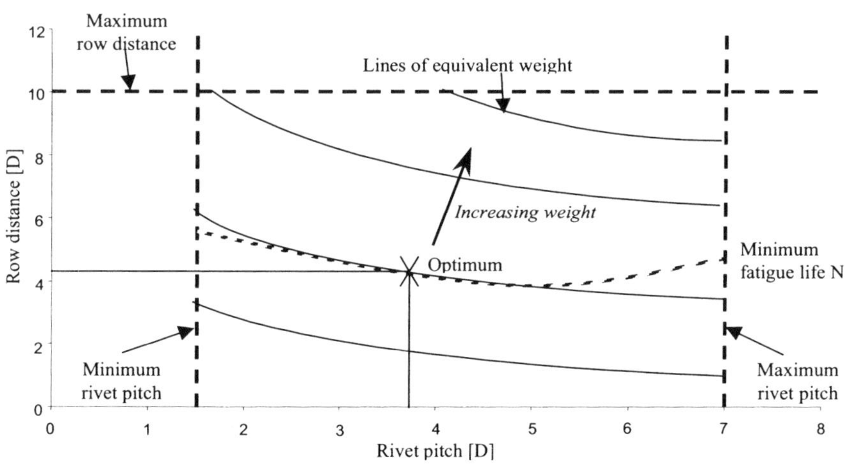

Figure 17.15   Design space of a lap joint.

In this case, the optimum is a rivet pitch of about 4.7 times *D* and a rivet row distance of about 4.2 times *D*. In practice this will be rounded to respectively 5*D* and 4.5*D*. With the aid of a computer tool much more complicated designs spaces can be analysed. It is also possible to implement FEM analysis [4]. Data generated this way can be converted for design handbook guidelines.

17.5 Concluding remarks

The design process of a structure can be divided into three steps: the first to determine the requirements imposed on the structure, then a global/conceptual design of the structure and third the detailed design. Of course this is a simplification of a complex, iterative design process. In this chapter only a few aspects and examples of detailed design in Glare were addressed, but the given examples indicate that it is possible to design lightweight, cost effective structures with Glare. The optimisation procedure described in section 17.4 is a very useful aid. However, thorough knowledge of the Glare properties <u>and</u> a thorough understanding of the underlying mechanics are necessary to come to an optimised design. It is also very important in this respect that the strong relationship between function and geometry of a structure, fabrication process and material properties is perceived. The splice design is an excellent example of this.

References

[1] G. Roebroeks, *The development of the splice geometry for the Daimler-Benz Aerospace Airbus Glare 3-3/2-0.3 barreltest panel 1&2*, Report TD-R-93-005B, Delft: Structural Laminates Company, 1995.

[2] G. Roebroeks, *The self forming technique; Recent progress in the development of a method to produce large aircraft fuselage panels of Glare*, Report TD-R-97-007, Delft: Structural Laminates Company, 1997.

[3] R.P.G.M. Muller, *An experimental and analytical investigation on he fatigue behaviour of fuselage riveted lap joints,* Ph.D. thesis, Delft: DUT, 1995.

[4] H. de Boer, *Shape optimization of fabric reinforced products. Design sensitivy analysis,* Shaker Publishing, ISBN 90-9014817-5, 2001.

18

Numerical modelling: delamination buckling

J. J. C. Remmers and R. de Borst

Abstract – Laminates can be sensitive to delamination buckling, which occurs when a partially delaminated panel is subjected to a compressive load. The interaction of local buckling and extension of the delaminated zone typically results in a decrease of the residual strength and, eventually, in a collapse of the structure. Fortunately, this phenomenon has never been observed in experimental tests with Glare. Although the delaminated layers buckle locally, the delamination front does not propagate within the range of compressive stresses that can be expected in typical aerospace structures.

Numerical analyses can give more insight into the mechanisms that prevent Glare panels from collapsing. In this paper, some experimental observations regarding delamination in Glare are discussed and, based on these observations, a numerical model is constructed at a meso-mechanical level. In this approach, solid-like shell elements are used to model the individual layers. They are connected by interface elements, which are capable of modelling delamination between the layers. These numerical techniques are used to simulate two classical delamination-buckling tests of a Glare laminate.

18.1 Introduction

There are still a number of open questions regarding the physical and mechanical behaviour of Fibre Metal Laminates. One of these concerns the behaviour of panels that are subjected to compressive and/or shear loading and which have initial delaminated areas, caused by fabrication or maintenance, for example. The compressive or shear stresses may lead to local buckling. This local buckling can subsequently cause growth of the delaminated area, which in turn will lead to progressive buckling. This process may eventually result in catastrophic failure.

Although delamination growth due to buckling has not yet been observed in experiments with Glare panels [1, 2], numerical simulations can be of assistance if we wish to arrive at a better understanding of the phenomenon. The numerical models must be chosen and handled with care, since the problem is a combination of stability and delamination (crack) growth. In this case, a basic issue is the level of modelling that is chosen. In the spirit of earlier research [3, 4, 5], we adopt a meso-level approach, in which the individual layers are modelled separately from the interface. Specifically, we use so-called solid-like shell elements [6] for modelling the individual layers. These elements share the advantage of conventional shell elements in that they can be used in applications as thin structures without exhibiting overstiff behaviour (locking). Intimately related to this is that no rotational degrees-of-freedom are needed, which makes them particularly suitable for application in layered structures, where we have a stack of shell and interface elements. The latter elements model the interfaces and can simulate debonding of the adhesive.

This chapter is ordered as follows. First, we give a more detailed description of the phenomenon of delamination buckling in Glare, starting from SEM-pictures. Then, the modelling strategy will be discussed, including a succinct description of the solid-like shell model [5, 6], the interface elements [4], the constitutive model for the interface [7, 8] and a contact model. The chapter concludes with two numerical examples of buckling and delamination growth of Glare laminates with initial delamination.

18.2 Delamination buckling

We consider a Fibre Metal Laminate panel with an initial delamination, see Figure 18.1(a). This delamination can be regarded as an imperfection, so that when the panel is compressed, the delaminated layer will buckle locally (b). As a result, the interface at the delamination front is subjected to a combined mode I / mode II stress [9]. When the stresses exceed the ultimate limit, the layers debond and the delaminated area grows, resulting in a reduction of the global stiffness (c). The process continues up to the point that the panel collapses (d).

Figure 18.2(a) shows a small specimen of Glare 2-3/2-0.3, which has been subjected to a three-point bending test [10]. The specimen was expected to delaminate as a result of interlaminar shear failure, but in this specific case - most likely due to production errors - the outer layer has buckled and delaminated, Figure 18.2(b). The prepreg layer has remained undamaged[1], apart from some single fibres that seemingly have left the prepreg, see Figure 18.3(a).

When we take a closer look at the crack front, Figure 18.3(b), we notice that the actual interface between the aluminium and the prepreg layers has remained intact and that the crack starts near a fibre in the prepreg layer. This observation can be explained as follows. Before fabrication, the aluminium panels are extensively treated. First, an etching process removes the existing oxidised surface. Then, the

[1]The epoxy matrix in the second prepreg layer shows some damage, but this is due to large tensile stresses in this layer

aluminium is anodised to create an oxide layer with a good morphology. After this, the surface is treated with a primer in order to enhance the durability of the bond. Finally, several hours after application of the primer, the laminate is stacked and put in the autoclave.

Consequently, the adhesive bonding between the aluminium and the prepreg layer is of excellent quality. In fact, the ultimate interface strength is much higher than the strength in the transition zone between fibre rich and resin rich prepreg. A crack is therefore expected to initiate in this area.

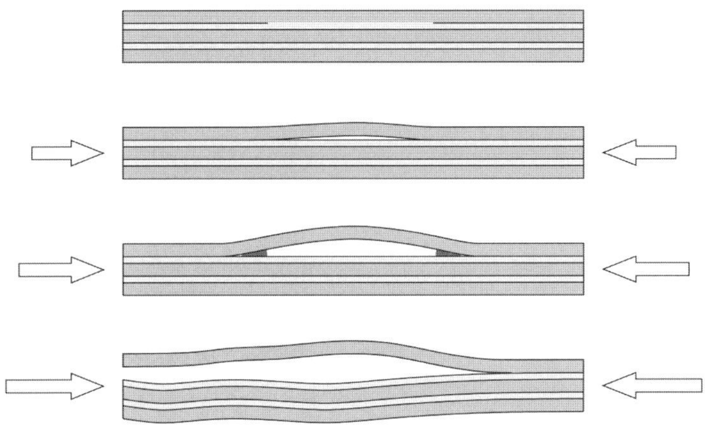

Figure 18.1 *Static representation of the delamination buckling failure mechanism; (a) initial delamination, (b) local buckling, (c) delamination growth, (d) failure.*

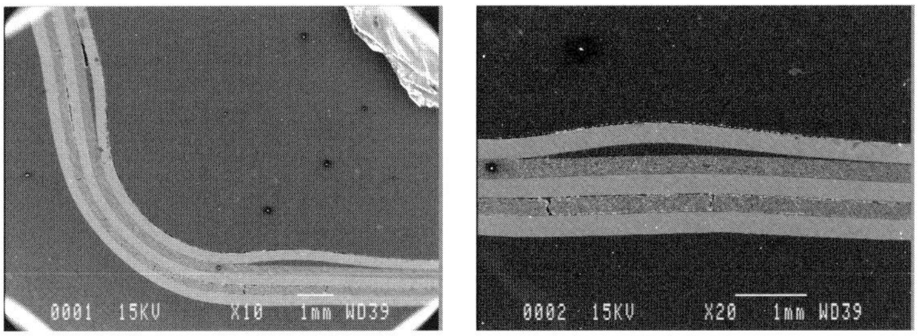

(a) overview (b) detail

Figure 18.2 *SEM picture of local buckling of the top layer of a Glare 2-3/2-0.3 specimen after being subjected to a three point bending test.*

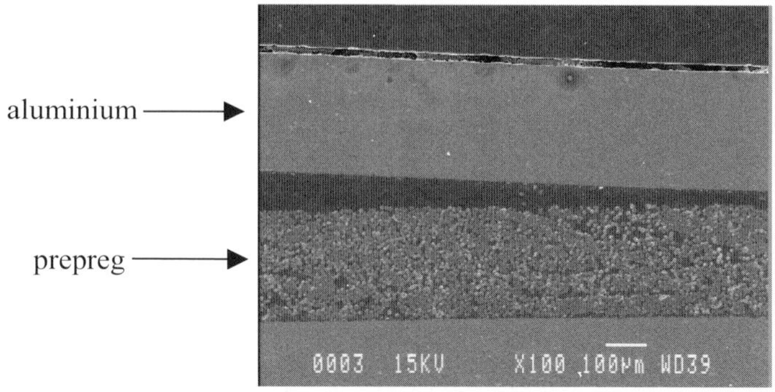

(a) overview (magnification 40×)

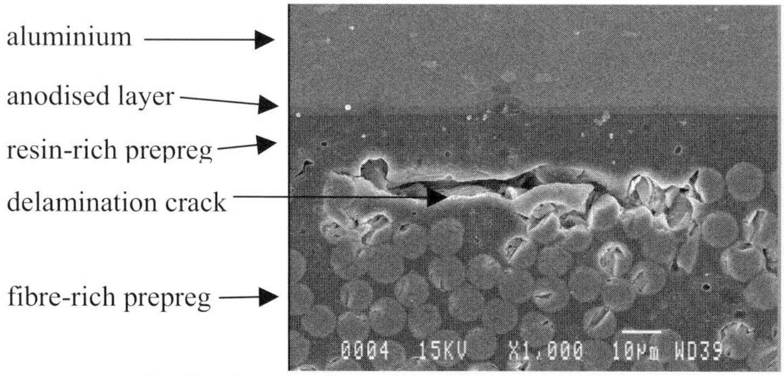

(b) detail at crack tip (magnification 400×)

Figure 18.3 *SEM picture of delamination front of a Glare 2-3/2-0.3 specimen due to delamination buckling.*

18.3 Numerical strategy

In order to be able to simulate the failure mechanisms on a macroscopic (buckling) and a microscopic level (delamination growth), we must investigate the phenomena with a mesoscopic model. This means that both the aluminium and the prepreg layers will be considered as homogeneous materials, with given constitutive parameters. The layers will be modelled with stackable solid-like shell elements, Figure 18.4. Furthermore, it is assumed that the prepreg layers remain undamaged. The crack growth in the transition zone of fibre rich and resin rich prepreg is simulated with interface elements that are placed between the solid-like shell elements representing the aluminium and the fibre prepreg layers. The crack characteristics of the transition zone are lumped in these elements.

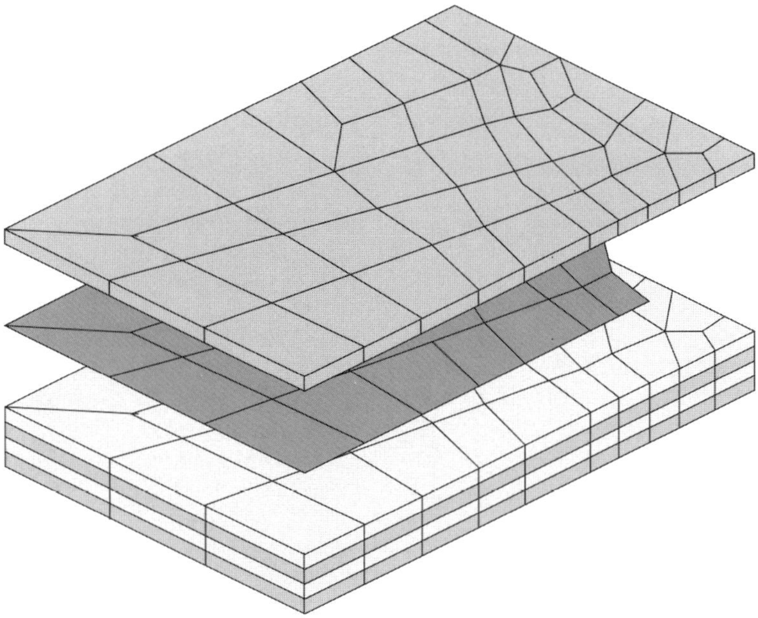

Figure 18.4 *Schematic representation of a finite element model of a Glare panel. The individual layers are modelled with solid-like shell elements. Interface elements (dark-grey) are applied between the top layer and the second layer.*

18.4 The solid-like shell element

The analysis of thin structures under compression requires special properties from finite elements. For this purpose, a whole range of geometrically non-linear, discrete shell elements have been derived. Most shell elements are two-dimensional in a three-dimensional space. In the present case, where we wish to build a laminate by stacking multiple elements, we cannot use standard shell elements.

Alternatives to shell elements are solid, three-dimensional elements. However, in order to avoid Poisson-thickness locking, the three dimensions must be of the same order of magnitude. This implies that in this application, the length of the elements must approximately be equal to the thickness of one layer, i.e. less than half a millimetre. Regarding the dimensions of an average panel in aerospace structures, the mesh must have a very large number of degrees-of-freedom.

The solid-like shell element derived by Parisch [6] combines the good characteristics of both families. It is a solid, consisting of eight external nodes, each with three translational degrees of freedom, see Figure 18.5. The four internal degrees of freedom are used to setup a quadratic displacement field in the thickness direction. Hence, the normal strain in this direction varies linearly instead of being constant as in classical shell elements. Poisson-thickness locking is then avoided and the element can be applied to thin applications. This element has previously been used by Hashagen and de Borst [11] to study failure mechanisms in Fibre Metal Laminates.

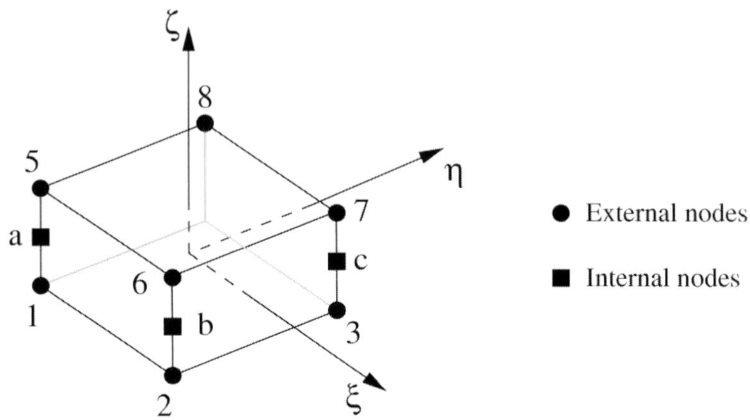

Figure 18.5 *Geometry of the 8-noded solid-like shell element. Each external node has three translational degrees of freedom. The internal nodes contain one degree of freedom*

18.5 The interface element

Delamination is modelled by means of interface elements that are placed between two layers. These elements consist of two adjacent surfaces, as shown in Figure 18.6.

The relative displacement $\boldsymbol{v} = [v_n, v_{s1}, v_{s2}]^{\mathrm{T}}$ between these surfaces is a measure for the internal strains[2], which are related to the tractions $\boldsymbol{t} = [t_n, t_{s1}, t_{s2}]^{\mathrm{T}}$ by means of the constitutive relation:

$$\vec{t} = \vec{t}(\vec{v}, \vec{\kappa}) \tag{18.1}$$

with: $\boldsymbol{\kappa}$ = a set of damage history parameters.

In this study, the interface element is equipped with a damage model for the simulation of delamination growth and a simple contact model to prevent delaminated layers from penetrating the core material.

---

[2]In the following, the strains and tractions in the interface in normal direction will be denoted with a subscript n and the quantities in transverse direction with the subscripts $s1$ and $s2$. Vectors are denoted in the text using **bold**.

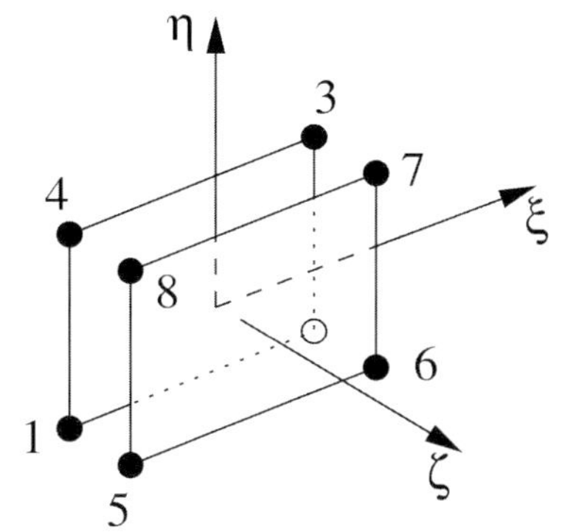

Figure 18.6 *Geometry of the 8-noded interface element.*

18.5.1 Delamination model

From Figure 18.3 it appears that delamination is a combination of two damage mechanisms on a microscopic scale: an adhesive debonding of the interface between fibre and matrix and a cohesive matrix crack. In order to capture both mechanisms, a damage model is used at a mesoscopic scale. In this model, the stiffness of the interface between the aluminium and the prepreg layers is reduced by means of damage parameters, which are initially equal to zero and equal one when the material is completely damaged [12]. The equivalent relative displacement of the interface is taken as the controlling parameter for damage growth. As soon as the equivalent relative displacement exceeds an initial or threshold value, the damage parameter starts to increase according to a damage growth relation.

Delamination of a Fibre Metal Laminate depends on the angle between the fibre orientation and the delamination front. Therefore, a single scalar damage parameter cannot be used. Furthermore, the delamination crack is a combination of mode I and mode II. Moreover, a difference can be observed between tension and compression in the mode I direction. In the latter case, the material is not damaged but behaves perfectly plastically. A model that incorporates these aspects has been derived by Schipperen [7, 8] and is based on the constitutive assumptions set by Schellekens and Hashagen in their plasticity-based delamination models [4, 5].

It is assumed that the stiffness of the interface element decreases when damage occurs. The constitutive relation, equation (18.1) can then be written as:

$$\vec{t} = (\vec{I} - \vec{d})\vec{C}\vec{v} \qquad (18.2)$$

with: $\boldsymbol{I}$ = the unit tensor
$\boldsymbol{C}$ = the initial elastic stiffness of the interface
$\boldsymbol{d}$ = a second-order damage tensor.

We consider the special case that only the diagonal terms of this tensor are non-zero. Damage occurs when the current equivalent relative displacement κ is higher than all previously attained values:

$$\kappa(\bar{v},t) = \max[\kappa(\bar{v},\tau) \mid \tau < t), \kappa_{init}] \tag{18.3}$$

with: κ_{init} = the initial or threshold value of κ.

The equivalent relative displacement is a function of the relative displacements in the three directions:

$$\kappa = \sqrt{\tfrac{1}{2}\bar{v}^{\mathrm{T}}\bar{P}\bar{v} + \bar{q}^{\mathrm{T}}\bar{v}} \tag{18.4}$$

where

$$\bar{P} = \mathrm{diag}\left[2\frac{\bar{v}_n^t}{\bar{v}_n^c}, 2\left(\frac{\bar{v}_n^t}{\bar{v}_{s1}}\right)^2, 2\left(\frac{\bar{v}_n^t}{\bar{v}_{s2}}\right)^2\right] \tag{18.5}$$

$$\bar{q} = \left[\bar{v}_n^t\left(1 - \frac{\bar{v}_n^t}{\bar{v}_n^c}\right), 0, 0\right]^{\mathrm{T}} \tag{18.6}$$

with: v_n^t = the relative displacement equivalent for the yield tension in the normal direction

v_n^c = the relative displacement equivalent for the yield compression in the normal direction

$v_{s1,2}$ = the relative displacement equivalents under the shear loading.

The softening is governed by the fracture toughness G_c. Experiments with Glare show a softening behaviour of the interface that resembles an exponential curve, see Figure 18.7. Inspired by this, the following damage relation is used for the diagonal components of ***d***:

$$d_{ii} = 1 - \frac{\kappa}{\sqrt{\bar{v}^{\mathrm{T}}\bar{v}}}\exp\left(-\frac{2C_{ii}v_n^t}{G_c}\left(\kappa - v_n^t\right)\right) \tag{18.7}$$

The damage parameter in this model will approach the limit value 1.0 asymptotically. This improves the numerical stability of the model, which is important, since we can expect large relative velocities in the case of delamination buckling. The use of a tangent stiffness matrix guarantees quadratic convergence in the nonlinear analysis.

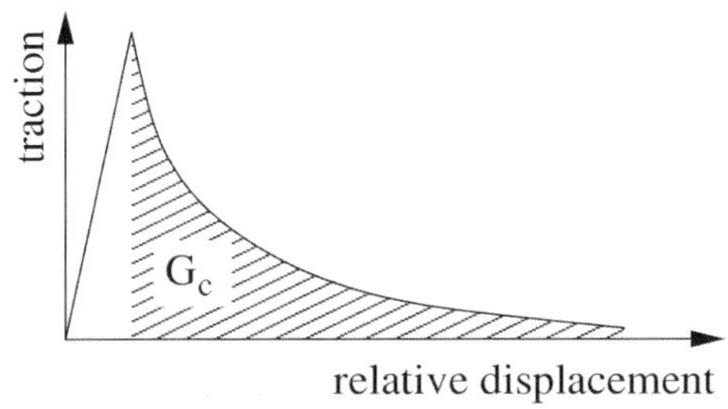

Figure 18.7 *Traction-relative displacement relation in mode I direction for the damage model.*

18.5.2 Frictionless contact model

The local buckling mode can change as a result of delamination growth. Consequently, the delaminated layer can move inwards and penetrate the core material. A contact algorithm is needed in order to prevent this in the numerical simulation.

In the current model, the interface element is equipped with a penalty contact algorithm. Initially, the interface element is traction free, i.e. the stiffness terms in the constitutive relation are equal to zero. As soon as the normal component of the relative displacement v_n of the two adjacent surfaces in the element is negative, a penalty stiffness D_{pen} is applied, according to

$$\vec{t} = \bar{C}\vec{v} \qquad (18.8)$$

with: $C = \text{diag}[D_{pen}, 0, 0]$ if $v_n < 0$
$C = \text{diag}[0, 0, 0]$ if $v_n > 0$.

Note that for reasons of simplicity, the additional stiffnesses in the transverse directions of the element due to friction are neglected, assuming that this will not have a large effect on the global behaviour of the structure.

18.6 Numerical examples

In the past, several studies have been carried out to investigate the behaviour of a partially delaminated Fibre Metal Laminate panel under compression. It appeared that the global stiffness of the panel was slightly reduced due to local buckling of the delaminated layer, but when the load was further increased up to 200 N/mm^2, no delamination growth was observed [1, 2]. The same conclusions could be drawn from fatigue tests [13] and experiments with moisture absorption [14]. Apparently, delamination growth due to delamination buckling under axial compression is not a major concern for Fibre Metal Laminates.

Nevertheless, it is interesting to predict the load levels for delamination growth although they may be significantly above those occurring in the aircraft. Therefore, two standard tests have been simulated; a quasi two-dimensional specimen and a panel with a circular initial delamination.

18.6.1 Thin, delaminated strip

The following numerical simulation is based on one of the first investigations into the residual strength of delaminated Fibre Metal Laminate panels in compression [1]. Consider a thin strip of Arall as shown in Figure 18.8. The adhesive between the outer aluminium layer and the aramid-prepreg material has partially delaminated over a length of 18 mm. The strip is attached to a thick substrate in order to prevent global buckling of the structure, which would blur the phenomena we wish to examine; local buckling in combination with delamination growth.

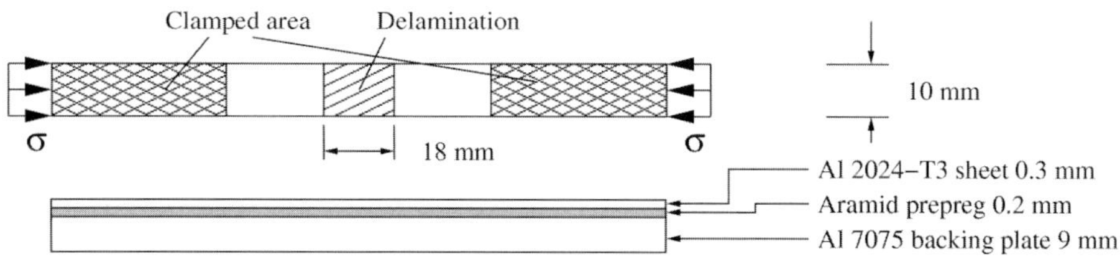

Figure 18.8 Geometry of delaminated Glare specimen.

Both the aluminium and the prepreg material are assumed to be elastic. The aluminium is isotropic with Young's modulus $E = 72000$ N/mm^2 and Poisson's ratio 0.3. The UD aramid is modelled as a homogeneous, orthotropic material. The material properties are obtained from Verbruggen [15] and are shown in Table 18.1. The initialisation stresses for fracture of the interface layer are obtained from previous numerical simulations with similar materials [5]. The ultimate traction in the normal direction is $t^t_n = 50$ N/mm^2, $t^c_n = 150$ N/mm^2, the ultimate traction in the two transverse directions equals $t_{si} = 25$ N/mm^2. The fracture toughness is $G_c = 1.1$ N/mm.

Table 18.1 Material properties of Aramid prepreg material

E_{11}	$5.769\ 10^4$ N/mm^2
E_{22}	$5.022 \cdot 10^3$ N/mm^2
ν_{12}	0.195
ν_{21}	0.032

An analytical solution for the local buckling load of this configuration was formulated by Kachanov [16]. The critical load for local buckling of a thin layer attached to a substrate of infinite thickness is equal to:

$$\sigma_{x,K} = \frac{\pi^2 E}{3(1-\nu^2)}\left(\frac{h}{l}\right)^2 = 72.3\,\text{N/mm}^2 \qquad (18.9)$$

for the given dimensions and material parameters of aluminium.

Because of symmetry, one quarter of the specimen is modelled in the finite element analysis, Figure 18.9. Symmetric boundary conditions are applied to the edges of symmetry. The correct local buckling mode is triggered by a small initial out-of-plane deformation of the top layer. The model contains 1920 solid-like shell elements, 540 interface elements and has a total of 10524 degrees of freedom.

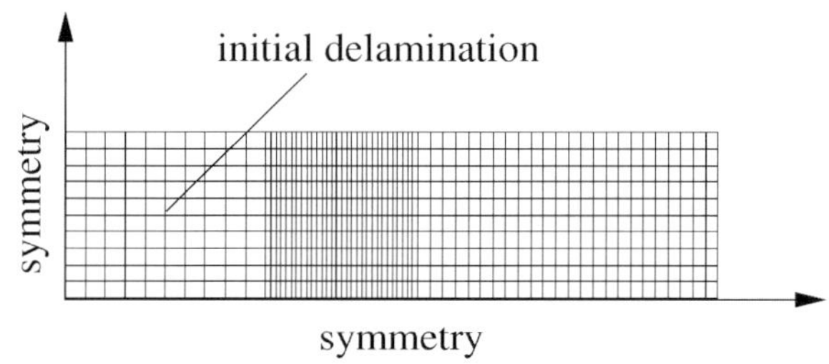

Figure 18.9 Finite element model for delaminated specimen.

The initial (dummy) stiffness of the interface elements is usually set to a value that is much larger than the Young's modulus of the surrounding structure. In this case however, a large dummy stiffness resulted in a bad convergence of the numerical model. An initial stiffness C_{ii} = 50000 N/mm^3, (i = 1, 2, 3) gives better results: the calculations remain stable and the interface elements do not overly influence the total stiffness of the model prior to cracking.

Figure 18.10 shows the out-of-plane displacement of the delaminated aluminium layer as a function of the axial compressive stress σ_x. The deformation can be divided into three parts, the pre-buckling path, the stable post-buckling deformation without delamination growth and the stable post-buckling deformation with delamination growth. The calculations were terminated after the top layer had completely debonded.

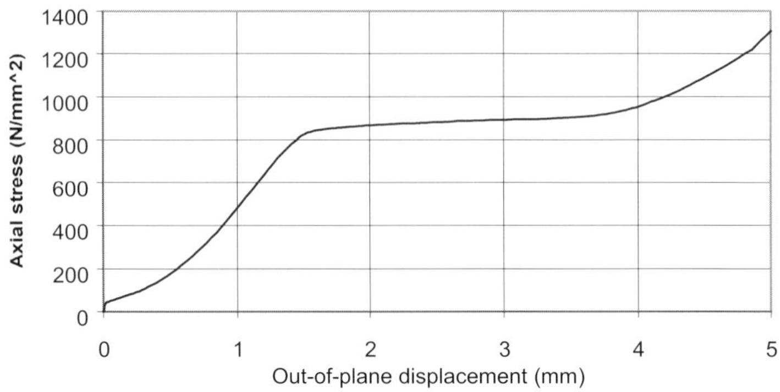

Figure 18.10 Out-of-plane displacement of the delaminated Arall strip as a function of the applied axial load.

A close up of the load-displacement curve is shown in Figure 18.11. Due to the small initial imperfection, it is not possible to pinpoint an unambiguous value for the buckling load. When extrapolating the post-buckling path, an axial load $\sigma_x \approx 50$ N/mm^2 can be regarded as the calculated buckling load, which is lower than the analytical solution but in agreement with the experimental values $\sigma_x = 40\text{-}45$ N/mm^2 obtained by Vlot. The load-displacement curve does not coincide with the curve obtained from the experiments, which may be caused by significantly large initial imperfections due to residual stresses in the specimens. The slope of the post-buckling path indicates the residual post-buckling stiffness. It can be seen in the figure that the calculated slope is in agreement with the one observed in the experiments.

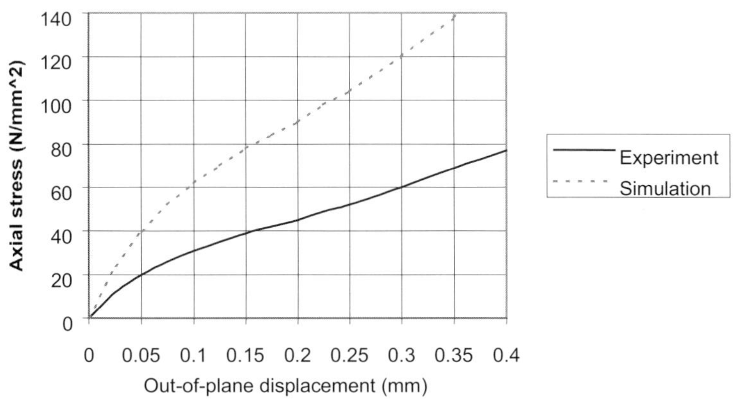

Figure 18.11 Out-of-plane displacement of the delaminated Arall strip as a function of the applied axial load for numerical simulation and experiment [1].

Progressive delamination growth starts at an axial load level $\sigma_x = 750$ N/mm^2. This delamination growth is triggered by the shear tractions in the interface exceeding the ultimate level of 25 N/mm^2. The calculations are continued up to the point that both layers have separated completely. The peculiar increase of stiffness at the end of the load-displacement curve can be attributed to edge effects, see Figure 18.12.

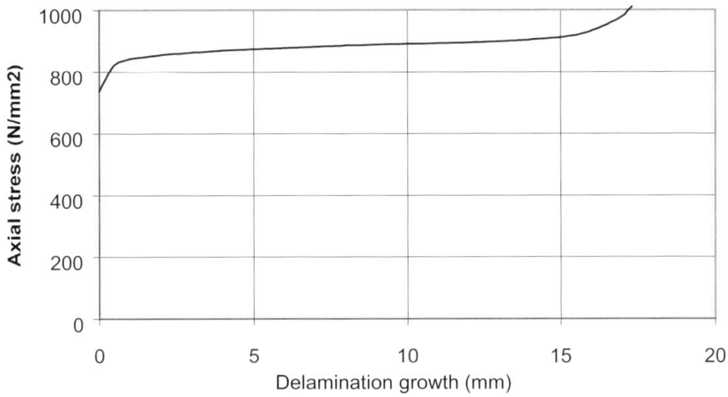

Figure 18.12 *Delamination growth of the delaminated Arall strip as a function of the axial load.*

18.6.2 Panel with circular delamination

The second example is somewhat more complicated, as the behaviour of a Glare 3 laminate with a circular initial delamination under uniaxial compression is examined, see Figure 18.13. In this example, the direction in which the delaminated zone will propagate is not evident. Similar studies with carbon-epoxy laminates [17] show that the circular delamination grows in a direction perpendicular to the loading direction; the delamination area transforms from a circular shape into an ellipsoidal one. Consequently, the buckling mode can change as well. Therefore, we have to consider the possibility of self-contact and must apply the contact algorithm in order to prevent this.

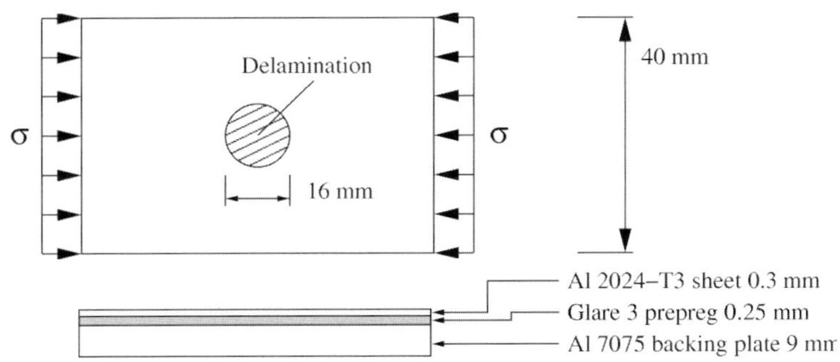

Figure 18.13 *Geometry of Glare laminate with circular delamination.*

The classical buckling load for a laminate with an ellipsoidal initial delamination can be calculated by using the Rayleigh-Ritz energy method, as given by Shivakumar and Whitcomb [1, 18]. For this configuration, the lowest buckling load is equal to 254.6 N/mm^2.

It is assumed that the thick substrate does not have a large influence on the mechanical behaviour of the specimen apart from preventing global buckling. Hence, in order to reduce the model, this layer has been left out. Global buckling is prevented by constraining the out-of-plane displacement of the lower boundary of the prepreg layer. The material parameters for this layer are given in Table 18.2. The mesh is shown in Figure 18.14 and consists of 1136 solid-like shell elements, 520 interface damage elements and 48 interface contact elements. The penalty stiffness for these elements is set equal to the dummy stiffness of the delamination elements, e.g. D_{pen} = 50000 N/mm^3.

Table 18.2 Orthotropic material parameters for 0° / 90° prepreg layer.

E_{11}	$3.317 \cdot 10^4$ N/mm^2	G_{12}	$5.5 \cdot 10^3$ N/mm^2	ν_{12}	0.195
E_{22}	$3.317 \cdot 10^4$ N/mm^2	G_{23}	$5.5 \cdot 10^3$ N/mm^2	ν_{23}	0.195
E_{33}	$9.4 \cdot 10^3$ N/mm^2	G_{31}	$5.5 \cdot 10^3$ N/mm^2	ν_{31}	0.06

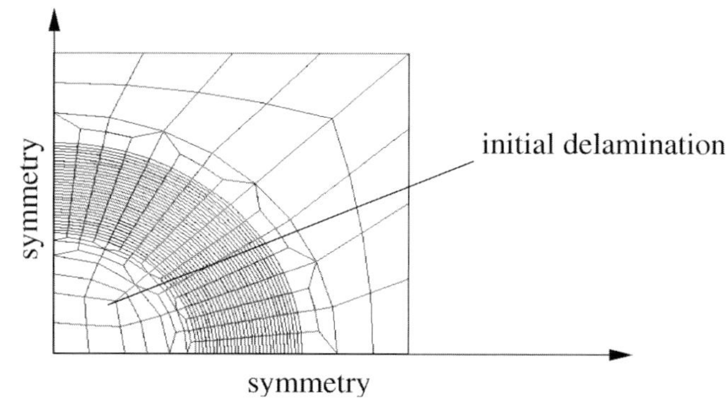

Figure 18.14 Finite element model of Glare laminate (top view).

The out-of-plane displacement curve of this model is shown in Figure 18.15. The local buckling load is in agreement with the analytical estimate. The delamination starts to grow at a compressive stress σ_x = 1010 N/mm^2. Again, the shear tractions in the interface appeared to be the critical factor for delamination growth. The delamination has progressed in a direction perpendicular to the loading direction, as can be seen in Figures 18.15(b) and (c). In the other direction, the aluminium layer moved inwards and made contact with the prepreg material, Figure 18.16.

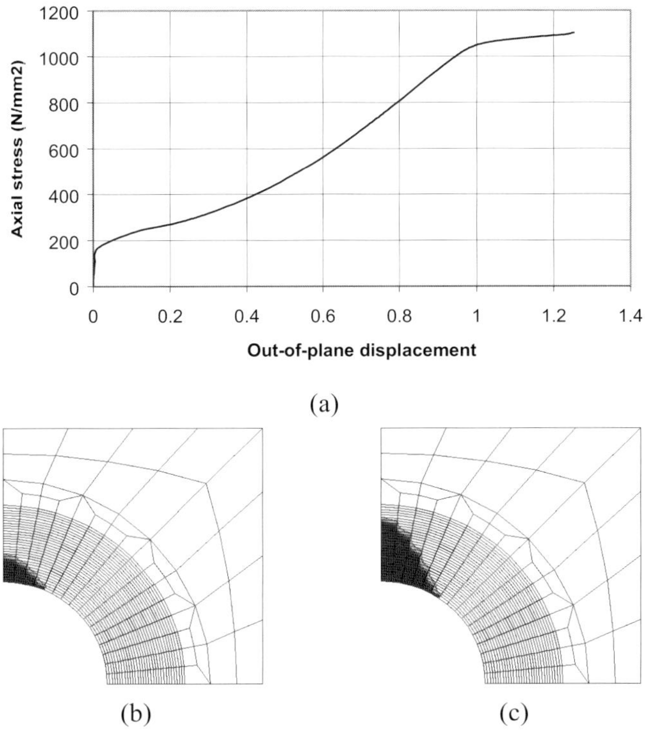

Figure 18.15 *Stress-deformation curve and delamination propagation of Glare laminate with an initial circular delamination.*

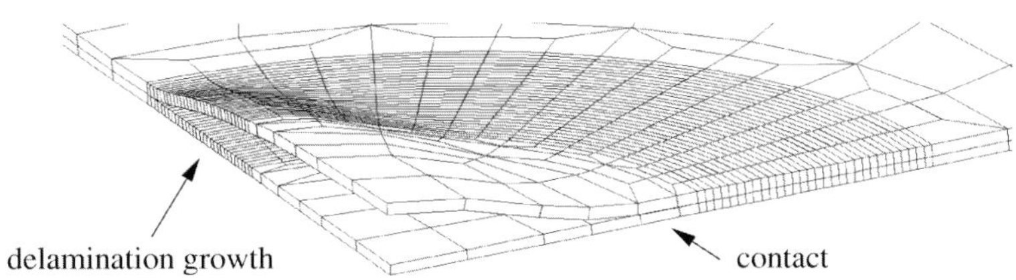

Figure 18.16 *Deformation of Glare laminate with an initial circular delamination.*

18.7 Closing remarks

In this chapter a numerical strategy has been derived for the simulation of delamination buckling of Fibre Metal Laminates on a mesoscopic level. The individual layers of the laminate are modelled by solid-like shell elements. The delamination progress is modelled using interface elements. SEM pictures of experimental tests are used to capture the failure mechanisms on a microscopic level and to translate them into a mesoscopic damage model.

The present numerical tools have been applied to the simulation of delamination buckling of an Arall strip and a Glare panel. The obtained results are in agreement with analytical calculations and experimental observations. Delamination growth does not occur within the range of stresses that are usual in aerospace structures. However, in our simulations the stresses have been raised to higher but less realistic levels in order to pinpoint the stress levels at which delamination growth starts and to determine the accompanying failure mechanisms.

In both cases delamination growth starts at a stress state where in reality both the aluminium and the prepreg material would be completely degenerated. Delamination growth was initiated when the shear tractions in the interface material exceeded the ultimate value. This is most likely the explanation for the fact that delamination growth is not observed in axial compression tests. Here, the shear tractions in the interface at the delamination front do not increase rapidly. This observation suggests that delamination buckling can be a point of concern when the specimen is loaded in shear.

Acknowledgements

This research is part of the multidisciplinary research programme 'Reliability of Materials and Constructions' (DIOC 10) at the Delft University of Technology. The authors would like to thank T. de Jong for providing the SEM-pictures of the delaminated specimen.

References

[1] A. Vlot, *Low velocity impact loading in fibre reinforced aluminium laminates and other aircraft sheet materials*. PhD thesis. Delft (NL): Delft University of Technology, 1991.

[2] O. J. Bosker, *Growth of debonds in glare, part I (no moisture absorption)*. Technical Report B2V-98-24. Delft (NL): Delft University of Technology, 1998.

[3] O. Allix and P. Ladevèze, 'Modelling and computation of delamination for composite laminates.' In: *Arch. Mech.*, volume 44: 1992, p. 5-13.

[4] J. C. J. Schellekens and R. de Borst. 'Free edge delamination in carbon-epoxy laminates: a novel numerical/experimental approach.' In: *Compo. & Struct.*, volume 28: 1994, p. 357–373.

[5] F. Hashagen, *Numerical Analysis of Failure Mechanisms in Fibre Metal Laminates*. PhD thesis. Delft (NL): Delft University of Technology, 1998.

[6] H. Parisch, 'A continuum-based shell theory for non-linear applications.' In: *Int. J. Num. Meth. Engng.*, volume 38 (11): 1995, p. 1855–1883.

[7] J. H. A. Schipperen, 'Mixed mode delamination in graphite-epoxy prepregs: A numerical analysis.' In: *Proceedings of ECCOMAS*, 2000.

[8] J. H. A. Schipperen, *Computational Modelling of Failure in Fibre Reinforced Plastic*. PhD thesis. Delft (NL): Delft University of Technology, 2000.

[9] J. W. Hutchinson and Z. Suo, 'Mixed mode cracking in layered materials.' In J. W. Hutchinson (editor), *Advances in Applied Mechanics*, volume 29: Academic Press, 1992, p.63–191.

[10] T. de Jong, *Interlaminar shear failure of GLARE*. Delft (NL): 2001. (in preparation).

[11] F. Hashagen and R. de Borst, 'Numerical assessment of delamination in fibre metal laminates.' In: *Comp. Meth. Appl. Mech. Engrg.*, volume 185: 2000, p. 141–159.

[12] J. Lemaitre and J. L. Chaboche, *Mechanics of Solid Materials*. Cambridge (UK): Cambridge University Press, 1990.

[13] P. A. Hooijmeijer and O. J. Bosker. *Debond growth in glare under compression and tension fatigue loading and elevated temperature.* Technical Report B2V-00-61. Delft (NL): Delft University of Technology, 2000.

[14] O. J. Bosker, *Growth of debonds in glare, part II (with moisture absorption).* Technical Report B2V-00-08. Delft (NL): Delft University of Technology, 2000.

[15] M. L. C. E. Verbruggen, *Aramid reinforced aluminium laminates: Arall, adhesion problems and environmental effects.* Technical Report LR 503. Delft (NL): Delft University of Technology, 1986.

[16] L. M. Kachanov, *Delamination Buckling of Composite Materials*. Kluwer Academic Publishers, 1988.

[17] S. Rinderknecht, *Delamination in Faserverbundplatten, ein vereinfachtes Berechnungsmodell.* PhD thesis. Stuttgart (D): Universität Stuttgart, 1994.

[18] K. N. Shivakumar and J. D. Whitcomb. 'Buckling of a sublaminate in a quasi-isotropic composite laminates.' In: *Journal of Composite Materials*, volume 19: 1985, p. 2–18.

19

Glare – from invention to innovation

F. Hashagen, C. Haack and M.Wiedemann

Abstract – In recent years, significant progress has been made in understanding the behaviour of Fibre Metal Laminates, especially Glare. Now, this material has been chosen to serve as skin material in fuselage sections of the Airbus A380. However, a very challenging task must still be solved. All the available knowledge and research results must be transferred into quality standards and procedures, design and sizing methods in order to meet the stringent requirements set up by clients, airworthiness authorities and Airbus quality standards. Therefore, the following paper describes major activities that are carried out to transfer Glare from an invention into a real innovation in the aerospace industry.

19.1 Introduction

Since the mid-eighties a lot of research has been carried out into the behaviour of Fibre Metal Laminates, especially Glare. Quite a number of open questions have been answered satisfactorily and promising experience has been gained from early in-service experience [1]. This knowledge has formed the foundation to decide to apply Glare in the fuselage of the Airbus A380 mega-liner. However, this decision is not the final step in the long lasting research activities. Now Glare must be transferred from an invention into an innovation.

In contrast to previous research activities where particular items, e.g. blunt notch behaviour [2], crack propagation [3] or quality aspects [4], have mostly been investigated independently of each other, now the different requirements of several functions in the design process must be met simultaneously. Therefore, the focus on the development activities shifts from phenomenological aspects to design aspects. The design with the final release of the definition dossier must incorporate all different aspects affecting the performance of the structure. Unfortunately, this implies conflicts between different targets in some cases, e.g. minimum production costs versus a highly optimised design. Additionally, to profit fully from the advantages of Glare additional actions or changes in the sizing procedures or the production process might be required. Therefore, the application of the new material in the fuselage structure is considered from a multi-disciplinary standpoint. Hereby, three main activities must be carried out: 1st the material qualification must be established; 2nd the actual design and sizing optimum must be met; 3rd production processes must be established on the work floor. Above all, the quality of the product must be assured.

In the following, case studies are presented which highlight some of the design aspects for establishing Glare within the fuselage design of the Airbus A380. However, for legal reasons and commercial aspects, only qualitative and not quantitative points will be discussed.

19.2 Panel design

19.2.1 Composite versus metal

As we highlight the research into Fibre Metal Laminates and take into account the recent discussions and findings about Glare we have to accept the fact that both metal and composite characteristics must be taken into account when designing for Glare [5, 6, 7]. The focus of the recent research has been on the three standard Glare layups, i.e. Glare 2, Glare 3 and Glare 4. Many material design allowables are available for the latter types, and these can be applied for sizing. The Metal Volume Fraction approach (MVF) [8] is a good example of how these characteristic Glare types can be described. However, to fully profit from the potential of Glare we have to use non-standard types of Glare. In contemporary designs, these types are applied as doublers in the door surrounding areas where the fibre orientation is set to best fit the load distribution around the door. Consequently, the finite element description of Glare is based on an orthotropic shell applying the classical laminate theory [9]. When certifying the structure this approach must find its equivalent in the qualification process for the structural element in which these special Glare grades are applied.

A similar line of reasoning requires severe changes of the justification procedure that is carried out as very extensive post-processing step after a finite element calculation. This post-processing is based on the analytical and half-analytical description of skin-stringer combinations and of the skin-frame combinations. For isotropic materials a lot of different sizing methods have been derived. A huge collection of these methods, e.g. the Euler-Johnson and Kuhn methods, is summarised by Niu [10] and Bruhn [11]. For sizing, the range of application of these methods must be extended to Fibre Metal Laminates for which their orthotropic character and temperature dependency must be taken into account. In particular, the stress-strain relation can be set up according to the method by Verolme [12] in which the direction dependent stress-strain relation can be accounted for by variation of the input material parameters.

On the other hand, typical metallic features can be observed in Glare, e.g. severe plastic behaviour and fatigue. A macroscopic approach is well suited for describing these effects in the material. Consequently, both approaches are incorporated into the justification methods currently applied in the programme packages ISSY and SCHADI [13] at EADS-Germany.

19.2.2 Design studies

Prior to the design of the fuselage, general aspects of the design have to be studied. Therefore, a typical part of the fuselage has been built for the megaliner. This should provide us with important initial experience for sizing and designing such a large structure.
A typical distribution of the bending moment is depicted in Figure 19.2.
In the rear fuselage the bending moment M_y is much higher than that of the front fuselage. Consequently, the longitudinal tension and compression forces both for fatigue and static load cases acting on a member in these areas are best suited to study the load bearing capacity of new materials and design concepts.
Following an extensive period of sizing studies, the final panel layup for the Airbus A380 has now been chosen. Glare offers a lot of sizing possibilities, i.e. application of Glare 3, Glare 4a or Glare 4b which must agree with other requirements (even non-technical ones) to obtain an optimised structure.

Two main design drivers can be observed when applying Glare. First, the two-bay-frame-crack criteria determine the thickness of the panels, Figure 19.1. Glare shows superior behaviour in this respect, which gives rise to some promising weight savings. Second, the blunt notch criterion is of major importance for the choice of the Glare type and panel thickness. One has to bear in mind that these two failure criteria represent a huge number of different phenomena that are checked during the design studies.

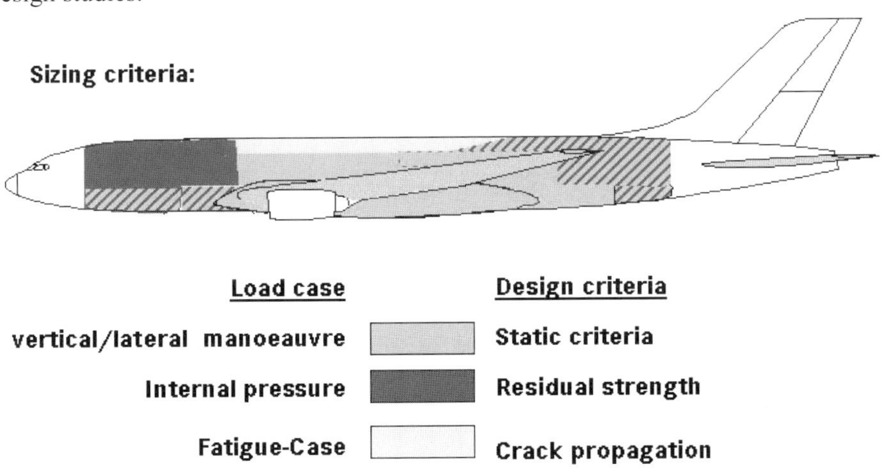

Figure 19.1 Sizing criteria of a typical fuselage and corresponding load cases.

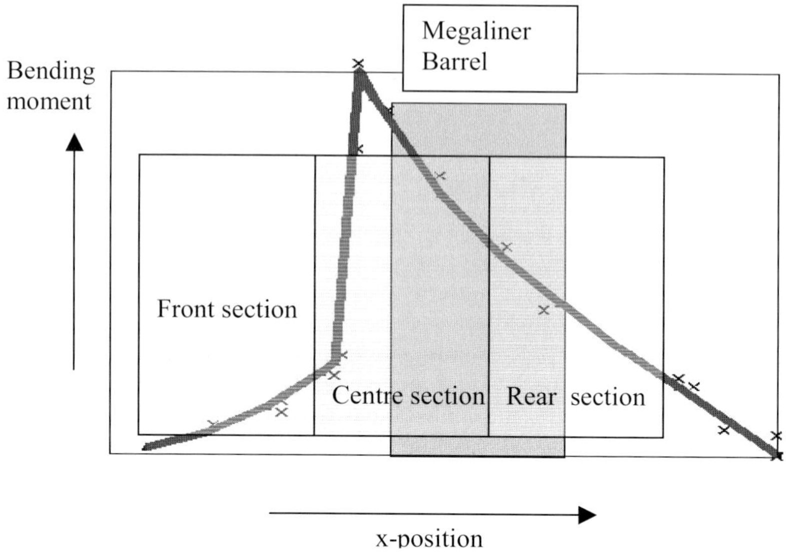

Figure 19.2 *Typical distribution of the bending moment M_y along the x-axis of a typical fuselage and location of the mega-liner barrel.*

19.2.3 Residual strength capability

Taking the residual strength capability of Glare as an example, we have to incorporate more than mechanical aspects into the analysis. Certain design requirements must be met if we are to profit fully from the very good residual strength allowables of the Glare skin. The frames of the back-up structure could be made of high strength aluminium. However, their allowable fatigue stress has to be compared to that of 2xxx-alloys in the corresponding load cases. Additionally, when sizing leads to design solutions we have to accept that some stress allowables are only valid with certain detailed design solutions, e.g. choice of rivets. These can be precisely determined in experiments, but may lead to problems for serial production. Therefore, to arrive at a highly optimised structure that fulfils the client and manufacturer requirements, the fuselage structure must be seen as a whole system consistent of all structural elements (frames, stringers etc., see Figure 19.3) with all sizing criteria, load cases and feasible design solutions.

For all these structural elements, approximately twenty different sizing criteria that result from fatigue, damage tolerance and static load cases will be checked. Additionally, load cases resulting from ground operations such as transport of individual panels before final assembly must be taken into consideration. Consequently, a continuous optimisation must be carried out. In this optimisation process no single criterion is considered independently from the others, but all sizing criteria and the corresponding load cases have to be taken into account. At EADS-Germany with ISSY [13] a highly specialised software tool has been developed in the last years in order to cover all these aspects.

Next to the phenomenological analysis that can be carried out in FEM studies, the experience from structural testing must also be incorporated into the design solutions. To meet the residual strength criterion a certain type of riveting is required when mounting the thin skin to clips and stringers. In this case, rivets must be qualified and applicable to the particular production process. Consequently, additional investments in new machines could be required, although this might conflict with the planning and possibilities at the production sites.
Therefore, the challenge of the design process lies in the combination of the different functional requirements. Turning Glare from an invention at laboratory level to an industrial innovation depends largely on the performance of this development process, see Figure 19.4.

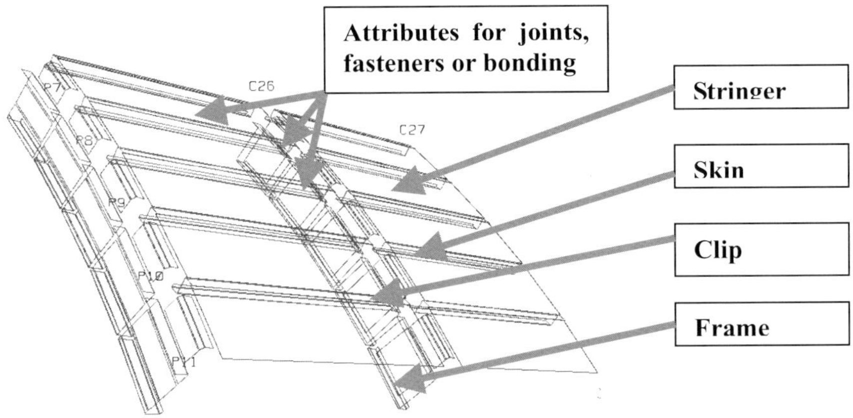

Figure 19.3 *Typical ISSY-model of a panel with stringers, frames, clips and skin.*

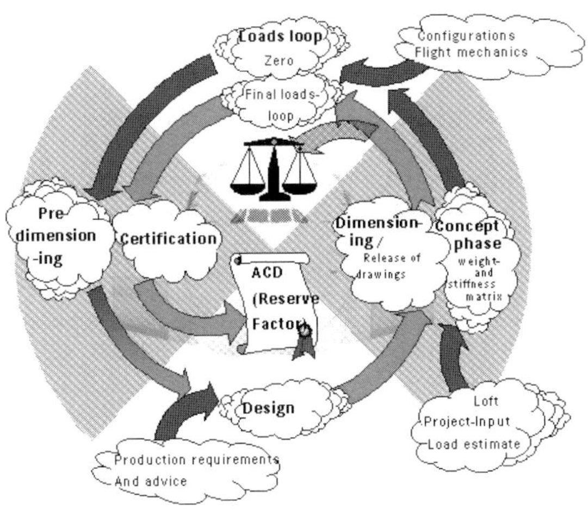

Figure 19.4 *Design process and major parameters of the design process.*

19.2.4 Blunt notch criterion

The situation is different in section 18 of the A380. The blunt notch criterion should be considered for sizing due to the high bending moment and resulting tension stress. The two-dimensional loading situation and the orthotropic character of Glare should especially be considered. Areas with high shear stress suffer from the chosen layup with fibres oriented only at 0 and 90 degrees. It is therefore crucial for establishing a sound design that sufficient design methods become available. For the blunt notch criterion, these design methods must take the orthotropic character into account, as well as the degradation of the material parameters as function of the temperature. Based on intensive research [2] in this field, a common approach has recently been established to account for this sizing criterion [5, 14].
Additional tests are still planned in order to assess how the blunt notch criteria will affect structural details such as skin-clip connections. The table of material design allowables should be extended, especially for the door surrounding with the special layup. In parallel to this study, investigations about surface protection systems and edge protection will show how the degradation of the blunt notch allowable due to temperature and ageing can be minimised.

Figure 19.5 shows the stress distribution. At the section interfaces, special attention should be paid to the blunt-notch sizing criterion when sizing the butt strap. As a result of the possible fibre orientation, Glare could lead to some new design possibilities.

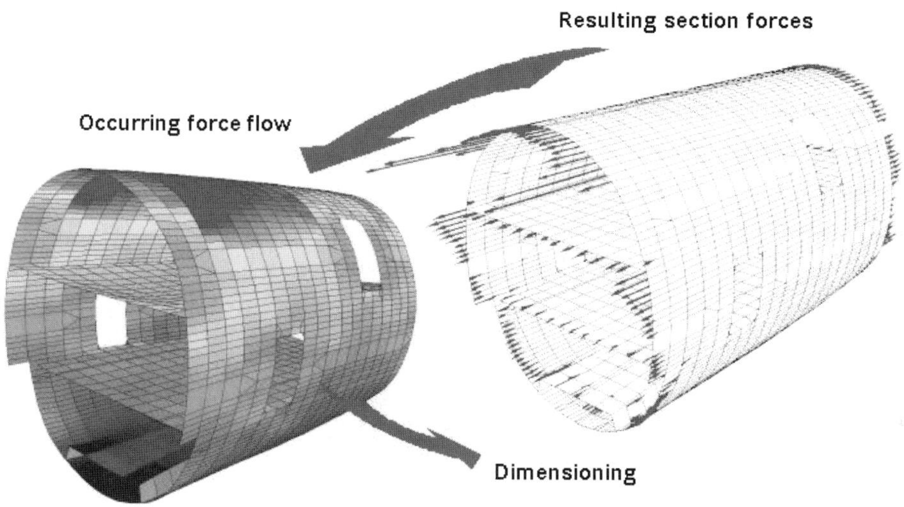

Figure 19.5 Stress distribution in the fuselage (not to scale).

19.3 Further investigation

Next to the current design work, future activities and a further development of Glare must also be taken into consideration. In this light, stringers become an attractive option for manufacture from Glare. Recently, many development activities have been initiated in order to set up the production process and assess the performance of Glare stringers in the fuselage design.

An important parameter for the performance of Glare stringers is the crippling stress. Crippling describes a local buckling phenomena of the flanges of the stringer. The critical crippling stress of the stringer enters the design methods as a sizing parameter. Therefore, tests have been carried out with Glare stringers of a thickness that could also be applied in the fuselage of the A380. The results obtained in buckling tests have been compared to results calculated via design rules for crippling stress [12]. The impact of elevated temperature on the crippling stress has also been studied. Figure 19.6 depicts a Glare stringer after loading. We observe a delamination in the buckled area of the flange in the post-buckling region as a secondary failure mode after crippling. In Figure 19.7, the calculated crippling stresses and the experimentally measured crippling stresses are shown. The impact of the elevated temperature leads to a reduction of approx. twelve percent and must be taken into account for sizing.

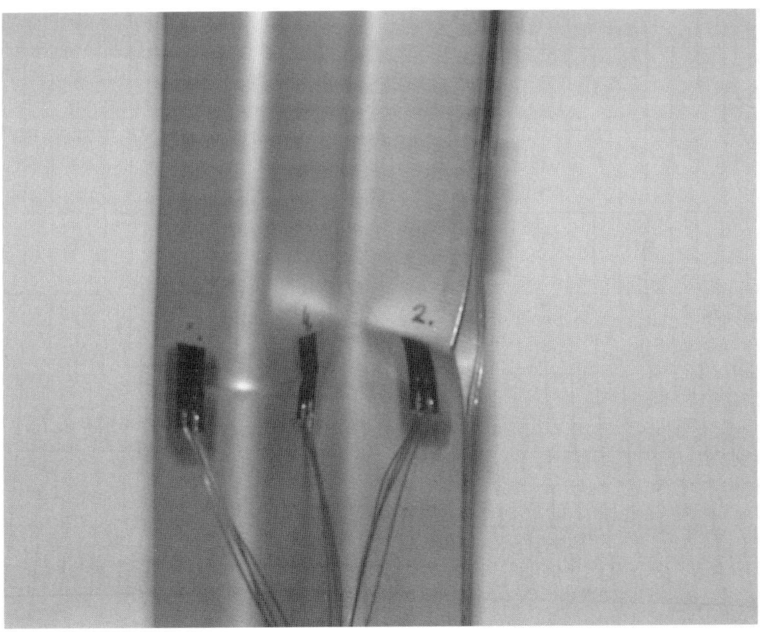

Figure 19.6 Destroyed Glare stringer, with delamination in the post-buckling region.

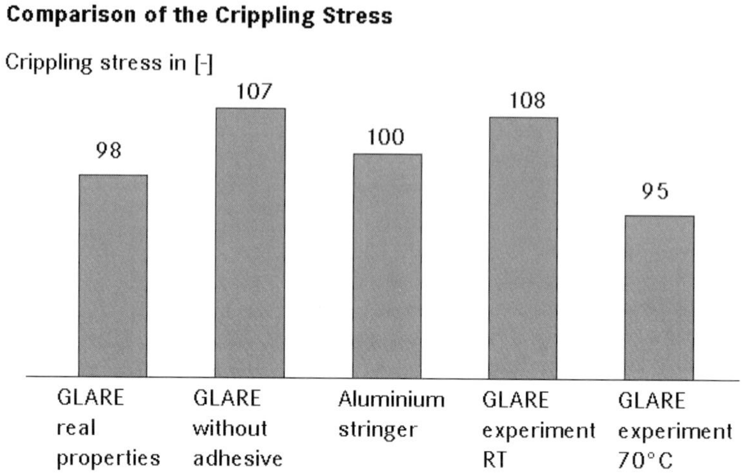

Figure 19.7 *Dimensionless crippling stress of Glare stringers based on theoretical calculations and experiments.*

19.4 Summary

This paper has emphasised that the current activities in the development of Glare focus on the transformation of the research result into an industrial process. The challenge of this transformation lies in the fact that all items of concern have to be considered conjointly and cannot be seen as separate, independent items. The example of the residual strength capability has shown that now even non-technical items such as investment into new machinery must be taken into consideration when new design solutions are proposed.
Furthermore, entering a new technology gives room for new basic changes in the design concept and offers the possibility to further increase the performance of the fuselage structure. Research should be carried out in parallel with the application of Glare on the A380 to allow us to fully profit from this potential.

References

[1] B. van Wimersma, *A3XX status reporting on in-service experience and full-scale testing of Glare*, AI/LE-T-presentation, 1998.

[2] T. de Vries, *Blunt and sharp notch behaviour of Glare laminates*, Dissertation, Delft University of Technology, 2001

[3] R.P.T. Müller, *An experimental and analytical investigation on the fatigue behaviour of fuselage riveted lap joints*, Dissertation, Delft University of Technology, 1995.

[4] R.A:M. Coenen, *Design of a quality assurance system for structural laminates*, Dissertation, Delft University of Technology, 1998.

[5] O.J. Bosker, *Tsai-Hill criterion to predict the bi-axial blunt notch strength of Glare*, Report: TD-R-0001A, Structural Laminates Industries, 2001

[6] T. Beumler, *Glare durability program, summary of the results*, Minutes of meeting EMF-932/99, 1999.

[7] F. Hashagen, *Numerical analysis of failure mechanisms in fibre metal laminates*, Dissertation, Delft University of Technology, 1998.

[8] G. Roebroeks, *The metal volume fraction approach*, TD-R-003-06, SLI report, Structural Laminates Company, 2000.

[9] R.M. Jones, *Mechanics of composite materials*, Mc-Graw-Hill Book Company, New York, 1980.

[10] M.C.Y. Niu, *Airframe structural design*, Hong Kong Conmilit Press Ltd, Hong Kong, 1988

[11] E.F. Bruhn, *Analysis and design of flight vehicle structures*, Jacobs Publishing, Inc, Indianapolis, 1973.

[12] K. Verolme, *The development of a design tool for fiber metal laminate compression panels*, Dissertation, Delft University of Technology, 1995.

[13] O. Erdmann, et al., *SCHADI Theory Manual*, EADS-Germany, 2000.

[14] O. Erdmann, *Letter from Mr. Erdmann (EADS-Germany) to Mr. Van Hengel (Structural Laminates Industries)*, EADS-Germany, 2000.

20

Secondary applications

J.W. Evancho

Abstract - Advanced Fibre Metal Laminates, Arall and Glare, were developed at TU Delft, The Netherlands. These patented, engineered materials were initially developed for application in the construction of fatigue prone aircraft components such as lower wing skins and fuselage skins. While ideal for these applications, commercial application in these primary structures was slow since new aircraft models were required in order to take full advantage of these materials and to avoid high additional costs of certification. Material characteristics of Fibre Metal Laminates are such that they also provide significant advantages over monolithic metals (aluminium, magnesium and titanium) and composites when used in the manufacture of secondary structural aircraft components. The materials are high-strength and lightweight. They are highly resistant to fatigue-crack propagation and impact damage. They are also highly resistant to flame penetration. Such characteristics make these materials ideally suited for flap skins, cargo bay liners, floors and firewalls. The materials are also ideal for use in the manufacture of speciality airline containers. This paper will review the characteristics of Fibre Metal Laminates and will give examples of significant improvements made in the service lives of aircraft components manufactured with these materials.

20.1 Introduction

Fibre Metal Laminates (FML), Arall and Glare, were originally developed for application in primary aircraft structures. These unique hybrid materials were configured to provide an optimum combination of resistance to fatigue crack propagation, high strength and excellent damage tolerance, all properties significant to the design of primary aircraft structural components such as the fuselage skin and lower wing skins. The unique attributes of these laminates are the result of a patented configuration that utilises straight, parallel fibres: aramid in the case of Arall and high strength S2-glass in the case of Glare. The optimum combination of properties result from the interleaving of these fibres, which are imbedded in an epoxy adhesive, between sheets of thin, aerospace grade aluminium [1, 2]. Figure 20.1 shows a schematic representation of a Fibre Metal Laminate with unidirectional fibres. Different configurations have been developed to provide optimum combinations of properties depending on the application of interest [3]. The literature abounds with papers documenting Arall and Glare's attributes towards this objective.

Arall and Glare meet their originally intended objectives of providing unique combinations of properties of interest to aerostructure design engineers. However, while considered a technical success, the introduction of these materials into commercial application has been considered by some a commercial disappointment. Commercial acceptance into primary aircraft applications has been frustratingly slow. This is not because the materials do not perform as they were designed to. The principal reason for slow application in primary components is because of the dearth of new aircraft programmes during the decade of the 1990's. The only new aircraft programme with timing to coincide with the development of the current family of Fibre Metal Laminates was McDonnell Douglas' C-17. Arall did earn its way onto this aircraft in a significant primary application, the outer skin on the aft cargo bay door Figure 20.2. The high cost of certification essentially blocked these materials from replacing aluminium in primary structures of existing aircraft models.

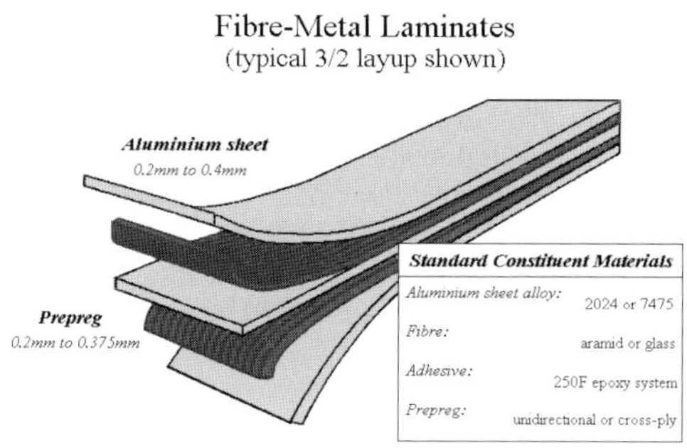

Figure 20.1 Schematic representation of unidirectional Fibre Metal Laminate.

During this time period, Boeing was already well into the development of its 777 aircraft. Unfortunately, the aircraft programme was at the end of its development phase by the time Glare was introduced and there was no possibility that it could be considered for application in a major primary component. Fortunately, Boeing recognised that Glare was a material that offered considerable benefit and they launched an intensive effort to find a possible use for it in the 777. Recognising the weight savings opportunity offered by Glare over aluminium, Boeing evaluated it for use in the bulk cargo bay floor. The weight savings provided by Glare was further magnified by the good thermal insulation properties of FMLs. This enabled a further weight reduction by reducing the insulation materials required in the bulk cargo area.

Figure 20.2 Arall aft cargo door skin on the C-17 (Arall 3-4/3).

Consideration of Glare for this application led to it being evaluated for impact and burn characteristics. Results of these tests revealed the excellent burn-through resistance and high impact strengths. Specification of Glare 3-3/2 for the bulk cargo bay floor of the 777 showed the materials viability for secondary applications. This not only created a new market opportunity for FMLs but, most importantly, it created a more immediate market opportunity in the airline after-market. This paper discusses the benefits of FMLs, particularly Glare, for secondary applications.

Attributes of FMLs for secondary applications are:

- low weight
- high strength
- high impact strength
- flame resistant
- corrosion resistant
- good thermal insulation
- traditional workshop properties
- easy reparability
- low maintenance.

These benefits make FMLs ideally suited for certain secondary aircraft part applications such as: control surfaces, non-pressurised doors, faring surfaces, cargo bay floors and liners, firewall liners, non-pressurised bulkheads and containers. Application of FMLs in certain of these applications will be discussed in further detail. The secondary applications can be grouped in two general categories: those that take advantage of an FMLs excellent fatigue resistance and those that utilise the good impact strength and flame resistance of Glare.

20.2 Control surfaces and flaps

The outstanding resistance to fatigue crack propagation, see Figure 20.3, make FMLs ideally suited for use as an outer skin on flaps and control surfaces.

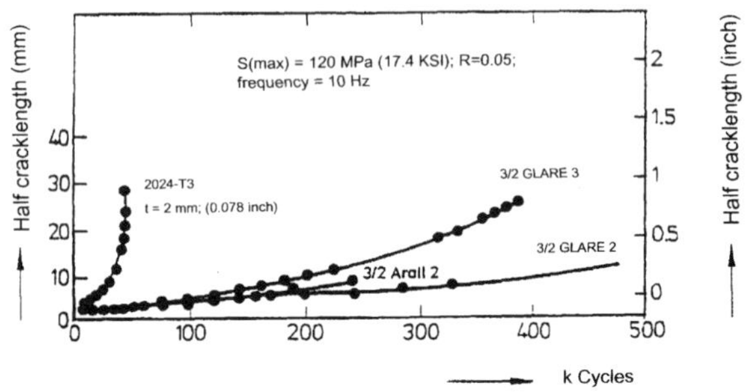

Figure 20.3 *Comparison of fatigue crack growth rate of Arall, Glare and 2024-T3 aluminium* [4].

Original work by Fokker Aircraft in the 1980's demonstrated Arall's outstanding resistance to fatigue crack propagation when applied to the lower wing skin of the Fokker 50. This work led to consideration of Arall in flap skins on the United States Air Force C-130 Hercules aircraft. These Special Forces aircraft experienced considerable fatigue damage due to their aggressive flight patterns and sonic waves from the ships side guns. Fatigue cracking in the lower aluminium skins typically occurred after 160 flight hours requiring continuous field repair and eventual removal of the flaps and replacement of the skins. The flaps also suffered sever impact damage from the rough runways. Under a programme to solve service problems by applying new technology, the Air Force funded evaluation of Arall as a replacement skin for the C-130 lower flaps. This programme demonstrated that replacement of the aluminium skins completely eliminated fatigue cracking in flaps that were flown for over 2000 flight hours. The flaps and skins were removed for detailed inspection. The Arall skins showed no fatigue cracking and no impact damage and ultrasonic inspection revealed no disbonding. The interior structure of the flaps was also in pristine condition. Figure 20.4 shows the flap with the Arall skin after flight testing and Figure 20.5 shows a similar flap with the conventional aluminium skin.

In 1991, DeHavilland demonstrated the advantage of using Arall for flap skins, but from a somewhat different perspective than what the industry had been focusing on [5]. Generally, Fibre Metal Laminates generated enthusiasm as an advanced aerospace engineered material since its high resistance to fatigue crack propagation provided an opportunity to either reduce weight during the design of a structure or drastically increase the life of a component. The focus of DeHavilland was to take advantage of improved properties of Fibre Metal Laminates and eliminate the waffle-doubler design in their flaps, see Figure 20.6.

Figure 20.4 Arall lower flap skin – C-130 (Arall 3-2/1).

Figure 20.5 Conventional aluminium C-130 lower flap skin.

This approach resulted in a simpler, and more importantly, cheaper component to manufacture. This was probably the first consideration when using Fibre Metal Laminates to reduce manufacturing costs. This approach was one of the factors that led to specifying Glare for the fuselage crown of the A380 aircraft.

Another Air Force demonstrator was initiated due to a fatigue-cracking problem present in T-38 dorsal covers, see Figure 20.7. Besides the severe fatigue-cracking, these covers, which were manufactured from magnesium, also presented a safety hazard as the magnesium would incinerate when struck by lightning. The five covers aft of the cockpit were redesigned in Arall and subjected to both static and in-flight testing. Both static and in-flight tests demonstrated Arall's outstanding resistance to fatigue cracking. Initially installed in 1994, the Arall covers are still flying with no fatigue cracking. Furthermore, because Arall and Glare have excellent lightning strike characteristics, a serious safety hazard can be eliminated.

Figure 20.6 *DCH-8 lower flap skin (Arall 3-3/2).*

Figure 20.7 *T-38 Dorsal cover (Arall 3-2/1).*

20.3 Impact properties

While the outstanding fatigue resistance of Fibre Metal Laminates provided an initial opportunity for improved performance in secondary aircraft components, the greatly enhanced impact strength of Glare generated interest leading to most of today's commercial applications. Boeing specified Glare 3–3/2 for the bulk cargo bay floor-material in the B777. This configuration of Glare had been optimised for strength and fatigue. Because of the interest in impact prone areas of aircraft, research was carried out by Structural Laminates bv to optimise a Glare configuration specifically for impact performance. Effects of aluminium alloy, glass fibre type, thickness of the fibre layers and unidirectionality versus weaves were investigated [6]. Results of this work are shown in Figure 20.8.

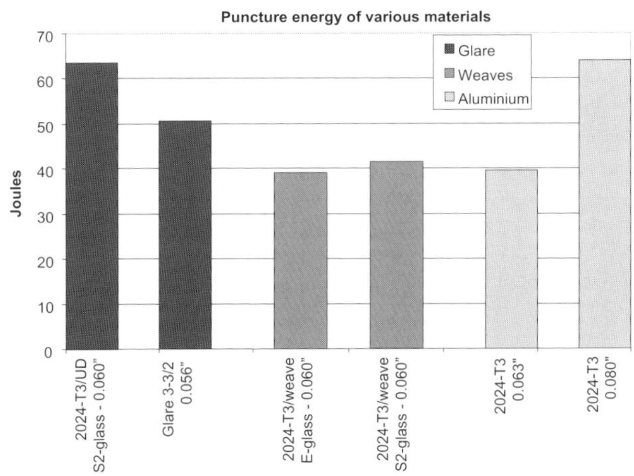

Figure 20.8 Impact performance of various materials.

The results revealed interesting findings. It is evident that FMLs produced using unidirectional S2-glass fibres provided maximum impact performance when compared to laminates made with either S2-glass or E-glass weaves. The Glare optimised with a fibre layer 0.060 inch thick provided maximum impact performance. This work supported the patent for Glare 5 [7]. Because impact strength of a component is also dependent on thickness of the component, it is more illustrative to compare impact strength on an equal thickness basis. Figure 20.9 compares different grades of Glare to 2024-T3 aluminium sheets as a function of thickness. The impact strengths of FMLs made from weaves are also shown. The lower impact strength compared to the patented Glare is significant.

More importantly for the airlines, a similar presentation can be made as a function of surface density, as shown in Figure 20.10. Both Figure 20.9 and Figure 20.10 show the outstanding impact strength of Glare. Depending on the preference of the airline, this improved performance can be used to reduce maintenance costs resulting from impact damage, to reduce weight and accept the same level of impact damage, or a combination of both. Details showing economic benefits that can be derived from replacing aluminium floors with Glare 5 are presented later in this paper.

Substitution of Glare instead of aluminium sheets for cargo bay floors is relatively straightforward in most Boeing aircraft. Bulk cargo bay floors in the Boeing 737, 747, 757, and 767 are simply sheets of 2024-T3 aluminium. The floors of the 777 are FML although not Glare. Glare 5 meets the requirements of FAR 25.855. A simple substitution of Glare 5 for the aluminium sheet will provide significant extension of service life of the floors, reduce maintenance costs, reduce weight and provide superior protection against flame burn-through compared to sheet aluminium. Potential damage from corrosion is also significantly reduced or eliminated. While other, lesser performing FML floor products offer many of the same advantages offered by Glare, impact strength is not improved and maintenance cost reductions are not realised.

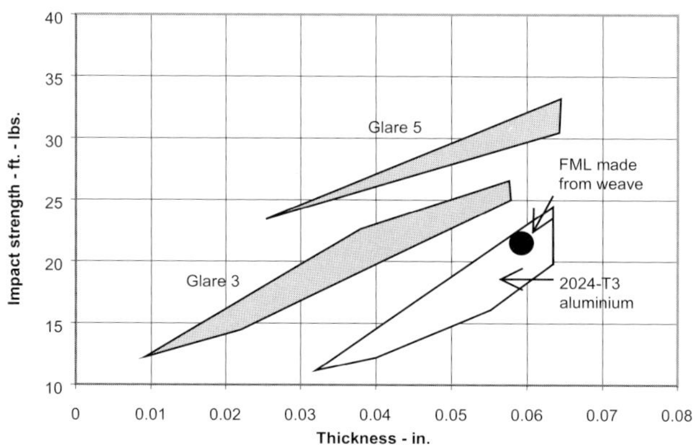

Figure 20.9 *Impact strengths of Glare versus thickness.*

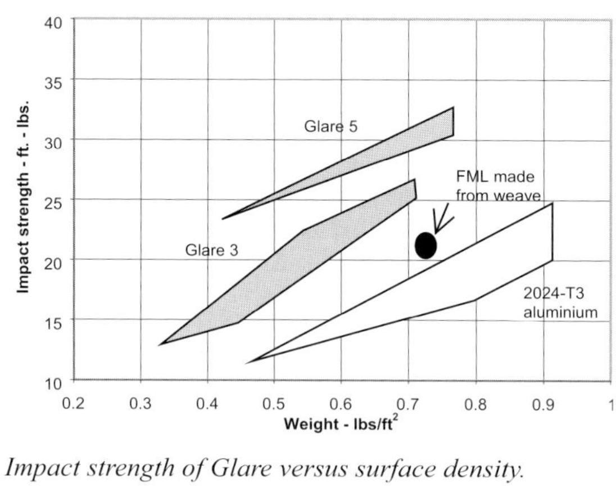

Figure 20.10 *Impact strength of Glare versus surface density.*

Several airline trials have been performed with Glare 5 as the cargo bay floor to support these claims: United Airlines (B757), US Airways (737), Qantas (737, 747), Midwest Express (MD-80 & 88 aft cargo bay), and Air Pacific (737). Of these, Qantas, Midwest Express and Air Pacific are now using Glare 5 as replacement material for cargo bay floors, while United is using a FML made from woven fibres which consequently has a lower impact strength. Figure 20.11 and Figure 20.12 show the original US Airways Glare 3–3/2 test panel. This panel was removed for evaluation after a service period of 2 years and 8 months. No through-penetrations were encountered. Installation of the Glare floor panels in the Qantas 737 is shown in Figure 20.13.

Figure 20.11 *Glare 3–3/2 floor panel after 2 years and 8 months service in US Airways 737.*

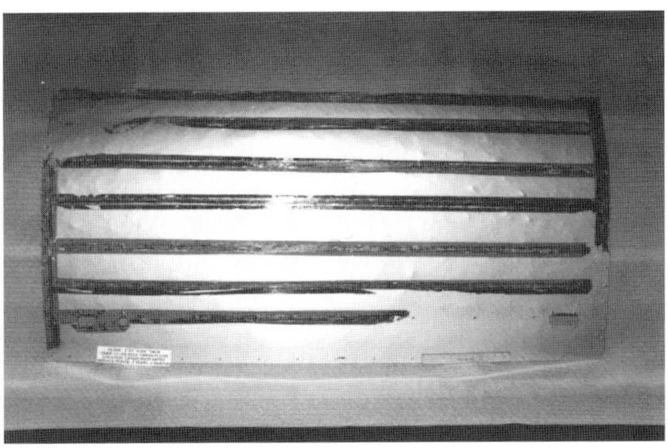

Figure 20.12 *Back-side of panel in Figure 20.11.*

Figure 20.13 Glare 5–2/1 installed in Qantas 737.

20.4 Glare in sandwich panels

Cargo bay floors in aircraft generally fall into two categories:

1. Single panels of metal or composite.
2. Sandwich panels consisting of metal or composite outer sheets and honeycomb or balsa wood cores.

While both concepts offer high impact strength and excellent corrosion resistance, each has their disadvantages. The single panel is less rigid and is susceptible to sagging, while the sandwich panel is considerably more expensive. On the other hand, the single panel is less expensive and the sandwich panel is much stiffer. Generally Boeing aircraft follow the single panel concept and Airbus aircraft use sandwich panels. Regional aircraft generally use sandwich panels.

Considerable work has been done evaluating the benefits of using Glare as the top sheet of honeycomb or balsa core sandwich panels. Tjahjono investigated using Glare 5–2/1 as a top sheet over a balsa core sandwich for the cargo bay of a MD-11 [8]. His evaluation showed significant improvements in impact properties and corrosion. Field-service trials of the sandwich panel extended the life of the floor by at least 100% compared to the original OEM floor. Service testing was stopped before the actual life could be determined in order to do laboratory testing of the service panel.

Figure 20.14 Glare floors and liners in cargo bay of Garuda A330.

Subsequent to the successful efforts of Tjahjono, Airbus Industries together with Structural Laminates Company, embarked on a major floor and liner development programme to improve the life cycle of Airbus sandwich panels used in cargo bay floors and liners. This work also showed that using Glare as a top sheet of the sandwich structure improved impact performance over monolithic aluminium or composite top sheets [9, 10]. Service trials have been going on since June of 1999 with excellent performance being reported for the Glare floors and liners. Figure 20.14 shows the floors and liners installed in a Garuda A330 aircraft.

20.5 Economics of Glare in aircraft floors

Improved properties in aircraft structures are of no interest if the improvements cannot be translated to either improved passenger safety or reduced operating costs. In the case of cargo bay floors, the primary incentive for the airlines is life cycle cost savings.

As an engineered material, Glare is more expensive than sheet-aluminium when compared on a unit purchase price basis. When compared on a lifecycle cost basis however, Glare shows a significant cost saving over aluminium. Analysis, based on actual service data, illustrates the savings provided by the use of Glare as bulk cargo bay flooring.

The analysis compares the use of 0.060 inch thick Glare 5, to panels 0.063 inch thick 2024-T3 aluminium floor panels. Differences in labour and material costs associated with floor panel removal and re-installation, perforation repair and fuel savings incurred from weight differences were calculated on an annual basis. Cost savings are reported as annual average cost savings, cumulative cost savings over a thirty

(30) year period and net present value of the cumulative savings using a 9% discount rate. Costs/savings are for a single 4 feet by 12 feet panel or 48 square feet of flooring. Larger or smaller areas would be affected proportionately. The analysis was conducted for a heavy use panel (entrance to bulk cargo bay).

Assumptions for this analysis are listed below in Table 20.1. While the actual service experience will vary for each individual airline, the assumptions used for this analysis are based on actual airline experience. The results indicated for the heavy-use panel are based on typical service data.

Results of the analysis are indicated in Table 20.2 and Figure 20.15. In this example, total cumulative cost savings over a 30 year period for one panel is $31,440, a net present value of $8,061 per year. For operators who prefer to use savings per flight hour or flight mile, the indicated results can be divided by the number of hours or miles per year.

Table 20.1 Assumptions for analysis.

Labour cost	$65 / hour
Glare 5 price	$26 / sq. ft
2024-T3 price	$5.00 / lb.
Fuel savings	$25 / lb. weight saved/year
Time required to replace floor panel	8 hours
Time required to repair perforation	1 hour
Repair patch size	6 sq. in
Aluminium heavy use panel life	1 year
Glare heavy use panel life	5 years
Aluminium heavy use repair frequency	1 / month
Glare heavy use repair frequency	1 / year

Table 20.2 Floor life cycle cost analysis heavy use panel.

	Aluminium	Glare
Thickness (in)	0.063	0.06
Total floor weight (lbs)	43.98	37.05
Total floor cost ($)	$ 219.91	$ 1,248.00
Repair frequency (#/yr)	12	1
Patch size (in^2)	36	36
Repair time (hr)	1	1
Replacement frequency (#/yr)	1.0	5
Total cumulative cost savings	$ 38,976	
NPV savings	$ 10,055	
Discount rate	0.09	

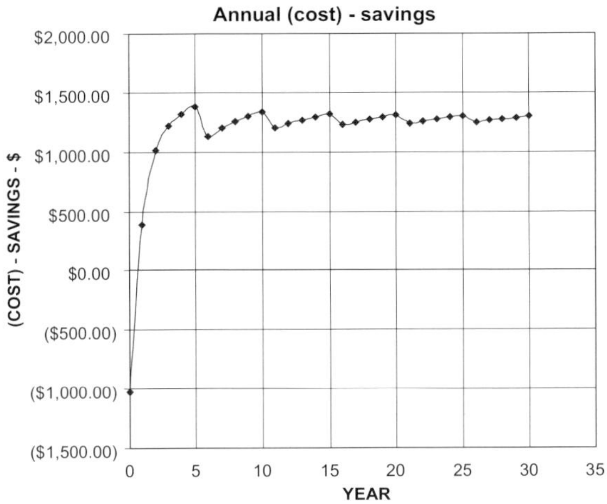

Figure 20.15 *Lifecycle cost savings for Glare cargo bay floor panels.*

20.6 Other impact prone Glare applications

20.6.1 Bulkheads

One of the big advantages of Glare imparted by the S2-glass fibres is the positive effect of strain rate on strength and strain to failure, see Figure 20.16. This effect results in even further increases in impact strength for higher velocity impact. Shorts Brothers plc took advantage of this attribute in development of the radome front bulkhead of the Lear 45, see Figure 20.17. This part has been in production using Glare 3-6/5 since 1996. Glare provided a 39% mass reduction compared to aluminium plate, a 19% mass reduction and a cost reduction compared to sandwich material consisting of composite and honeycomb and, a 23% mass reduction and a cost reduction compared to sandwich material consisting of aluminium and honeycomb.

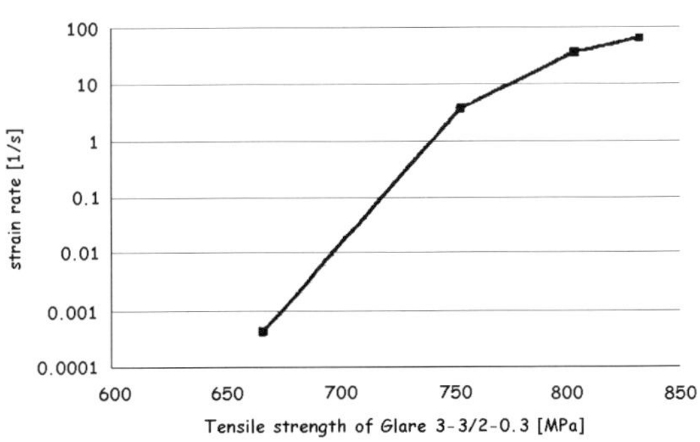

Figure 20.16 *Strain rate effect on tensile strength of Glare 3* [11].

Figure 20.17 Shorts Brothers Lear 45 radome front bulkhead made from Glare.

20.6.2 Airline and speciality containers

Glare's attributes that make it ideal for use in aircraft also make it an ideal material for construction of speciality airline/transportation containers. Its high impact strength leads to significant reduction in handling damage and consequently reduced repairs and reduced lifecycle cost compared to containers made from more conventional materials. As in aircraft applications, Glare's low weight and excellent corrosion resistance provide additional advantages. Because of the positive strain rate effect and Glare's resistance to flame penetration, it is most well known in this application area as the material used in the manufacture of the only airline luggage container meeting the blast mitigation requirements of the FAAs Draft Appendix A of ISO 6517.

The FAA, with participation of the department of Defence, US aircraft manufacturers, and the aviation industry, has been investigating ways to increase an aircraft's resistance to blast effects. As a participant in this programme, Galaxy Aviation Security, L.L.C., Egg Harbor Township, New Jersey developed the ECOS$^{3®}$ Secure Luggage Container (LD-3 configuration).

The ECOS$^{3®}$ is constructed of panels of Glare 5 and Glare 7, which are joined to an aluminium frame system. FAA testing has demonstrated its ability to withstand the harmful effects of a blast and protect the surrounding aircraft. Any flame resulting from the blast is fully contained by the ECOS$^{3®}$. Photographs of the ECOS$^{3®}$ before and after FAA explosives testing are shown in Figure 20.18 and Figure 20.19.

A big advantage provided by the use of Glare in the ECOS$^{3®}$ is that weight of the container is held to a minimum; it is only about 100 lbs heavier than a conventional LD-3 and it meets all LD-3 specifications. The ECOS$^{3®}$ has been in service tests since 1999 and units have been commercially deployed. The success of this concept has expanded to incorporate Glare in other airline container configurations and in luggage overhead bins for blast mitigation.

Figure 20.18 *ECOS$^{3®}$ before explosives test.*

Figure 20.19 *ECOS$^{3®}$ after explosives test.*

Reference

[1] J. Schijve, et al., *Laminate of Aluminium Sheet Material and Aramid Fibers*, US Patent No. 4.500.589, 1985.

[2] L.B. Vogelesang, et al., *Metal-Resin Reinforced with S2-Glass Fibers,* US Patent No. 5.039.571. 1991.

[3] Aviation Equipment, *Glare and Arall Laminates,* Inc. Product Brochure.

[4] A. Vlot, et al., *Towards Application of Fibre Metal Laminates in Large Aircraft,* Paper provided by www.glareconference.com.

[5] L. Kok, et al., *Evaluation of Arall Laminates for Fuselage Applications – Fatigue and Residual Strength Performance,* Presented at AEROMAT Conference, Long Beach, CA. 1991.

[6] Structural Laminates Company internal report, TD-R-94-006. *Test Data During the Development of Glare Floor Material and Glare Fire Walls*. 1994.

[7] G.H.J.J. Roebroeks and A.C. Mattousch, *Resistant Laminate,* US Patent No. 5.547.735, 1996.

[8] S. Tjahjono, *Consequences and Challenges of Glare for Structural Repair and Newly Designed Fuselage Structure,* Structures and Materials Laboratory, Faculty of Aerospace Engineering, Delft University of Technology. 1997.

[9] A. van Ginkel, *Optimization of Glare Floor Panels with Respect to Durability,* Pre-thesis Report, Structures and Materials Laboratory, Faculty of Aerospace Engineering, Delft University of Technology. 1999.

[10] Ferdinandus, *Impact and Corrosion Tests on New Sandwich Structures,* Structures and Materials Laboratory, Faculty of Aerospace Engineering, Delft University of Technology, 1997.

[11] G.H.J.J. Roebroeks, *A Structural Material for Fire Resistant Aircraft Fuselages*. Presented at the PEP Symposium on "Aircraft Fire Safety", Dresden, Germany: Published in CP-587, Oct. 1996.

D. Production

21

Machineability

P. Broest and J. Sinke

Abstract – Different machining processes have been tested for their applicability when machining Glare, including drilling, milling, water-jet cutting and laser-jet cutting. For machining processes in general, three phenomena are important when a particular process is applied: the wear of the cutting tool, the delamination of the laminate, and the introduction of heat into the cutting tool and the laminate. Some changes in cutting tool geometry, tool material and machining procedures may be required for successful application of a particular machining process for Glare.
This chapter describes the milling and drilling process in more detail. The influence of lubrication, laminate thickness and other parameters is also discussed.

21.1 Introduction

When manufacturing Glare components for large structures, the laminates have to be machined to the right dimensions, holes have to be drilled, and other machining processes have to be performed.
As a result of the layup and constituents of Glare, machining processes should be performed with care and one should know what parameters are important.
In this chapter a brief overview of the most important phenomena will be given. Some of the phenomena will be illustrated for the most common machining processes, i.e. milling and drilling.

21.2 General Comments on Machining of Glare

Fibre Metal Laminates are laminates with alternating layers of metal alloys and prepregs. In the case of Glare the metal alloys are aluminium alloys and the prepregs are unidirectional glass-fibre prepregs placed in one or two directions.

During extensive testing of Glare laminates, three phenomena proved to deserve the most consideration during machining operations. The three phenomena are:

- the wear of the cutting tool
- the delamination of the laminate
- the introduction of heat effected zones at the edges of the laminates.

We shall discuss these effects in this section.

The wear of the cutting tools is due to the hard and abrasive glass fibres embedded in the prepreg layers of the laminate, see Figure 21.1 for an example. Since the hardness of these fibres is relatively high with respect to most cutting tool materials, only a selected number of cutting tool materials are applicable for Glare laminates. The High Speed Steel (HSS) that is commonly used for the machining of aluminium alloys is not suitable for Glare laminates. Even HSS-tools with a Titanium-Nitride or Cobalt coating do not improve the tool life satisfactorily. The options that proved to give the best results were tools based on Solid Cemented Carbides (Hard Metal (HM) tool bits) and Poly Crystal Diamond tipped (PCD) tools. However, most of these tools are rather brittle, and a failure of the tool by shock or other accidental events is very likely. Therefore, these tools should be used on stable machines only. ***Note***: When fibre metal laminates are made from carbon fibres the same wear problem will be encountered. In the case of aramid fibres the cutting tools will not deteriorate by wear, but a slight decrease in "sharpness" of the cutting tool will result in incomplete cutting of the aramid fibres and "hairy" edges of the laminates.

Of course, tool wear is not applicable for machining processes such as water jet cutting and laser jet cutting. For these processes the edge quality is an issue: both processes produce a rough edge; 'sandy' in the case of water jet cutting and 'burred' after laser jet cutting. The edge quality is poor for both processes and can only be accepted for a limited number of applications.

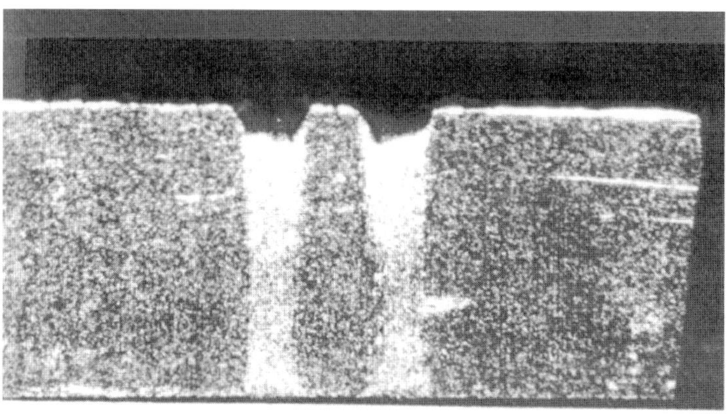

Figure 21.1 *Wear marks on the clearance plain of a Tungsten carbide mill, after 30 meters routing at 4000 rpm and a feed per tooth of 0.1 mm (used for milling of a Glare 3-3/2-0.3 laminate).*

Figure 21.2 *Schematic presentation of delamination during milling (left) and drilling (right).*

The second phenomenon is delamination. Delamination occurs when the peel forces, the shear forces, or a combination of these two exceed the maximum values.
During machining operations the peel forces are the most critical. When the machining forces perpendicular to the layers of the laminate become too high, the layers will delaminate. An example is related to the helix angle of a drill bit or a milling tool. When the helix angle becomes too large, the top layer of the laminate can be peeled from the laminate. Therefore, particularly for milling operations, the helix angle of the milling tool should be kept as small as possible. Another example of potential delamination is the feed force during drilling. When the drill bit exits the laminate the feed force may push the outside layer away and cause a delamination. The danger of such a defect is increased when the drill bit becomes dull, and a high(er) feed force is required to drill the hole. Both examples are sketched in Figure 21.2.

The last phenomenon that will be described is related to the heat generated during a machining operation. For some processes such as water jet cutting and for thin laminates this temperature increase of the laminate does not exist or is negligible. For thick laminates and for processes such as laser jet cutting the heat introduction into the laminate is an important issue. Although the heat effected zone in laser jet cutting is small (only a few tenths of a millimetre inside the material) it may reduce the applicability of laser jet cutting to rough cutting operations only. For machining operations such as milling the heat can be counteracted by the use of lubricants, but the heat introduction into the laminate may limit the feed rate and the cutting speeds.

21.3 Milling

During the milling process there are a number of parameters which can be varied such as the mill diameter, type of mill, speed, feed, milling condition, etc. A particular combination of those parameters is selected depending on the type of laminate and its thickness. During the development of Glare, a limited number of combinations have been examined to determine the influence of these manufacturing parameters on the quality of the milled edge. Especially the feed, the milling condition (lubrication) and material thickness showed a considerable influence on the quality of the milled edge. The trends that have been discovered will be discussed in this section.

The feed in the milling process is very important. With a high feed it is possible to manufacture a product in less time, which saves costs. A high feed also results in high milling forces on the milling tool and on the laminate edge. High forces on the milling tool can cause fracture of the (brittle) tool and increase of the wear of the tool, while high forces on the laminate edge can cause delamination, which is a typical failure mode for Glare laminates.

Figure 21.3 shows the influence of the feed on the edge quality. An increase of the feed leads to a lower edge quality. The figure shows two Glare 2B-11/10-0.3 laminates. The top picture has been milled with a feed of 1200 mm/min and the lower one with a feed of 1500 mm/min (for both tests the following parameters were used: mill diameter = 8 mm, speed = 8000 rpm and milling condition = semi-dry). The laminate milled with the highest feed shows delamination and edge deformation that is unacceptable.

Figure 21.3: Influence of feed on edge quality (mill diameter = 8 mm, speed = 8000 rpm and milling condition = semi-dry).

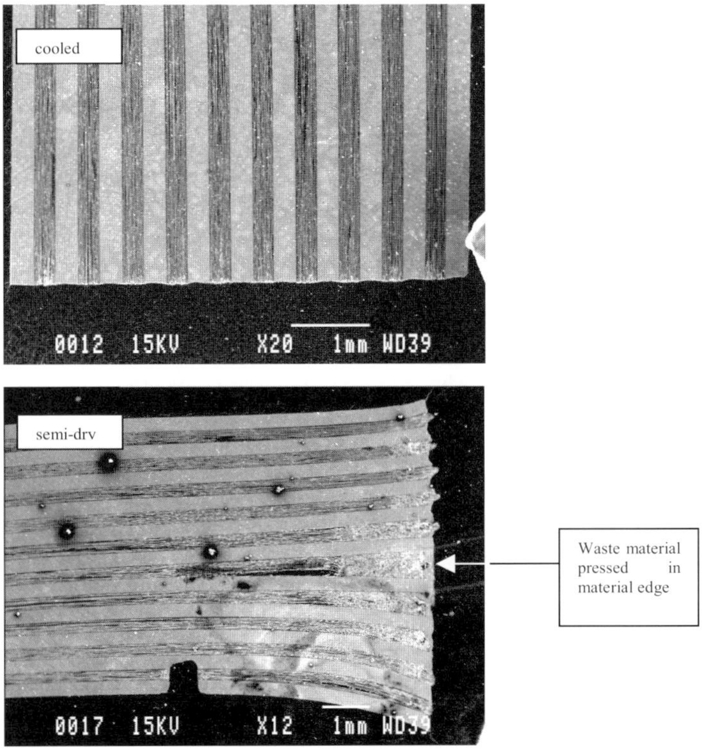

Figure 21.4: Influence of the milling condition on the edge quality.

Milling of a laminate can be performed under different conditions. For Glare two conditions with respect to lubrication have been examined: cooled milling and semi-dry milling. Semi-dry milling does not use a coolant, which saves an extra step in the manufacturing process: cleaning the product. However, during semi-dry milling higher temperatures are reached which can lead to unacceptable edge qualities (edge delamination and/or deformation). Therefore, cooled milling is an option when semi-dry milling gives unacceptable results.

The effect of the milling condition on the edge quality can be viewed in Figure 21.4. Both Glare 2B-11/10-0.3 laminates have been milled with the same parameters (speed = 8000 rpm, feed = 1500 mm/min, mill diameter = 8 mm), but under different conditions. Delamination and deformation of the edge is visible with the semi-dry milled specimen. However, semi-dry milling was possible for thin laminates, so the thickness of the laminates is also of influence on the edge quality and is related to the milling condition. The thickness of the laminate influences the milling forces on the edge and the tool. High forces cause delamination and/or deformation of the edge. A larger thickness also gives a larger contact area that produces more heat. At higher temperature the chips and fibres adhere to the milling tool. This waste material can be forced into the laminate edge and result in delamination/deformation of the material. This phenomenon can be viewed in the lower picture of Figure 21.4. When cooling is applied, this phenomenon is not present because the coolant transports the waste material and reduces the heat.

For semi-dry milling the laminate thickness is also important for the quality of the milled edge. When laminates of different thickness are milled under semi-dry conditions and the same parameters, the thick (~6 mm) laminates give less acceptable results. Figure 21.5 shows two examples of the influence of the thickness for semi-dry milling. The top photograph shows that waste material has been forced into the laminate edge. This phenomenon has been explained above. With thinner laminates the temperature will be lower and the transport of chips is easier. Heat production is lower due to the smaller contact area and the chips will not adhere to the tool.

To conclude this section: milling Glare laminates is possible when the proper parameters, conditions and tools are applied. The thickness of the laminate is also important. In general, it can be stated that for thin laminates (t ~ 3 mm) feeds between 600 and 1500 mm/min are possible for both cooled and semi-dry milling. For thick laminates (t ~ 6 mm) these feeds are only possible with cooled milling. For semi-dry milling only a feed of 600 mm/min is possible.

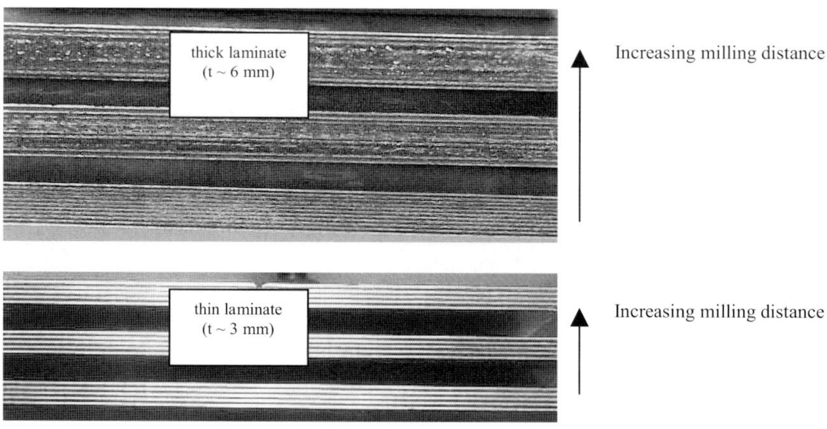

Figure 21.5 *Influence of thickness with semi-dry milling of Glare.*

21.4 Drilling

Drilling holes in Glare laminates requires other process parameters than those for aluminium due to the layered structure of the laminate. Delamination is a specific failure mode for Glare laminates and delamination of the top (helix angle) and bottom layer (feed force) is especially possible when drilling. Figure 21.6 shows an example of delamination of the top and bottom layers due to drilling of a hole with the wrong set of parameters and tools.

Drilling a hole can be done automatically or manually. Automatic drilling appeared to be an appropriate method for drilling holes in Glare. After investigation of the drilling process parameters for automatic drilling good results were obtained with specific speed/feed ratios and specific tools. There were no delaminations or deformations of aluminium layers present and the hole diameters met the tolerances. Figure 21.7 shows drilled holes of good quality (no delamination or deformation). Following tool-life tests, it has been proved that this quality could be maintained up to 700 holes. Even at a higher number of holes the tool still did not show wear and the holes had a good quality.

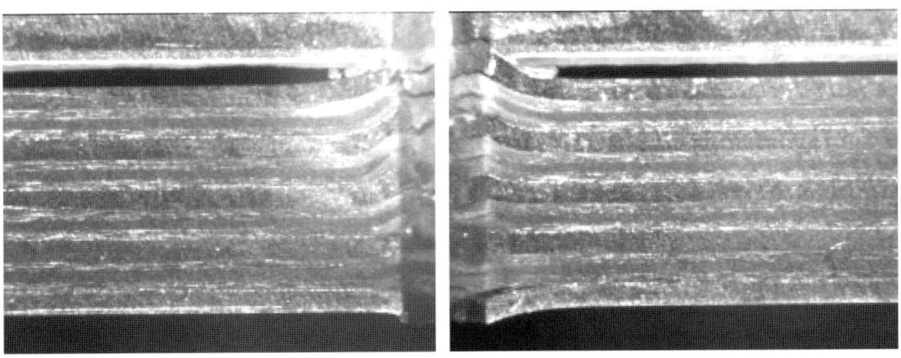

Figure 21.6: Delamination of top and bottom layers (manually drilled specimen).

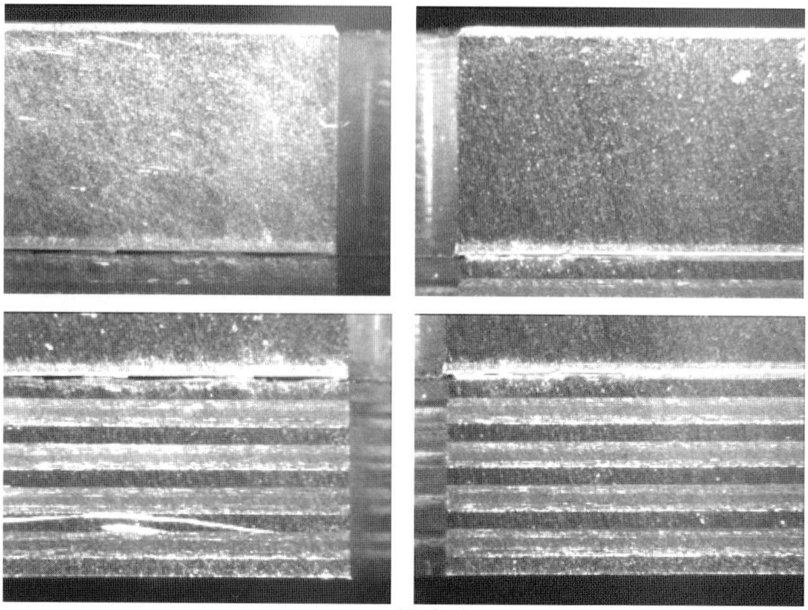

Hole 2

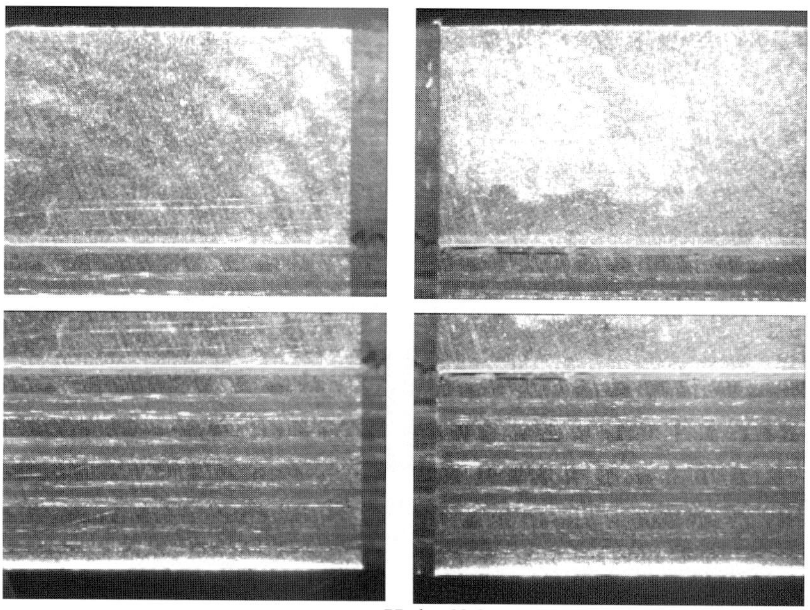

Hole 696

Figure 21.7: *Good hole quality (tool-life tests with 6.26 mm MBBN 624 drill, tool material: HM, speed/feed = 6000/0.15 rpm/mm).*

However, during the assembly of substructures, manual drilling of Glare is required for drilling holes in positions that can not be pre-drilled or have poor accessibility. Manual drilling tests showed unacceptable results, see Figure 21.6. This was caused by the wrong tool material and wrong drill geometry. For manual drilling, hard metal (HM) drills cannot be used because the tool material is brittle and susceptible to shocks and high loads (Hard Metal tools can only used on stable machines as with automatic drilling). After adjustment of the drill geometry and the tool material, manual drilling will also result in good results. No delaminations or deformations of the aluminium layers were visible and the hole diameters were according the required accuracy. Figure 21.8 shows a picture of a manually drilled hole with the right drill geometry.

As a final conclusion it can be stated that drilling of Glare laminates is possible, both for automatic and manual drilling. With the proper processing parameters, drill geometry and tool materials a high quality of the holes can be obtained. During drilling the layered structure of Glare laminates has to be taken into account and will set limits to the feed force. For automatic drilling HM tools can be used and for manual drilling other tool materials (e.g. cobalt coated drills) have to be used.

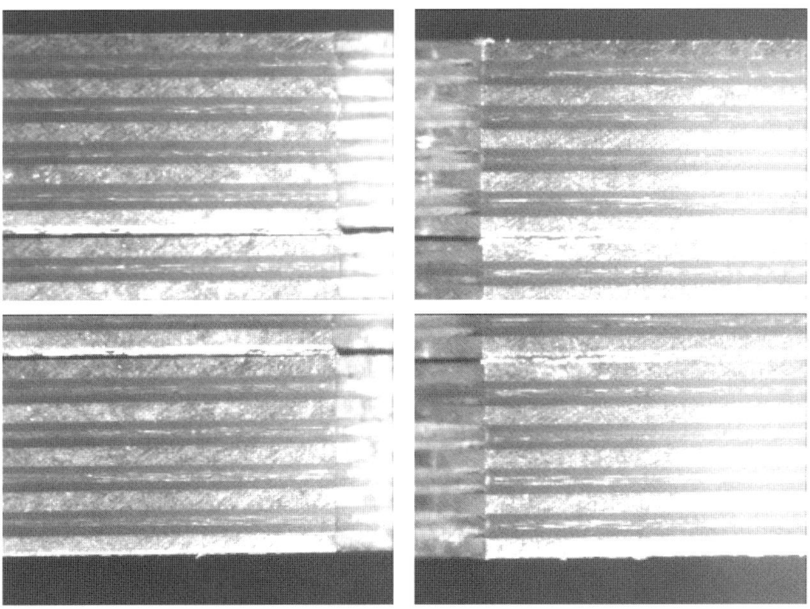

Figure 21.8: Manual drilled hole with proper drill geometry (4.8 mm 3.3/4.478-x drill, tool material: HM, speed ~ 2500 rpm).

21.5 Safety Requirements

During chip-forming machining of glass fibres or related fibre reinforced materials, small fibre particles may be released. It is well known that small asbestos fibres have proved to cause cancer. The glass fibres used in fibre metal laminates have diameters of 10 μm, much larger than those of asbestos fibres. Fibres of these sizes are not expected to reach the lungs. In addition, no splitting of the fibres in longitudinal direction during machining operations, the results of which may become dangerous, was observed (see chapter 33). However, to exclude any risks of spreading and inhalation of fibre particles proper safety precautions should be taken, such as wearing protective clothing, masks and application of lubricants when possible.

As regards the chemical agents that are released, no unacceptable risks have been observed with respect to these agents when the evaporation of gases is compared to international standards. During the investigation the release of gases have been compared to Dutch, German and French standards, and for each standard the release of chemicals was well within the prescribed margins (see chapter 33).

22

Formability

T.W. de Jong, E. Kroon and J. Sinke

Abstract – Despite the limited formability of Fibre Metal Laminates (FMLs), most structural components for a stiffened wing or fuselage structure can be manufactured. In general the applicable manufacturing processes are a combination of conventional processes used for sheet metal parts and laminating principles used for composites. The formability of Glare is related to the different failure modes that may occur. These failure modes are related to the constituents of the laminates and the layup of the laminates. Despite the limited formability, the manufacture of stringers does not pose a problem, as has been demonstrated by extensive research. On the other hand, in-plane deformations of Glare laminates and the related forming processes, such as stretch forming, are affected by the limited failure strains of Glare.

22.1 Introduction

The formability characteristics of Fibre Metal Laminates, and Glare in particular, have been tested extensively. The main reason is that for a successful application of new materials, adequate production methods for parts or structural elements should be developed.

As a result of its excellent damage tolerance properties, the fuselage is the most obvious application for Glare. In the fuselage structure, the potential of Glare can be exploited in the skin panels, consisting of reinforced skins and stringers. The skins that can be reinforced locally by doublers can also be laminated, as will be explained in chapter 23. The stringers can be manufactured by a combination of bending and laminating. In some cases even double-curved parts can be made, although the stretch forming capability of the laminates is limited.

In this chapter a brief overview of the bending of laminates, the manufacture of stringers and the stretch forming capabilities of FMLs are presented.

22.2 Bending of Glare

Bending is one of the major deformation modes during plastic deformation of sheet metals and laminates. In all forming processes some amount of bending takes place. Without any bending a flat sheet would remain flat.
At the same time bending provides the opportunity to illustrate the differences between monolithic materials and laminates such as Glare. The major differences can be found in the failure modes and the failure limits.

22.2.1 Failure modes

The first difference between the forming properties of Glare and monolithic aluminium is the number of failure modes for each material. Glare can have two failure modes: failure of one of its constituents or failure of its coherence. A monolithic metal only fails due to cracking.

Failure of constituents

A fibre-metal laminate has three main constituents: the fibres, the resin or matrix material and the metal layers. The fibres are embedded in the resin.
The glass/epoxy layers in Glare have a low failure strain in both directions. Therefore, fracture of a prepreg layer is one of the most common failures in Glare. In the fibre direction the deformation is limited by the small failure strain of the fibres.

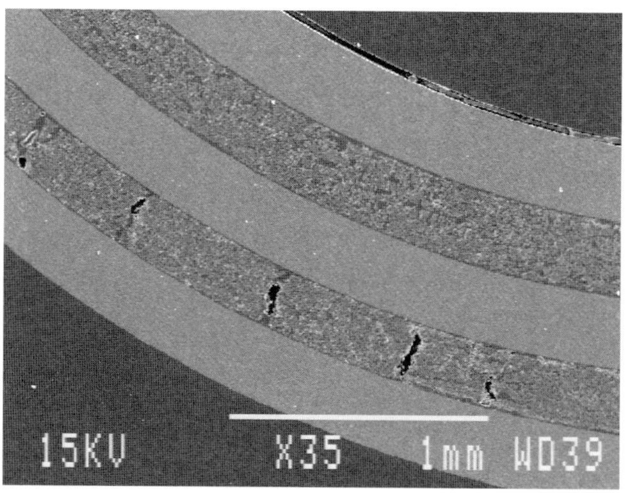

Figure 22.1 *Cracks in the matrix along the fibres (cross-section perpendicular to fibre direction).*

The failure strain perpendicular to the fibres is also small. This is due to the low failure strain of the epoxy in combination with the stress concentrations induced by the embedded fibres. When the resin fails, the cracks are parallel to the fibres, as is shown in Figure 22.1 for Glare 2-3/2-0.3. These cracks will not occur in a 2/1 laminate since for this laminate the prepreg layer is close to the centre line (neutral axis) of the bend.
A problem with this failure mode is that it is very hard to detect, since it is not visible from the outside and the C-scan process may not detect these cracks. On the other hand, the influence of the cracks on the mechanical behaviour has proven to be insignificant.
If the bending radius is reduced further, the metal layer will crack when the failure strain of the metal alloy is exceeded.
Another failure mode is fibre buckling during compression loads, which is shown in Figure 22.2. In this case the fibres fail at the inside fibre layer. When these fibres fail, they push the outer metal layer outwards, causing a rim or a buckle. This rim or buckle is visible at the inside of the curvature.

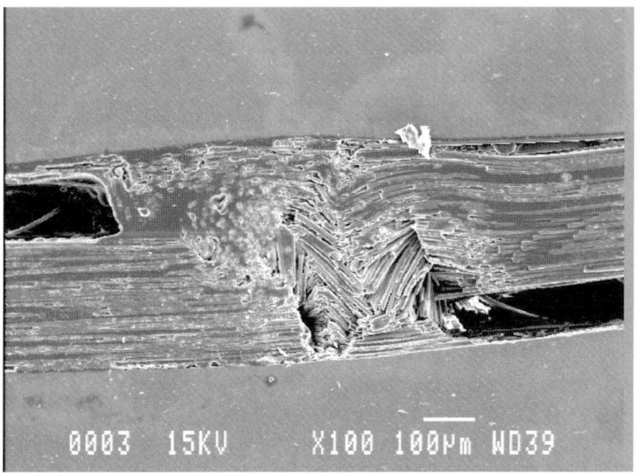

Figure 22.2 Fibre failure in compression inducing buckling of the metal layer.

Coherence Failures

The interlaminar shear failure is one of the failure modes related to the composition of laminates in general. When the laminate delaminates, one of the interfaces between the layers fails by debonding. This failure mode is also known from other formable laminates [1].
When the interlaminar shear failure occurs, it is clearly visible from the outside due to the "seagull" shape, see Figure 22.3. This shape is directly caused by the failure of one of the interfaces and the slippage between the compression and tension sides in the laminate.
With the naked eye it is possible to confuse this failure with the delamination buckling failure. But the difference is clear when the specimen is seen through a

microscope. It is then clearly visible that for the interlaminar shear failure, the delamination occurs near the centre line (neutral axis) of the laminate, see Figure 22.4, while for delamination buckling the failure occurs at the outside. However, there are clues that both failure mechanisms are related to each other, since they both have a dependency on the transverse shear force at failure.

The location of the delamination in an interlaminar shear failure is always in one of the resin rich layers close to the centre line of the laminates, because at that location the highest transverse shear stresses are present. The interface between the resin rich layer and the fibre rich layer is often the weakest interface in the laminate.

Delamination buckling, which is shown in Figure 22.5, is a failure mode for which the metal layer buckles when the interface with the prepreg layer delaminates. The failure mode is clearly visible when the buckle is large. As noted above, the overall appearance of this failure mode looks like the overall appearance of the interlaminar shear failure. However, the cause of the failure is totally different. During bending the inside of the laminate is compressed. When the compression stresses in the inside layer become too high, notwithstanding the adhesion and the stabilising curvature of the bend, the inside layer buckles as the peel stresses on the prepreg layer exceed their ultimate values.

It is obvious that the quality of the adhesive bonds plays a large role in this failure mode. As with most stability problems, there is a large influence of small deviations, test results show a large amount of scatter for this failure mode.

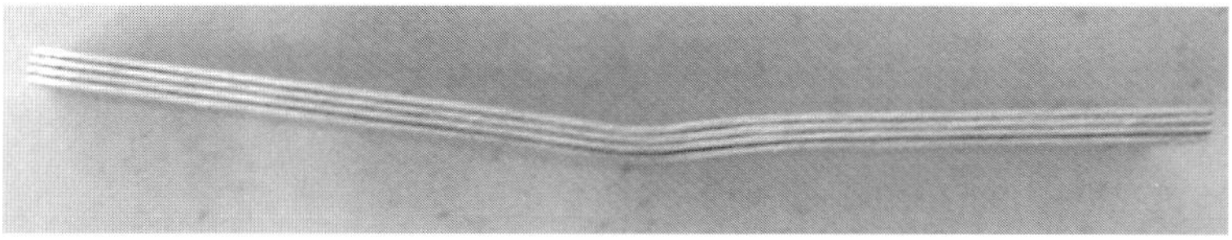

Figure 22.3 *Characteristic "seagull" shape after interlaminar shear failure.*

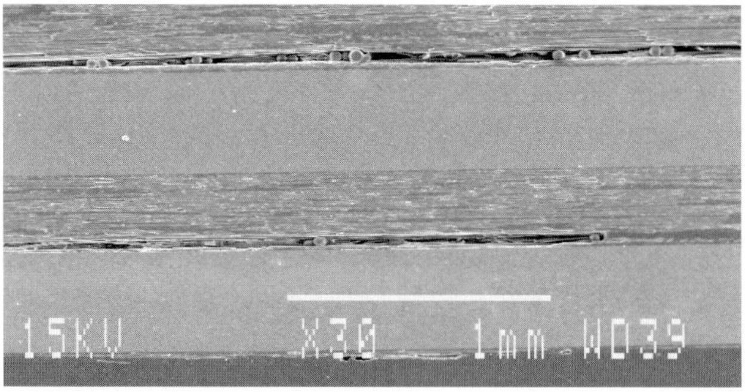

Figure 22.4 *Cracks due to interlaminar shear failure.*

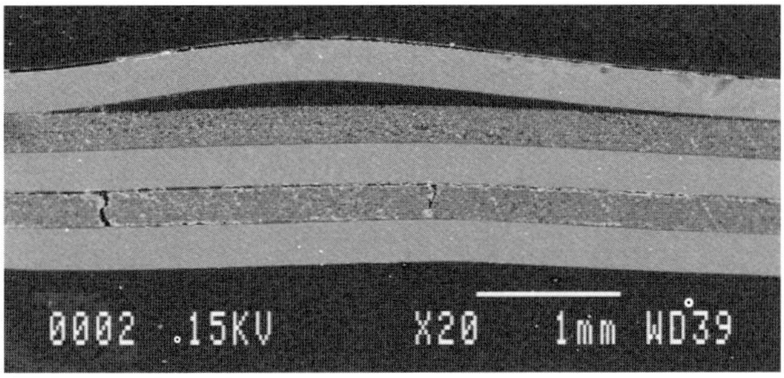

Figure 22.5 *Delamination buckling (with some secondary cracks).*

22.2.2 Failure limits

For pure bending of simple, symmetrical laminates, the strains in the different layers of a laminate can be calculated. When the failure strain of each layer is known, the formability limit of the whole laminate, i.e. the minimum bending radius, can be predicted.

This calculation method is based on the classical laminate theory, although some additional routines have been added to incorporate the plastic behaviour of the metal layers. However, in the classical laminate theory the transverse shear stresses are not included while these stresses are responsible for the interlaminar shear failure. Therefore, a new method is under development that takes care of both the transverse shear stresses and the plasticity of the metal layers. The method is based on a combination of so called "Zig-Zag" theories and higher order deformation methods [2].

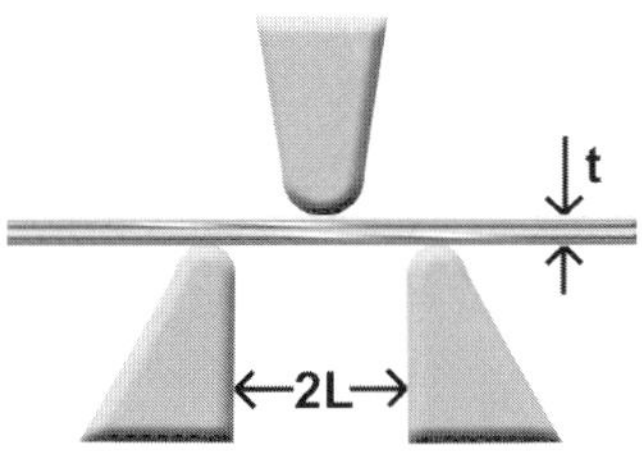

Figure 22.6 *Definition of parameters in the three-point bending test.*

The failure modes described above can be examined with the relatively simple 3-point bending test, see Figure 22.6. By changing the span between the supports of the test rig, the ratio between the bending moment and the shear force can be changed. Different ratios result in different failure modes.

During the tests, the forces and displacements are continuously measured. Since there is no relation between the punch displacement and the shape of the specimen (and thus the strains) at failure, only the force at failure can be used.

The best visualisation for the change in failure mode is generated if the maximum bending moment in the specimen is plotted against the shear force in the specimen.

Both parameters are presented per unit width of the specimen. This maximum bending moment in the specimen is equal to the shear force multiplied by the distance between the punch and one of the supports (L), which is a variable in the tests. The shear force is half of the load on the punch.

Such a plot is given in Figure 22.7. From this plot it is obvious that the failure type where one of the constituents is deformed beyond its limit, is independent of the shear force in the specimen.
Also, from this plot it can be concluded that the interlaminar shear failure not only depends on the shear force in the specimen, but that there is also a clear relationship with the maximum bending moment. The fact that the maximum shear force decreases when the bending moment increases indicates that the failure should start where the bending moment is the highest, i.e. at the punch. The fact that there is a relationship between the bending moment and the shear force can be accredited to the plasticity of the metal layers, which changes the internal transverse stress distribution in the Glare laminate.

The laminate forming prediction model, developed at the Structures and Materials Laboratory of Delft University of Technology, can also be used to calculate the minimum value for the bending radius for different Glare types and layups, see Figure 22.8.

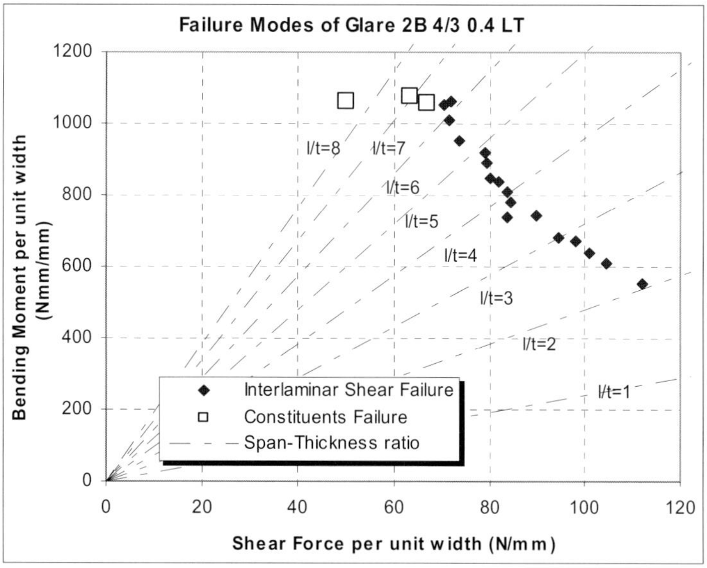

Figure 22.7 *The relationship between the failure mode and the loading for Glare 2B-4/3-0.4 with the bend-line parallel to the rolling direction.*

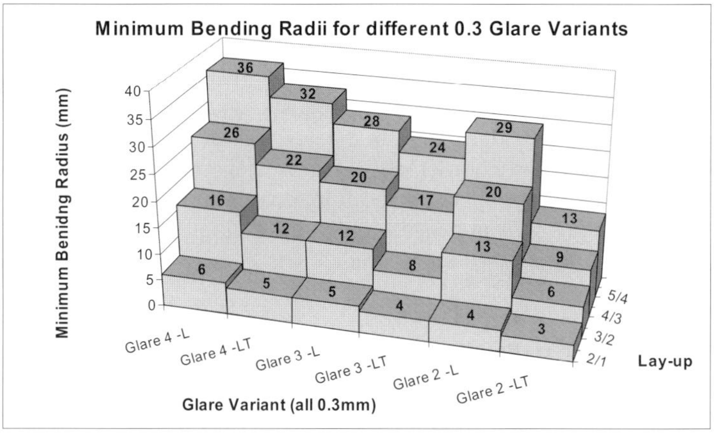

Figure 22.8 *Minimum bending radii for three different Glare variants (in two bending directions) with 4 different layups, but all using 0.3mm aluminium layers.*

The strain distribution in a cross-section for the case of pure bending is given by:

$$\varepsilon_{ij}(z) = \begin{Bmatrix} U_0 + \kappa_x \cdot z & 0 & 0 \\ 0 & 0 & 0 \\ 0 & 0 & - \end{Bmatrix} \qquad (22.1)$$

with: z = thickness co-ordinate originating at the centre of the laminate
U_0 = generalised stretching strain
κ_x = curvature

The curvature can be related to the bending radius by:

$$R = \frac{1}{\kappa_x} - \tfrac{1}{2}t \qquad (22.2)$$

The strain ε_{33} is not prescribed, since the laminate is assumed to be loaded in plane stress. U_0 is changed to keep the resulting stretch force of the laminate zero, although this is not necessary for symmetrical laminates. The model searches for the largest curvature for which none of the failure criteria is met at all positions in the laminate. The results given in Figure 22.8 do not have the same failure mode.

The figure shows that the minimum bending radius depends on the angle between the rolling (=prime fibre) direction in the bend-line. The suffix L means a bend-line perpendicular to the rolling direction and LT means a bend-line parallel to the rolling direction. For the Glare 2 variants this goes without saying, since the fibres are oriented only in one direction, for the other laminates the direction of the outside

fibre layers is dominant since the strain in those layers is the largest (largest distance to the neutral line).
Since the glass fibres deform only elastically, the spring-back of the laminates will usually be much larger then for sheet metals.
For 2/1 layups and for thin Glare-laminates bent in the LT-direction, the spring-back angles are of the same order as of the metal constituents. However, when the thickness (layup) is increased or the fibre content in the L-direction is increased the spring-back angles will increase rapidly, up to angles of 20°-30° instead of the 5°-10° spring-back for aluminium.

22.3 Manufacturing Glare stringers

Stringers stabilise the fuselage skin of an aircraft. In general, stringers are made of formed aluminium sheets or made from extrusions. Because of the excellent fatigue characteristics of Glare, the allowable stresses in a Glare fuselage skin will be higher than in an aluminium fuselage skin. The application of aluminium stringers on a Glare fuselage skin will therefore presumably lead to early fatigue failure of the aluminium stringers at some locations. In addition, differences in material stiffness may increase this problem. As a result, the use of Glare stringers in combination with a Glare fuselage skin will be the only option at those locations.

To date, three concepts have been developed for the manufacturing of Glare stringers:

1. forming of separate aluminium sheet and applying prepreg between the formed aluminium followed by curing
2. forming of non-cured Glare followed by curing
3. bending and bonding of seperate cured Glare 2-2/1 sections into a thicker layup using adhesive and curing.

For reasons of handling, stringer homogeneity, constant flange thickness, stringer joggling and quality control the third concept, i.e. bending and bonding of Glare 2-2/1, is considered to be the most feasible option for mass production of Glare stringers.

22.3.1 Manufacturing concept of bending and bonding Glare 2-2/1 laminates

Though feasible, forming of Glare has an important drawback in that small bending radii are not possible for thick laminates, since this will lead to interlaminar shear delamination or to bending cracks in the aluminium and/or prepreg layers. However, by bonding thinner Glare stiffeners to form a thicker stiffener we can avoid this disadvantage.
The basic consideration for these kinds of stringers is to bond two or more thin Glare 2-2/1 profiles together to produce a thick stringer, see Figure 22.9. Since each

separate profile has a relatively small bending radius, the bending radius of the built-up stringer will be relatively small as well [3].

The direction of the fibres in the prepreg layers of the Glare 2 laminates is parallel to the bend-line since this is the loading direction of the stringer.

From a design point of view, the stringer radius is important because of the minimum allowable distance between rivet or Hi-Lok and stringer radius. This allowable distance may lead to large flanges. To realise sufficient structural design possibilities, the Glare stringer should preferably be produced with the smallest radius possible.

From a production point of view, all separate profiles of the total stringer should be formed with one and the same bending radius, otherwise a different bending radius would be necessary for each profile. One typical aspect of applying the same bending radius to each profile is that the distance between the separate profile radii will be larger than the distance between the straight flanges, see Figure 22.9. However, this variation in distance is not a problem because the adhesive will flow from the straight flanges to the corners during the curing process and will fill all the space between the separate profiles.

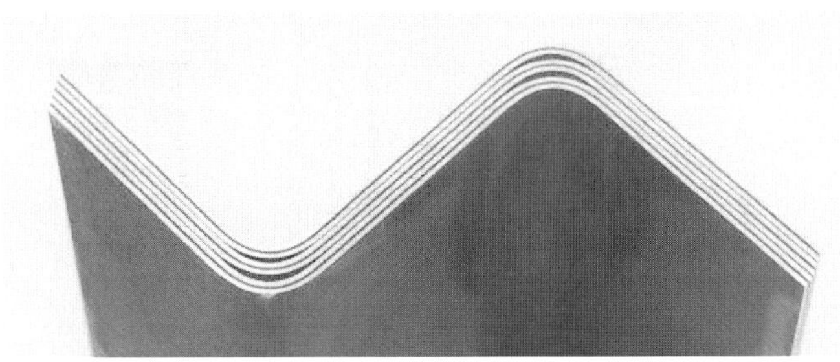

Figure 22.9 *Cross-section of a Glare stringer.*

Figure 22.10 *Conventional brake form machine.*

To ensure proper bonding, the separate Glare 2-2/1 profiles need to match very well, i.e. the geometrical tolerances of each Z-profile should be within narrow bounds [4]. Up to a certain profile length, the accuracy and reproducibility of a conventional brake form machine, see Figure 22.10, enables identical profiles within the required dimensional tolerances. The brake-formed profiles remain straight after each bending operation; lateral curvature and twist after the first bend are negligible.

There are two options for the bonding of Glare 2-2/1 profiles, see Figures 22.11 and 12: bonding with adhesive or with adhesive and one or two additional prepreg layers. Although both methods are feasible, laminating with adhesive only is easier since the adhesive film can be cut into strips of one meter length, which can easily be laminated. However, in contrast to adhesive, the prepreg can not be cut into separate strips, because loads have to be transferred by the fibres in the final product. To facilitate the laminating of long prepreg plies on Z-profiles, flat prepreg and adhesive plies are first laminated on top of each other before being placed on the Z-profiles.

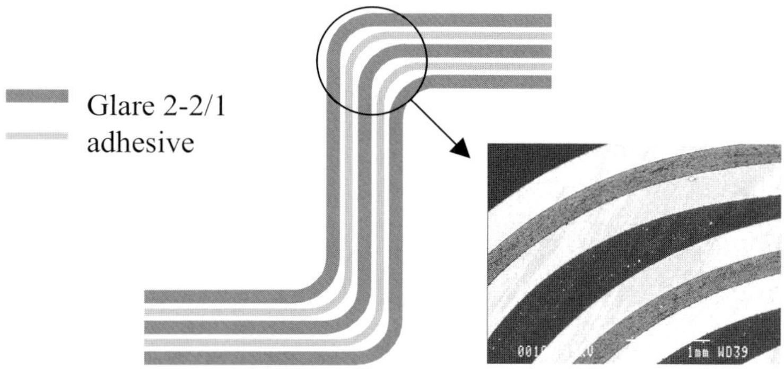

Figure 22.11 Bonding of Glare 2-2/1 profiles with adhesive only.

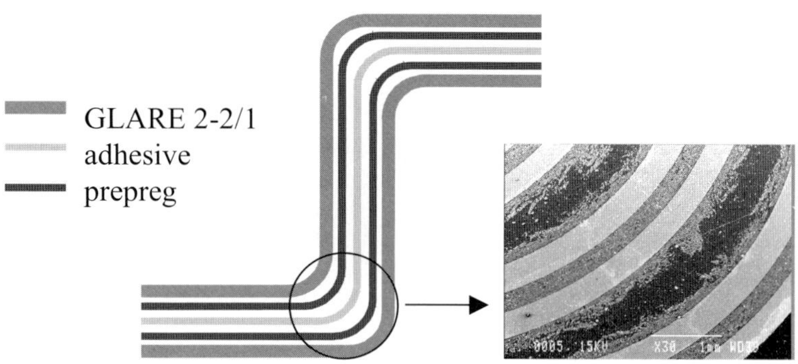

Figure 22.12 Bonding of Glare 2-2/1 profiles with adhesive and two prepreg layers.

When the Glare 2-2/1 Z-profiles are laminated with the adhesive film (and additional prepreg layers) the whole package is vacuum bagged and cured with a standard Glare autoclave cycle. The vacuum bag method assures equal pressure on all sides of the stringer, resulting in the best bond quality.

22.3.2 Joggling of Glare stringers

Inevitably, the aircraft skin will contain all kinds of local thickness variations, e.g. doublers, splices, crack stopper bands, etc. There are three options to guide the stringers over these thickness variations:

- conventional joggling
- bonding straight stringers over thickness steps, forcing them to follow the contour by autoclave pressure
- application of a staggering concept

Figure 22.13 Conventional joggling of a Glare stringer.

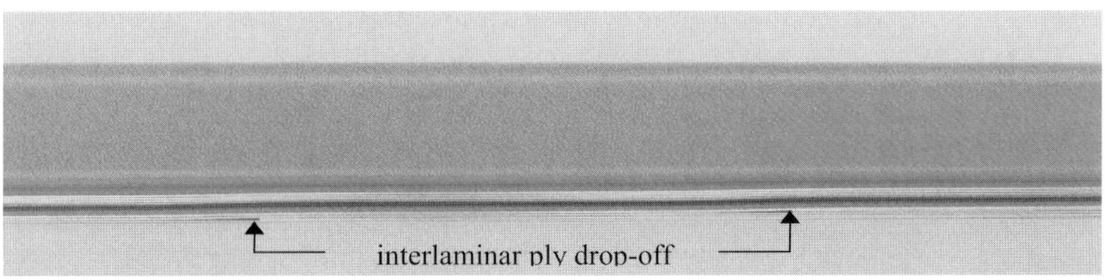

Figure 22.14 Glare stringer bonded over two thickness steps in a Glare skin.

Although feasible for small ratios of joggling depth to joggling length, the possibility of conventional joggling, see Figure 22.13, is limited with Glare stringers. The high transverse forces used induce delaminations in the Glare stringer.

Fortunately this disadvantage can be avoided by bonding straight Glare stingers over small thickness steps such as interlaminar ply drop-offs, see Figure 22.14. The lower stiffness of the Glare stringer compared to the aluminium stringer, the high autoclave pressure and the softening of the prepreg at raised temperatures during the cure cycle enables straight Glare stringers to be bonded over thickness steps in the aircraft skin.

This feature avoids costly conventional joggling and is therefore an important cost saving aspect of Glare stringers.

For large thickness steps, which can occur in stringer couplings for connecting two stringers over a butt strap, we can apply the staggering concept. In this concept the stringer coupling is milled to the required dimensions after which additional Glare 2-2/1 L- or Z-shaped profiles are bonded to the coupling to compensate for the locally reduced thickness, see Figure 22.15.

22.3.3 Quality control of Glare stringers

The flat Glare 2-2/1 panels that are used for stringer production are C-scanned – i.e. ultrasonically inspected – according to standard Glare quality procedures. The panels are then cut to strips and brake-formed using the appropriate brake form radius without inducing delamination or cracks. Subsequently, only the quality of the adhesive and possible additional prepreg layers used for bonding the separate Glare 2-2/1 profiles need to be checked.
For the Glare 2-2/1 profiles that are bonded with adhesive only, this can be done by testing of the bond line the straight stringer flanges according to today's standards.
C-scanning can also be used to check the Glare stringers, which may be required for the variant with the additional prepreg layers, see Figure 22.16.

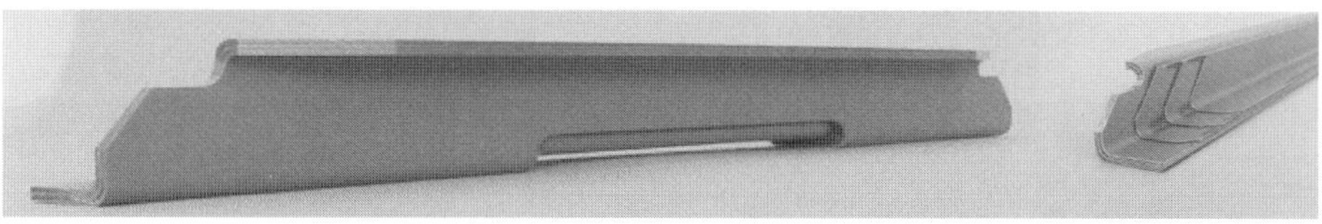

Figure 22.15 Glare stringer coupling with staggering of L-shaped Glare 2-2/1 profiles.

Figure 22.16 C-scanning of Glare stingers.

Procedures for the quality control of Glare stringers by C-scanning are currently being developed. Recent studies indicate that C-scanning of the Glare stringers is less time-consuming and therefore less expensive than bond testing.
Only the bond layers need quality control, since the separate Glare 2-2/1 profiles in the Glare stringer are C-scanned when the material is still flat and when the appropriate brake form radius is applied without inducing delamination or cracks. Therefore C-scan procedures may be reduced to the detection of local defects in the bond layers of the Glare stringer, cutting down costs on quality control.

22.3.4 Future developments

To date, Glare stringers have been produced by cutting Glare 2-2/1 panels to strips, brake-forming the strips to profiles, bonding the profiles to form Glare stringers and then milling the stringers to the required dimensions. Recently a new bending machine has been developed which, besides greater lengths, is also capable of bending up and down without rotation of the sheet. This feature enables the manufacture of corrugated Glare panels, see Figure 22.17, which facilitates much faster bending operations and implies a simplification of the laminating and bonding process.

Furthermore, a possible cost reduction can be obtained as quality control can be performed on one stringer after the stringers are milled from the corrugated Glare panel. This stringer is a kind of 'witness' and represents the quality for the entire panel.

Figure 22.17 Corrugated Glare 2-2/1 panels.

22.4 Stretch forming

Stretch forming of monolithic materials such as aluminium alloys is the most frequently used process for the manufacture of slightly double-curved shells and curved profiles. These shells and stringers are applied in the wing and fuselage structures of aircraft, i.e. in the cockpit area, the tail sections, and the wingtip areas.

Stretch forming is a method that combines controlled stretching and bending of flat sheets, roll-formed sections, and extrusions around form blocks to produce accurately contoured parts without wrinkles. Usually, two opposite ends of the workpiece are gripped in jaws and pulled or wrapped around the die. This can be accomplished in several ways. One of the most common processes of stretch forming of shells is presented in Figure 22.18. The stretch forming of profiles, such as stringers and frames, is performed in a similar way.

During stretch forming of monolithic materials, the entire cross section of the sheet is plastically deformed in tension, resulting in a negligible spring-back of the imposed curvatures. At the same time, the residual stresses in the formed sheet are almost zero.

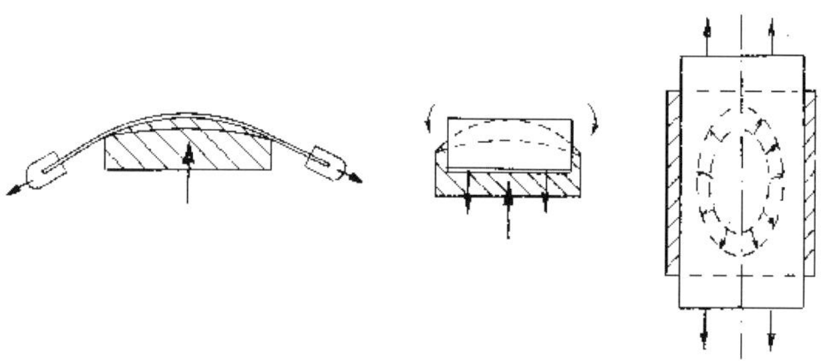

Figure 22.18 The principle of stretch forming (side, front and top view).

22.4.1 Stretch forming of Glare

Stretch forming of Glare laminates is much more difficult than stretch forming of monolithic materials. The significant differences in stretch forming capabilities of Glare can be related to its material properties:

- the failure strain of Glare in the fibre direction (L-direction) is very small (< 5%). The plastic strain that remains after stretching is even smaller (< 3%) due to the large spring-back of the material.
- although the failure strain in transverse direction (LT-direction) of UD laminates is larger (approx. 10-12%), the stretch forming capability does not improve since the material should be strained plastically in all directions, see the top view of Figure 22.18.
- the strain cannot be increased by heat treatments; Glare laminates can not be heat-treated.

- the fibres in the Glare laminates will deform elastically under all circumstances. These elastic stresses will cause considerable spring-back when the loads are released from the panel.

On the other hand, there are a few beneficial characteristics in the behaviour of Glare:
- the high strain hardening results in an even distribution of the strains over a large surface area.
- cross-ply laminates can be stretch-formed by tension in the bias directions ($\pm 45^0$), resulting in a maximum strain of 5-7%.

22.4.2 Stretch forming tests

During the development of Fibre Metal Laminates, several stretch forming processes have been investigated and demonstrator panels have been made.

One of the first applications of stretching of Arall and Glare was to reverse the residual stress system. When Arall or Glare are cured at an elevated temperature, the residual stresses in the metal layers are tensile stresses and the residual stresses in the fibre layers are compressive stresses. When the laminate is stretched just beyond the yield point and released, the residual stress system will be reversed resulting in compressive stresses in the metal layers. The compressive stresses in the metal layers are favourable with respect to crack initiation and growth. However, because the process is expensive, cannot be used for spliced laminates and can only be applied in one direction this so-called 'post-stretching' has only been considered as optional so far.

Several test series have been performed for the stretch forming of Glare sheets. These tests have been conducted in the industry as well as on a laboratory scale at the Delft University of Technology and have revealed the limited stretch forming capability of Glare. Because of the embedded fibres, the curvatures achievable in Glare are very small (large radii). At the same time the spring-back is considerable and related to the anisotropic nature of the laminates, and therefore it will be difficult to design a proper mould for a particular part. The best results have been achieved by stretching in the bias direction (fibres diagonal to the stretch directions). The low strain limit and the spring-back are less pronounced in this direction.

Other processes for the manufacture of double-curved shells, like peen forming and explosive forming, have also been tested, although the same problems will arise: a limited strain, and a spring-back that is large and difficult to predict.

The same is true for the stretch forming of stringers and frames: the curvatures achieved are small and the spring-back is significant. The spring-back phenomena are even worse than for stretch forming of shells, since stringers are made of UD-laminates; the fibres are in longitudinal direction of the stringers and the fibre direction coincides with the direction of the stretch force.

The height of most stringers is limited, allowing the manufacture of slightly curved stringers, but the height with respect to the curvatures for frames is too large to be stretch-formed. Besides, the fibre orientation along such a frame will change along the curvature. Therefore, only small frame sections can be assembled by a rubber-formed part (including the stretch flange) and a stretch-formed profile at the location of the shrink flange of the frame. Such an assembly has been made for demonstration purposes and is shown in Figure 22.19. However, an assembly like this one will result in additional assembly costs and weight and will be disadvantageous in most cases.

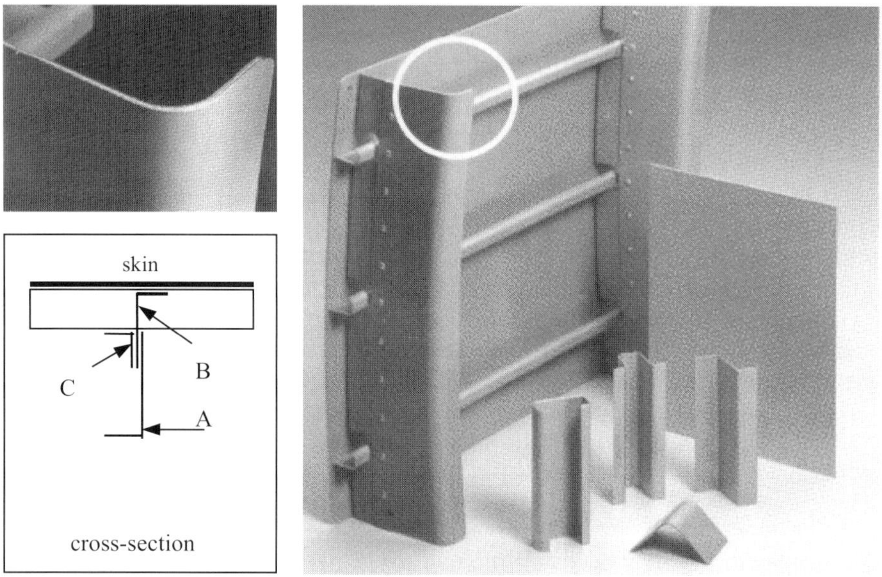

Figure 22.19 *Demonstrator panel with frame assemblies: rubber-formed web and stretch flange (A), rubber-formed castellation (B) and stretch-formed shrink flange (C) of the frame. Note the large bending radius in the top left picture.*

Figure 22.20 *Beads in a Glare 3-3/2-0.3 laminate (back) and Glare 3-2/1-0.3 laminate (front).*

22.4.3 Manufacture of beads

Biaxial stretching of the material can be used to make some details in Glare laminates, e.g. beads, see Figure 22.20. Beads, made by rubber-pad forming here, can be used to increase the stiffness of a flat surface. However, the geometry of a bead in Glare is limited when compared to aluminium alloys. The layup of beaded Glare laminates is limited to 2/1- and 3/2-layups.

22.4.4 Conclusion

Stretch forming processes for the manufacture of double-curved shells or stringers are usually not suitable for Glare, due to the small limit strain and the large spring-back. The most appropriate way to manufacture double-curved shells and stringers is by laying up separate layers or thin laminates, as will be explained in chapter 23 for double-curved shells.

References

[1] E. Atzema, *Formability of sheet metal and sandwich laminates,* Master Thesis, University of Twente, 1994.

[2] N.J. Pagano ed., *Interlaminar response of composite materials,* Elsevier, 1989.

[3] E. Kroon, *Manufacturing Glare stringers, status report,* report B2V-00-05, Delft University of Technology, Delft, January 2000.

[4] E. Kroon, *Brake form concept for Glare 2A laminates*, report TD-R-00-008, Structural Laminates Industries, Delft, September 2000.

23

Curved panels

J. Sinke and N. Jalving

Abstract - The Self Forming Technique and the splice concept were developed for the manufacture of large single or double curved skin panels. This technique is very similar to the layup processes used for full composite structures. The application of the spliced concept enables the designer to design large skin panels, the only limitation being the size of the autoclave used for curing. Other details such as doublers, stringers, and thickness steps can be made during the same or subsequent autoclave cycle. The layup tool for the manufacture of those large skin panels can be a single curved or a double curved tool. The layup tools, which should have a high accuracy, have been designed as welded aluminium structures. For both single and double curved skin panels a few demonstrator panels have been made, which have proven the concepts for the large skin panel manufacture.

23.1 Introduction

Suitable manufacturing processes are required in order to apply Glare laminates in real and large structures such as fuselage skin panels. The investigation and development of manufacturing processes for Glare has been part of the development of Glare for structural applications.

Many different manufacturing processes were investigated for their applicability with Glare during the development of suitable production processes of Glare parts. When all manufacturing processes of Glare parts or structures are evaluated, we find that some of the processes that can be applied are related to the manufacturing processes of metal sheets, such as bending and machining processes. Some of these processes have been described in chapters 21 and 22. Other processes, including the layup processes described in this chapter, are related to the manufacturing processes of full composite structures. Using layup technology, large and complex shells can also be made cost-effectively.

This chapter describes a few concepts with respect to the manufacture of large (fuselage) skin panels. First, the principle of the Self Forming Technique will be described in section 23.2, followed by a brief description of the splice concept in section 23.3. In section 23.4 the manufacture of the layup tool, the most important tool necessary for the manufacture of large skin panels, will be discussed. In the last two sections, 23.5 and 23.6 the manufacture of the demonstrators, both single and double curved panels, will be described.

23.2 The principle of the Self Forming Technique

During the early years of Arall and Glare development in the 1980s, only flat laminates or flat panels were fabricated. These panels could, like aluminium sheets, be formed by processes like roll-forming, bending and even stretching. After 20 years of development a lot of progress has been made in the manufacturing processes of Glare skin panels. The current standard to manufacture those panels is the application of the so-called Self Forming Technique.

The Self Forming Technique is based on the laminating of layers like when manufacturing full composite structures. A single or double curved layup tool or mould is required for a skin panel and the layers are stacked in the preferred orientation and layup. The splice concept provides the possibility to fabricate large panels. The size of the panels is no longer limited by the width of the metal sheets but by the size of the autoclave used to cure the panels instead. In the same process cycle many details such as the splices mentioned previously, see section 23.3, thin doublers and ply-drop-offs can be incorporated. The layers will be bonded together during the autoclave cycle and at the same time the pressure will form the metal layers over the (small) thickness steps in the panel. Thick doublers and stringers are bonded to the skin panel in subsequent autoclave cycles.

Figure 23.1 A panel with bonded stringers, door and window cut outs.

Figure 23.1 shows a large fuselage skin panel that has been made using a single curved layup tool. The stringers and doublers around the door cut-out have been bonded to the panel in a second autoclave cycle using the same tool.

When the layup tool has accurate dimensions and a proper surface finish, i.e. roughness, it can be directly used to laminate the skin panels to the right contour.
The Self Forming Technique is applicable for single curved panels as well as for double curved panels, the latter extends the application of Glare to a larger part of the airframe structure.

23.3 Splices and doublers

The splice concept is one of the key elements in the design and manufacture of large skin panels. A splice is a discontinuity in a metal layer, resulting from the limited width of the metal sheets; the maximum width of metal sheet being about 1.5 metres. The gaps in the different layers are not located at the same cross-section, but are staggered, although the splices are located in the same region. Splices in the prepreg layers are not required since the resin will fuse the different layers together.
The application of splices allows the width of a panel to be limited only by the dimensions of the autoclave used for curing the panels. During the development of the manufacturing concepts over the past years, various splice concepts were considered. The first concepts were based on "butt-like" splices, laminating the edges of the metal layers close to each other with only a small gap in between. As the tolerances for these types of splices are very small, the splice concepts have since changed. The concept used today is an overlap splice. The tolerances are much larger for these splices and the ease of manufacturing has been increased. Both types of splices are presented in Figure 23.2.

A second feature of large skin panels are doublers, additional layers added to locally reinforce the laminate, e.g. in regions of window and door cut outs.
Thin doublers (one or two additional layers) can be placed in-between the layers of the laminate. These doublers are known as internal doublers. The other possibility is to place the doublers on the outside of the Glare, usually on the inside of the fuselage.

On some single and double curved demonstrator panels the placement of doublers was investigated. Thick outside doublers were bonded in a second autoclave cycle. Such a reinforced panel that was later milled to simulate the door surround area is shown in Figure 23.3.

The progress that has been made in manufacturing technology, using the spliced concept, the use of doublers and the application of secondary bonding cycles, has led to the ability to design and manufacture large, stiffened (fuselage) skin panels.

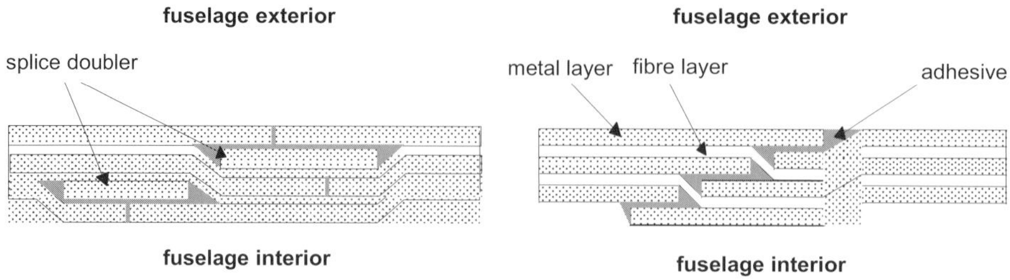

Figure 23.2 *Old (left) and new (right) splice concept.*

Figure 23.3 *Double curved panel with doublers and doublers in close up.*

23.4 Mould design

Before the manufacture of single or double curved panels can start, the required tooling has to be designed and fabricated. For the single curved demonstrators that were made first, the most important tooling consisted of the layup tool, the carriage riding over the layup tool and the handling system for the metal sheets.

For the fabrication of the tools the following list of technical requirements was made:

- The layup tool should be made of aluminium to limit the build-up of internal stresses as much as possible. When the difference in coefficients of thermal expansion of the layup tool and the laminate is too large, additional internal

stresses will be introduced in the laminate. The curing temperature for the panels is 120 ^{0}C.
- The top surface of the layup tool is representative for the A380-configuration in order to obtain representative demonstrator panels. The tool may be used also for the production of panels in a later stage of the A380-development.
- The dimensions of the layup tool should be as large as possible. The dimensions of the autoclave will set the boundaries.
- The layup tool should be durable: the tool should be designed so that it can be used 1000 times.
- The top surface of the layup tool should be airtight since the laminate will be vacuum bagged before curing.
- With the carriage one should be able to reach every spot on the layup tool.

Based on these requirements (and the requirement of low tooling costs) a set of tools were designed and fabricated as presented schematically in Figure 23.4.

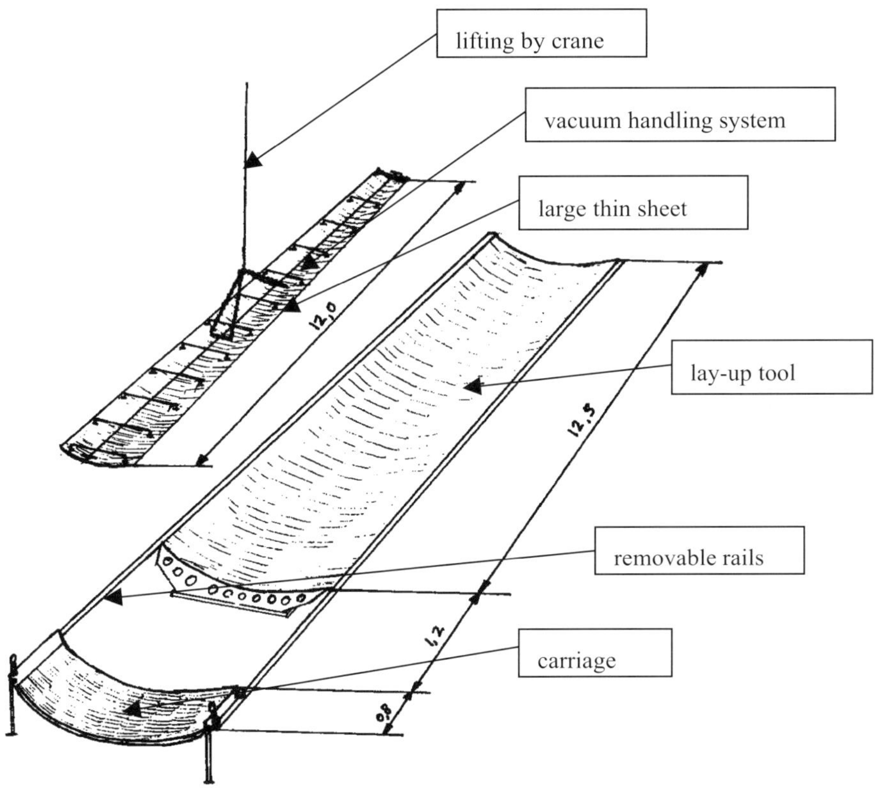

Figure 23.4 Schematic overview of the tools developed for the manufacture of large panels.

23.4.1 The layup tool

The layup tool was made of thick, single curved shells welded to each other and welded to a girder and web substructure. The substructure was also welded. The top surface was covered with aluminium sheets with a thickness of 2 mm, placed side by side.

For the construction of the layup tool several options were investigated:

- The first option was to manufacture the curved shell by bolting thick plates together thus creating one large surface for the layup tool. The joints, running in the width direction of the tool, would be butt-joints. In the butt straps, seal cords would be incorporated in order to achieve an airtight top surface. The top shell would be bolted to the substructure, which could be a structure made from plates bolted or welded together.
  The major disadvantages of this construction were the problems obtaining an airtight and smooth top surface and the durability of the seal cords. The seal cords cannot be used 1000 times (as required), but should be replaced frequently, which is a time-consuming and difficult job.
- The second option was a construction based on a top shell made by bonding of metal sheets. The substructure could be made of plates welded or bolted together.
  The difficulties for this type of construction were the possible leakage of the bond lines after repetitive use of the layup tool and the risk of creep, which may change the dimensions of the tool.
- The third option was a top shell made by welding of plates. At first, the welding was expected to result in inaccurate tolerances of the top surface (distortions due to the heat of the welding process). To overcome this problem it was decided to use thick plates that could be machined afterwards. It turned out that the distortion was small enough that milling of the complete tool was not necessary and grinding at a few places was sufficient.
  The substructure was made of welded plates. In Figure 23.5 a side view of the final layup tool is presented, showing the welded plates of the substructure.
  In addition to the top shell, exchangeable sheets were placed side by side on top op the shell. The main reason for this is the easy replacement in case of surface damage.
  The final dimensions of the layup tool were 12.5 by 3.3 metres, which just fits into the autoclave of Fokker Aerostructures (Papendrecht).

Figure 23.5 The welded substructure of the layup tool.

23.4.2 The carriage

A carriage that can ride over the layup tool is required to be able to reach any location on the tool. The carriage, as sketched in Figure 23.4, is also curved in order to decrease its height to the layup tool. The carriage is powered by an electromotor. The carriage rides over two rails mounted at the longitudinal edges of the layup tool. These rails have to be removed when the layup tool is moved into the autoclave. The length of the rails is two metres longer that the length of the layup tool, so that the carriage can be placed outside the layup area for unobstructed access to the laminate if necessary.

23.4.3 The Vacuum Handling System

A special vacuum handling system was developed for the layup of large thin sheets. The metal layers, oriented in the longitudinal direction of the layup tool, are too long to be handled manually. The handling system consists of a long and stiff girder in the longitudinal direction and lateral profiles, and is equipped with two rows of suction-cups at 1.1 metre intervals. This system, which is moved by a crane, allows easy handling of the metal sheets. When a sheet is moved to within a small distance above the correct location, the vacuum is removed and it can then be accurately placed.
This handling system is not necessary for the prepreg layers, the prepreg can easily be moved and located by hand.

23.4.4 Manufacture of tooling for the double curved panels

After some experimental tests on an existing double curved lay up tool, it was also decided to manufacture a large double curved lay up tool according to the same construction principles as for the single curved layup tool.

The size of the 3D-layup tool is 7 x 3.3 metres. The radius in the longitudinal direction varies from ∞ to 20 metres, in the transverse direction between 1.5 and 3.4 metres.
The top shell is made of a double curved plate with a thickness of 20 mm. After assembly of the tool by welding, the tool was stress relieved (1 hour at 200 ^{0}C). Subsequently the top surface was milled to its final dimensions and surface finish, see Figure 23.6. The minimum required thickness after milling was 8 mm.

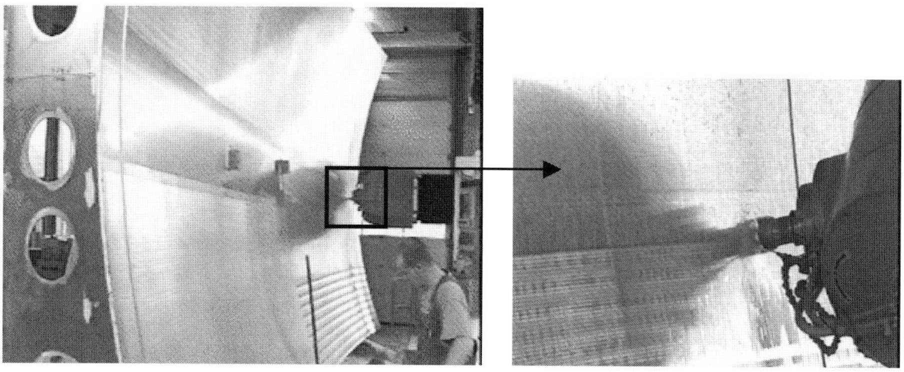

Figure 23.6 Milling of the 3D-lay up tool and close up of milling head.

The carriage was changed for its use with the 3D-layup tool. It was not only necessary to ride the carriage over the U profiles, but it has also to move up and down in order to reach the surface of the lay up tool as close as possible.
The use of the vacuum handling system was not a success because the sheets could not be bent in two directions without damaging them.

23.5 The manufacture of single curved panels

Using the tooling as described in section 23.4 several single curved panels were fabricated with the dimensions of 11.8 metres length and 3.2 metres width. In Figure 23.7, the panel (at the left side of the picture) and the layup tool are presented.

Figure 23.7 The layup tool and the first panel.

The panels that were manufactured were Glare 3-4/3-0.4 laminates, having two splices in the longitudinal direction and two doublers (with a thickness of 0.4 mm each) positioned at the centre of the skin panel.

The panels were manufactured at the Fokker Aerostructures (Ypenburg). They were prepared and laminated in the clean room, after which the completed panels were vacuum-bagged and cured in an autoclave. Although the lay up tool was 12.5 metres long the maximum length of the panel was 11.8 metres because the maximum length of the metal layers for pre-treatment was limited to 11.8 metres.

For the first panel the carriage frequently had to be moved because the prepreg needed to be smoothed to eliminate gaps, see Figure 23.8. Since the vehicle was operated at a speed of 6 metres per minute, it took quite some time to laminate this panel. Additional experiments revealed that a small gap between two adjacent prepreg layers is acceptable. The allowed gap results in a considerable reduction of time for prepreg positioning, since the smoothing of the prepreg could be eliminated.

Figure 23.8 Moving with the carriage (left) and smoothing of prepreg (right).

The experiences gained from the positioning problems of the aluminium sheets, the prepreg and the adhesive film, resulted in the need of a long lightweight ruler.
A 12.5-metre ruler has been made and used instead of thin ropes that were initially used for positioning. The ruler is made as a sandwich: a honeycomb (Nomex) core, aluminium U-profiles on both sides, and for the top and lower facings aluminium sheets with a thickness of 0.4 mm. The complete ruler weighs about 20 kilos and proved to be a very useful tool.
The vacuum handling system was designed for manual manipulation. During the production of the system some changes were made to the design, e.g. adding the possibility of rotation of the sheets from a horizontal to a vertical position and vice versa. These changes increased the weight of the handling system and manual operation was no longer feasible. Instead a small crane that was in the clean room of Stork Aerostructures was used, and three people operated the handling system. Two guided the system, one at each end, while the third person controlled the crane movements. For series production it is recommended to integrate the control of the crane and the vacuum handling system, to reduce the required number of workers.

During the evaluation of the manufacturing process of the panels the following recommendations were made:

- The replaceable sheets of the layup tool should be covered with an anti-stick material, so the adhesive film and prepreg will not adhere to them.
- The carriage should move over permanent rails, which are not attached to the layup tool ("dock principle"). All devices should be assembled at one side of the layup tool; the other side can be used to move the sheets over the tool. In addition, there should be a possibility to climb onto the carriage at any place along the layup tool.
- During lamination, the vacuum handling system proved to be too flexible, therefore, the (bending) stiffness of the vacuum handling system should be increased, i.e. by changing the design of the main aluminium girder.
  The distances between the suction cups should be reduced from 1.1 metres to about 0.5 metre. The reduction of these distances will also improve the positioning of the sheets with regard to the ruler.

To force the sheet to the correct curvature during lifting, additional pushing pads are necessary. These should be placed in between the other suction cups. These pads should have the potential to move the sheet about 10-centimetres downwards.
The valves of the suction cups should be controlled electrically instead of manually. When small sheets have to be transported some valves have to be shut otherwise the system does not work properly.

- To manipulate the crane efficiently the control panel (switches) should be integrated in the vacuum handling system.
- A laser projection system can be used to assist the layup of the sheets and prepregs at the right position. Such a laser system projects the exact boundaries of each layer on the layup tool.
- During the layup of a panel a lot of waste materials such as protective foils of the prepregs have to be removed. Particular provisions like dustbins are required for the waste materials.

After evaluation of the manufacture of the single curved panels it could be concluded that the manufacture of those large panels is not a major problem, see Figure 23.9. The tooling as described in section 23.4 was adequate. The most important changes of the process and tooling were the manufacture of the ruler and the 2 mm allowable prepreg gap, which reduced the labour time considerably.

Figure 23.9 The second single curved panel (Glare 3-4/3-0.4) including 3 doublers.

23.6 Double curved panels

Before the double curved panels were made using the tooling described in section 23.4.4, a few panels were made using other available tooling. In Figure 23.10 a small double curved panel (2.5 times 1.5 metres) with bonded stringers is shown. This panel was made on a Fokker 50 bonding tool.

The specially designed and built tooling was used to manufacture two double curved panels. The size of these panels is 6.5 x 3.0 metres, which is considerably larger than the panel presented in Figure 23.10. The material produced was a Glare 3 4/3-0.4 laminate. In length direction of the panels there are three splice locations. Two photos of the manufacture of the panels are presented in Figure 23.11.

The first panel was made from metal layers that were only roughly contoured therefore the accuracy of the splice areas was not very high. This is also due to the slight waviness that occurs due to the layup of flat, undeformed sheets in a double curved mould. Compressive stresses are introduced at the edges of those sheets, resulting in mild wrinkles. The size of the wrinkles is related to the width of the sheets. For double curved panels the width of the sheets is reduced in order to limit the wrinkling. The wrinkles will be removed during the autoclave cycle.
In the second panel, designing and machining accurate sheets increased the accuracy of the spliced areas. For the design part of these cut-outs the Catia system was used. An accuracy of about 0.5 mm was achieved.

Figure 23.10 A small double curved panel made on a Fokker 50 bonding tool.

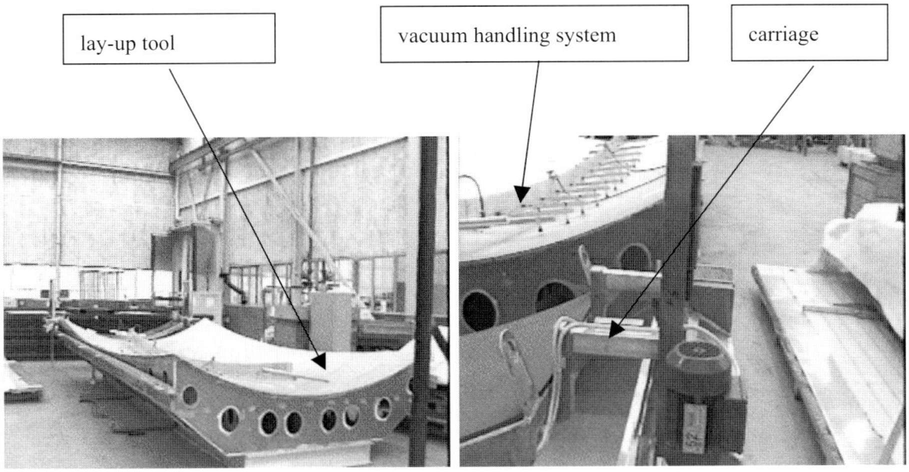

Figure 23.11 The assembly tool and its subsystems.

During layup, the sheets were positioned with position pins at the short ends of the layup tool. These pins were used to prevent slipping of the sheets.

The overall production process of the double curved panels is similar to the manufacture of the single curved panels, except for the use of machined sheets and positioning pins as previously described.

The first panel, see Figure 23.12, was cut into sections in order to evaluate the quality through the C-scan method. No delaminations or other manufacturing defects could be found. The only comment that could be made was about the inaccurate splice geometry.
Stringers were bonded onto the second panel in order to evaluate the second curing cycle and the positioning of stringers that were formed before bonding. No difficulties were encountered for either.

Figure 23.12 Layup of the first double curved skin panel (left) and finished panel (right).

23.7 Conclusion

In the previous sections (23.5 and 23.6) the manufacture of large skin panels were discussed. It has been proven through the manufacture of several large demonstrators for each configuration that the manufacture of those large panels is fairly easy and straightforward using the Self Forming Technique and the splicing concept (sections 23.2 and 23.3). The time for the layup of those panels is a few hours.
Through the application of these techniques, large panels including doublers and other details can be manufactured in a manner similar to the manufacture of panels made from full composites (layup technology). This method of manufacturing large, integrated panels will reduce the costs of those panels considerably.
The constituents, such as the prepregs, metal sheets and adhesive tapes can be used as fabricated. Particularly for the metal sheet, no expensive stretch forming is required to bring the sheets to the proper curvatures. For the single curved panels the width of the sheets is limited to the width that can be supplied by the material manufacturer. For the double curved panels the width of the metal sheets is limited by the curvatures of the tooling. The width is usually in the order of 0.5-1.0 metres. For the prepregs and adhesive tapes there are no limitations.
The tooling required for the manufacture of the large skin panels, as described in section 23.4, can be made by simple techniques such as welding. For the tooling for production series the handling system, the positioning tools and the layup procedures can be improved.

24

Quality control

S.C.H. van Meer and R.A.M. Coenen

Abstract – The non-destructive ultrasonic C-scan method is an effective tool to check the quality of Glare panels and components. This automated method measures the ultrasonic attenuation through the panel and images it over its entire surface. Local defects and porosity give a significant increase in attenuation and can therefore be located within the component. To determine the significance of these indications, the scope of the inspection method has to be defined and a reference system developed. An "effect of defects" programme is needed to provide suitable rejection levels.

24.1 Glare fabrication process

24.1.1 Production process sequence

The Glare manufacturing process sequence, see Figure 24.1, is very similar to that of bonded metal products and composite products. In fact, it is a combination of the two. The most important difference with composites is that the adhesive is already combined with the fibre layer in the raw materials process.

Raw materials

The raw materials for Glare consist of aluminium sheets, glass fibres and epoxy adhesives. The glass-fibres and adhesive film are put together in one assembly, i.e. the so-called prepreg (pre-impregnated fibres). Some separate adhesive is necessary for doublers, at splice locations and to bond stringers to the skin.

The aluminium sheets are anodised to obtain a uniform and homogeneous oxide layer and then primed to achieve additional corrosion protection. The quality of the oxide layer is checked for each batch of produced metal sheets. Every batch of prepreg is locally inspected for the fibre volume fraction and the thickness of the adhesive film.

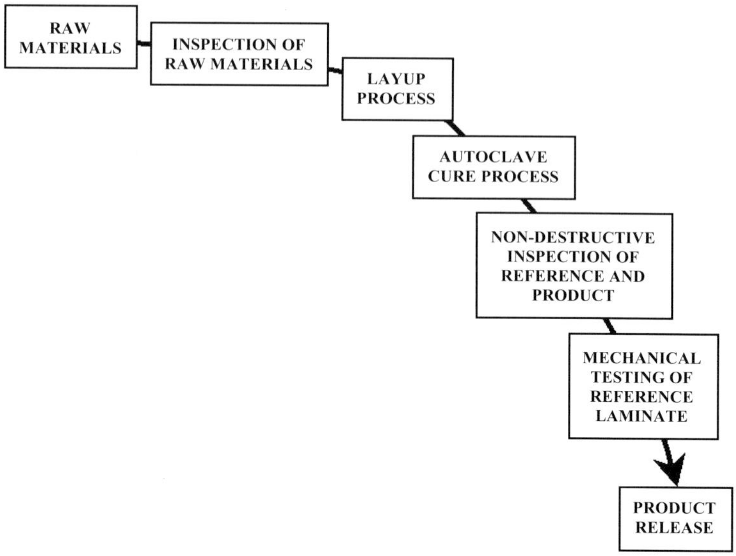

Figure 24.1 Different phases for fabrication of Glare products.

Layup

During the layup process the aluminium and prepreg layers are positioned in the right stacking order and orientation. Layup simulations connected to a monitor are required to assure correct stacking for complex panel configurations. On this monitor the person carrying out the layup can see the next layer that he has to put on the stack. This reduces possible mistakes. Contamination must be prevented and the room temperature must be kept below approximately 25 °C to prevent premature adhesive curing. The moisture content of the air should not exceed a well-defined level.

Cure

During the cure process the different layers are bonded together. The activity of the two-compound adhesive reaction is thermally controlled and pressure is applied to bond the layers. This process is performed in an autoclave where pressure and temperature can be accurately controlled. A Batch Witness Panel (BWP) is cured together with the panel batch. This BWP will later be mechanically tested to verify the general quality of used materials and the overall result of the cure process. The purpose of reference panels like BWP is further explained in section 24.2.5.

Quality control

After the production process, the BWP and the panel are inspected in an automated inspection system, the so-called ultrasonic C-scanner. The panel is ultrasonically scanned with a small index coverage resolution of one square millimetre (1 mm^2).

Ultrasonics is very sensitive for detecting disbond discontinuities. The position of any discontinuity is accurately registered.

A part of the BWP is mechanically tested to determine the general quality of the raw materials and the quality of the bonding process. If the test results are outside the specified boundaries, the whole batch of panels must be rejected for further use. A troubleshooting action is immediately started to find the process error.

Additional processes such as bonding an external stringer on a skin-panel will follow similar process steps as described above.

24.1.2 Choice of a non-destructive inspection method

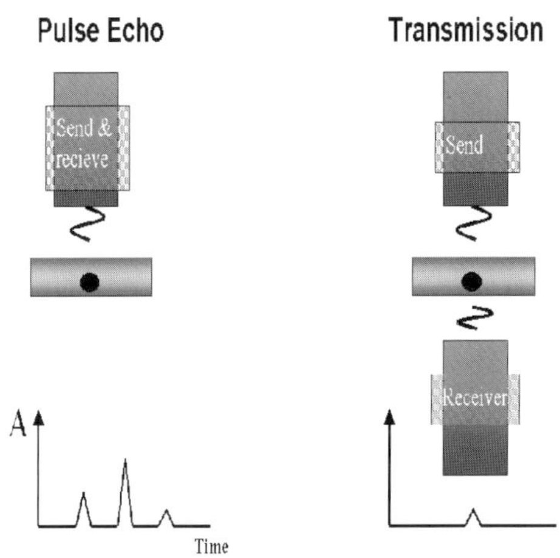

Figure 24.2 Ultrasonic Pulse-Echo and single through Transmission method.

Ultrasonics is the most suitable method to detect disbonds and porosity in laminated products. This type of discontinuity generally contains a gas that has a very different density and sound velocity in comparison to the aluminium and prepreg layers. Ultrasonics will detect these types of discontinuities with an excellent sensitivity. The ultrasonic inspection method can be easily automated. Two basic principles can be adopted, see Figure 24.2, i.e. the Pulse-Echo and the single through Transmission method. As with composites, the single through Transmission method was chosen for inspection of the Glare laminates.

24.1.3 Definition of single through Transmission method

This method requires two separate ultrasonic transducers, i.e. a transmitter and a receiver, directly facing each other and separated by the specimen. The principle of operation is the evaluation of the change in transmission of ultrasound through a specimen. Defects in the specimen will attenuate or block the transmitted ultrasonic signal because of the difference in acoustic impedance; thus a reduction in the signal amplitude or a total loss of signal occurs. Water-jet (squirter) systems can be employed to provide ultrasonic coupling.

In the C-scanner, a transducer-couple scans the specimen following a raster pattern and presenting a view of the total area including possible discontinuities as if seen from above and comparable to an X-ray image, see Figure 24.3. Discontinuity location and size data are available from changes in amplitude as a function of position. These are displayed on a screen. The raw data is digitised and stored on magnetic media.

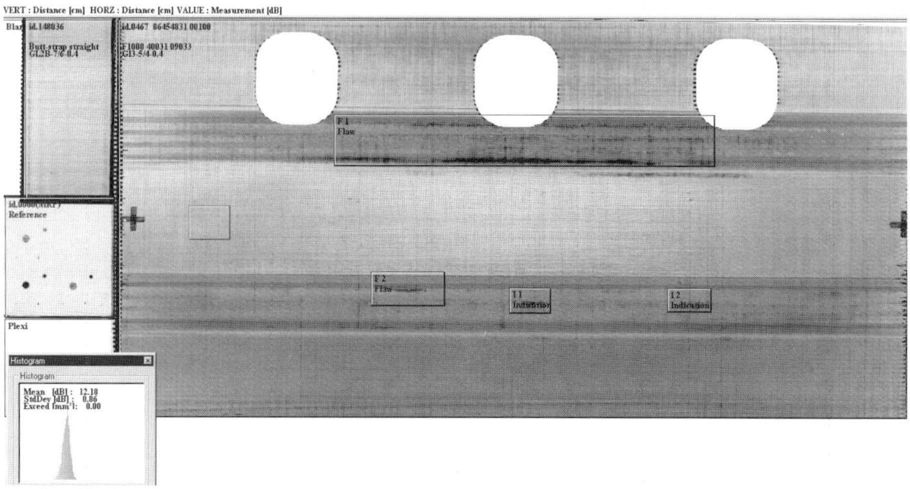

Figure 24.3 Ultrasonic C- scan image of curved Glare aircraft panel.

Figure 24.3 is an ultrasonic image of a slightly curved Glare panel with three window sections. Notice on the left starting from the top: a Butt strap panel, a Master Reference Panel (MRP) and a plexiglass panel. The MRP contains several circular artificial defects. The purpose of reference panels like MRP is further explained in section 24.2.5.

The ultrasonic attenuation scale is proportional with the (gradual) grey scale, some defects in the MRP appear as black spots meaning that maximum attenuation is exceeded. The window openings give no attenuation because there is no material to attenuate the signal. Plexiglass attenuation is also close to zero due to the low density of plexiglass. In the large panel there are two longitudinal doubler sections having significantly more attenuation. Looking more in detail, the upper doubler has one flaw section appearing as a black bold line while the lower doubler has three black spots corresponding with one flaw and two indications. The histogram (lower left) corresponds with the data of the square-area in the middle-left part of the skin panel. It shows a normal distribution with a mean value of 12.2 decibel, the standard deviation is 0.86 decibel.

The Batch Witness Panel (BWP) was scanned prior to this scan and passed the ultrasonic C-scan test. Economic considerations prevented the manufacture of a Component Reference Panel (CRP) as reference for this skin panel, since this skin panel is strictly to be used for exhibition purposes.

For more detailed coverage of this subject see Ness [1], upon which this section is based.

24.1.4 Defect detectability scope

The objectives of the ultrasonic C-scan method are:

- To detect local defects caused by foreign material contamination, e.g. piece of wrapping foil accidentally inserted into the laminate.
- To detect local defects caused by raw material contamination, e.g. glass splinters from cutting action of prepreg.
- To detect high concentrations of defects or porosity, e.g. caused by an over-aged adhesive layer, insufficient adhesive film thickness or a faulty autoclave process (e.g. insufficient pressure).
- To detect positioning errors in configuration discontinuities such as splices, doublers or thickness steps.
- To detect an incorrect fibre direction (in average through the thickness, since the various layers cannot be separated). The difference between unidirectional (Glare 2), crossply (Glare 3, Glare 4, Glare 5) and +/-45° crossply (Glare 6) can be determined, see Figure 24.4. Notice that the difference between the crossplies Glare 3, Glare 4 and Glare 5 cannot be determined.

Figure 24.4 *Examples of C-scan images of Glare laminates. First row, from left to right showing: crossply, +/- 45° crossply, unidirectional. Second row from left to right: local flaw, porosity, light porosity.*

24.1.5 Defect sizing

Naturally occurring discontinuities in monolithic metals are rarely neither smooth nor parallel to the inspection surface, i.e. the entry surface of the sound. Therefore the detected discontinuous area does not always match exactly with an artificial defect with which the image is compared. The size of real flaws is often underestimated, see Figure 24.5.

Well-defined laminated products such as Glare and Arall have the benefit that occurring discontinuities, e.g. delamination, are in general limited to one layer. Since the layers are thinner than 0.2 mm, the discontinuity will also nearly always be parallel to the entry surface. Consequently the sound beam will be perpendicular to the discontinuity. The reflection of sound will be close to optimal and the area relationship will show a high degree of validity, see Figure 24.5.

For a more detailed coverage of this section, see Roebroeks [2], upon which this section is based.

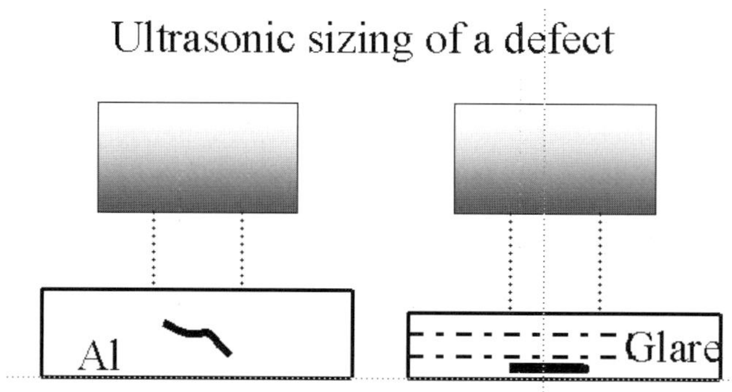

Figure 24.5 Defect sizing in aluminium and in Glare.

24.2 Reference system design for Glare

24.2.1 Introduction

Most NDT methods require a known reference to compare with the obtained data in order to achieve a reproducible result. In general, the reference material consists of a panel with artificial defects of known sizes and shapes. It is important that the reference material resembles the actual component that needs to be inspected.

When the general response of the material to ultrasound is a clearly defined function of the thickness, the results from the reference system can be extrapolated to an identical material with another thickness. This applies to metals in general and holds for many composites as well. If the thickness of a Carbon Fibre Reinforced Plastic (CFRP) component increases with for example a factor 2, the attenuation will double. Tables are available for metals with their attenuation coefficient in terms of dB/mm.

However, because of resonance effects in Fibre Metal Laminates a simple linear relationship between thickness and attenuation is not valid, see Figure 24.6. It is possible, for example, that a laminate with 0.3 mm thick aluminium layers has a higher attenuation than a thicker laminate with 0.4 mm aluminium layer. This occurs due to a more favourable resonance characteristic of the inspection frequency with the 0.4 mm thick layer.

A special reference system for Fibre Metal Laminates has therefore been developed. The objective of the following section is to describe this system and to provide an insight into its logic.

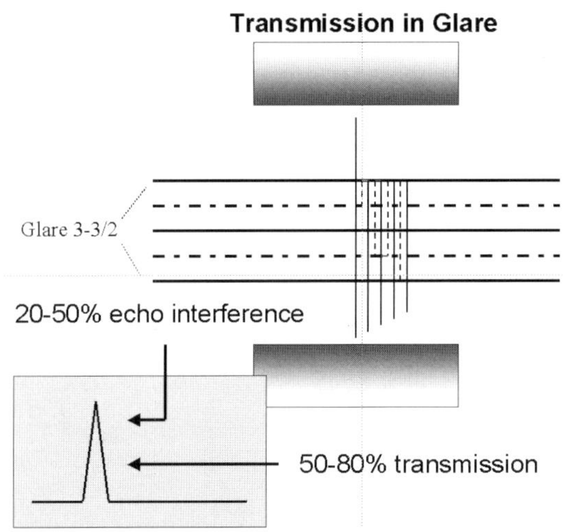

Figure 24.6 Echo interference pattern in Glare for ultrasonic waves.

24.2.2 Reference system design

The reference system is the core of the developed inspection method for Glare. A complicating factor, compared to normal full-composite structures, is the interlayer constructive and destructive interference effects of ultrasonic waves in Fibre Metal Laminates, see Figure 24.6. These effects occur because of the thin layers and the large acoustic impedance contrast between the metal and the composite layer and lead to a highly non-linear attenuation characteristic. Until a fully functional ultrasonic forward model has been developed and tested, a reference system will be necessary.

Objectives

The primary objective of the reference system is: "*To provide all the required reference information for the inspection of a batch of laminates.*"
The system should achieve this objective without having to test all the laminates and configurations present within this batch of material.

Local defects

The reference system should provide, either directly or indirectly, a reference level for every layup present on the primary component, because a Glare structure can be fairly complex and built up from different layups. From this reference level an offset (generally 6 dB) is taken. This level can then be used to determine whether defects, larger than a pre-set criterion, exceed this level.

General quality

For the general quality assessment of a batch, the reference system should provide an extremely sensitive method to detect general quality changes, e.g. porosity content or insufficient cure. This deviation will be expressed by statistics of the ultrasonic transmission response of a laminate, such as average attenuation, standard deviation and mechanical testing values as average interlaminar shear strength and other variables.

Configuration

The reference system should provide comparative material to verify the correct locations of splices and the presence of fibre-layers at the right orientation.

24.2.3 Assumptions

The system is based on the following assumptions:

- quality is independent of layup or configuration
- raw materials can be assumed of constant quality within one batch
- equal raw material in the same process will lead to equal results

Some details are provided in the sections below.

Layup independence

The first assumption is extremely important because the entire reference process depends on the acceptance of the fact that a panel may have a different performance and still has the same general quality. The difference in performance for a laminate is then caused by the difference in layup and not by a difference in quality. This means that if a reference laminate of a given configuration performs as expected, all the other configurations will also perform as expected but at different levels. It is important to note that the quality assurance tests do not therefore establish design mechanical performance levels.

Because the above system functions very well, a qualification of all the different layups can be avoided. This is especially valuable since there may be several hundreds of layups present in one panel!

State of control

If one of the base materials or processes is not in a state of control, relatively large variations may occur within one batch. This means that statistical construction of suitable reference values is difficult, or even impossible. This may lead to a relatively high rejection rate of large panels even if only small patches of low quality material have been used in a panel.

24.2.4 Reference system layout

The reference system has been split up into three control lines of defect classes:

1. general quality (porosity)
2. local defects
3. configuration mismatch

The aim of this distinction is to allow the separate performance of more sensitive, general quality inspections and tests. This is primarily desired because destructive tests have to be performed to verify the general quality. This means that the configuration of these panels should be fixed (see previous subsection).

Another advantage of this system would be the possibility to perform the general quality control for the BWP on a much smaller C-scan-system, using low noise amplifiers, immersion testing and low acquisition speeds compared to the rough evaluation on the large panels for local defects and configuration errors.

24.2.5 Types of reference panels

This section describes the essential elements of the reference system, see Table 24.1.

Table 24.1 Types of reference panels.

Ref. Panel	Description	Goal
MRP	Master Reference Panel Glare-3-3/2-0.3	Check instrument and serves as reference for BWP
BWP	Batch Witness Panel Same configuration as MRP	Check batch on general quality problems
CRP	Component Reference Panel Contains all the configurations as the component it is used for.	Check local defects in special Glare configurations, e.g. splices and doublers

Master Reference Panel

The Master Reference Panel (MRP) is used to assess the C-scan capability and serves as a direct reference for the other witness panels. The MRP is a laminate with a fixed layup (Glare 3-3/2-0.3). Just as with the witness panel, a significant part of the MRP is mechanically tested, e.g. tensile test and interlaminar shear test. However, it is possible to replicate the MRP for use during inspection procedures. The MRP also serves at a cross-batch reference level, to enable the construction of a useful database.

Batch Witness Panel

The Batch Witness Panel (BWP) is used to check the general quality of the batch and should contain the same batches of raw materials as the other panels in the batch. The BWP must have had the same autoclave cure cycle as the other panels in the batch. Until now many general quality problems are related to porosity in the cured laminate.

The witness panels have to be of the same configuration as the MRP. The ultrasonic statistic results, such as mean attenuation and standard deviation, are compared to those of the MRP. This statistical information is compared to database recordings which relates this data to certain quality level. This statistical data has been collected during the last decade and will be updated with information generated during the Glare qualification process. The BWP is also mechanically tested during the lot release-testing phase.

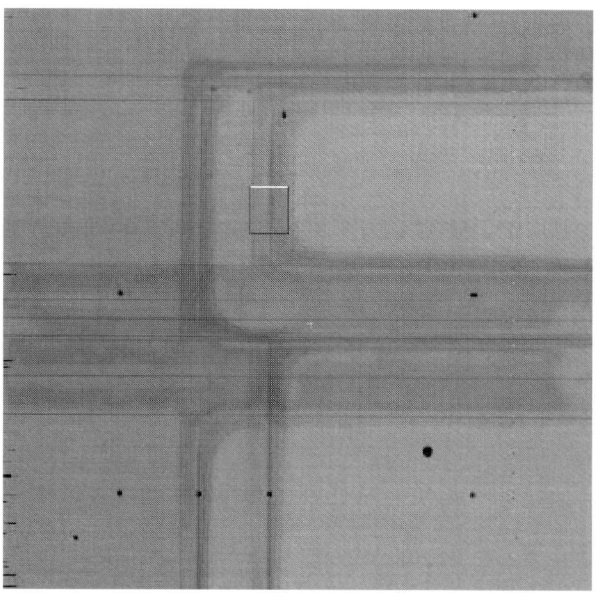

Figure 24.7 Ultrasonic C-scan image of a CRP for a A310 skin panel.

The part of the panel that is not used for destructive testing is stored in a material library. It can then be used later for future reference and to evaluate the performance of new NDT methods or technologies.

Component Reference Panel

The Component Reference Panel (CRP) is used to verify the C-scan gauge capability for local defects in special Glare configurations such as splice-area's, internal doublers, thickness steps, etc., see Figure 24.7. Due to the effects of constructive and destructive interference of the internal echoes, identical sections on CRP and the components cannot be assumed to generate identical responses (mean attenuation). Small thickness differences of the layers can result in significant differences of ultrasonic attenuation. The CRP should be stored in the material library for future reference and comparisons.

For a more detailed coverage of this section, see Coenen [3], upon which this section is based.

24.3 Conclusions

The ultrasonic C-scan through Transmission technique is very effective for inspecting fibre metal laminates. The disbond-type discontinuities are very good reflectors of ultrasound. Depth information is not provided by this non-destructive method. Local defects and porosity are detected with excellent sensitivity. Fibre-layer orientation is presented in the C-scan image as a projected image because of the through Transmission principle. Positions of special configuration elements such as splice areas, internal doublers and thickness steps are easily detected and can be accurately defined.

The ultrasonic attenuation does not have a straightforward relationship with the panel thickness due to the effect of constructive and destructive interference of the interlayer echoes. Therefore comparison of ultrasonic attenuation is often not direct between different configurations and sometimes is even difficult between identical configurations. Small changes in layer thickness can give significant attenuation differences, especially around the resonance points.

A reference system has been developed for Glare panels and components that take the interference effect typical for thin-laminated materials into account. The reference system uses only one key reference panel to assure the batch quality. The *effect of defects* programme will soon provide more information for even more accurate rejection levels.

References

[1] S. Ness, C.N. Sherlock, P.O. Moore and P. McIntire, *Nondestructive handbook*, volume 10, 2nd ed., American Society for non-destructive testing, 1996.

[2] G. Roebroeks, E. Kroon, S.C.H. van Meer, *Quality Control Specification for primary Glare structures*, Report TD-R-98-006, Rev.04, March 2001.

[3] R.A.M. Coenen, *Design of a quality assurance system for structural laminates*, PhD-thesis, Delft University Press, 1998.

E. Safety, Maintenance and Inspection

25

Inspection and maintenance

F.O. Taddey

Abstract – This chapter shows the after-sales related activities for Glare structures within the Airbus consortium. For the Glare demonstrator panel on the MRT A310 MSN 484 of the German Air Force a supplemental SRM has been developed. This SRM is the starting point for the activities related to the A380 design and development. Allowable damage limits and repairs on Glare parts will be tested as part of the "Megaliner Barrel" research project.

25.1 History of the Glare demonstrator panel

In 1999 the Glare demonstrator panel for an Airbus A310 at position Fr. 35-40 and between STGR 5RH and STGR 13 RH (Section 14) was redesigned, manufactured and installed. The panel consists of internal doublers at the riveted butt and lap joint run-outs, which are all covered by the internal aluminium layer, and a splice configuration in the middle of the panel. The stringers are mainly bonded, but in the first and last frame bays, riveted stringer/skin joints have been used. Clips are riveted, as are butt and lap joints.
The installation was performed during the conversion working party of this aircraft from a PAX to a Multirole Transport A/C (MRT) at Lufthansa Technik in Hamburg.

25.2 Design of the Glare panel

The original production design has been rebuilt, taking Glare design principles into consideration. Around the panel edges, the original design requires reinforcement doublers. These reinforcement doublers are all located at the inner side of the panel.

In addition to the doublers a Glare splice was designed, which interferes with the reinforcement doubler. This was a complicated design that required a lot of harmonisation work with manufacturing. Fokker Aerostructures performed the redesign.

25.3 Manufacturing of the Glare demonstrator panel

The panel was manufactured without having an original template for its shape. The manufacturing template or mould was built especially for this panel. To avoid any sizing problems, a detailed measurement of the original panel was carried out. These data were transferred to the mould of the new panel and the panel was delivered with these final dimensions. Fokker Aerostructures undertook the manufacturing of the panel.

25.4 Installation

The installation procedure was defined by Service Bulletin A300-53-021 under modification number GAF06698A [1] and was approved by the LBA (German airworthiness authority).

The installation was performed like a normal panel replacement after major damage. Only two concessions / abnormalities occurred:

1. Problem: one countersunk hole was damaged due to a broken tool.
   Corrective action: the hole was enlarged to the next nominal diameter
2. Problem: clips P/N was changed.
   Corrective action: interchangeable parts were installed.

The installation procedure is illustrated by a series of photos. Figures 25.1 through 6 show in succession: the removal of the original skin panel, the Glare demonstrator panel and the installation of the Glare demonstrator panel on the fuselage.

Figure 25.1 *Removal of the original skin panel.*

Figure 25.2 *Fuselage after removal of the original skin panel.*

Figure 25.3 *The Glare demonstrator panel.*

Figure 25.4 *Lifting of the Glare panel.*

Figure 25.5 *Positioning of the Glare panel.*

Figure 25.6 *The Glare demonstrator panel in place.*

25.5 SRM activities related to the Glare demonstrator panel

A specific supplemental SRM [2] has been developed for the Glare demonstrator panel, which contains the identification of the installed parts, the Allowable Damage Limits (ADL) and specific skin repairs, see Figure 25.7. The defined repairs and ADL are based on test results from Delft University of Technology, National Aerospace Laboratory NLR and Fokker Aerostructures.

DaimlerChrysler Aerospace
Airbus

STRUCTURAL REPAIR MANUAL
SUPPLEMENT

DaimlerChrysler Aerospace Airbus
Customer Services
Technical Publications Department
Kreetslag 10
21129 Hamburg
Germany
Telefax 49 (40) 7437–6605

STRUCTURAL REPAIR MANUAL SUPPLEMENT
FOR
AIRBUS A310–304 COVER E.O. FOR CONVERSION PROGRAM GAF 035 96A
(COMBI CONVERTIBLE AIRCRAFT)
FOR THE
GERMAN AIR FORCE (GAF)

Effectivity MSN 484

This Supplement must be attached to the AIRBUS INDUSTRIE Model A310–304 Structural Repair Manual Supplement when the Glare Panel has been installed.
The information contained herein supplements or supersedes the information of the basic Structural Repair Manual only in those areas covered herein. For limitations, procedures and performance Informations not contained in this supplement, consult the basic Structural Repair Manual.
When the Glare Panel is removed and the original state is restored this Supplement has to be completely removed.

<u>LBA Compliance Statement</u>

We have reviewed this Structural Repair Manual Supplement for Airbus A310–304 with Change GAF 06698A. This is to certify that this document is LBA accepted for showing compliance with the certification basis. This document is approved by the DaimlerChrysler Aerospace Airbus, Office of Airworthiness under the LBA granted privileges of the German design organization approval I–EA 1.

Printed in Germany

Figure 25.7 Introductory page of the structural repair manual supplement [2].

25.5.1 ADL definition – rework, lightning strike and dents

Figures 25.8 through 12 show five pages of the SRM supplement [2]. On these pages the Allowable Damage Limits (ADL) are defined along with the corrective measures that should be taken in case such damage occurs.

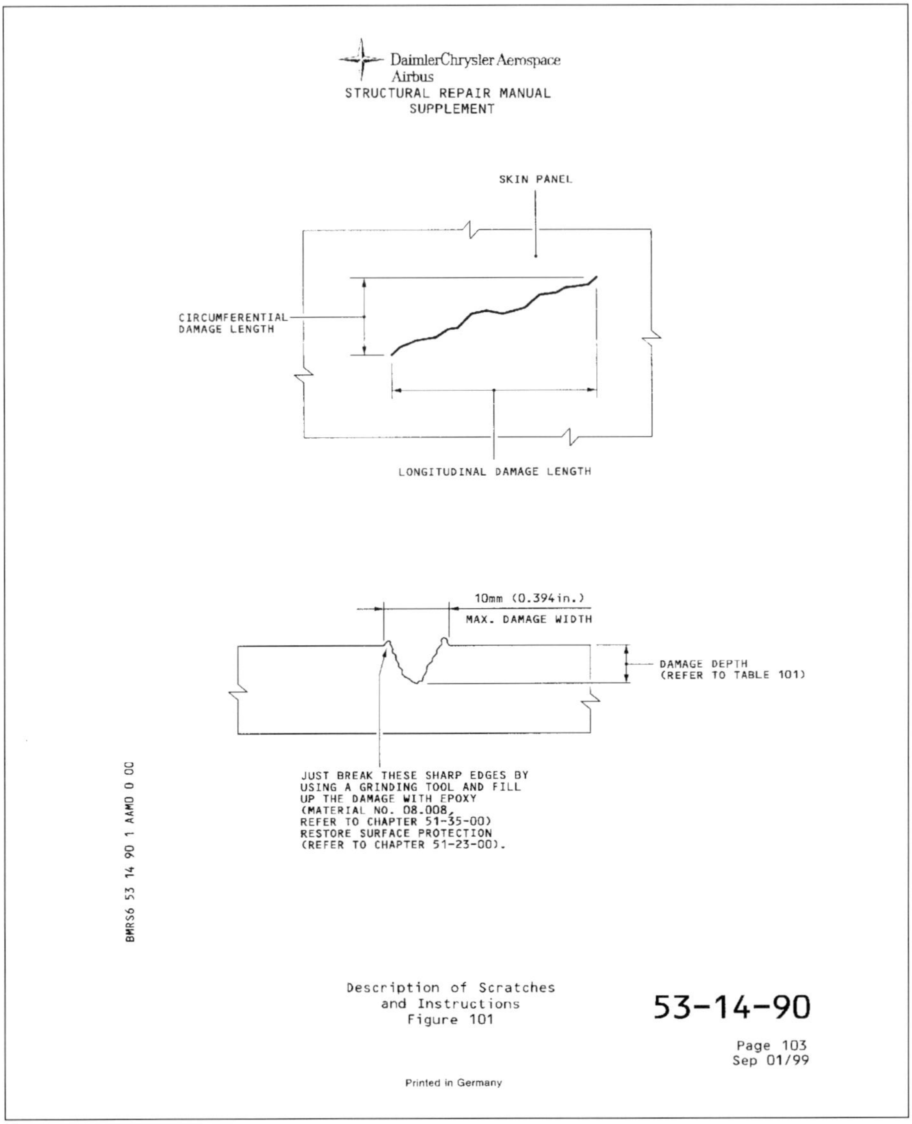

Figure 25.8 Page of SRM supplement [2] *describing scratches and instructions.*

DaimlerChrysler Aerospace
Airbus
STRUCTURAL REPAIR MANUAL
SUPPLEMENT

CAUTION: FOR DAMAGE DEPTH GREATER THAN 0.65MM (0.026IN.) IT IS ALLOWED TO HAVE A FIRST CRACK AND A CRACK THRU. IN CASE OF A HOLE THRU ALL LAYERS WITH THE RISK OF PRESSURE LEAKS, A TEMPORARY SKIN REPAIR IS RECOMMENDED (REFER TO PAGE BLOCK 201).

DAMAGE DEPTH	CIRCUMFERENTIAL DAMAGE LENGTH	LONGITUDINAL DAMAGE LENGTH	LIMITATIONS
less or equal 0.4mm (0.016in.) (glass fibers must be undamaged)	maximum 375mm (14.76in)	maximum 500mm (19.69in)	no limitation
greater than 0.4mm (0.016in.) and less than 0.65mm (0.026in.)	maximum 250mm (9.84in.)	maximum 250mm (9.84in.)	6000 FC
greater than 0.65mm (0.026in.)	maximum 37.5mm (1.47in.)	maximum 37.5mm (1.47in.)	3000 FC

NOTE: The maximum damage width is 10mm (0.394in.) (Refer to Figure 101).

NOTE: The minimum distance between adjacent damages is 200mm (7.87in.).

NOTE: These limits are valid for unriveted area or riveted area with a maximum of three adjacent rivets.

Scratches
Table 101

53-14-90

Page 104
Sep 01/99

Printed in Germany

Figure 25.9 Page of SRM supplement [2] *defining scratches.*

DaimlerChrysler Aerospace
Airbus
STRUCTURAL REPAIR MANUAL
SUPPLEMENT

This table contains requirements and restrictions for burn marks on skin caused by lightning strikes at a distance of more than 15mm (0.591in.) from an existing fastener, measured from the fastener center line.

	DAMAGE DEPTH	
	< 0.4mm (0.016in.)	< 0.65mm (0.025in.)
Permissible number of burn marks per stringer / frame bay	5	3
Minimum distance between two damages	100mm (3.94in.)	
Maximum damage size per burn mark	50mm (1.97in.) x 50mm (1.97in.)	
Live limit	No	
Inspection method	External FOKKER bond test to determine the damage size (to be performed by DA)	
Corrective action - permanent	Remove damaged or burned aluminum and fiber layer portions by blending out. Material removal of the second aluminum layer is not allowed.	
Corrective action - conversion	Remove temporary applied sealant or cold curing adhesive. Material removal of the second aluminum layer is not allowed.	
Surface pretreatment	Prepare surface for a cold bonding procedure of aluminum sheets (Refer to Chapter 51-77-12). Additional requirement: dry the surface for a minimum of 1 hour at 80°C.	
Surface protection	Install a doubler (Material 2024T3 thickness 0.4mm (0.016in.) up to 0.6mm (0.024in.) according to procedure given in Chapter 51-77-12 Paragraph 4.C. For surface protection refer to Chapter 51-23-00.	

Conversion and Permanent Repair of Lightning Strike Damage
Table 104

53-14-90

Page 110
Sep 01/99

Printed in Germany

Figure 25.10 Page of SRM supplement [2] *describing conversion and permanent repair of lightning strike damage.*

DaimlerChrysler Aerospace
Airbus
STRUCTURAL REPAIR MANUAL
SUPPLEMENT

	DAMAGE DEPTH	
	< 0.4mm (0.016in.)	< 0.65mm (0.025in.)
Permissible number of burn marks per stringer /frame bay	2	1
Minimum distance between two damages	150mm (5.91in.)	
Maximum damage size per burn mark	100mm (3.93in.) x 20mm (0.79in.)	
Live limit	No	
Inspection method	External FOKKER bond test to determine the damage size (to be performed by DA)	
Corrective action - permanent	Remove damaged or burned aluminum and fiber layer portions by blending out. Material removal of the second aluminum layer is not allowed.	
Corrective action - conversion	Remove temporary applied sealant or cold curing adhesive.	
Surface pretreatment	Prepare surface for a cold bonding procedure of aluminum sheets (Refer to Chapter 51-77-12). Additional requirement: dry the surface for a minimum of 1 hour at 80°C.	
Surface protection	Install a doubler (Material 2024T3 thickness 0.4mm (0.016in.) up to 0.6mm (0.024in.) according to procedure given in Chapter 51-77-12 Paragraph 4.C. For surface protection refer to Chapter 51-23-00.	

Conversion and Permanent Repair of Lightning Strike Damage at Bonded Splice between Stringer 8RH and 9RH
Table 105

53-14-90

Page 111
Sep 01/99

Printed in Germany

Figure 25.11 *Page of SRM supplement* [2] *describing conversion and permanent repair of lightning strike damage at bonded splice between stringer 8RH and 9RH.*

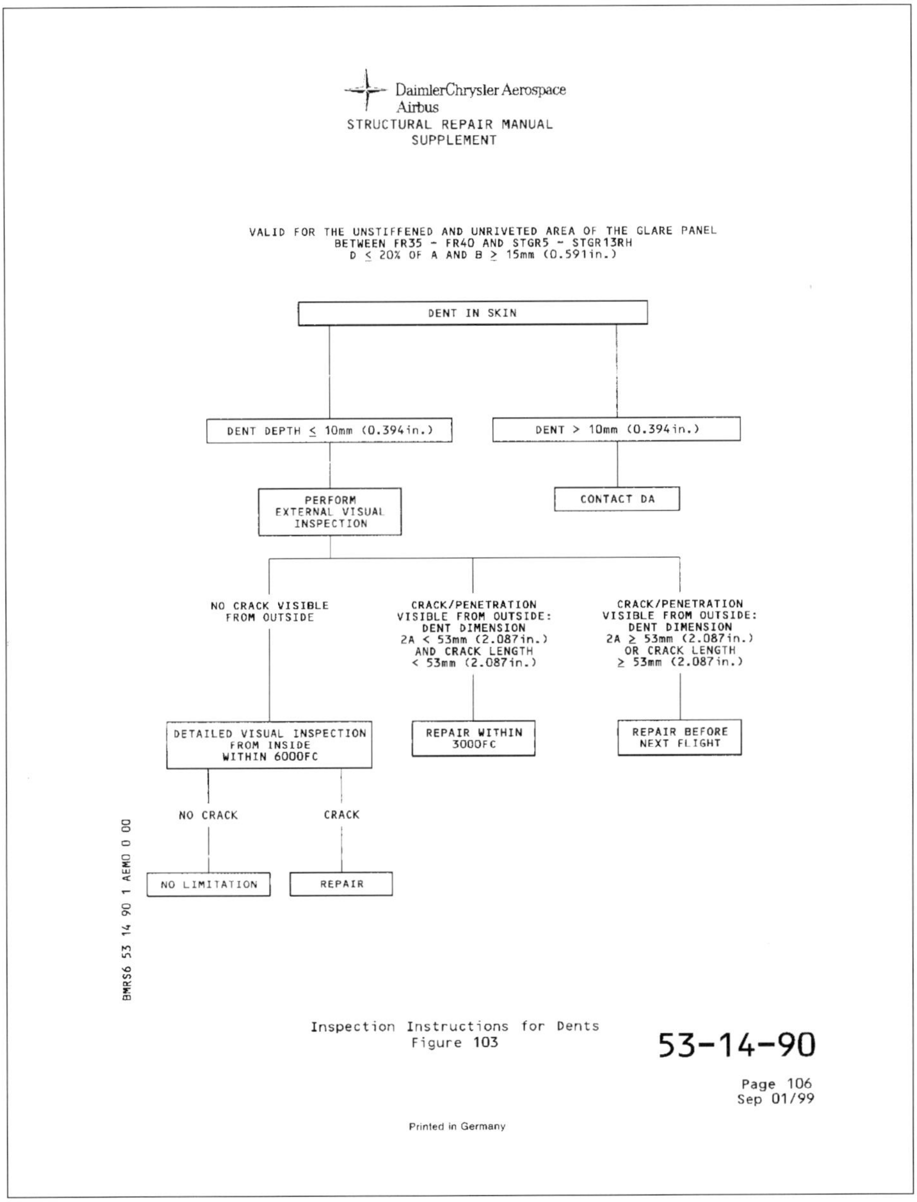

Figure 25.12 Page of SRM supplement [2] describing inspection instructions for dent.

25.5.2 Definition of skin repairs

The skin repairs for the A310 are fully comparable to the skin repairs of monolithic aluminium structures. For this demonstrator panel it was possible to define 1.8 mm thick 2024 T3 and 1.7 mm thick Glare 3-3/2-0.4 as doubler material with the same inspection threshold and interval. As a result of the low fatigue stress level in this panel, the standard SRM skin repair principles do not impose additional inspection requirements until the design service goal. Figures 25.13 through 15 show three pages of the SRM supplement [2] about the repair of damaged skin panels.

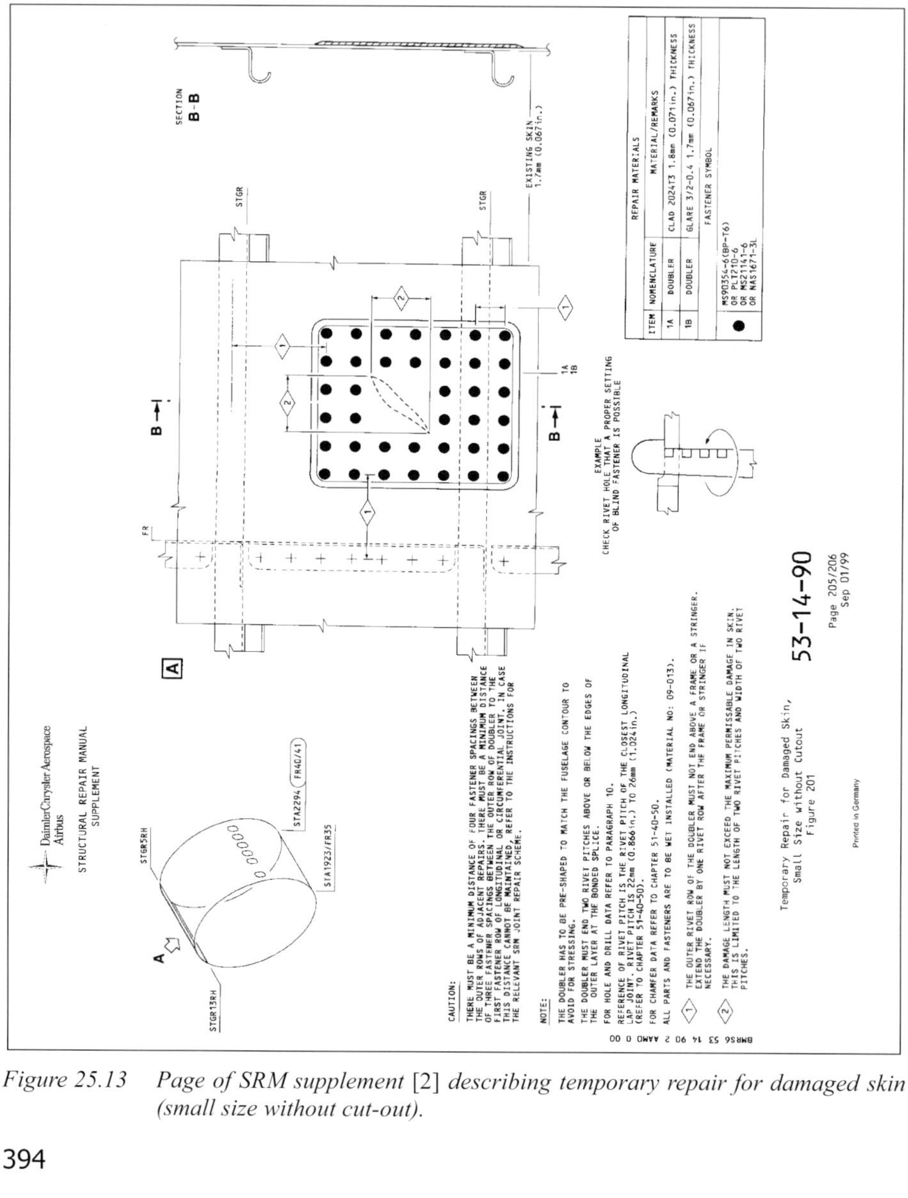

Figure 25.13 *Page of SRM supplement* [2] *describing temporary repair for damaged skin (small size without cut-out).*

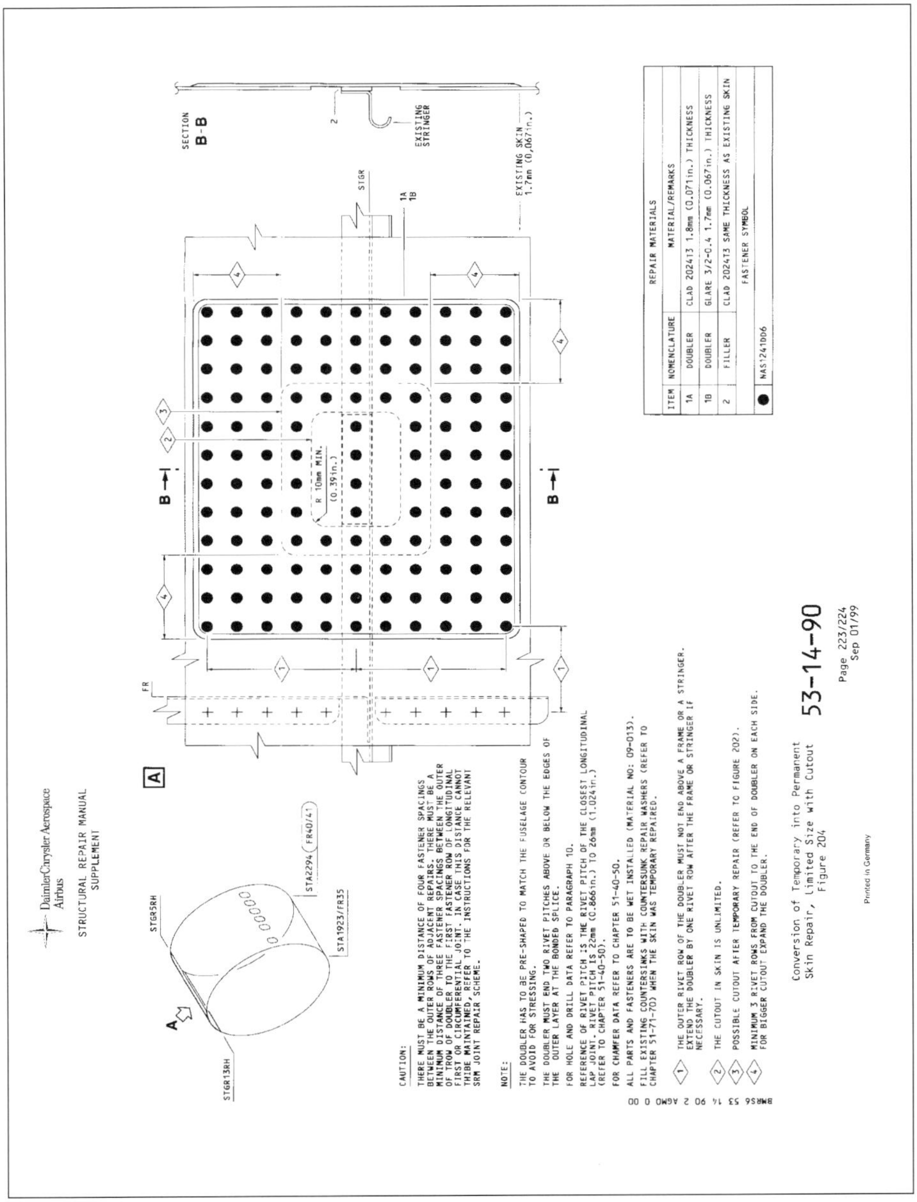

Figure 25.14 *Page of SRM supplement* [2] *describing conversion of temporary into permanent skin repair (limited size with cut-out).*

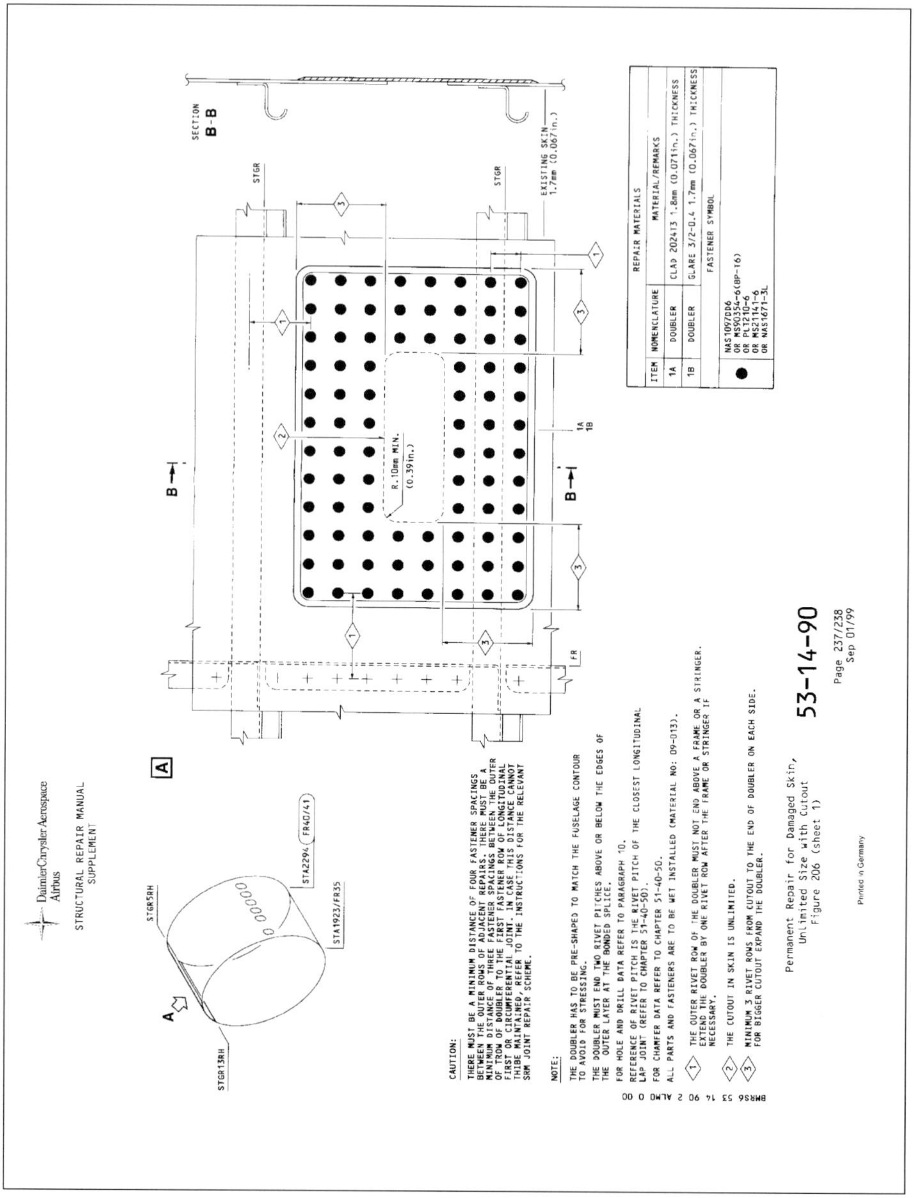

Figure 25.15 *Page of SRM supplement* [2] *describing permanent repair for damaged skin (unlimited size with cut-out).*

25.6 Glare related SRM development and testing for the A380 and Megaliner Barrel

The basis for the relevant ADLs and repairs of Airbus' newest aircraft programme is the work and development for the A310 MRT Glare demonstrator.
The ADLs will be comparable to A310 supplemental SRM and the skin repairs will in principle be comparable, but adapted to the new design and material thickness, as well as for the different stress levels.
Repairs for stringers (longitudinal stiffeners) made from Glare will also be developed by taking into account well-known repair principles for the other A/C programmes.
Different repair principles will be tested on the Megaliner Barrel.

25.6.1 Skin repairs

- maximum doubler thickness 5.0 mm
- doubler materials
  - Glare 3-X/Y-0.4 – permanent repair
  - 2024 T3 – interim repair (life limit until next overhaul)
  - 7075 T6 – interim repair
  - AMS5518 (high strength steel) – interim and permanent repair
- use of titanium fasteners, minimum diameter 5.6 mm
- flush skin repairs
- external skin repairs
- double shear skin repairs
- repair principles will be developed for maximum skin thickness of 20 mm (this is a goal, which depends on the definitive design)
- additional inspections will be required.

25.6.2 Stringer repairs

- new specific splice and partial replacement repair principles have to be developed due to different design of the A380 and Megaliner Barrel – reduced clearance between stringer and frame
- U and C-profiles as splice or replacement profiles
- several repair materials will be offered with different life limits
  - Glare
  - monolithic aluminium
  - steel
- use of titanium fasteners, minimum diameter 5.6 mm.

References

[1] Service Bulletin No.: A300-53-021; Modification No. GAF06698A; Title: *Fuselage – Install Glare panel demonstrator*; Revision No. 1; March 1999

[2] A310 MRT GAF – *Structural Repair Manual – SRM Supplement Glare Panel German Air Force – Registration 10+25 – Manufacturer serial number 484 – for change 06698A*. As PDF file available. Issue Sep01/99.

26

Burn-through and lightning strike

P.A. Hooijmeijer

Abstract – Burn-through tests showed that Glare is very fire resistant. Depending on the Glare grade and thickness, only the outer aluminium layer will melt, whereas the other layers will remain intact. Carbonisation of the fibre layers and delamination of the laminate will insulate the non-exposed side of the material from the flames, which results in relatively low temperatures at the non-exposed side. Burn-through tests meant for baggage compartment liners were performed for more than 10 minutes, which did not result in burn-through or flame leakage of the Glare sheet.

Lightning strike testing revealed a similar behaviour. Due to arc attachment of the lightning strike, heat was created and consequently the outer aluminium layer locally melted and/or vaporised. Also in this case the underlying fibre layers carbonised and were partially damaged, depending on the severity of the strike. In all cases, all layers underneath the first prepreg layer remained intact, whereas in a monolithic aluminium sheet the material melted through the thickness. In contrast with aluminium, Glare consequently showed a much better fatigue behaviour after a lighting strike, since most of the layers remained intact.

26.1 Introduction

For civil passenger aircraft a significant number of fatalities, in otherwise survivable accidents, are associated with the effects of fire. Two major accidents can be mentioned as examples.

The first is an accident at Manchester airport in 1985. External fire, which occurred during take off, entered the cabin by burning or melting through the fuselage. During this accident the amount of fuel involved was only limited and the rescue and fire-fighting services responded rapidly. However, many lives were lost and this was attributed to the rapid burn-through of the fuselage skin followed by flames entering the passenger cabin.

The second example is the accident with a Saudi Arabian Airlines L-1011 Tristar in 1980. The aircraft landed safely after an in-flight emergency. A delay in the evacuation process resulted in the deaths of all 301 passengers aboard, before the doors could be opened, see Figure 26.1.

Figure 26.1 *Wreckage after burn-out of a Saudi Arabian Airlines L1011 Tristar.*

The frequency of accidents with jet aircraft in which fire is involved is about 1.5 per ten million flying hours. The number of fatalities due to fire in these survivable accidents is about 19 per accident [1]. Although the frequency of accidents involving fires has been reduced, improvements in aircraft fire safety have become increasingly more important for future aircraft design. One of the reasons is that larger aircraft are currently being designed and the importance of the safety of these aircraft is evident. Therefore the protection of passengers against post-crash fires has become an important area of aviation safety.

To meet the airworthiness requirements of a maximum of 90 seconds evacuation time (FAR 25.803), an intact fuselage with sufficient integrity is of major importance. If the structural integrity of a fuselage is poor, part of the structure and/or thermal and acoustic insulation material may fall on the passengers during the evacuation after a post-crash fire.

Aluminium has relatively short burn-through times and the structural integrity of a fuselage rapidly diminishes during the melting of the skin and other structural members. Therefore, new materials such as composites and Fibre Metal Laminates are subjected to burn-through testing. The layered composition of Glare might provide a better burn-through behaviour compared to aluminium. In section 26.2 the behaviour of Glare during burn-through tests is described.

A comparable subject, however far less dangerous, is the strike of an aircraft by lightning. All aircraft in service will suffer from lightning strike. The amount of damage done to the structure depends on the structural material and also on the severity and the type of the lightning strike. Because the aircraft fly far above the clouds nowadays, the hazard of being struck by lightning is most pronounced when the aircraft is on the ground and during climb and descent.

Metallic structures have a good electrical conductivity and therefore they will only suffer from very local and minor damage due to a lightning strike. Composite materials on the other hand, have a much lower conductivity and will therefore suffer from severe structural damage if the material is not protected properly, see Figure 26.2.

Figure 26.2 *Typical damage due to lightning strike on an unprotected carbon component* [2].

In the case of Glare, which is a hybrid material, a different behaviour might be expected. Section 26.3 will give insight into the behaviour of Glare when exposed to a realistic lightning strike.

26.2 Glare exposed to fire

The aviation authorities carried out a long-term investigation into the behaviour and properties of current aerospace materials and structures in post-crash pool fires. In addition, new materials such as composites and Fibre Metal Laminates have been studied. The studies were performed to find a way of improving the burn-through resistance of aircraft fuselages. Apart from extending the time for the fire to penetrate the fuselage, the associated toxic gasses and smoke in the cabin should also be minimised and the degradation of the structural integrity of the fuselage delayed to increase the possible evacuation time. The ultimate goal for the American airworthiness authorities is to reduce the fatalities in the United States resulting from fire accidents to zero.

26.2.1 Regulations and tests

The available Federal Airworthiness Regulations (FAR) do not impose burn-through requirements for fuselage skin structures. The requirements only treat the fire resistance of interior materials: seats and cushions, thermal acoustic insulation materials and baggage compartment liners but not the fuselage structure. Glare was developed as a material for the primary structure of a fuselage such as the skin. Since no regulations are provided for the skin, burn-through tests on Glare were mostly performed using the test setup for cargo compartment liners.

Regulations prescribe a burn-through test for cargo compartment liners consisting of a test setup with a 930 °C flame with a heat flux of 91 kW/m^2. In comparison with the possible characteristics of a large post-crash fuel fire - a temperature of 1150 °C and a heat flux of 160 kW/m^2 [3] - these values are rather low. The apparent difference from the conditions during a post-crash fuel fire indicates that an improvement in burn-through tests is preferable. However, the present test is very useful for comparison purposes between the behaviour of traditional materials and Glare.

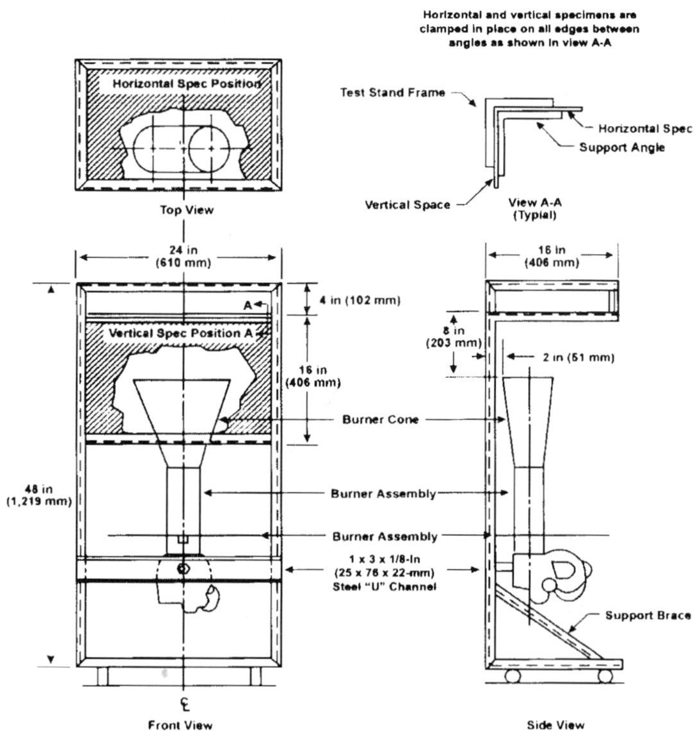

Figure 26.3 *Baggage compartment liner test setup* [4].

In the test for cargo compartment liners, the test panel is a setup with either a horizontal or vertical frame. The panel is exposed to the flame for a certain time and flame leakage and burn-through times are recorded together with the temperature on both the exposed and the non-exposed side. A Bunsen burner that represents large fires produces the flame. Figure 26.3 gives a schematic representation of the test.

26.2.2 Burn-through system

Glare is a material consisting of two materials with different burn-through characteristics. The aluminium layers show behaviour identical to that of a monolithic aluminium sheet when exposed to fire. Aluminium of 1 to 2 mm thickness will burn through within a minute in post-crash pool fire conditions, depending on the emissivity. The latter parameter is determined by the amount of soot deposited on the specimen.

In Glare, the aluminium layers are very thin and will burn through very quickly when exposed to fire. Within only seconds the aluminium layer with a maximum thickness of 0.5 mm burns through. In Figure 26.4 this typical burn-through behaviour is shown.

After burn-through of the outer aluminium layer, the glass-fibre prepreg layer is exposed to flames of around 1100 °C. The fibres in Glare are S2-glass fibres in an epoxy matrix. The melting temperature of the glass fibres is higher than the flame temperature, therefore the glass fibres are able to keep the fire from passing the layer and thus create a fire barrier.

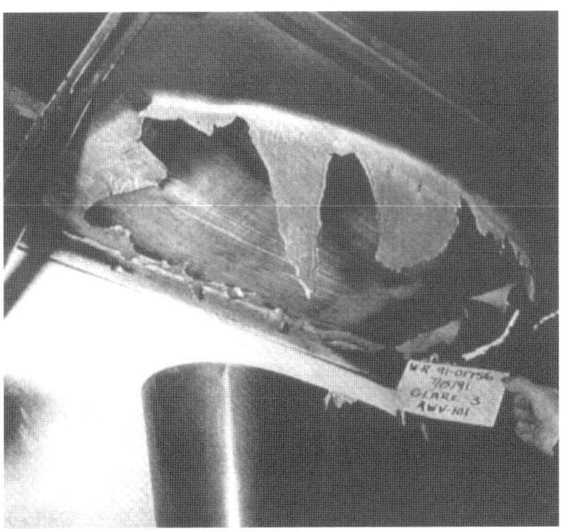

Figure 26.4 *Typical damage on Glare panel after 10 minutes of exposure to 1100 °C flames (courtesy Boeing)* [5].

The epoxy matrix cannot withstand this heat and will carbonise throughout the whole thickness under this condition. The carbonisation causes complete delamination of the laminate, starting from the location of exposure. The delamination implies separation of the layers. This separation allows air to flow in between the layers, which consequently acts as an insulator.

The remaining aluminium layers in the laminate are protected by the insulating air barriers between the delaminated sheets and by the carbonised epoxy in the fibre layers and remain intact. In this way, the non-exposed side of the laminate is insulated from the fire. Delamination is a very unpredictable feature and scatter in the burn-through times of Glare can be attributed to the different amounts of delamination in the specimens. The temperature on the non-exposed side will remain relatively low compared to the temperature on the non-exposed side of a monolithic aluminium specimen.

The excellent burn-through properties of Glare, as a result of the phenomena described above, were verified in many test series. For example, in 1994 fireproof tests were done by the Boeing Company according to FAA standards on Glare 3 and Glare 4. The tests were performed according to Advisory Circular 20-135 [6]. Although the exposed side of the laminate was damaged, the high melt temperature of the glass fibres prevented penetration of the flame. The relatively low temperatures on the non-exposed side of the laminate proved the good insulation behaviour. The results of these tests showed that Glare can be qualified as fireproof and can consequently be used for firewall applications. The results of the tests are summarised in Table 26.1.

For comparison purposes, the burn-through times of aluminium roughly 1.5 to 2.0 mm thick (comparable to the thickness of the materials given in Table 26.1) are about 1.5 minutes, whereas Glare did not show flame penetration after 15 minutes, at which point the tests were stopped.

Table 26.1 *Results of burn through tests on Glare by Boeing [5].*

Test	Fire-proof and Fire Resistance tests
Standard	Federal Aviation Administration Advisory Circular, 20-135 (2/6/90). ANM-110
Performed by	Boeing, Seattle
Flame temp	1100 ±25 °C

Type FML	Temperature non-exposed side (°C)
Glare 3-2/1	After 5 min: 220
Glare 3-3/2	After 5 min: 160 After 10 min: 215
Glare 4-2/1	After 5 min: 220

26.3 Glare exposed to lightning

26.3.1 Lightning strike testing

When an aircraft is flying through a thunderstorm, it will most certainly be hit by lightning. This is mainly because the aircraft will create a shorter path over which the lightning can discharge. Lightning on aircraft can result in burn marks and pitting. Penetration of the skin can even occur when lightning strikes thin-walled structures.

Lightning strikes of class 2A [7] on the fuselage and cockpit are of prime interest for a fuselage skin. A 1B strike simulates the most severe lightning load possible, but is only found at the rear of the aircraft and engines, locations that are not so important when considering Glare. However, this severest strike was also used as a worse case scenario in the investigation of the behaviour of Glare. Table 26.2 gives a classification of the severest lightning strike 1B used in testing.

Table 26.2 Definition of class 1B lightning strike (MIL-STD-1757a).

Class 1B lightning strike, pulses A+B+C+D	
A-pulse (Initial stroke)	Peak amplitude = 200 kA (±10%); Action integral = $2x10^6$ A^2s (±20%); Time duration ≤ 500 μs
B-pulse (Intermediate current)	Maximum charge transfer = 10 Coulombs; Average amplitude = 2 kA (±20%); Time duration ≤ $5x10^{-3}$ s
C-pulse (Continuing current)	Charge transfer = 200 Coulombs (±20%); Amplitude = 200-800 A; Time duration: $0.25 \leq t \leq 1$ s
D-pulse (Restrike)	Peak amplitude = 100 kA (±10%); Action integral = $0.25x10^6$ A^2s (±20%); Time duration ≤ 500 μs

Two different lighting-strike tests can be distinguished: Direct Hit and Current Flow tests. In a Direct Hit test an actual bolt of lightning will hit the material, whereas a Current Flow test sends a large current through the skin from the point of current introduction to the point where the current can be released to the environment or other structural components.

Since Glare is not a monolithic material and also consists of non-conductive materials, it is interesting to investigate the influence of a lightning strike on Glare. Improper protection of composites has shown that a lightning strike can cause severe damage to the structure.

Lightning-strike tests were performed at the KEMA (a Dutch research institute for electrical components and electrical power supplies), simulating severe but realistic lightning in flight. In a test setup, see Figure 26.5, lightning strikes hit several configurations of Glare material. In the case of an insufficient current flow from the location of the arc attachment of the lightning to the position where the current is guided out of the material, the structure will be damaged. This damage was studied after the tests.

Figure 26.5 *Test setup for lightning strike at the KEMA company, Arnhem, The Netherlands.*

26.3.2 Glare behaviour

The behaviour of Glare when hit by lightning can be explained by the layered composition of the material. The fibre layers in the laminate are non-conducting, therefore the current must find its way through only the aluminium layers. If those layers are not connected, the current can only be transmitted through the thin outer aluminium layer.

When a lightning strike hits the surface, a lot of heat is created in the material. Similar to Glare behaviour when exposed to fire, the outer aluminium layer melts and vaporises locally. The underlying fibre layer carbonises and, depending on the severity of the hit, the fibre layer will be partially damaged. Examination of the first fibre layer showed that in most cases the first prepreg layer will be damaged and that only in case of very severe hits will the second prepreg layer be damaged as well.

In all cases, the underlying layers of the laminate will not be damaged. Figure 26.6 gives shows a detail of damage caused by a Direct Hit. As can be clearly seen, the damage to the laminate is concentrated on the outer aluminium layer and the first fibre layer only. The damage will not penetrate through the thickness of the sheet.
C-scan investigations have shown that no invisible damage is present outside the visually damaged area of the lightning strike. From this it can be concluded that no expensive detection methods are necessary to identify the size of the damage that needs to be repaired.

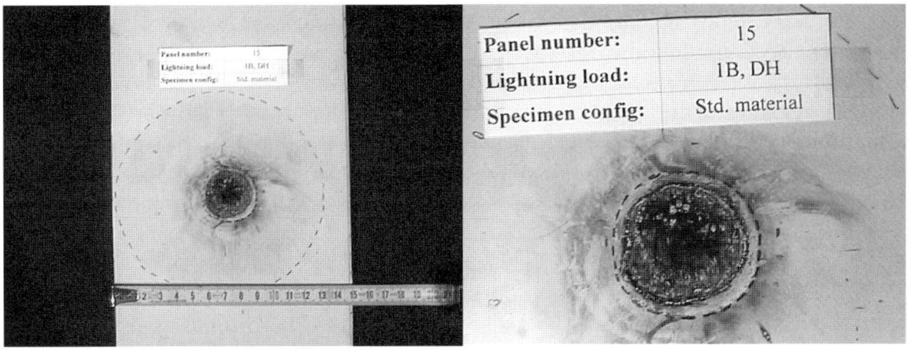

Figure 26.6 *Details of damage due to lightning. The outer aluminium layer is melted and part of the first fibre layer is damaged after a 1B Direct Hit.*

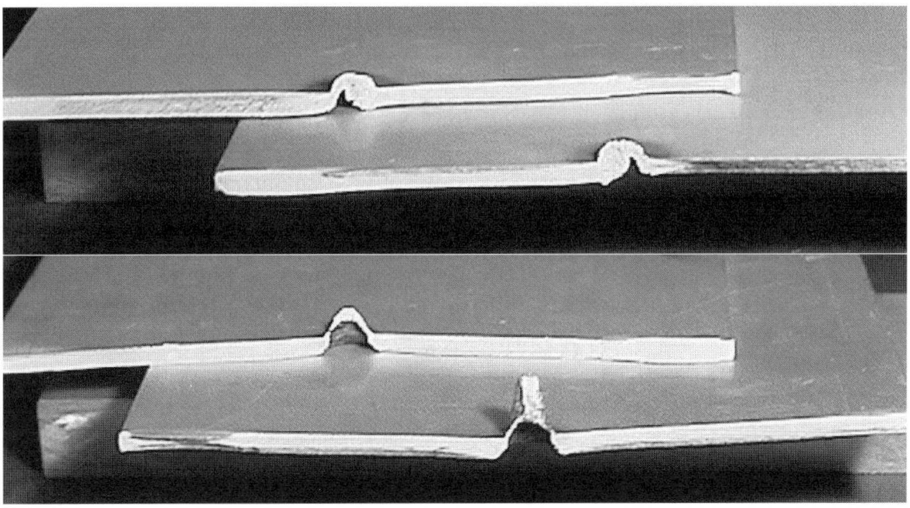

Figure 26.7 *Damage of lightning strike in monolithic aluminium layers after a 2A hit (top) and a 1B hit (bottom).*

In Figure 26.7 an example is given of the damage due to lightning strike in a monolithic aluminium panel. As can be seen from the cross-sections of the specimens, the aluminium is melted by the lightning strike, but because the panel is monolithic, the material is melted though the thickness and even causes a protrusion on the non-exposed side. In an exceptional case a small hole was caused by the lightning strike.

From the lightning strike tests, we can conclude that the resistance of Glare against lightning strike is better than for aluminium. The combination of conductive aluminium layers and isolating fibre layers results in lightning damage that only affects the outer two layers (one aluminium and one prepreg layer), leaving the underlying layers undamaged. For the aluminium panels, however, it was found that the lightning affects the complete thickness of the panel, even when exposed to the less severe 2A zone lightning strike.

The damage due to the lightning strike directly influences the fatigue life of the aluminium specimens. Fatigue cracks will start from the damage with fatigue crack propagation extending over the thickness of the panel. The reduction of fatigue life is not only due to the visible damage, but also the disturbed heat treatment caused by the lightning strike.

In the Glare specimens the damaged layer is fully supported by the intact underlying layers. In Glare a fatigue crack will also occur but will be limited to the damaged outer aluminium layer. The other layers will remain intact. Part of the load on the specimen will be transferred over the damage and the stress intensity at the tip of the fatigue crack is reduced. Crack growth will consequently be very slow.

References

[1] G. van der Kevie, *The modification of Glare to develop a more fire-resistant and non-delaminating fuselage skin for the Airbus A3XX*, Master thesis, Delft University of Technology, Faculty of Aerospace Engineering, 1997.

[2] G.W. Reid, 'Mechanical damage to aircraft structures from lightning strikes', *Journal of Aerospace Engineering Part G*, Vol. 207, pp.1-14, 1993.

[3] D.C. Dodd, C.T.M. Hall, J. Pollard, M.A. Snell, *Burn through resistance of fuselages: Initial findings*, Civil Aviation Authority CAA Paper 94002, London, July 1994.

[4] FAA, *Aircraft Materials Fire Test Handbook, Final Report*, DOT/FAA/AR-00/12, US Department of transportation, April 2000

[5] G.H.J.J. Roebroeks, 'Glare; a structural material for fire resistant fuselages', *Aircraft Fire Safety*, AGARD Conference Proceedings 587, October 1996, p.26-1 - 26-13.

[6] FAA, AC 20-153, *Powerplant installation and propulsion system component fire protection test methods, standards and criteria*, US Department of Transportation, June 1990.

[7] FAA, AC 20-53A, *Protection of airplane fuel systems against fuel vapor ignition due to lightning*, (paragraph 10: Lightning attachment phenomena), US Department of Transportation, December 1985.

27

Impact properties

M. Hagenbeek

Abstract – An aircraft structure should be able to tolerate a certain amount of impact energy and damage, dependent on the location and probability of impact. The structural design and material choice must guarantee the structural integrity after a realistic impact event. For Glare the impact damage resistance is related to the aluminium and glass fibre/epoxy properties. Glare is stronger than aluminium at higher impact velocities due to the strain rate dependent behaviour of the fibres. The dent depth after impact is comparable to aluminium with the same thickness. The residual properties for Glare are comparable to (e.g. compressive strength), or better than (e.g. fatigue) monolithic aluminium. The inspection of a Glare structure is no more complicated than for aluminium, since Glare has the same indentation and because the damage size due to impact is smaller than the visible dent size.

27.1 Mechanisms

27.1.1 The impact regime

Low velocity impact (< 11 m/s) occurs through damage from for example service trucks, cargo containers and dropped tools during maintenance operations. High velocity impact (>11 m/s) can occur when debris from the runway hit the fuselage during take-off or landing, or when ice from the propellers strikes the fuselage. Different behaviour is observed during ballistic impacts (> 500 m/s), which are a concern for military application, and for hypervelocity impact (> 2000 m/s) of spacecraft by space debris. For Glare a strain rate effect can be observed, i.e. Glare is stronger at a higher strain rate at a higher impact velocity. This effect is shown in Figure 27.1.

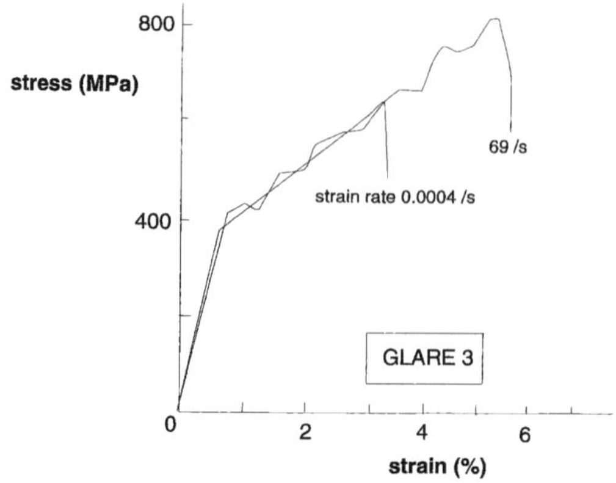

Figure 27.1 *Strain rate effect for Glare (0.0004 /s; quasi-static and 69 /s; high velocity).*

27.1.2 The impact process

Impact, in its physical context, is defined as the deformation and failure process during the collision between projectile and structure. Impact is of a dynamic nature: mass effects have to be taken into account and forces and displacements are time dependent. Large forces are exerted and removed during the short period of impact and initiated stress waves travel away from the region of impact.

The energy-absorption capability of the structure is decisive for the damage that may be created by the impact. The damage done during impact depends on: how the stresses build up, how they compare with the various dynamic material strengths, how they are redistributed as the fracture progresses, and the fracture energies and the energy available to drive the fracture process. These in turn depend on the material properties and geometry and energy and momentum of the projectile. Impact at low energies can result in surface damage and delamination. Impact at higher energies can produce cracks through the whole thickness or complete perforation.

Since impact is a very complicated phenomenon, predicting impact damage requires a thorough understanding of the basic damage mechanisms and mechanics governing the impact event. The shape of the dent, see Figure 27.2, is controlled by these deformation mechanisms.

Figure 27.2 *Shape of the dent.*

In general the deformation mechanism can be split up in three parts, see Figure 27.3:

1. Membrane strain deformation: caused by in-plane normal stresses which is constant over the thickness.
2. Flexural deformation: pure bending deformation only takes place for small deformations and is described as bending without membrane strain.
3. Transverse shear deformation: these deformations can cause shear cracks from which delamination can start to grow.

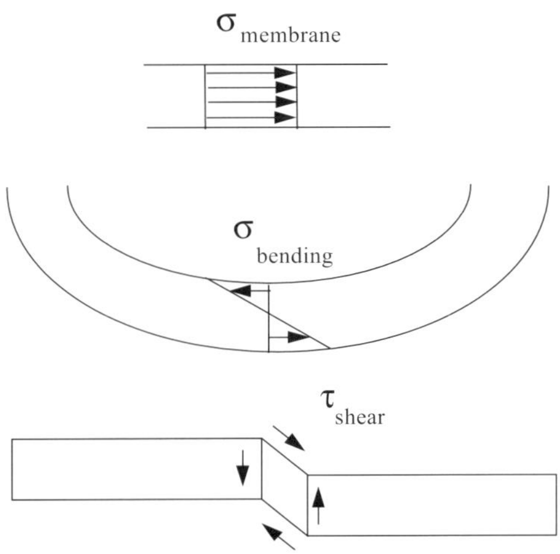

Figure 27.3 Three deformation mechanisms during impact.

27.1.3 Failure modes of Glare

For Glare there are several material properties related to the impact damage resistance. The damage modes of aluminium are combined with those of the glass fibres and matrix. Damage in Glare is therefore complex. It is a combination of cracks in the aluminium layers, crack growth and composite fibre breaking, matrix cracking and debonding of fibre/epoxy and epoxy/metal. Vlot [1] states that Glare has a fibre critical or an aluminium critical behaviour. For the fibre critical behaviour a crack occurs perpendicular to the outer fibre direction, see Figure 27.4(a). After an aluminium critical failure the crack runs in the rolling direction of the aluminium layers and in general this is the fibre direction of the first fibre layer, see Figure 27.4(b). Due to the relatively high strain to failure of S2-glass both types of failure can occur depending on the layup of the material.

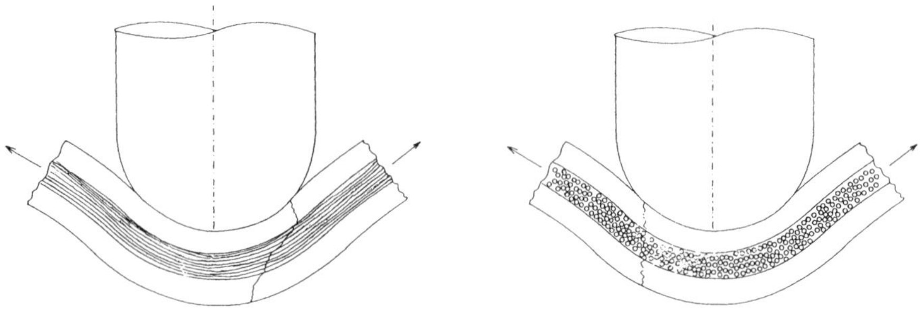

Figure 27.4 *(a) Fibre critical behaviour (left) and (b) Aluminium critical behaviour (right).*

For increasing impact energy level of the projectile, the following damage can be found in Glare:

1. Dent due to plastic deformation of the aluminium layers.
2. Epoxy cracking and minor delamination in the dented area.
3. First crack: combined fibre and aluminium cracking at non-impacted side.
4. Through crack: also aluminium cracking at impacted side.

27.1.4 Damage quantification

To quantify the severity of impact damage there are four major damage criteria:

1. Maximum permanent deflection.
2. Energy restitution coefficient.
3. Damage width.
4. Damage grade energy level.

1) Maximum permanent deflection.

This is the dent depth after impact. Figure 27.5 shows a comparison of dent depth versus impact energy for low velocity impact. The dent depth is strongly dependent on the thickness of the tested material. The dent depth decreases with increasing specimen thickness. To be able to compare the specimens regardless of their thickness, the dent depth is multiplied by the specimen thickness. The result is shown in Figure 27.6. The Glare 2B, Glare 4A and Glare 4B specimens now show identical results to aluminium specimens concerning dent depth. The same results have been found for high velocity impact [2].

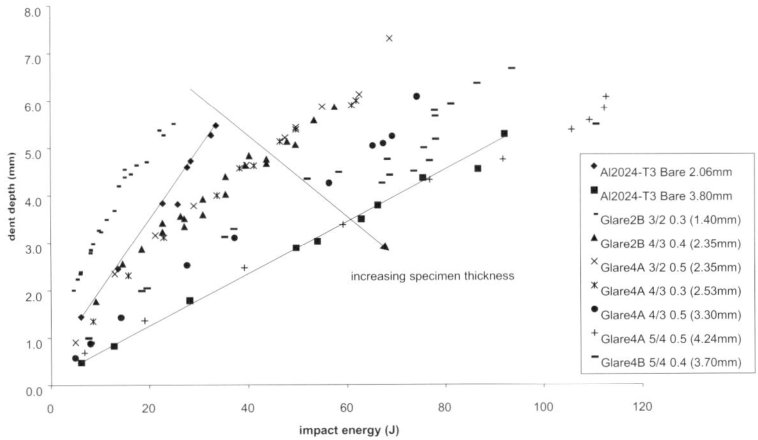

Figure 27.5 Dent depth versus impact energy for low velocity impact.

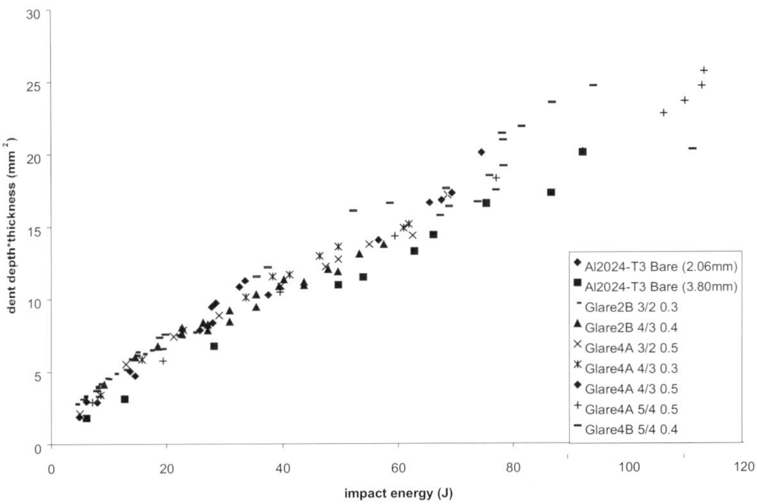

Figure 27.6 Dent depth times thickness versus impact energy, for low velocity impact.

2) Energy restitution coefficient.

When the projectile rebounds, the energy restitution coefficient is defined as:

$$e_E = \frac{E_{rebound}}{E_{impact}} \qquad (27.1)$$

With: $E_{rebound}$ = kinetic energy of the impactor immediately after impact
E_{impact} = kinetic energy of the impactor before impact.

This parameter is also a measure for the damage in the material. When $e_E = 1$, the collision of projectile and plate was fully elastic and there will be no remaining damage. If e_E becomes smaller more energy is absorbed by damage of the specimen.

3) Damage width.

To detect delamination in Glare specimens the outer aluminium layers have to be removed chemically. Delamination of the inner fibre/epoxy layers cannot be observed this way. However, the maximum damage occurs at the non-impacted side. For Glare the delamination zone is smaller than the visible damage at the non-impacted side. Therefore, the delamination damage width can be defined as in Figure 27.7(a).
The damage width in a carbon fibre-epoxy specimen cannot be defined in the same way. The damage at the non-impacted side is larger than the delaminated area and much larger than the damage at the impacted side, see Figure 27.7(b).

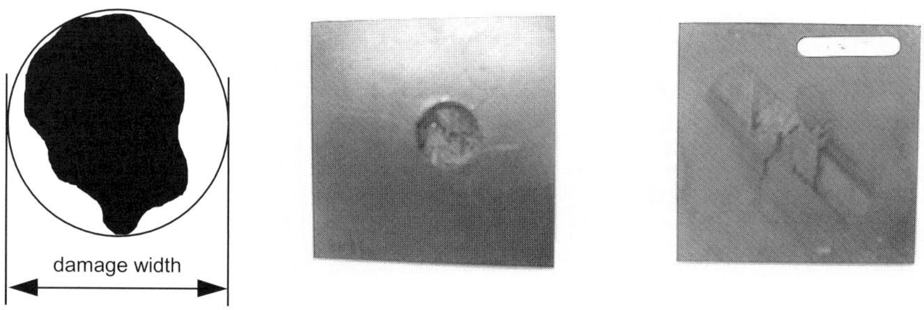

Figure 27.7 *Definition of damage width for Glare compared with damage in carbon fibre-epoxy specimen after impact: (a) Definition for Glare (left) and (b) Carbon specimen; impacted side (middle) and back side (right).*

4) Damage grade energy level.

This is the minimum cracking energy to create a first crack or a through crack. The minimum cracking energy is a measure of the ability to withstand impact damage. With fibre metal laminates, for first crack the impact energy needed for a visible crack in the specimen is taken. For a through crack the impact energy needed for a visible crack on both sides of the specimen is determined. Figure 27.8 and Figure 27.9 show a comparison of damage grade energy level per unit weight for respectively low and high velocity impact. The energy is divided by the specimen's areal density, i.e. mass per square metre material, for a better comparison. The difference in energy level for first and through crack is even more evident at high velocity impact testing. At both low and high velocity impact all Glare specimens, except the Glare 2B-3/2-0.3 specimens, perform better than the aluminium specimens by comparison with the damage grade energy levels.

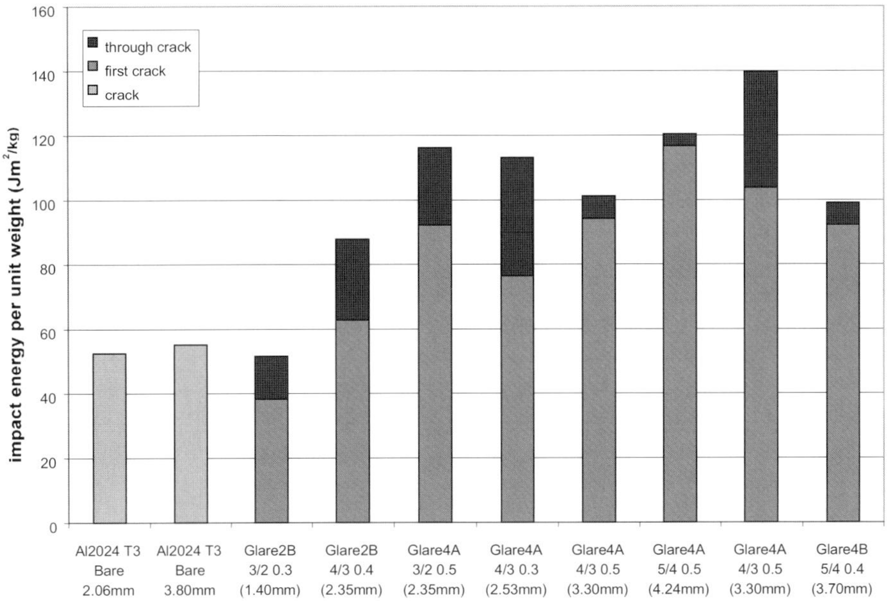

Figure 27.8 *Damage grade energy level per unit areal density for a crack in aluminium 2024-T3 and for a first and a through crack in Glare, at low velocity impact (~10 m/s, drop weight).*

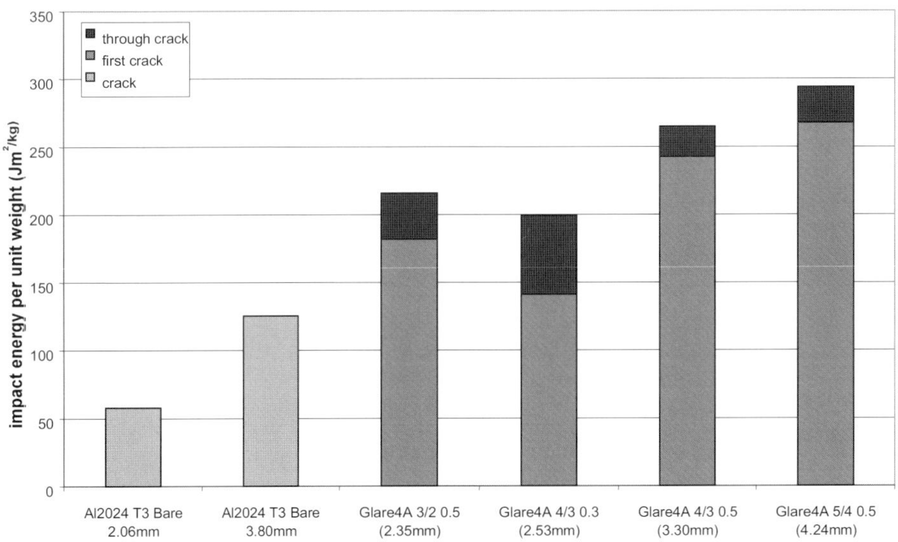

Figure 27.9 *Damage grade energy level per unit areal density for a crack in aluminium 2024-T3 and for a first and through crack in Glare, at high velocity impact (~100 m/s, gas gun).*

27.2 Impact damage resistance

27.2.1 Environmental effects

Environmental effects can degrade the properties of composites, especially hot/wet conditions. Moisture absorption can reduce the strength and stiffness of the material. The higher temperatures and moisture absorption tend to plasticize the resin. Laminates may even have improved impact resistance and residual strength under hot/wet conditions, because of the improved peel toughness. It is currently assumed at the certification of civil aircraft that the environmental degradation is caused only by moisture absorption under service temperatures. Other effects like thermal cycling, solar radiation and freeze/thaw thermal cycling of moistened laminates are assumed not to degrade the material [3].

An important advantage of Glare is that the aluminium sheets protect the glass fibre from moisture absorption. For Glare the moisture ingress rate around holes or damaged areas must be determined. The static strength has to be demonstrated with measurements for this moisture content. The environmental effects on the static and residual strength are simulated in the test setup to determine the effect. For fatigue testing the temperature cycles that the aircraft endures during every flight combined with the load cycles are used. Moisture ingress is slow and will not affect the impact damage resistance of Glare.

27.2.2 Design aspects

According to the damage tolerance requirements the design limit for compressive strain is limited to 0.003 to 0.004 for 'classical' composites with an epoxy matrix. The design strain for compression members is based on compression tests after impact, with or without loading during impact. There are several ways in which the design of a stiffened panel in composites can have an improved damage tolerance. The first one is related to the ply orientation (layup) of the skin. A skin of a damaged panel with a 'soft' orientation, i.e. with a low percentage of plies oriented parallel to the loading direction, will have a higher strain to failure than a skin with a stiff ply orientation. Second, the outer layer of a composite layup has to be ±45° or 90° to assure good damage tolerance. Also areas subjected to runway debris, propeller ice, engine fragments etc. can be shielded with impact protection material. For Glare the same design aspects are important, though the aluminium sheets largely improve the damage resistance and damage tolerance behaviour of the material.

27.3 Damage tolerance

27.3.1 Definition

Damage tolerance is defined as the ability of a structure to tolerate a reasonable level of damage and defects that might be encountered during production or service. The literature refers to 'damage tolerance' as a property of the material or the structure related with damage. However, it is important to realise that there are four damage properties [3]:

1. damage resistance (influence of environmental conditions, material choice, design aspects, etc.)
2. damage tolerance (effect of damage on the residual (mechanical) properties)
3. inspectability (ease of detection of damage)
4. reparability (ease of repair)

Two more aspects that need to be considered in the damage tolerance method are:

5. damage sources (cause of damage and defects)
6. damage types (form of damage)

27.3.2 Damage sources

An important part of the damage tolerance evaluation is the establishment of impact damage criteria. The type, probability of occurrence, and damage level should therefore be defined and damage sources known.

Bird strike

For commercial transport aircraft Davis and Sakata [4] state: To ensure that the flight crew is afforded the same level of protection from bird strike provided by the windshield, certain areas of the cockpit structure must be capable of withstanding impact of a 1.8 kg bird. Glare can sustain more damage per unit weight than aluminium 2024-T3, especially at high velocity, and could well be applied in the cockpit structure. In the crown section the relative ease of production of double curved Glare panels and the good impact properties are a fine weight-saving combination.

Hail impact

Hail damage has a low probability, because only few hailstones grow dimensions that are big enough to create damage. In case these hailstones do occur, they can severely damage the aircraft.
At high aircraft speeds small hailstones can also create considerable damage. Hail should be considered separately, both on the ground and in flight. The impact locations are different, which can be seen in Figure 27.10.

The pilot should try to avoid hailstorms and a reasonable prediction of the expected hail intensities must be made. On the ground damage can occur on the upper surfaces and in flight on the upper, side and lower surfaces of the wing and fuselage.

The size, terminal velocity, and probability are important factors that determine the severity of the damage for both ground and flight operations.
In addition, the number of hailstones impinging on an aircraft fuselage per unit area as a function of the duration is important. Hail encounters are characterised by multi-impacts of various size hailstones and the number of impacts is influenced by geographical location. A summary of the frequency of encounter and the impact density of hailstones is presented in Figure 27.11.

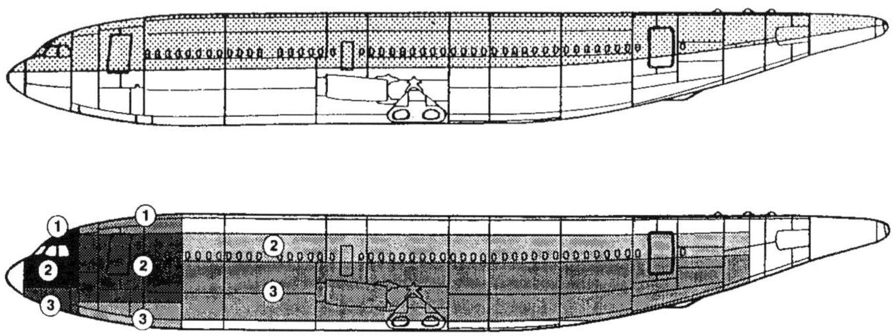

Figure 27.10 *Hail impact locations: above: on the ground; and below: in flight (darker areas indicate more severe hail impact damage).*

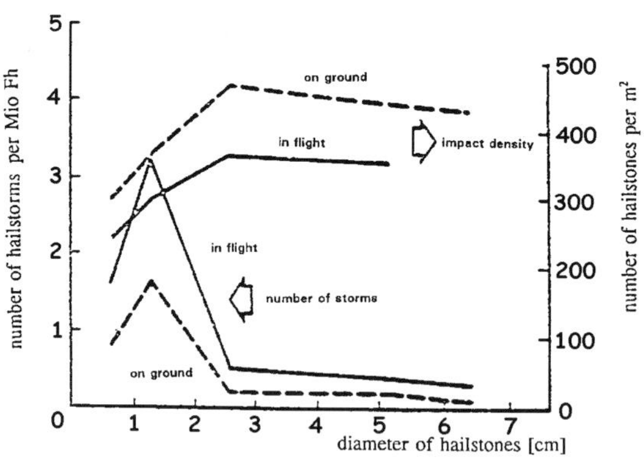

Figure 27.11 *Frequency and density of hailstorms [4].*

The application of Glare in the upper fuselage section is therefore prone to hail impact damage. Figure 27.12 shows a 3.2 mm thick Glare 4B-4/3-0.5 specimen after etching, impacted with a 5.7 cm ice ball at a speed of 140 m/s. The frequency of encounter and impact density of this size of hailstones is extremely small, which can be seen in Figure 27.11. The speed corresponds with the approach of an aircraft. This corresponds to the fact that a hailstorm encounter is most likely to happen during approach, due to the flight altitude and due to the fact that it is more difficult to avoid. A gas gun is used to accelerate the ice ball to this speed. A high speed camera captures the ice ball during flight and at impact, and determines the speed and the integrity of the ice ball. No damage or delamination was found [5].

Figure 27.12 *No damage or delamination shows after etching away the outer aluminium layers of a Glare 4B-4/3-0.5 specimen, impacted with a 5.7 cm ice ball at a speed of 140 m/s.*

Takeoff and landing area (runway) debris

The size and weight of the impactor is based on a standard representation of gravel. From the likeliness of a structure hit by runway debris (gravel, bolts and ice) per 10 flight hours an inspection period for this type of damage can be defined. The impact always occurs on ground with aluminium spheres of diameter 1.27 cm, with an impact velocity of 60 m/s and impact angles of 18 to 34 degrees. The impact energy of approximately 5 Joules is of no concern for Glare. The probability of impact on the upper fuselage, where Glare is applied, is also very small.

Tools

To simulate impact of tools, a weight is dropped from a representative working height. For tools an impact with 0.453 kg, a diameter of 2.54 cm and a velocity of 6 m/s at 90 degree angle should be considered. For Glare this condition is of no specific concern.

Engine fragments

This type of damage is realistic for turbine engines. Cruise power is used to determine the velocities imparted to the engine fragments when a motor explosion occurs in flight. The sizes of the fragments are determined from experience (generally 1/3 disk is taken). These fragments will always completely penetrate the fuselage if the engine cowling does not contain them. The hole size should at least be expected to be 2.6 times greater than the area of the fragment. For Glare a two bay crack criterion is used for residual strength considerations.

Nose gear tire shrapnel

The size and weight of tyre shrapnel is mostly dictated by thrown thread. Of the tyre failures that lead to shrapnel, 80% are thrown threads and the remainders are carcass ruptures. This type of failure rate is based on airline experience. It is expected that in some cases the crew will be unaware of impact from tyre shrapnel during takeoff, which indicates that the structure should be able to withstand operational loads until possible damage is detected. For Glare this condition is of no specific concern.

Cargo and service trucks

The impact caused by collision of a service / cargo-handling truck with the fuselage can be considerable. The damage occurs especially at door-surroundings. Severe damage will be easily detected and must be repaired prior to the flight. The damage is of general concern and does not place special restrictions on the application of Glare.

27.3.3 Damage types

Damage and defects may be induced during the fabrication of the part and the material, e.g. voids and porosity, or may occur during the service life of the aircraft. A damage tolerance concept has to deal with the types of damage for the design and certification of the structure. Table 27.1, Table 27.2 and Table 27.3 give a summary of respectively the in-service mechanical damage, the manufacturing defects and in-service environmental damage. In-service damage can be divided in four categories: fatigue damage, environmental degradation, accidental damage (impact) and wear and erosion (by rain, sand and dust). These damages do not act independently: impact damage causes moisture ingress and fatigue damage accumulation. Damage can be caused during the manufacturing process, e.g. delamination induced at the edge of a hole by drilling or by handling. Quality control, e.g. ultrasonic inspection or coin tapping, is needed to detect whether the area of defects exceeds a certain threshold level.

Table 27.1 *Typical in service mechanical damage.*

defect	typical causes
cuts and scratches	Mishandling
abrasion	rain/grit erosion
delaminations	impact damage
disbonds	impact damage overload
hole elongation	overload/bearing failure
dents (with delamination and crushed core)	impact damage walk in no-step regions runway stones
edge damage	mishandling of doors and removable parts
penetration	battle damage severe mishandling, e.g. fork lift

Table 27.2 *Typical manufacturing defects.*

defects	typical causes
voids	poor process control
delaminations	inclusion of release film poor process control faulty hole formation procedure
disbonds (in bonded joints)	poor fit of parts inclusions of release film poor process control
surface damage	poor release procedure bad handling
misdrilled holes	faulty jigging

Table 27.3 *Typical in-service environmental damage.*

defect	typical causes
surface oxidation	lightning strike overheat battle damage, e.g. laser
delamination	freeze/thaw stressing (due to moisture expansion) thermal spike (causing steam formation)
disbonds (in honeycomb panels)	same as for delaminations
core erosion	moisture penetration into honeycomb
surface swelling	use of undesirable solvents, e.g. paint stripper

For Glare moisture ingress is possible at the edges or after damage, when the aluminium has cracked. The edges should therefore be sealed to protect the panel from moisture and solvents (paint strippers etc.). Special attention must also be paid to the process control, which can be done by C-scanning the Glare panels. Overheating can degrade the material, but is not likely to occur in the fuselage and can be detected by the discolouring of the paint.

27.3.4 Residual properties

The damage tolerance of a material is related to its residual strength in the presence of damage. This is governed by the damage grade in the material after impact. Even the lowest grade, a dent left induced by impact, causes a small stress concentration around the dent and consequently a lower strength, especially in fatigue.

Compression strength after impact

Glare 4-5/4-0.5 (3.2 mm thickness) and aluminium 2024-T3 (3.8 mm thickness) show a similar compression strength after impact, i.e. 2% strength reduction due to a dent, and 10% strength reduction due to a through crack [6]. The compression test setup with anti-buckling guides is shown in Figure 27.13. Global buckling of the specimens was initiated by local buckling in the dented area. No delamination buckling (critical for composite materials) was observed in the Glare specimens.

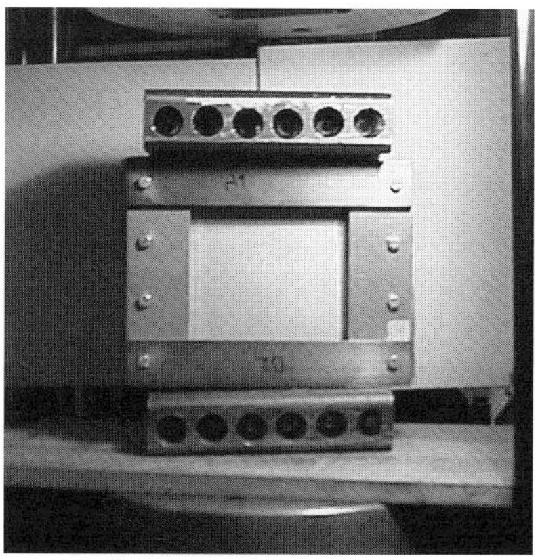

Figure 27.13 Compression test setup.

Fatigue properties after impact

The fatigue properties of Glare versus aluminium 2024-T3 after impact are shown in Figure 27.14. The Glare specimen only shows crack growth in the outer aluminium layer at the impacted side and no failure after 200.000 cycles (CA, 120 MPa, R=0.05). The crack initiation time in aluminium 2024-T3 is much longer, but the crack grows through the thickness very fast resulting in final failure after approximately 130.000 cycles.

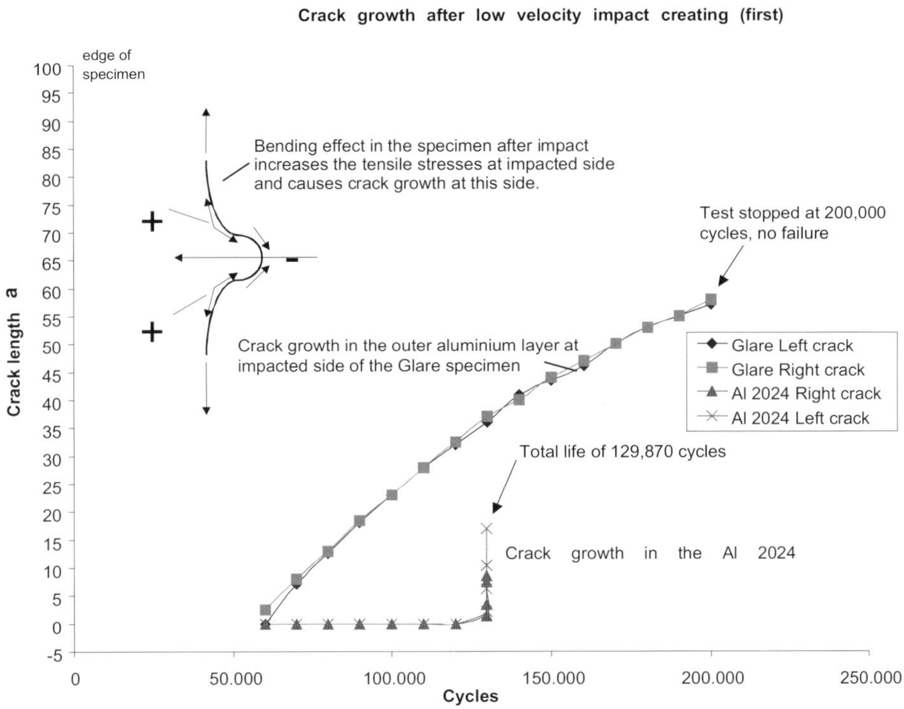

Figure 27.14 *Crack growth curve for Glare 4 (outer aluminium layer at impacted side) and Al 2024-T3.*

Due to a bending effect in the dent area, the crack starts to grow at the higher loaded impacted side. This is favourable since the impacted side is usually the inspectable, outer side of the structure.

27.4 Inspection

27.4.1 Detection of damage

Accurate non-destructive testing (NDT) methods are needed to detect damage in composites. The aircraft should be able to sustain detectable damage for a certain period before its detection. In Table 27.4 the flaws and damages, that are assumed to be non-detectable, are listed.

Table 27.4 *Assumed non-detectable flaw/damage (U.S. Air Force) for composites.*

type	size
scratches	a surface scratch that is 100 mm long and 0.50 mm deep
delamination	an interply delamination that has an area equivalent to a 50 mm diameter circle with dimensions most critical to its location
impact damage	damage caused by the impact of a 25 mm diameter hemispherical impactor with 135 J kinetic energy, or with that kinetic energy required to cause a dent 2.5 mm deep, whichever is least

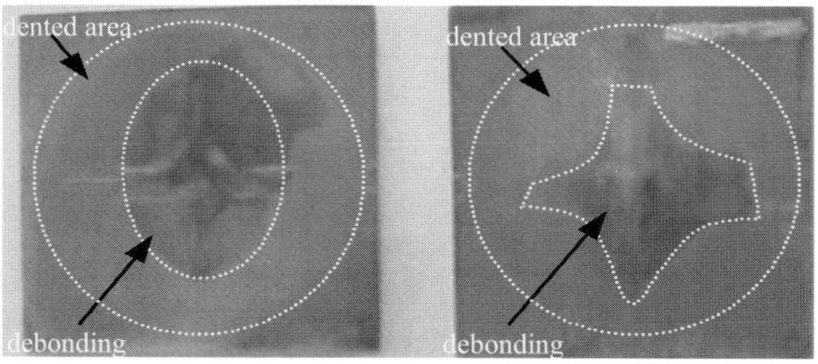

Figure 27.15 *Etched Glare 4A-4/3-0.5 specimens impacted with 165 Joules at high velocity. The dented area is larger than the debonding: the impacted / concave side (left) and the back / convex side (right).*

It is difficult to C-scan Glare specimens because not only the delamination but also the plastically deformed dent in the aluminium layers will reflect the ultrasounds since it is in general not perpendicular to the signal. However, Glare has a visible dent after impact and the damage size is always smaller than the dent size. This can be seen in Figure 27.15, which shows high velocity impacted Glare 4A-4/3-0.5 specimens after etching [7]. Thus, only the surface scratches in the outer aluminium layer need to be considered for Glare. Inspection can be limited to the dented spots and scratches where the dent depth and respectively scratch length and depth are the decisive factors for eventual repair.

27.4.2 Inspection intervals

For Glare the same approach as for metals can be followed. An initial small crack is assumed in the material, which grows during the service life of the aircraft. The main damage source is fatigue and not impact. The residual strength of the construction gradually decreases. Inspection of the damage and repair restore the structure to the initial strength. The inspection interval is such that uncontrolled crack growth will not occur. Strength reduction below limit load is avoided, see Figure 27.16.

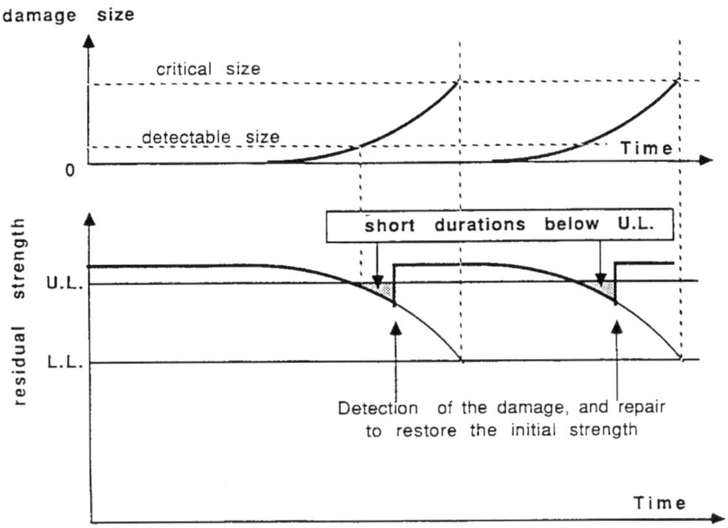

Figure 27.16 *Residual strength as a function of time of a damage tolerant metal structure with developing fatigue damage.*

27.5 Conclusion

Glare is stronger than aluminium at higher impact velocities due to the strain rate dependent behaviour of the fibres. The dent depth after impact is comparable to aluminium with the same thickness. The residual properties for Glare are comparable to (e.g. compressive strength), or better than (e.g. fatigue) monolithic aluminium. The inspection of a Glare structure is no more complicated than for aluminium, since Glare has the same indentation and because the damage size due to impact is smaller than the visible dent size.

References

[1] A. Vlot, *Impact loading on Fibre Metal Laminate*, Int. J. Impact Engng., Vol.18, no.3, 1996, p. 291-307.

[2] M.N. van der Voort, *Overview impact testing and damage assessment,* Report B2v-99-23, Delft University of Technology, Delft, February 2000.

[3] A. Vlot, *Low-velocity impact loading - on fibre reinforced aluminium laminates (Arall) and other aircraft sheet materials*, PhD-thesis, Delft University of Technology, 1991.

[4] G.W. Davis and I.F. Sakata, *Design considerations for composite fuselage structure of commercial transport aircraft*, NASA CR 159296, 1981.

[5] M. Hagenbeek, *Impact of ice, Impact properties of Glare*, Report B2v-98-15, Delft University of Technology, Delft, February 1999.

[6] J.J. Homan, C*ompression after impact*, Report B2v-98-11, Delft University of Technology, Delft, December 1998.

[7] M. Hagenbeek, *Impact comparison between Glare, carbon and aluminium. Impact properties of Glare,* Report B2v-99-03, Delft University of Technology, Delft, February 1999.

28

Corrosion

B. Borgonje, M.S. IJpma and W.G.J. 't Hart

Abstract – Corrosion is a problem that is hard to avoid in an aircraft structure. The different types of environments in which an aircraft operates and the large variety of materials used create a spectrum of different types of corrosion. Aluminium alloys are the most commonly used materials in aircraft structures. Aluminium itself is not so sensitive to corrosion, but the addition of other elements to the alloy makes it much more prone to corrode. In Glare the outer layers always consist of a sheet of an aluminium alloy, therefore the same type of corrosion-resistance can be expected as for a monolithic aluminium sheet. A difference may exist due to thickness effects. The thin sheets of aluminium that are used in Glare have a different microstructure compared to thicker sheets. This may have a positive or negative influence on the occurrence of different types of corrosion. If corrosion does occur, it will be limited to the outer aluminium layers, since the prepreg layers act as a corrosion barrier. Therefore Glare has better corrosion resistance than monolithic aluminium sheets.

28.1 Introduction

Environmental attack usually starts from the outside of a material. Since the outer layers of Glare always consist of aluminium, it is important to have a good understanding of the durability properties of this material. Fortunately, aluminium has been the most widely used material in the aerospace industry for many years now. Therefore the environmental effects, e.g. corrosion, on aluminium are already well known. Corrosion attacks the material, which leads to a decrease in structural performance of the metal. Severe corrosion may even penetrate and create holes in thin sheets of material, making good protection systems necessary. The theory of corrosion will be discussed in section 28.2. Section 28.3 will focus on the impact of corrosion on aircraft structures. The typical corrosion aspects of Glare are discussed in section 28.4.

28.2 Corrosion of aluminium

28.2.1 The corrosion process

When metals are exposed to a humid environment the metal changes from the metallic state into an oxidised state in the form of metal ions or hydrated metal oxides. This process is called corrosion and is the result of chemical reactions that occur between the metal and an aqueous solution containing different ions and/or dissolved oxygen (electrolyte solution). These chemical reactions involve charge transfer in the form of metal ions or electrons passing the interface between metal and solution. They are therefore called electrochemical reactions, or as the solid metallic phase is usually termed an electrode, electrode reactions. An electrochemical corrosion process always involves two or more electrode reactions, from which at least one causes oxidation of the metal. The oxidation reaction is accompanied by one or more reduction reactions [1].

Aluminium is a very reactive material with a high affinity for oxygen. The metal is nevertheless highly resistant to most atmospheres and to a great variety of chemical agents. This resistance is due to the formation of a protective layer of aluminium oxide on the surface. However, to obtain higher strength and stiffness properties, aluminium is often alloyed with other elements such as copper, magnesium, manganese, zinc, silicon, etc. These elements can accumulate around grain boundaries forming metallic bonds, which can cause potential differences between precipitate and grain. At the same time, precipitate-free zones can be formed creating potential differences between these zones and the rest of the crystal. As a result, the corrosion resistance of aluminium is lowered.

Influence of aluminium type

Standard types of Glare contain either a 2024-T3 or 7075-T6 aluminium alloy. These types of alloy have reasonable corrosion properties.
The heat treatment usually applied to these alloys changes the structure of the materials, and so therefore changes the corrosion behaviour. The heat treatment process commonly used for 2024 is indicated by T3. This includes a solution heat treatment and quenching followed by ageing at room temperature. When ageing is conducted at higher temperatures (~190 °C) a T8 condition is obtained, creating a structure with better corrosion properties.

28.2.2 Types of corrosion in aluminium

Corrosion can appear in different forms. A division can be made into general corrosion and local corrosion. General corrosion usually only takes place on unprotected metals exposed to a corrosive environment. In metals with a protective layer (which can be a natural oxide layer or an artificially applied metallic, organic or inorganic layer) corrosion is usually local. The most important types of local corrosion in aluminium will be discussed in this sub-section. Only a brief description will be given of the way each type of corrosion occurs and under which circumstances this will happen. More detailed information about the underlying chemical processes can be found in Shreir [1].

Pitting corrosion

Pitting corrosion is the most common form of aluminium corrosion. A pit is probably initiated by the adsorption of chloride ions at imperfect spots in the oxide layer. The reaction taking place causes the solution within the pit to become acidic, which makes the forming of new oxide impossible. The cathodic reaction will occur on the adjacent aluminium layer over a relatively large surface area, causing the anodic pitting reaction to take place rapidly.

Intergranular corrosion

Intergranular corrosion occurs because of heterogeneity in the alloy structure. In the aluminium-copper type alloys, like 2024, precipitation of $CuAl_2$ particles at the grain boundaries makes the adjacent solid solution anodic and therefore sensitive to selective corrosion. The degree of sensitivity to intergranular attack depends on the nature, amount, size, distribution, etc. of the precipitates. These are factors that are influenced by the heat treatment of the alloy. In general, precipitation-hardenable alloys are more likely to be sensitive to intergranular corrosion when aged to peak hardness, and less susceptible in the overaged condition.
Chloride-rich environments in particular are likely to cause severe intergranular corrosion of certain aluminium alloys. However, this can be avoided if correct manufacturing and heat treatment conditions are observed. The amount of minor elements that facilitate intercrystalline attack should be controlled to a safe level. The presence of copper needs special attention, since relatively small additions can cause a noticeable increase in intercrystalline attack in some alloy systems.

Galvanic corrosion

If different metals are in direct contact with each other in the same structure exposed to a corrosive environment, the corrosion rates may be different from those of the single metals in the same environment. The metal with the lowest corrosion potential will show an increased corrosion rate when it is connected to a metal with a more positive corrosion potential. The metal with a more positive corrosion potential will have a decreased corrosion rate. This type of corrosion is called galvanic corrosion, and must be avoided.

Stress corrosion

Stress corrosion is a failure mechanism that requires the cooperation of a corrosive environment and a sustained tensile stress. The stress corrosion process occurs in two phases. First there is a crack nucleation period, followed by the crack propagation period. In the crack nucleation period corrosion is the predominant factor. When corrosion develops on the material surface, penetration on a micro level occurs. This will eventually lead to a microscopic notch. At that point the crack propagation period begins. The micro-notch will induce a stress concentration at its tip, changing the local anodic potential and thereby often accelerating local corrosion. The tensile stress will also open the micro-notch, thus giving further access to the electrolyte. Under these conditions the crack will penetrate deeper into the material, possibly leading to structural failure.

Crevice corrosion

Corrosion can initiate in crevices due to the development of an aeration cell, which is caused by differences in the amount of dissolved oxygen in the environment. Usually, oxygen transfer takes place from the atmosphere into a liquid, but in crevices this is hampered. Corrosion processes use the dissolved oxygen, which can lead to a reduction of the oxygen concentration in these areas. In turn, this leads to an increased salt concentration and lower pH value. These two effects both limit the formation of an oxide layer in the crevice. This explains the rapid rate of deterioration, once the process has been initiated.
Factors that influence the occurrence of crevice corrosion are the shape and dimensions of the crevice. In general, for crevice corrosion to occur, the crevice has to be wide enough to enable the penetration of liquid, but narrow enough to limit the amount of diffused oxygen. The depth of the crevice also plays an important role in this respect.

Filiform corrosion

Filiform corrosion is also a type of corrosion that is caused by differential aeration cells. It is probably initiated by formation of a drop of highly concentrated electrolyte solution at the location of a hygroscopic, i.e. water-attracting, salt crystal underneath a surface coating. Oxygen diffuses through the coating towards the drop. Underneath the drop an oxygen-deficient region is formed, surrounded by an oxygen-rich area.
Filiform corrosion appears as a random white tunnel of corrosion products beneath thin surface coatings. This behaviour is the result of capillary forces, which move the drop around under the surface coating. The increase in pH of the solution at the edges of the drop is thought to accelerate this process.
This type of corrosion is more often detrimental to appearance than to strength, although thin foil may be perforated and corrosion-attack of thin clad sheet (as used in aircraft structures) may expose the less corrosion resistant aluminium alloy core. Filiform corrosion often initiates from countersink bore holes, where the protective layer is missing.

Layer corrosion

Layer corrosion is also known as exfoliation or lamellar corrosion. It occurs on materials that have a marked "fibre" structure caused by rolling or extrusion. The attack is rapid and very selective, forming partly detached layers of relatively uncorroded material. Layer corrosion mainly occurs in copper-bearing aluminium alloys in a number of environments, some of which are regarded as only mildly corrosive.

Aluminium alloys are susceptible to layer corrosion because of the pronounced grain structure after rolling or extrusion. The attack proceeds along a number of narrow planar paths (usually but not necessarily intergranular) parallel to the working directions. The corrosion products are voluminous and press the layers apart, causing the metal to swell and, in severe instances, to disintegrate into separate sheets of metal.

28.3 Corrosion in aircraft structures

28.3.1 Introduction

In aircraft structures a wide spectrum of different types of corrosion can be found. This has partly to do with the many different materials that are used. The structure largely consists of aluminium alloys, but also different types of steel, titanium and magnesium alloys are often used. The last few decades have seen the introduction of composite materials, making this issue even more complicated. This large variety of materials is used because strength and weight often have the highest priority for an aircraft material. However, the downside is that the occurrence of corrosion can be increased.
The extensive use of riveted, bolted and bonded joints in aircraft structures also promotes corrosion attack, since they create areas where moisture can accumulate. This is also the case when sandwich structures are used.

Also important is the fact that an aircraft is exposed to a large variety of environments during its life. The outside of the fuselage has to withstand salt, mud, “chipping” of stones, temperature changes, erosion, UV radiation, chemicals, etc.
The environments inside the aircraft can vary just as much as the outside. Materials can be exposed to condensed water (with or without contaminants), hydraulic liquids and aggressive environments near toilets and galleys.

28.3.2 Critical locations

Corrosion usually takes place in certain areas of the structure. These areas will be described next, together with possible solutions that can be incorporated in the design.

- *Locations where moisture and dirt can accumulate*
  The lower part of the fuselage is one of the most corrosion sensitive areas in an aircraft structure. This is caused by the accumulation of condensed water, which can contain several different chemicals, such as nicotine, tar, perspiration and hydraulic fuel, as well as aggressive liquids coming from toilets and galleys. In order to avoid the accumulation of water, drain holes can be introduced.
  Another example of an area that is exposed to a harsh environment is the landing gear bay.

- *Crevices in joints*
  Moisture can remain present due to capillary forces, which can cause corrosion. A sealing compound should be applied to the contact areas of a joint.

- *Sharp edges, due to sudden changes in thickness*
  Surface protection is minimal, because the protective layer cannot attach well to sharp edges.

- *Components under permanent stress*
  Stress corrosion can take place due to internal stresses or assembly stresses.

- *Holes in riveted or bolted joints*
  The area between fastener and sheet can give access to moisture, after which crevice corrosion may take place. Therefore fasteners need to be 'wet' installed, i.e. with use of a sealant.

- *Contact areas of metals with different corrosion potentials*
  As described earlier in this chapter, galvanic corrosion can occur when two metals with different corrosion potentials are connected to each other. The increased use of carbon fibre reinforced plastics in aircraft can have a negative impact on corrosion properties, because the fibres are electrically conductive.
  In order to avoid galvanic corrosion, the flow of current needs to be blocked. This can be done by inserting an insulating layer between the different materials.

28.3.3 Protection methods

There are several ways to eliminate or decrease the amount of corrosion on a material. The most obvious solution is to choose a corrosion insensitive material. However, many other aspects influence the choice of a material.
The operational life of current civil aircraft is at least 30 years and therefore application of an adequate protective coating on the outer surface is essential for corrosion prevention. Different ways of doing this will be described next [2, 3].

Cladding

In the aviation industry one of the most widely used corrosion protection methods is the application of a so-called 'clad layer' on aluminium alloy sheets. This layer usually consists of high purity aluminium, which is relatively insensitive to corrosion attack. The aluminium can also be sacrificial with respect to the aluminium alloy core, which is cathodically protected by the clad layer due to their different corrosion potentials. Clad layers can be applied by pressing, rolling or extrusion, and is done by the aluminium supplier.

In Glare, only the exposed sides of the outer aluminium layers may have a clad layer. The outer aluminium layers, in this case, have a single sided clad layer. The inner aluminium layers do not have a clad layer applied to them, since this may cause bondline corrosion, i.e. corrosion at the aluminium-epoxy interface.

Aluminium oxide films

Different methods can be used to obtain an oxide layer. Acid etching creates a thin oxide film, which is mechanically weak. A thicker and more durable oxide layer can be obtained by anodising, which is therefore widely applied on aerospace aluminium.

Anodic coatings are formed by oxidation of the metal in certain electrolytes. Hence, the metal being oxidised is the anode in the system. Usually the electrolyte consists of an acidic solution. In Europe, the most common solution used is chromic acid (Chromic Acid Anodising – CAA). Other methods are phosphoric (PAA) and sulphuric acid anodising.

The anodic coatings produced on aluminium are sub-microscopically porous. Therefore, to achieve lasting corrosion resistance the porous coating must be sealed, which is usually done by immersion in boiling water. Unsealed anodic coatings also serve as effective surfaces for adhesive bonding and for the application of organic coatings. Corrosion resistant anodic layers are also present on the aluminium layers in Glare. For uncladded Glare these layers are also present at the outside of the material.

One of the shortcomings of anodic coatings is their lack of elasticity. This causes the coatings, particularly the thicker ones, to develop small cracks when the substrate material is subjected to tensile stresses. When the substrate metal is bent through a small radius the coating will often crack. These fine cracks can act as stress raisers, lowering the fatigue strength of the protected material. This can however be compensated by shot peening.

Conversion coatings

Conversion coatings consist of chemical treatments such as those known by their brand names Alodine and Iridite. Originally these coatings were sometimes used as an anti-corrosion treatment for internal surfaces of aircraft structures that were not exposed to a corrosive environment. Nowadays they are mainly used as a pre-treatment for paint systems. Alodine is also applied to protect the aluminium edges of Glare since those edges are usually not anodised.

Paint systems

One of the oldest methods used to prevent corrosion is application of a paint coating on the metal surface. The paint system usually consists of an epoxy chromate containing primer and a polyurethane topcoat.

Water displacing corrosion inhibitors

A more recent type of corrosion protection is provided by the so-called water displacing corrosion inhibitors (WDCIs). Some well-known brand names are LPS-3 and Boeshield T-9. These liquids can be applied onto a wet surface. After evaporation of the solvent a thin wax film is left behind on the surface and protects the underlying material from corrosion attack. WDCIs are usually applied on aircraft inner surfaces.

Corrosion inhibitive sealants

The primary function of a sealant is to exclude moisture, dirt, dust and chemicals, or to help to contain a liquid or gas. Sealants may be used as elastic surface coatings to protect a substrate from an aggressive environment, thereby acting as a corrosion inhibitor. They are mainly used in joints, to seal off rivets and bolts ("wet" fastener installation).

28.4 Corrosion of Glare

28.4.1 Corrosion of Glare vs. monolithic aluminium

Although from the outside fibre-metal laminates seem to have similar corrosion characteristics compared to monolithic aluminium, there are several aspects which distinguish Fibre Metal Laminates (FML) from conventional, monolithic (thick) metal sheets, which could lead to different corrosion properties. These FML-specific aspects will be discussed next.

Electrical conductivity of the fibres

The use of electrically conductive fibres could lead to the creation of potential differences and therefore an increase in corrosion sensitivity. This is the main reason why carbon fibres have not been used in FMLs with aluminium layers. Since glass fibres are electrical insulators, this is not a problem for Glare.

Heat treatment during production

Aluminium 2024 usually receives a T3 treatment before use in aerospace structures. The T3 treatment consists of solution heat treatment, quenching, stress-relief by stretching and natural ageing. This gives the alloy high ductility, excellent fatigue crack propagation resistance and high fracture toughness. The good combination of these properties is mainly due to the lack of coarse grain boundary phases and the absence of precipitate free zones.

During the production of Glare the aluminium undergoes another heat treatment, which is meant to cure the adhesive layer that keeps the fibres in place and joins the metal layers together. For Glare an epoxy system is used that is cured at a temperature of 120 °C. This can act as an ageing process for the aluminium alloy of Glare, although it is a relatively short period. It could therefore have some influence on the microstructure and subsequently on the corrosion behaviour [4].

Thickness of the metal layers

In Glare the aluminium layers usually have a thickness of 0.3 to 0.5 mm, which is much thinner than that of conventional aluminium sheets used in aircraft.
SEM investigations of 2024-T3 alloy of 0.3 and 1.6 mm showed different microstructures. While sheets normally have a fibre structure due to the rolling process, this was not observed for sheets of 0.3 mm thickness. Thin sheets were found to have a lower ductility ratio, which decreases the resistance to stress corrosion [5].
Another cause for the difference in microstructure is the quenching process after heat treatment, because thin sheets have much higher quenching rates. Corrosion experiments have indicated that thin sheets (0.3-0.4 mm) have better corrosion resistance than sheets of conventional thickness (2-4 mm). It can therefore be concluded that the microstructure of thin aluminium alloy sheets has a beneficial effect on the corrosion properties [6].

Fibre layers acting as corrosion barriers

A major advantage that Fibre Metal Laminates have over monolithic aluminium is the fact that corrosion damage is limited to the outer aluminium layer, since the fibre layers act as corrosion barriers, see Figure 28.1. Therefore corrosion will have limited impact on the mechanical properties of the laminate, with the effect linearly decreasing with the number of layers of the laminate. For Glare with more than two aluminium layers this would mean that corrosion, if it does appear, would have minor consequences for static strength, due to the large number of aluminium layers remaining intact.

Bondline corrosion

Corrosion can occur at the interface between aluminium and prepreg. Usually this happens after debonding has taken place, but when a clad layer is present the corrosion can be the cause of debonding. A clad layer forms the anode in combination with the alloy base material, and is consumed first in a corrosive environment, thereby protecting the aluminium alloy. In an adhesive bond however, it is undesirable to have a clad layer, because it results in bond-line corrosion and ultimately leads to delamination [7]. Therefore in Glare no clad layers are used inside the laminate.

28.4.2 Experimental work

Corrosion tests

Extensive corrosion tests with Glare have been performed by Bommer [4]. Glare 3-3/2-0.3 specimens have been alternately immersed in an aqueous solution, salt sprayed and exposed to different humid environments. After these tests the specimens were mechanically tested and investigated for corrosion. The aluminium showed intergranular corrosion, which started mainly from the specimen edges, due to the absence of surface protection there. However, pitting corrosion was expected, not intergranular corrosion. This result may have been caused by the 120 °C production temperature during the adhesive bonding of the metal to the prepreg layers, which could have changed the microstructure of the aluminium. However, the detected intergranular corrosion would not be critical for a structure with the investigated corrosion protection (anodising and chromate-containing primer).

Investigation of the stress corrosion behaviour of Glare and thin sheets of aluminium has been carried out as part of the development of sandwich floor panels with Glare facings. Since impact (indentation) is very common in this type of structure, this has also been included in the research. Impact damage areas showed an increased rate of corrosion attack. A large difference was found between bare and anodised sheets, showing the benefits of surface treatment [8].

Unnotched specimens of Glare 3-3/2-0.3 did not show any stress corrosion cracking (SCC) after 30 days of alternate immersion and a constant tensile load of 250 MPa. Specimens did show pitting corrosion and intergranular attack. The notched specimens saw the occurrence of SCC, caused by local stress concentrations [4].

Influence of corrosion on fatigue and residual strength

A number of tests have been performed to study the influence of corrosion on fatigue and residual strength performance of Glare and monolithic aluminium.
After 175 hours of accelerated exfoliation corrosion testing (EXCO), the corrosion depth in 4 mm thick solid aluminium 2024-T3 was slightly higher than in Glare 4B-5/4-0.4, see Figure 28.1. Aluminium and Glare blunt notch specimens were subsequently fatigue tested, using a maximum stress level of 80 MPa for the aluminium and 130 MPa for Glare. The aluminium specimens failed after 48 to 62 kcycles, while testing of the Glare specimens was stopped after 100 kcycles, without failure having taken place. Residual strength of these Glare specimens was approximately 340 MPa, compared to 390 MPa for uncorroded specimens [6].

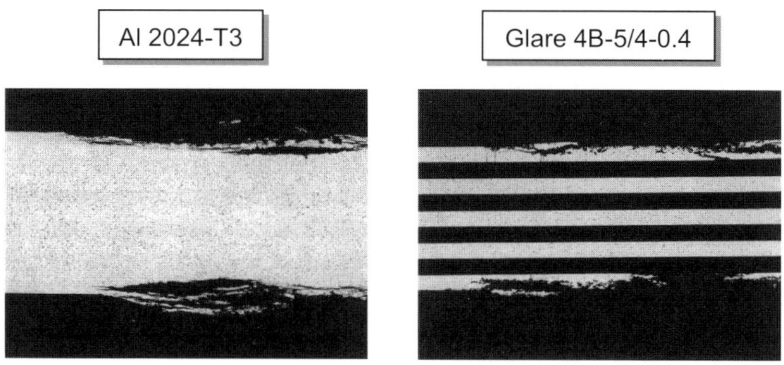

Figure 28.1 *Corrosion depth in 4 mm thick aluminium 2024-T3 sheet and Glare 4B-5/4-0.4 after 175 hours of EXCO testing* [6].

Fatigue testing of corroded Glare lap joints did not result in a reduction of residual strength after four lifetimes (100 kcycles). After eight lifetimes (200 kcycles), a difference was visible. The uncorroded specimens reached a residual strength of 193 MPa, while the corroded specimens failed at 177 MPa [9].

28.4.3 Maintenance aspects

Glare protection methods

Because of the problem of bondline corrosion, only the outer aluminium layers in Glare have a clad layer, on the outer side. The rest of the aluminium layers are attached to the prepreg layers and have an anodised surface, which has been treated with a bond primer (BR127). This enhances the durability of the bond between aluminium and epoxy, but it also provides an extra form of corrosion protection. This will be useful in case the inner aluminium layers are exposed to the environment due to damage of the prepreg layers, see Figure 28.2.

Painting is usually not considered during corrosion testing, thus when it is applied in practice the corrosion resistance will increase. Rivets and bolts in joints are wet installed, which is standard practice for aircraft structures. In this way Glare can therefore be treated like conventional aluminium. Only where there are unprotected free edges special attention is needed, because the cross-sections of the aluminium layers are not protected and moisture intrusion could take place in-between the different layers. Application of a sealant should however be sufficient.

Corrosion removal

Tests have been performed using conventional corrosion removal methods on Glare. Abrasive corrosion removal techniques, like Al_2O_3 blasting and glass bead blasting were considered inappropriate, because improper removal of corrosion and the possibility of damaging the underlying prepreg layer. The use of conventional corrosion removal methods for aluminium like air tools with grinding discs however, turned out to be acceptable for use on Glare [10].

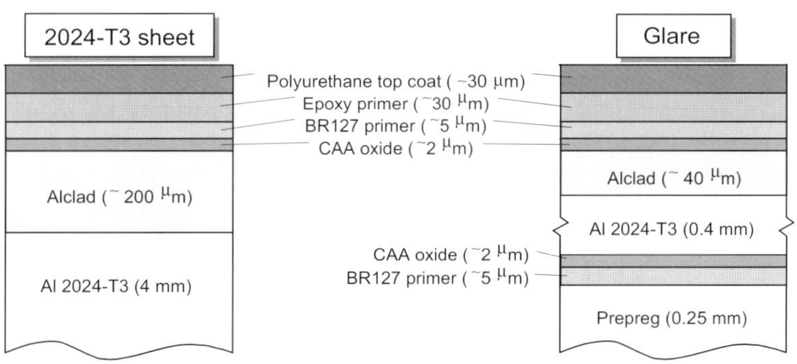

Figure 28.2 Schematic overview of corrosion prevention systems in conventional aluminium 2024-T3 sheet and Glare.

28.5 Conclusions

Since the outer aluminium layers of Glare consist of the same type of aluminium alloy that is used in most aircraft today, not many differences in initial corrosion behaviour are to be expected. In fact, the thin aluminium layers in Glare shows corrosion behaviour that is slightly better than that of monolithic aluminium sheets. The thinner sheets used in Glare have better corrosion characteristics due to a higher rate of quenching after heat treatment. If corrosion does occur, it is limited to the outer layers, since the prepreg layers act as corrosion barriers.

Surface protection methods are identical to those used for solid aluminium sheets. The standard corrosion protection of Glare consists of a clad layer on the exposed side of the outer aluminium layers, followed by anodising and coating with a bond primer. In case of corrosion attack, conventional grinding tools can be used to remove the corrosion products. However, care must be taken to avoid damage to the underlying prepreg layer.

References

[1] L.L. Shreir, R.A. Jarman and G.T. Burstein (ed.), *Corrosion. Volume 1: Metal/environment reactions.* Third edition, Oxford: Butterworth-Heinemann, 1994.

[2] L.L. Shreir, R.A. Jarman and G.T. Burstein (ed.), *Corrosion. Volume 2: Corrosion control.* Third edition, Oxford: Butterworth-Heinemann, 1994.

[3] W.A.J. Moonen, *Protective coatings for aircraft structures – A review.* Report LR-413, Faculty of Aerospace Engineering, Delft University of Technology, 1983.

[4] H. Bommer, *Environmental effects on the material Glare 3 3/2.* F4K/K-95-035 Technischer Bericht, Daimler-Benz, Ottobrunn, 1995.

[5] Ferdinandus, *Stress corrosion cracking susceptibility of unanodized and anodized thin aluminium alloy 2024-T3 sheet.* Thesis report, Faculty of Aerospace Engineering, Delft University of Technology, 1997.

[6] W.G.J. 't Hart, and L.C. Ubels, *Fatigue and residual strength performance of Glare 4B-5/4-0.4 and 4 mm thick 2024-T3 after accelerated corrosion testing.* NLR-CR-98514, National Aerospace Laboratory NLR, 1998.

[7] R. Oosting, *Toward a new durable and environmentally compliant adhesive bonding process for aluminium alloys.* Ph.D. thesis report, Faculty of Aerospace Engineering, Delft University of Technology, 1995.

[8] Ferdinandus, *Impact and corrosion test on new sandwich structures*, TZ-report, Faculty of Aerospace Engineering, Delft University of Technology, 1997.

[9] W.G.J. 't Hart and L.C. Ubels, *Effect of corrosion on the fatigue and residual strength performance of Glare 4B-5/4-0.4 lap joints*, NLR-CR-99358, National Aerospace Laboratory NLR, 1999.

[10] W.G.J. 't Hart, *Evaluation of conventional corrosion removal methods on Glare*, NLR-CR-2000-037, National Aerospace Laboratory NLR, 2000.

29

Riveted repairs

P.A. Hooijmeijer

Abstract – For the repair of Glare skins it is important that the current repair methods and tooling for aluminium structures can be used in order to avoid an increase of the operational costs for airliners. It has been shown that Glare can be repaired with normal riveted repair procedures, which are described in the Structural Repair Manual (SRM) of existing aircraft. Workshop operations like drilling, cutting and deburring could be performed on Glare without any problems. The excellent fatigue properties of Glare provide the riveted repair on a Glare skin with very good damage tolerance properties. Fatigue cracks in the first rivet rows of a repair will only occur in one or two layers of the skin. Stresses can be transferred to other layers in the laminate. Depending on the Glare grade, the residual strength results were equal or better to identical repairs on aluminium structures, even when the fatigue loading was performed at a much higher stress level.

The repair of a Glare stringer could also be performed according to an existing SRM. Again, all normal operations, which were also performed on the Glare skin repair, did not lead to problems when performed on a Glare stringer.

29.1 Introduction

All aircraft in practical use will suffer damage. For the application of Glare as a new structural material, the maintainability of a Glare (fuselage) structure is of prime importance for airlines. It should be at least as good as the maintainability of aluminium structures so that the operating costs do not increase. It is important that current inspection and repair methods for aluminium structures can be applied to Glare structures and that no new equipment and/or extensive training are necessary.

The most important damage causes for an aircraft fuselage structure are fatigue, corrosion and accidental damage. To maintain the airworthiness of the aircraft, these damages must be repaired. The common repair procedure for aluminium structures is to remove the damaged area and apply a riveted aluminium patch. An example is given in Figure 29.1. This repair method can be performed quickly and is not complicated and can be performed on any airfield with relatively simple tooling.

Note that the subject of the present chapter is riveted repairs only. Bonded patch repairs will be the subject of chapters 30 and 31.

Several aspects are important when considering riveted repairs on a Glare skin. Many of these aspects are related to the mechanical properties of Glare, but the workshop properties must also be considered. In section 29.2 the relevant workshop properties are discussed. The question is whether Glare can be repaired with standard tooling and without special training for personnel.

A repair has to be designed using damage tolerance requirements. In the past, repairs were often only designed according to static strength requirements. However, this design principle can lead to a repair without sufficient fatigue life. Therefore, the repair design philosophy has changed and now the fatigue behaviour is addressed; the effect of patch stiffness on the attraction of load to the repair area is now taken into account, for example. In section 29.3 this fatigue behaviour of actual riveted repairs on stiffened and unstiffened Glare skin panels is discussed.

Finally, section 29.4 treats a special type of repair, i.e. the repair of a Glare stringer.

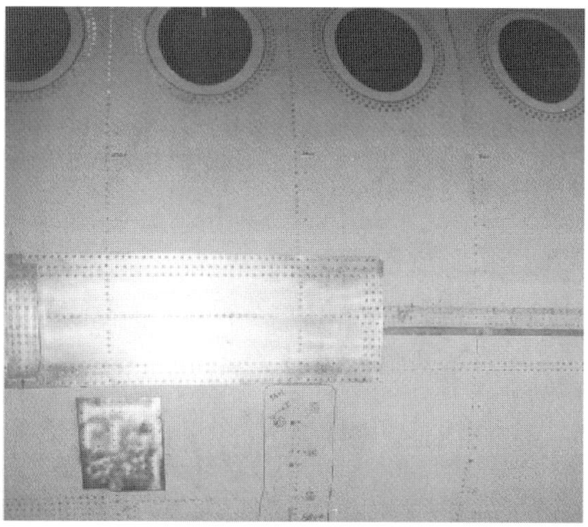

Figure 29.1 *Example of a riveted patch repair on an aluminium structure.*

29.2 Repair workshop properties of Glare

When the size of the damage in a structure is determined, the normal procedure is to drill or cut out the damaged area. The edges of the cut-out will then be re-worked with sanding disks and deburring knives to create a smooth cut-out. If necessary, a part of the stringer near the cut-out will also be repaired.

A Glare skin with bonded aluminium stringers was repaired at the location of one of the stringers. In order to be able to investigate a wide variety of aspects of the repair, not only was the skin repaired; a part of the bonded aluminium stringer was also removed. Note that the stringer to be removed was bonded directly onto a splice in the Glare skin. Due to the layered character of Glare it could be possible that during the removal of the bonded stringer from the Glare skin the laminate would be damaged as well. To create the worst-case situation the stringer was bonded onto a splice location.

In order to perform a repair with normal tooling, procedures like drilling, cutting, grinding and deburring must be applicable to Glare. Therefore, a small cut-out was made in the Glare skin by drilling out the corners of the cut-out, followed by a complete removal of the cut-out from the skin using a cutting disk. The procedure is illustrated in Figure 29.2. These operations created no problems such as delaminations. However, to drill holes with a good quality, High Strength Steel drills have to be used otherwise delaminations in the laminate around the drilling holes may occur. Cutting and deburring can be performed just as for aluminium structures.

Bonded stringers can be removed by cutting away the upper flange and the stringer web using a cutting disc. Removal of the bonded, lower flange of the stringer from the damaged area can be accomplished in two ways.
The first method is to sand it away from the Glare skin. In that case attention must be paid to the temperature of the material, which will increase by the sanding action. A high temperature may cause damage to the laminate. However, in practice it was observed that the temperature during normal operations was not extremely high. Workshop personnel could still touch the sanded structure by hand.
The second method to remove the lower flange of the stringer is by using a filler knife as a wedge between the lower flange and the skin to peel the stringer from the Glare skin.

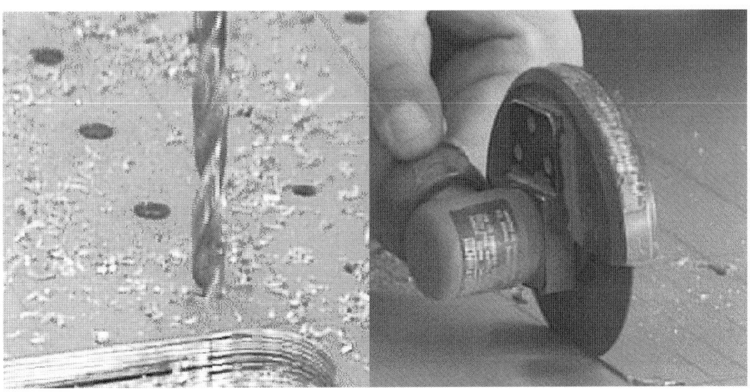

Figure 29.2 Drilling and cutting operations on a Glare skin.

Both methods were tried out on a Glare skin and did not reveal any problems. Even the splice area directly underneath the bonded stringer remained intact. The skin was not damaged, see Figure 29.3. When riveted stringers are applied they can be removed even easier by drilling away the fasteners.

Since all workshop operations could be performed on Glare, the repair could be finished without any problems. The riveted patch could be positioned and holes for the rivets could be drilled without damaging the Glare skin. The patch was riveted to the skin and a stringer coupling was riveted to the aluminium stringer.

The final result of the repair is shown in Figure 29.4. It can be concluded that a riveted repair can be performed on a structure with a Glare skin. No special or expensive modification to the operations or training of personnel is needed.

It has been shown that Glare can be repaired using traditional materials and tooling. Besides these good workshop properties the existing repairs must have equal or better mechanical properties than the riveted repairs on aluminium structures. This is the subject of the next section.

Figure 29.3 *A clean and undamaged Glare skin surface after removal of the lower flange of the aluminium stringer. The splice line underneath the stringer is clearly visible.*

Figure 29.4 *Finished repair on Glare skin.*

29.3 Mechanical properties of Glare repairs

The mechanical properties of a repair are of course even more important than the workshop properties. Airworthiness of the aircraft after the repair must be ensured. As was stated in the introduction (section 29.1) a repair must not only fulfil static requirements but must also be damage tolerant. To check the damage tolerance, stiffened and unstiffened Glare panels were tested under fatigue loading and after that the residual strength was determined. No optimised repairs were developed especially for Glare but the normal patch geometry and rivet pattern according to the SRM was applied. Optimisation of the repair was beyond the scope of this initial test programme.

The test specimens consisted of 800 x 1200 mm large repair panels with cut-outs of about 200 x 300 mm. Different patch materials and fastener types were used. As patch materials both Glare and aluminium were applied. If it could be demonstrated that it is possible to use an aluminium patch, it would reduce costs for the airlines for the storage of Glare material. These patches were fastened to the skin, e.g. with Hi-Loks and Visu-Loks. Glare skins were tested, and aluminium skins were also investigated as reference. In all cases, riveted patch repair methods obtained from the SRM of an existing aircraft were used. In Figure 29.5 an example is given of a riveted repair over a cut-out that was tested.

Figure 29.5 *Glare repair panel in test setup at National Aerospace Laboratory NLR, the Netherlands.*

The panels were fatigue loaded at constant amplitude with a minimum stress level of 0 Mpa and a maximum stress level of 120 MPa for about 100.000 cycles, after which the panels were tested for their residual strength.

All repaired Glare specimens showed the typical damage tolerance capability of the material. The outer rivet rows of a repair transfer the largest part of the load between the patch and the skin. These rows are critical from a damage tolerance point of view and showed crack initiation.

Fatigue cracks also occurred in the outer rivet rows of a repair in an aluminium structure. These cracks emanated from the rivets and progressed through the thickness of the skin. Eventually, this caused a fatigue failure of the repair or resulted in a low residual strength during a residual strength test.

However, for Glare a different behaviour was observed. Of course, fatigue cracks also initiated in the Glare skin but these cracks only occurred in the first one or two aluminium layers of the laminate, see Figure 29.6. The fibres in the laminate remained intact, as did the remaining aluminium layers in the laminate. A part of the load could still be transferred over these intact layers. For the aluminium skins, the cracks grew through the thickness.

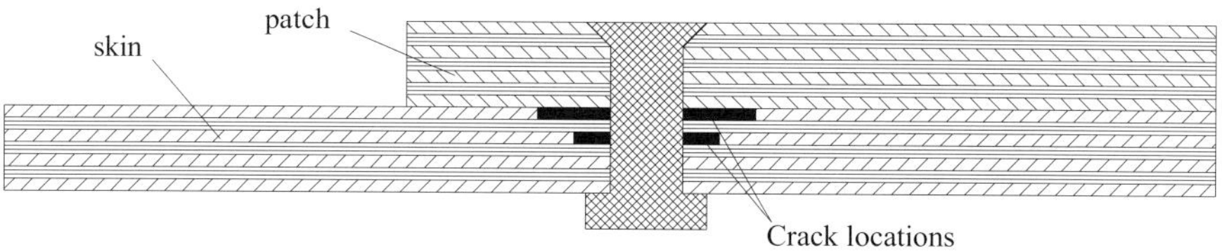

Figure 29.6 *Fatigue cracks in first critical rivet row in Glare repair only occurred in one or two layers of the laminate (view in the loading direction).*

As an example, several residual strength results are given here for repair panels with riveted patches. The results are obtained from the tests on unstiffened panels with a large cut-out of approximately 200 x 300 mm. The residual strength relative to the strength (no fatigue) of an aluminium patch on an aluminium skin is given in Figure 29.7.

This figure shows the superior behaviour of the Glare repair panel in comparison with the aluminium panel when the damage tolerance requirement is applied to the repair. Even though the maximum stress level for the Glare 4A panel (120 MPa) is much higher than the maximum stress level used for the aluminium panel (85 MPa), the residual strength after 120.000 cycles is only slightly lower. Because the result of the static strength test on a Glare 4B panel is lower than for aluminium, we also see that the result is dependent on the Glare grade used. In the Glare 4B skin and patch the largest part of the fibres is orientated in a direction perpendicular to the loading direction. The amount of fibres 'bridging' the fatigue cracks will consequently be lower.

Different fastener and materials combinations showed that the mechanical properties were influenced by the choice of fastener. The difference in clamping force of the fasteners influenced the rivet flexibility and the secondary bending. These are both of importance for the fatigue behaviour and the subsequent residual strength behaviour of the repaired structure.

For repairs over a much smaller cut-out these typical results for Glare were also found. Combinations of a Glare skin with an aluminium patch were also investigated. The static test results of a repair on a Glare skin with a small cut-out showed that the repair will be able to carry loads which can be as high as three times the equivalent of an operational internal pressure load. This value can even be reached after a fatigue loading of 200.000 cycles at a maximum stress of 120 MPa (R=0), see Figure 29.8.

Although the results shown in Figure 29.8 are already impressive, they were limited by the capacity of the used testing machine, i.e. none of the panels failed. The actual residual strength was therefore higher than shown in Figure 29.8.

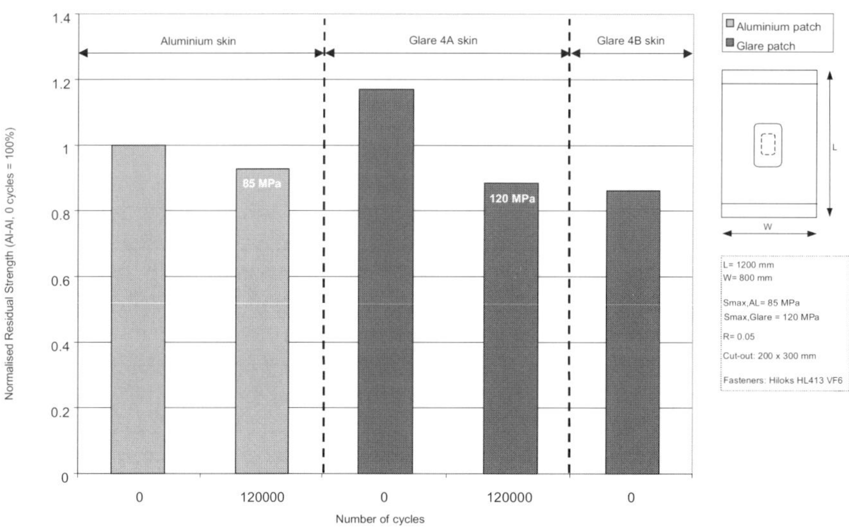

Figure 29.7 Residual strength results for aluminium and Glare repair panels.

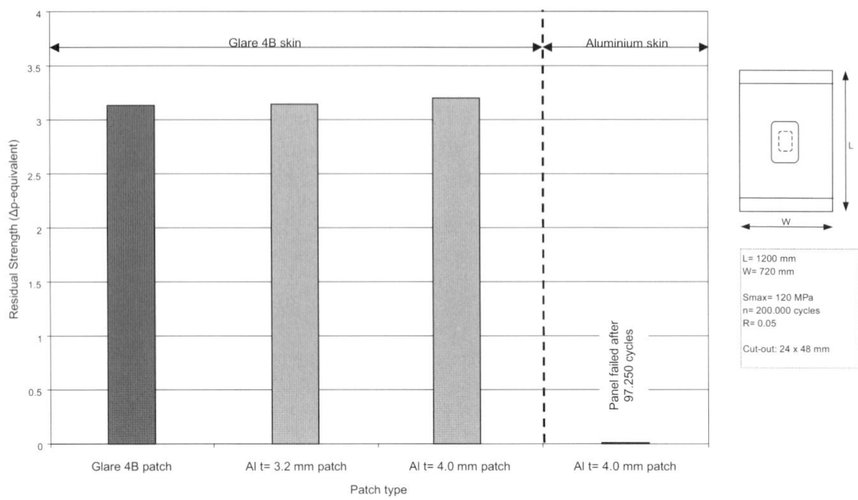

Figure 29.8 Residual strength results for repairs on small cut-outs (24 x 48 mm).

29.4 Repair of Glare stringers

Extensive research has led to the possibility of the production of Glare stringers. The use of Glare stringers on a Glare skin has the advantage that they have the same stiffness as the skin material. In some cases aluminium stringers would attract too much load when used on a Glare skin.
Of course, for Glare stringers the same requirements concerning repair are valid as for a Glare skin. Ideally, it would be best if existing aluminium stringer repair methods could be applied to Glare stringers.

A Glare panel with bonded Glare stringers was repaired using a standard method, existing tools and personnel without extensive training. A part of the stringer was removed from the panel using a cutting disk. In section 29.2 we have already shown that the cutting of Glare does not cause any problems. The same is true for the cutting of Glare stringers. The upper flange and the web were cut from the stringer. The remaining lower flange could again be removed by one of the two methods mentioned in section 29.2. The flange was removed by using a grinding disk. One after the other, the layers of the Glare stringer lower flange could be removed, see Figure 29.9. The final aluminium layer of the stringer, which was bonded to the Glare skin, was removed with a filler knife used as a wedge. Some of the adhesive remained on the skin surface but this could be removed by sanding. No damage was done to the skin although the anodising and primer layer of the Glare were sanded away. This is not a problem typical to Glare, however, since the same would happen when sanding an aluminium skin. The anodising layer can be restored by applying

Alodine to the surface. This is also a standard procedure in repair operations on aluminium structures.

The cutting edges of the Glare stringers could be treated in the same way as the edges of a repair cut-out were sanded and deburred.

A Glare stringer coupling can be used as a replacement for the damaged part of the stringer. This part can easily be machined from a Glare stringer. Glare stringers are all identically produced and will have an identical shape within a certain tolerance. Using a Glare stringer part as a stringer coupling would not lead to any problems with dimensions. An aluminium strip is bonded to the area where the lower flange of the damaged part of the stringer has previously been ground away. The stringer coupling is then installed and fastened to the remaining, intact parts of the original stringer. Drilling and reaming operations on Glare have been shown not to lead to any problems. In Figure 29.10 the final result of the stringer repair is shown after sealing and finishing.

Figure 29.9 *Grinding of lower stringer flange. The Glare layers are removed from the skin one by one.*

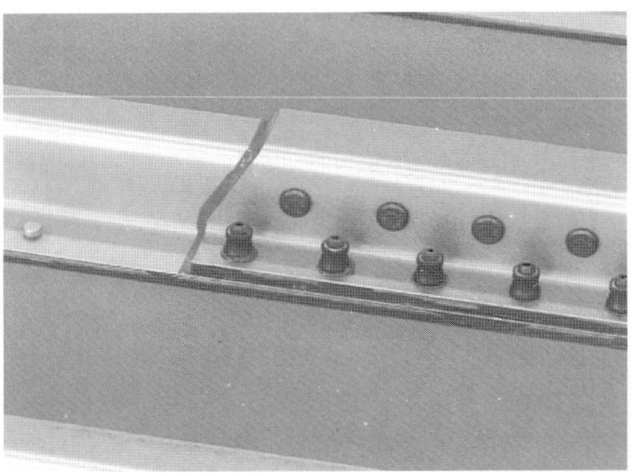

Figure 29.10 *Stringer coupling fitting nicely to the damaged stringer after fastening and sealing.*

30

Bonded repair patches

H.J.M. Woerden, W.J. Mortier, C.B. Guijt and S. Verhoeven

Abstract – The growing number of ageing aircraft around the world, both military and civil, has led to an increasing demand for safe, damage tolerant, and cost-effective repairs to solve the inherent fatigue damage and corrosion damage problems. Adhesive bonding of Glare repair patches can provide such a solution. While Glare can be used for conventional repair methods like riveting and bolting, bonded patches provide a uniform and efficient load transfer into the patch and reduce the risk of high stress concentrations caused by mechanically fastened repairs. The advantage of Glare repair patches compared to other patch materials lies in the small mismatch in the coefficient of thermal expansion (CTE) with the aluminium skin, the excellent fatigue properties, the high strength, the moderate extensional stiffness and high bending stiffness. This makes Glare a promising candidate for safe and cost-effective bonded repairs for thin skin structures, not only for ageing aircraft, but also for repair of in-service damage of relatively new aircraft (high request of repair performance). Extensive research has been performed in a joint research programme of the Faculty of Aerospace Engineering of Delft University of Technology and the Center for Aircraft Structural Life Extension (CAStLE) of the United States Air Force Academy. An overview of this research will be presented here.

30.1 Introduction

The number of ageing aircraft flying, both civil and military, has been increasing significantly over the last decades. Economic restrictions force operators to use their aircraft much longer than intended by their designers. To illustrate this problem: in the year 2000, more than 5700 commercial aircraft were over twenty years old. In 1993, 51% of the USAF fleet was over fifteen years in age, and 44% was over twenty years in age [1]. As a consequence, fatigue and corrosion problems become an important topic for the maintenance of these aircraft. A well-known, yet dramatic example of an ageing aircraft that proved the need for inspection and safe repair techniques is the Aloha Airlines Boeing 737-200. Multiple site fatigue damage, bonding and corrosion problems caused an explosive decompression on April 28, 1988 when a large portion of the crown section just aft of the main cabin door separated on flight 243 at cruise altitude near Maui.

Advanced bonded repairs with a fatigue, corrosion, and impact resistant material like Glare can help to ensure a damage tolerant solution for ageing aircraft as well as new aircraft (being at the start of their service life, repairs to these aircraft will demand the highest performance of a bonded repair). While Glare can be used for conventional repair methods like riveting and bolting, adhesive bonding has the advantage of providing a more uniform and efficient load transfer into the patch, thereby reducing the risk of high stress concentrations caused by mechanically fastened repairs. A well-designed bonded repair (i.e. bond line loaded in shear and stronger than the bonded members) applied in a properly controlled environment using stringent pre-treatment and processing steps, will result in a practically maintenance free, safe and durable structure.
The history of the crack patching bonding technique goes back to the 1980s, when it was successfully applied to the fracture-critical D6AC steel wing pivot fitting of the F-111 bomber [2]. Since then, mainly boron/epoxy and carbon/epoxy composite patches have been studied and used by the Australian and United States Air Force. Aircraft repaired with these materials include the C-141B transport aircraft, the F-111 and F-16 fighter-bombers and the B-1B strategic bomber. The high stiffness of these materials makes thin repair patches possible and the prepregs can be draped and co-cured to repair complex structures. However, during the adhesive cure cycle and at low operating temperatures these materials have the disadvantage of a large mismatch in coefficient of thermal expansion (CTE) with the skin material, which can result in increased thermal stresses in the patch, adherend and bond line. The CTE of Glare is much higher than that of boron/epoxy or carbon/epoxy. As a patch material, this results in a much smaller difference in CTE with the parent (metal) structure. Consequently, high residual thermal stresses due to the curing process and operating temperatures are prevented.

The application of bonded Glare repair patches has been studied in a joint research programme of Delft University of Technology and the United States Air Force Academy since 1994. For thin skin applications, Glare patches are superior to boron/epoxy and carbon/epoxy patches. For boron and Glare this is illustrated in Figure 30.1. It can be clearly seen that bonded patch repair provides for an impressive improvement in fatigue life, where Glare performs even better than boron/epoxy repairs. This behaviour is the pillar of the joint research programme. In this chapter an overview is given of the research performed since Fredell initiated the work on this subject [3]. To start with, we shall discuss the used processes and materials for a better understanding of the bonded repair process in general. Subsequently, the most important stresses for bonded repairs will be pointed out, as well as some stress-related problems encountered in bonded repair. Then the work done in the field of analytical model validation and finite element modelling for a variety of repair configurations will be discussed. The last section before the conclusions then goes into the subject of durability aspects of bonded patch repair, which is crucial for certification and public acceptance of the concept.

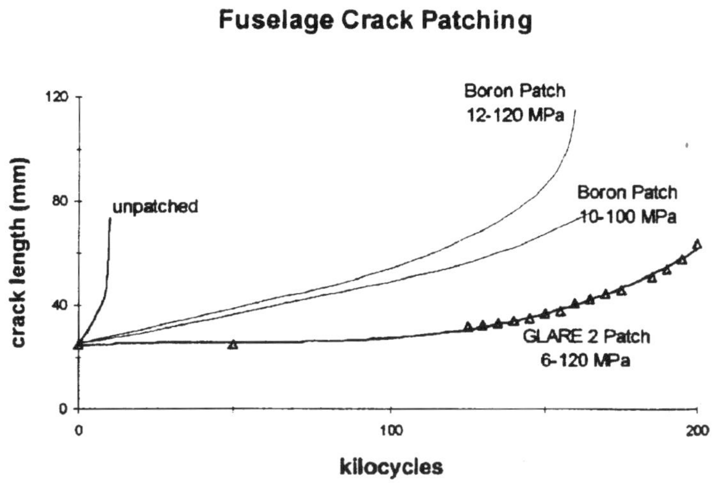

Figure 30.1 *Crack growth rates of specimens with different patch repair materials compared to unrepaired specimen, 1 mm thick aluminium 7075-T6 adherend* [4].

30.2 Processes and materials

In this section, we discuss the processes and materials as used in the bonded Glare repair research. The complete patch repair process from individual materials to repaired structure is outlined. The repair process involves many different steps, and strong, durable adhesive bonds will only be achieved when the entire process is carefully controlled.

30.2.1 Materials: adherend, adhesive and patch

The research is focused on large military transport aircraft where the fuselages are primarily constructed from 7000 series aluminium. Aluminium 2024-T3 is still the most commonly used material for commercial aircraft. Both of these aluminium alloys are used within this research programme as the adherend simulating the aircraft structure or skin. However, the design and analysis methods that emerge from this research will be generally applicable to other materials as well, provided that the material properties are known. For example, Glare repair patches on Glare skin structures, which will become important for (future) aircraft like the Airbus A-380 where large portions of the aircraft are constructed of Glare material.
The adhesives used in this research programme are epoxy adhesives especially tailored to provide the high strength needed and to withstand the operating environments encountered in aircraft operation. Two adhesives are commonly used (with comparable properties): FM-73 produced by Cytec Industries and AF-163 produced by 3M Aerospace. Both are film adhesives with a carrier to assure a minimum bond line thickness. The advantage of using film adhesives is that they do

not require mixing prior to use, they are easy to handle and they have a very uniform composition.
The Australian and United States Air Force mostly use composites (such as boron/epoxy and carbon/epoxy) as a patch material. The high stiffness of these materials makes thin patches possible and consequently these repair materials are extremely suitable for highly loaded or thick structures, or for repairs where aerodynamics play an important role (e.g. aircraft wings). This research programme focuses on repairs of aircraft fuselages where the Glare material seems to perform much better. This is due to the previously mentioned differences in coefficient of thermal expansion between adherend and patch.
Typical properties of the most commonly used patch materials and adherend materials are given in table 30.1.

Table 30.1 *Typical properties of some repair and parent structure materials* [5].

Properties	2024-T3	7075-T6	Glare2 3/2 0.2	Boron/epoxy	Carbon SP 500-2
E_L (GPa)	72.4	72	69	207	186
E_{LT} (GPa)	72.4	72	54	19	12
α_L (10^{-6}/°C)	23.2	23.6	16.3	4.5	-0.9
α_{LT} (10^{-6}/°C)	23.2	23.6	24.5	20	26
$\sigma_{yield\ L}$ (MPa)	345	505	383	n/a	2913
$\sigma_{yield\ LT}$ (MPa)	345	505	242	n/a	n/a
$\sigma_{ultimate\ L}$ (MPa)	485	570	1187.5	1585	n/a
$\sigma_{ultimate\ LT}$ (MPa)	485	570	313.4	62.7	n/a

30.2.2 Material pretreatment

The key to a strong and durable bond is the presence of a chemically reactive surface with good wettability on both adherend and patch. To achieve such a surface, a good surface pre-treatment is necessary. Pre-treatment of the patches is very simple, both for boron/epoxy and for Glare patches. The aluminium layer to be bonded on the outside of the Glare patch is already pre-treated (PAA: Phosphoric Acid Anodising or CAA: Chromic Acid Anodising) and primed. Only a solvent wipe is needed to remove surface contamination from the patch. Pre-cured boron/epoxy patches need an abrasive step before the solvent wipe to expose the epoxy en roughen the surface microscopically.
The aluminium adherends need a more intensive pre-treatment due to their reactive nature with oxides. Commonly used pre-treatments are gritblast-silane and PAA or CAA, all followed by a priming step to preserve the freshly created active oxide layer. The primer layer also acts as a barrier to corrosion through the small amount of corrosion inhibitors (chromates) in the primer. The primer used within this research is BR-127 from Cytec Industries.

30.2.3 Cure cycle

After the controlled pre-treatments, the repair patch and adhesive film are positioned on the damaged adherend. The stack-up can then be cured under pressure and elevated temperature to bond the patch to the adherend. The applied adhesive restricts the cure cycle, but in the case of FM-73 and AF-163 the same cure cycle can be used. Curing can be performed in an autoclave (for specimen-type structures) or with a heat blanket that locally heats the repair area (on real aircraft). In both cases pressure is provided by a vacuum bag layup over the repaired area, which is needed to evacuate air and to prevent the formation of voids in the adhesive. The cure cycle used for the epoxy adhesives in this research is depicted in Figure 30.2.

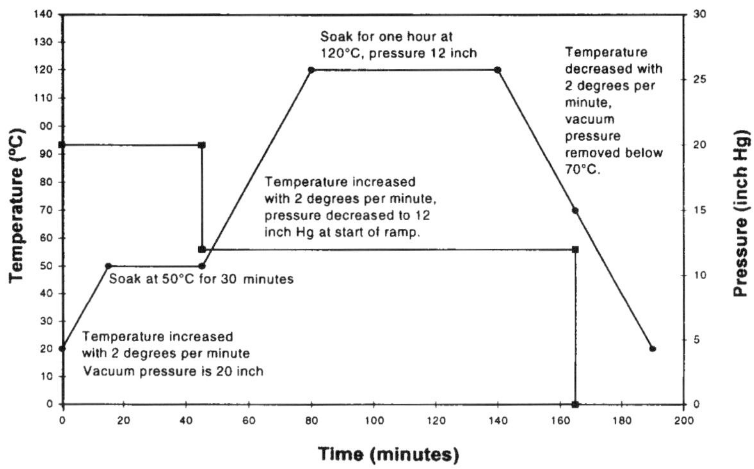

Figure 30.2 *FM-73 and AF-163 cure cycle used for bonded patch repair.*

30.2.4 Inspection

After the patch has been bonded to the adherend, an inspection step is needed to check the integrity of the bond line. The cured adhesive is checked for voids and delaminations using a C-scanning device, which will generate an image of the decibellic material-response compared to the response of the undisturbed material. Though voids and delaminations can be found using this technique, the absence of voids and delaminations is only a necessary condition for bond line integrity; it does not imply a strong bond. Non-bonded surfaces where the adhesive is in contact with the material, for example, cannot be found using C-scanning or any other NDI technique. Until such an NDI technique is found, the only guarantee for a strong and durable bond is the validation of the bonding process [6].

30.3 Repaired stress intensity factor

Bonded repair effectiveness is measured in this research by the ability of the bonded patch to reduce the crack growth rate of the repaired crack by reduction of the stress intensity factor of this crack. Adequate design of the bonded repair can accomplish this without creating new areas prone to crack initiation. The repaired stress intensity factor, however, depends on both material choice and repair configuration.

The patch material is often different from the adherend material creating thermal residual stresses in the repaired structure. One-sided repairs where double-sided repairs are impossible can cause severe secondary bending. When multiple patches are needed in close proximity (multiple site damage), this can have a considerable influence on the repaired stress intensity factor. It is also possible that the amount of load transferred from adherend to patch is such that the patch becomes susceptible to crack initiation. These problems have been studied and are discussed in this section.

30.3.1 Thermal residual stresses

The difference in coefficient of thermal expansion (CTE or α) between patch material and adherend material, see table 30.1, plays a crucial role in patching effectiveness. When materials with different CTEs are bonded, both the cure cycle and the operating temperature affect the thermal residual stresses in the patched area. For specimen-type repaired structures, which are small enough to be completely heated (in an autoclave or under a heat blanket) during the cure cycle, the thermal stress situation is relatively simple. The CTEs of the patch materials in this research are lower than the CTE of the aluminium adherend. During the cure cycle the repaired structure is heated to 120 °C (up to which the patch and adherend are assumed to be able to freely expand) and the structural bond between patch and adherend is formed. Then the repaired structure is cooled down to ambient temperature, but now adherend and patch act as mutual restraint on contraction through the difference in CTE. The aluminium adherend wants to contract more than the patch, resulting in tensile stresses in the adherend (crack opening) and compressive stresses in the patch. For boron/epoxy patches these detrimental tensile thermal stresses in the adherend are much higher than for Glare patches because the CTE of boron/epoxy is significantly lower than the CTE of Glare.
The thermal residual stresses involved in the actual repair of an aircraft are much more complex. The effective CTE of the aluminium fuselage skin in the heated zone is lower than the CTE of the aluminium alloy itself due to the restraint of the surrounding cool structure. During the cure of the adhesive, the patch is free to expand whereas compressive stresses will be induced in the repair area due to the thermal constraint produced by the unheated zone. When the effective CTE of the skin is smaller than the CTE of the patch, cooling to room temperature will thus create tensile stresses in the patch, which are balanced by a favourable crack closing compression in the skin. When the CTE of the patch is smaller than the effective CTE of the skin, the situation is reversed. For most composite patches, the effective CTE of the metallic skin will be higher than that of the patch and detrimental tensile stresses will be present at the crack tip after curing, balanced by compression in the patch. Because of the small CTE mismatch between aluminium and Glare, the effective CTE of the skin can become comparable or even lower than that of the patch, resulting in favourable compressive stresses at the crack tip after curing.
Besides the thermal curing effects, which lead to permanent, static stresses, there is also a cyclic thermal effect due to the operating temperature of the aircraft. At cruising altitude, the complete fuselage of an aircraft cools down to temperatures as low as –55 °C. When the CTE of the patch is considerably lower than the CTE of the parent

material (e.g. boron/epoxy), high tensile thermal stresses will act on the crack, reducing the effectiveness of the repair. These ground-air-ground (GAG) cycles impose a cyclic loading in the skin and patch and consequently a shear stress in the adhesive layer, which might lead to thermal fatigue problems. Minimising the difference in CTE and lowering the cure temperature of the adhesive are favourable for the static internal curing stresses as well as for the cyclic thermal GAG stresses. However, a minimum cure temperature is required, and a lower cure temperature may change adhesive properties and will increase cure time. An example of the influence of the difference in CTE between adherend material and patch material is given in Figure 30.3 for a Boeing 737-200 fuselage at cruise altitude [3].

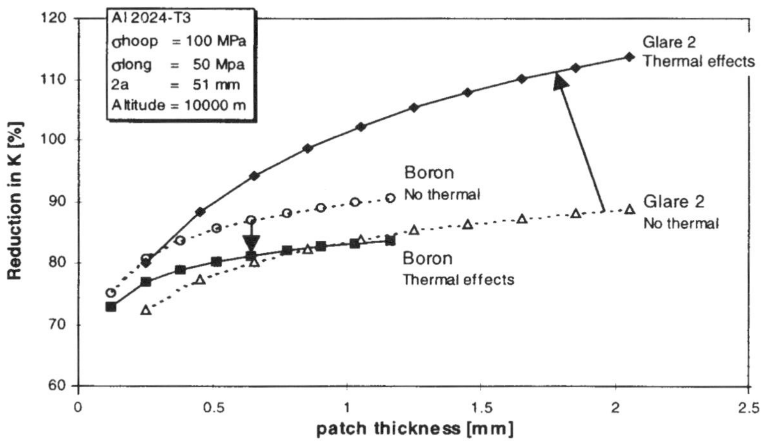

Figure 30.3 *Reduction in stress intensity factor for bonded Glare 2 and boron/epoxy patches, with and without thermal effects for B-737 fuselage* [3].

30.3.2 Secondary bending

A single-sided patch will induce bending due to eccentricity. Secondary bending will be more pronounced in unsupported structures, but even in real stiffened aircraft structures some out-of-plane bending will occur if a single-sided patch repair is performed see, Figure 30.4. This secondary bending can considerably increase the stresses at critical locations such as the patch tips and crack faces. At the patch tips it will increase the peel stresses, while secondary bending at the crack faces can stimulate crack growth and the formation of debonds.

Secondary bending is inevitable when using single-sided patch repairs, but the patch can be designed to minimise the detrimental effects. Secondary bending is largely influenced by the load attraction of the repaired area, the geometry of the patch, the shift in neutral axis and the bending stiffness of the repaired area.

Figure 30.4 *Secondary bending for single-sided (left) and double-sided repair (right).*

Load attraction to the repaired area is minimised by using patches with an extensional stiffness (Young's modulus times thickness) equal or close to the adherend material. For the high-stiffness boron/epoxy this results in relatively thin patches and therefore low bending stiffness values for the repaired structure. The Young's modulus of Glare 2 is below that of the aluminium adherend, resulting in relatively thick patches and therefore high bending stiffness values for the repaired structure. The negative influences of load attraction (like high patch tip stresses) can be further reduced by using elliptical patches with limited aspect ratios.
Tapering of the patches is one of the most important methods to reduce the bending stresses in the skin and the adhesive stresses at the patch tip. Tapering of the patches gradually changes the neutral axis position of the patch-skin combination, thereby decreasing the bending stresses. The influence of the taper ratio (i.e. taper length over patch thickness) is depicted in Figure 30.5 for Glare 2 and boron/epoxy patches with a length of 100 mm as calculated with a finite element model [7]. It can be clearly seen that a higher taper ratio will result in lower bending stresses. Realistic taper ratios in the range from 1:10 to 1:20 still result in bending stress reductions between 10% and 35% depending on patch material.

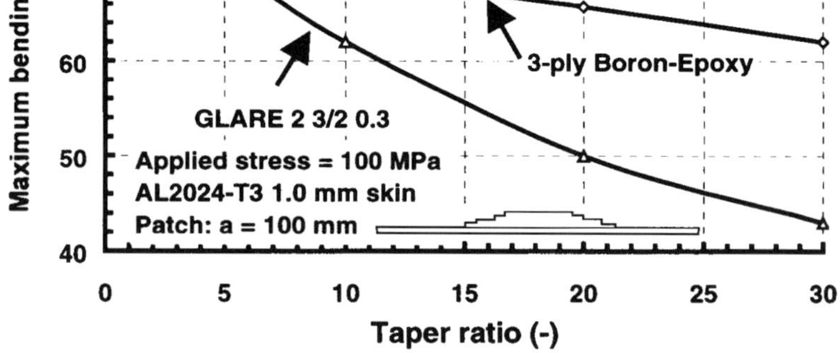

Figure 30.5 *The influence of taper ratio on the maximum skin bending stress* [7].

30.3.3 Patch interaction

Finite element modelling was also used to investigate the interaction effect of multiple patches in close proximity, which might be used in the case of multiple site damage [7]. Only uniaxial loads and stresses in the loading direction were considered for two different configurations: a "row" of patches (oriented perpendicular to the applied load) and a "column" of patches (oriented parallel to the applied load).
For a row of two circular patches it is evident from Figure 30.6 that load attraction as an interaction effect is determined by the ratio of absolute patch separation relative to patch width, S_p/b. The skin stresses at the patch tips, which are critical for this case, are slightly lower with reducing relative patch spacing. When elliptical patches are used, skin stresses at the patch tips show a slight increase with decreasing relative patch spacing, although the change in magnitude remains as marginal as for the circular patches.

In the case of a column of two patches, four elliptical patches with different aspect ratios were considered. Figure 30.7 shows the finite element analysis indicating a clear increase in stresses at the facing patch tips due to load attraction. For relative patch separation ratios (S_p/a larger than 0.5) there is hardly any patch interaction. The analysis also shows that elliptical patches with high aspect ratios should be avoided. In general one should strive to combine the lowest possible aspect ratio a/b with a patch separation ratio of at least 0.5 to achieve minimal skin stresses.

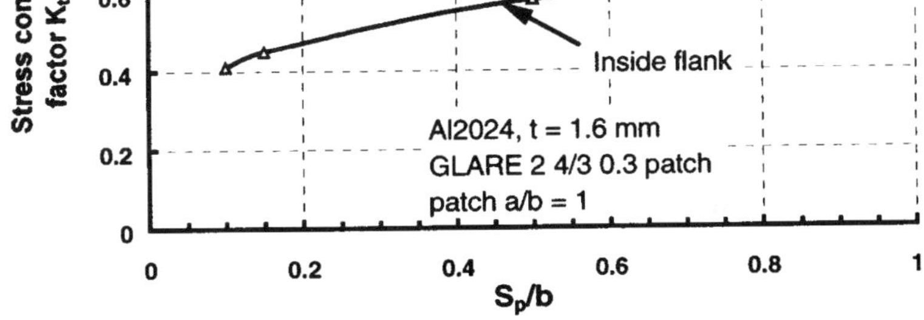

Figure 30.6 *Stress concentration for a row of two circular patches (a=half patch length, b=half patch width)* [7].

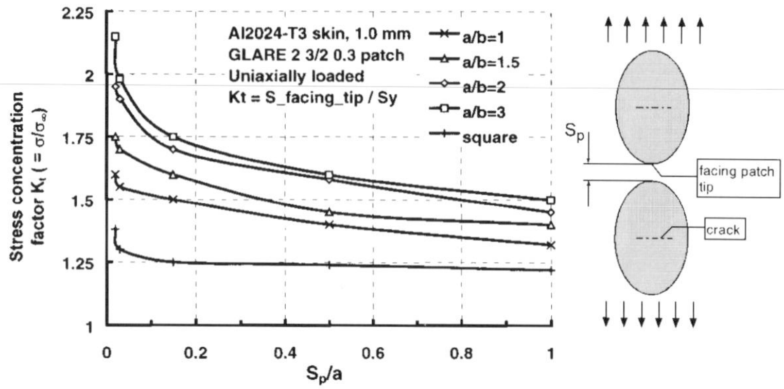

Figure 30.7 *Stress concentration at facing patch tips for a column of two patches (a=half patch length, b=half patch width)* [7].

Analysis of columns and rows of patches under biaxial loading in a barrel setup showed the skin tip stresses to be significantly lower than for the uniaxial loading discussed above.

30.3.4 Patch cracking

For single-sided bonded Glare patches, the secondary-bending component due to the jump in neutral axis can be significant. Added to the other stresses present in repaired structures, these high bending stresses can cause crack initiation in the most highly loaded part of the bonded Glare patch: the aluminium layer on the faying side of the patch. This aluminium layer is the first patch layer into which the adherend loads are transferred by the adhesive. This part-through crack in Glare patches was found in highly loaded single-sided test specimens after destructive disassembly of the bonded repair. This crack in the patch aluminium layer also appeared to be the failure location during residual strength tests after fatigue. Some research was done to investigate the severity of this patch-cracking problem, resulting in a simple prediction model. It was found that the patch crack always grows in the wake of and far behind the skin crack, having only minor influence on the repaired stress intensity factor for skin crack lengths up to 75% of the patch width. Also, the fibre layers of the Glare patch stay intact, thereby effectively bridging not only the skin crack, but also the patch crack. In addition, bending stresses will be reduced in real aircraft structures because of the stiffening elements present and crack initiation in the patch might not occur.

The prediction model uses finite element results for the stresses in the patch layers. These stresses are then corrected (assuming equal strain at the calculated stress level) to account for real material stress-strain behaviour, see Figure 30.8, since the finite element model assumes linear, elastic behaviour and calculates the stresses directly from the experienced strains.

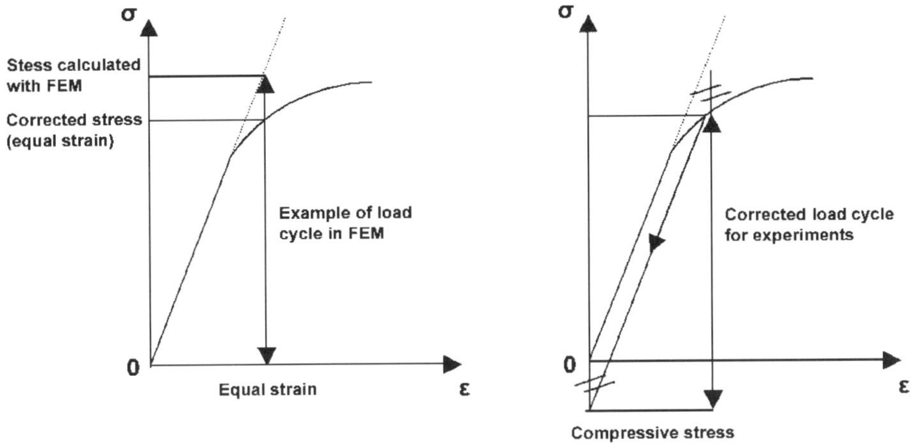

Figure 30.8 *Correction method of finite element stress to real stress (left) and corresponding corrected load cycle in experiments (right).*

Using the corrected stress cycle that the aluminium patch layer on the faying side is in reality subjected to, the corresponding corrected stress ratio can be calculated. This stress ratio, calculated for different adherend crack lengths, is then used together with the Miner rule (rule of linear damage) and an aluminium fatigue life curve (S-N) to predict the time to crack initiation in the Glare patch, expressed in terms of the adherend crack length and corresponding stress intensity factor at the crack tip of the adherend crack [8]. After the initiation point has been calculated, the aluminium patch layer is then assumed to be completely cracked (worst-case scenario) for an adherend crack length equal to the patch width (which is seen as the end of the life of the repair, even though the patch will be effective far beyond the point where the adherend crack reaches the patch edges). A new stress intensity factor can be calculated for this situation. A linear increase in stress intensity is assumed between patch crack initiation and the completely cracked aluminium patch layer.

The stress intensity factor range versus adherend crack length diagram constructed in this way, see Figure 30.9, can now be used to predict the adherend fatigue crack growth rate using a fatigue crack growth rate curve (da/dN-ΔK). In this way the increasing crack growth rate seen in experiments at adherend crack lengths approaching the patch width can be quite adequately predicted.

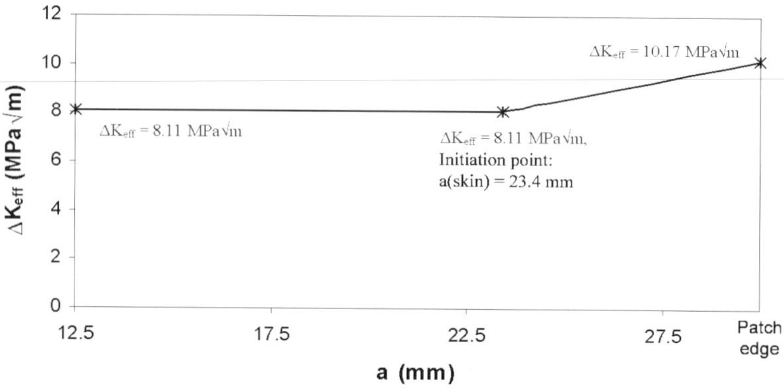

Figure 30.9 *Calculated stress intensity factor range versus adherend crack length diagram for Glare 2-3/2-0.2 patch at Constant Amplitude loading 6-120 MPa, used to predict adherend fatigue crack growth rates under influence of a part-through crack in the patch.*

30.4 Analytical and finite element modelling

To be able to predict the fatigue life or the fatigue life enhancement of bonded patch repaired aircraft structures or test specimen, accurate calculation of the stress intensity factor at the repaired crack tip is needed. Preferably analytical models are used, while numerical calculation techniques are far more expensive due to the computer time needed. Additionally, analytical models can be easily converted to a fatigue crack growth prediction computer programme. Numerical models, like finite elements are more often used as validation of experimental results and to quickly gain insight into the effect of slight changes (i.e. material properties or configuration) on the repaired stress intensity factor. In this research, analytical modelling as well as finite element modelling is used.

30.4.1 Analytical models

The analytical foundation of bonded patch repair is the model developed by L.R.F. Rose [9]. This model covers the stress intensity solution for the repaired crack, the adhesive shear strain in the bond line and the load attraction into the stiffened area. The Rose model is a continuum analysis based on the theory of elasticity. It considers an infinitely wide, centre-cracked, isotropic plate with a one-sided, bonded, orthotropic, elliptical patch. After calculation of the load transfer from the plate into the patch [10], a two-stage analysis is performed, see Figure 30.10. The first stage involves an analysis where the repair is modelled as equivalent inclusion and without the presence of a crack. The crack (with stress-free crack faces) is introduced in the second stage. This leads to a detailed stress solution with a repaired stress intensity factor that asymptotes to a constant value for increasing crack length.

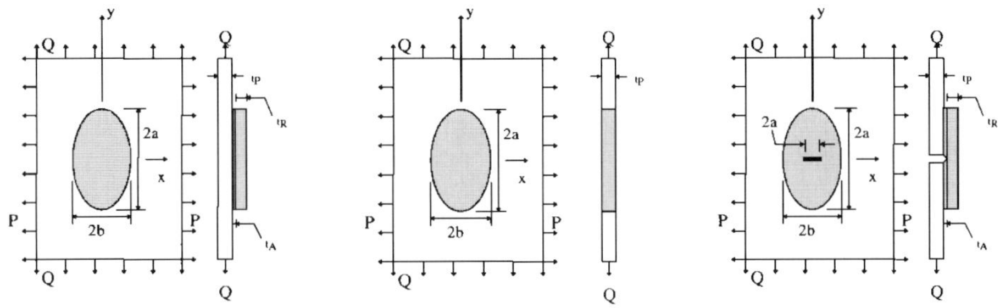

Figure 30.10 *Two-stage analytical Rose model showing first stage without crack and modelled as inclusion (left and middle), and second stage where crack is introduced (right)* [9].

The original Rose model can also take into account the effect of thermal residual stresses due to curing and operating temperatures, but out-of-plane bending is not modelled. Over the years the original Rose model has been improved (thermal stress calculations) and extended (secondary bending) to a model that quite accurately calculates the stress field for a bonded patch repaired structure. It is therefore one of the models incorporated in the CalcuRep2000 computer programme, which has been developed within this research programme as a design and prediction tool for patch repair. CalcuRep2000 is also briefly discussed in chapter 31b.

30.4.2 Finite element models

Finite element modelling has been used to gain insight on changes in bonded repair configurations and bonded repair material properties (for example the patch interaction effects discussed above). It also served as validation technique for experimental work. Some finite element modelling examples [11, 12] will be discussed.

Repaired stress intensity factors with out-of-plane bending

The previously mentioned influence of secondary bending on the repaired stress intensity factor can be calculated using analytical models such as Rose-based models extended to account for secondary bending [13, 14], or finite element models. The original Rose model [9] neglects out-of-plane bending and defines a constant, limiting value for the stress intensity factor. This constant value is also predicted by the Rose-Wang-Callinan model [14], but this model does incorporate secondary bending. The Ratwani model [13] leads to an increasing stress intensity factor for increasing crack length. Finite element modelling has been used alongside some experiments to investigate the accuracy of the analytical models.
The ABAQUS finite element model is based on the specimen types used for the experiments. The model consists of a cracked aluminium 2024-T3 layer, an adhesive layer and a layer representing the Glare patch over the full adherend width. The adhesive layer was modelled using 3-dimensional 20-node brick elements (3D20R), while adherend and patch were modelled with 8-node reduced integration shell

elements (S8R). The patch-adhesive and adherend-adhesive boundary conditions were according Mindlin plate theory [15]. Due to double symmetry, a quarter model was sufficient. The crack in the adherend was modelled by not restraining boundaries along the crack face. The modified crack closure method was used to determine the stress intensity factor at the crack tip [16]. Plain stress conditions and pure bending were assumed because of the thin adherends (1.2 mm). Initial thermal residual stresses due to curing were not modelled. Bending was taken into account by using the root-mean-square value of the stress intensity factor as defined by Young and Sun [17]. Debonding and plasticity effects were neglected. The adhesive used in the model was AF-163-2K with a thickness of 0.13 mm.

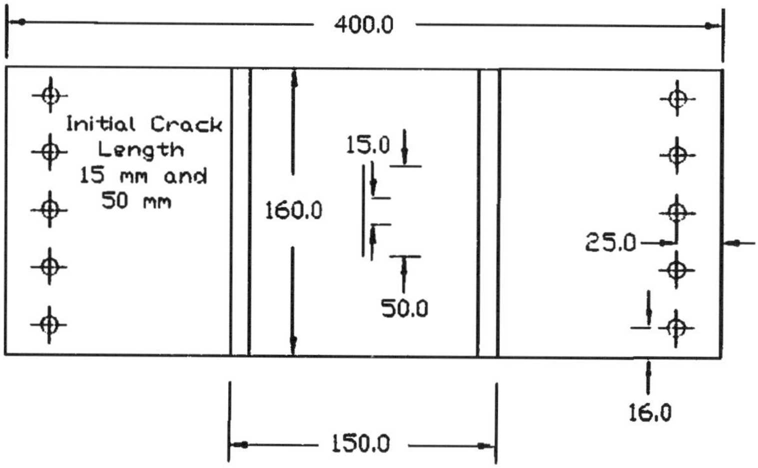

Figure 30.11 *Test specimen configuration for experimental work with full-width, tapered Glare patch* [18].

The same materials modelled with FEM were used during the experimental part of the research [18], which also used the same specimen configuration. The configuration is shown in Figure 30.11. The specimens were repaired with full-width Glare 20 3/2-0.2 patches with the fibre direction oriented perpendicular to the crack. The patch had an extensional stiffness slightly higher than the extensional stiffness of the adherend.

Stress intensity factor calculations (without thermal residual stress effects due to curing for analytical and FE models) and experiments were performed, both incorporating secondary bending. Stress intensity factors from the experiments were deducted from the strain gauge data. These strain gauges were applied to the adherends after the cure cycle and therefore readily comparable with finite element results. Figure 30.12 shows the results, where it can be clearly seen that the finite element calculations are close to the experimental results (two experiments), particularly at larger crack lengths.

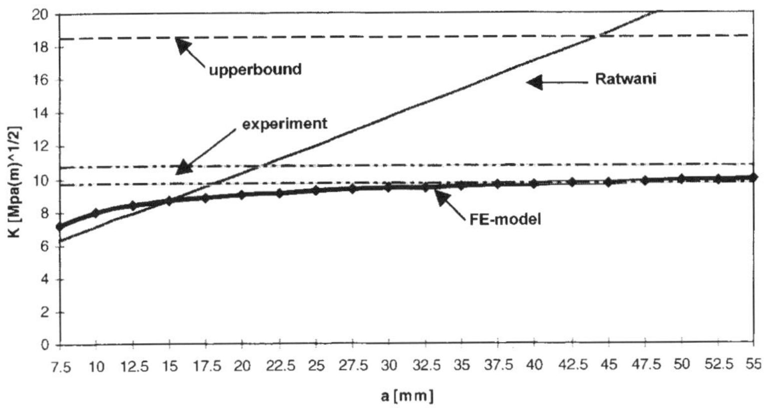

Figure 30.12 *Results of analysis with bending comparing analytical models with a finite element model and experimental results* [11].

The difference between finite elements and experiments might be that plain stress at the crack tip is assumed in the FE-model. Assuming plain strain would result in a slightly higher (around 5%) stress intensity factor. In reality, the situation at the crack tip will be in between plain stress and plain strain. Also, the fact that debonds are not accounted for in the FE-model will result in lower stress intensity factors than in reality, where debonds will grow. The analytical upperbound from the Rose-Wang-Callinan model is much higher than the other results. This might be explained by the fact that the analytical model assumes quasi-isotropic material and plain strain conditions, although the Glare 2 and boron/epoxy patches are highly anisotropic. The Ratwani model seems to agree better with the experiments and finite element calculations, but only for small crack lengths. The assumption of an increasing stress intensity factor for increasing crack lengths in this model is physically questionable though.

Influence of debonds on stress intensity factor

During (slow) fatigue crack growth of the crack in the aluminium after bonded repair, delaminations will grow around the crack due to the peel stresses in the adhesive caused by out-of-plane bending of the crack faces. Debonds around the crack result in a decreased crack closure, which in turn causes an increase in crack growth rate. The natural debond sizes are rather small and have limited effect on bonded repair efficiency. However, delaminations due to errors in the bonding process (pre-treatment, cure cycle, etc.) can occur even without being detected with NDI techniques. Environmental influences can also cause increased debond sizes. The effect of larger debonds has been investigated with finite elements.
Two different adherend crack lengths have been modelled: a=22.5 mm and a=50 mm. Debonds were assumed to be rectangular of shape, reaching as far as the crack tips. Four debond heights were examined: b=0.5 mm to b=2.0 mm with a step size of 0.5 mm. The results are shown in Figure 30.13 depicting the maximum stress

intensity factor ($K_{max}=K_{mid}+K_b$), the stress intensity factor due to pure bending (K_b), and the stress intensity factor midway the adherend thickness (K_{mid}). It can be seen that an increase in debond height only affects the K_{max} through K_{mid}. The stress intensity due to pure bending is unaffected by the debond height.

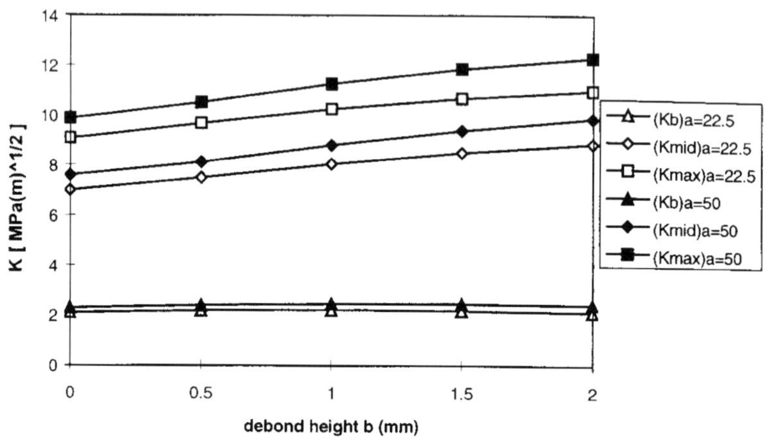

Figure 30.13 *Influence of debond height on stress intensity factors (with bending) at two crack lengths: a=22.5 mm and a=50 mm* [11].

Influence of material property changes and design variables on patch efficiency

Finite element modelling is very well suited for quickly investigating the effect of material property changes or other design variables on bonded repair efficiency. Material properties for example, can change under the influence of temperature and moisture, or can even vary from material batch to batch. Defects in the patches or adhesive system (disbonds, voids, etc.) can also affect the material properties. Finite element modelling has been used to investigate the effect of changing material properties. The basic FE-model is the same as discussed earlier, although the patch does not cover the full adherend width now and is octagonal (boron/epoxy) or elliptical (Glare 2) in shape. These specimen configurations are in accordance with the majority of test specimens used within this research. Figure 30.14 shows the relative influence of a change in Young's modulus of the adhesive or boron patch on the stress intensity factor. It can be concluded from this figure that even extensive changes in elastic modulus of the adhesive do not significantly affect the repaired stress intensity factor. A change in patch modulus however, has a dramatic effect on repaired stress intensity factor. A 20% increase in elastic modulus already results in a 10% decrease in stress intensity.

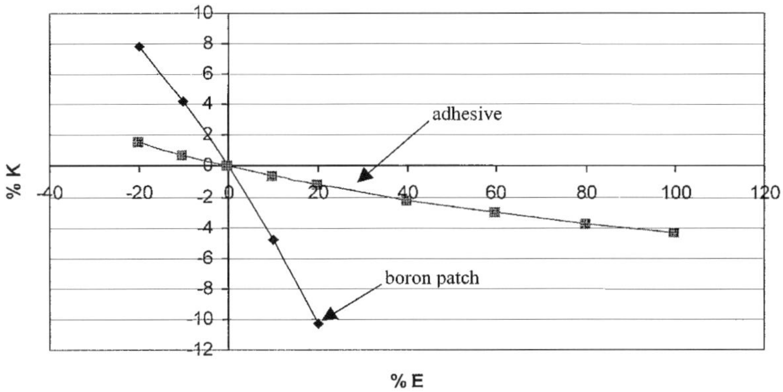

Figure 30.14 *Relative influence of changes in elastic modulus of the adhesive or boron patch on the repaired stress intensity factor* [12].

As example of the influence of design variables on patch efficiency operating temperature was taken. Operating temperatures for bonded patch repairs might differ depending on the application considered. Finite element modelling was used to investigate the influence of the operating temperature on the actual stress intensity factor at the repaired crack tip for the two patch materials used in this research.

Figure 30.15 shows that the effect is much larger for the boron/epoxy patch than for the Glare 2 patch. This was expected, because of the earlier discussed large difference in coefficient of thermal expansion of boron/epoxy and aluminium, whereas this difference is small for Glare compared to aluminium. Particularly at cruise altitudes where temperatures of –55 °C are quite common; this results in a much higher repaired stress intensity factor for boron/epoxy repairs, and consequently higher crack growth rates of the repaired crack.

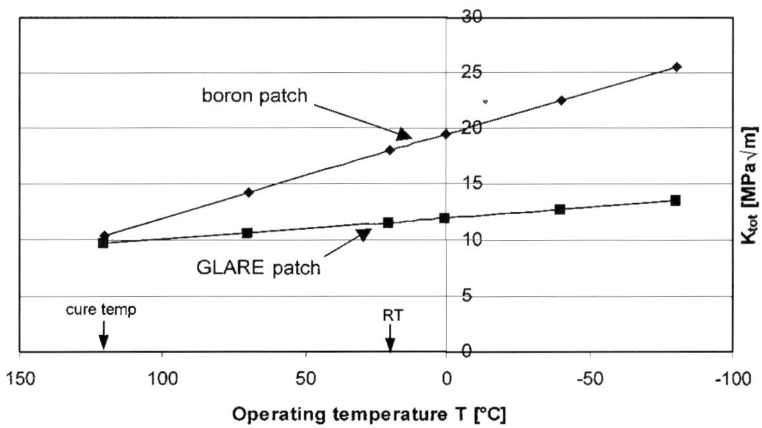

Figure 30.15 *Influence of operating temperature on the actual stress intensity factor at the repaired crack tip. Far field stress: 120 MPa, crack length: 30 mm* [12].

30.5 Durability of bonded patch repairs

Bonded patch repair is an economic, safe and damage tolerant option compared to riveting or reskinning. Life enhancements of several times the original design life of the repaired structure can be achieved without the explicit need for additional maintenance. Drawback of adhesive bonded patch repair is the fact that no NDI technique is currently available to check bond line integrity after curing. For example disbonds where the adhesive is in contact with the substrate cannot be found using NDI. Using NDI techniques can therefore in no way guarantee bond strength and repair durability. Therefore only structures able to demonstrate limit load capability (1.2 x Limit Load) in absence of the repair are certified when bonded repairs are used. Knowledge about the long-term behaviour of bonded patch repair is needed to provide a solution to this problem. Durability research and process control of the entire bonding process (instead of bond integrity checking techniques) are the key to general acceptation and airworthiness certification, even for critical damages. This section shortly discusses the durability research in this project.

30.5.1 Material property degradation

A bonded patch repaired structure consists of a build-up of different (material) layers that are susceptible to environmental attack: the aluminium adherend, an oxide layer, a primer layer, the adhesive layer, again a primer layer and an oxide layer, and then a patch. Each part of this bonded structure reacts differently to moisture and temperature. Focusing on material degradation through moisture ingress, the adhesive layer and prepreg layers in the patches are most sensitive to this kind of environmental attack. Simple moisture diffusion tests have shown that the epoxy adhesives used in this research, as well as the epoxy adhesive used for the Glare patch prepreg layers, behave according to Fick's diffusion laws [19]. Diffusion constants have been determined with Fick's laws for all epoxies used and the apparent saturation level was calculated. This apparent saturation level was then used for all further experimental research into adhesive and patch degradation due to moisture.

Property degradation of the epoxy adhesives used was experimentally investigated using dynamic mechanical analysis (DMA). For both dry and saturated adhesive specimens the glass transition temperature (T_g) was monitored and additionally for the saturated specimens the elastic modulus with increasing temperature. The glass transition temperature is a measure for the operational temperature envelope of the adhesive while the elastic modulus (in the elastic region directly proportional to the shear modulus) is a measure for the load carrying capabilities. Dry specimens were tested in an oscillating 3-point bending setup where the probe amplitude was monitored. Saturated specimens were tested in an oscillating tensile setup under water while the increasing temperature during the test would otherwise cause moisture to diffuse out of the specimens. In the case of fully reversible alteration of adhesive properties due to moisture, this would then yield the exact same results as the dry specimens. In the test setup under water the loss modulus was monitored. Table 30.2 shows the decreases in glass transition temperature for the epoxies tested (FM-94 is the epoxy used in the Glare prepreg layers, and AF-163 and FM-73 the epoxies used in the bond line) to be around 20%. All three saturated epoxies sharply deteriorate at around 80 °C. Although this drop in T_g is severe, the saturated T_g is still above the maximum operating temperature of an aeroplane of about 70 °C.

The decrease in Young's modulus with increasing temperature for the saturated specimens is depicted in Figure 30.16. This figure also gives an indication of the glass transition temperature at around 80 °C.

Table 30.2 *Glass transition temperature T_g (°C) comparison between dry and saturated epoxy specimens* [19].

Adhesive	T_g (dry)	T_g (saturated)	ΔT_g
FM-94	103.0	80.0	-23.0
FM-73	96.0	77.5	-18.5
AF-163-2K	106.0	85.0	-21.0

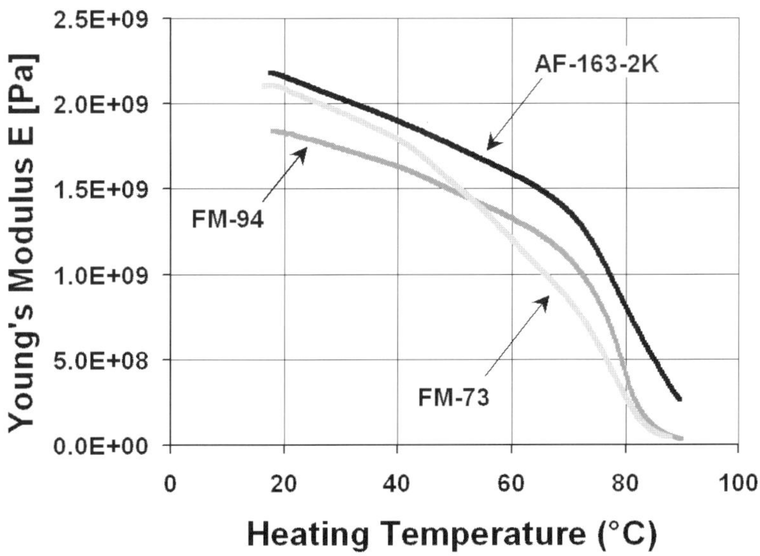

Figure 30.16 *Elastic modulus as function of temperature for saturated specimens* [19].

Up to the maximum operating temperature all three epoxies degrade. FM-73 shows a modulus decrease of about 60%, the other two adhesives a decrease of around 40%. This decrease in modulus will have its impact in load carrying capabilities of the epoxies. This was indeed seen during fatigue testing of adhesive CCT-specimens where saturated specimens showed significant reductions in fatigue life.
The effect of these dramatic reductions in adhesive properties on patch efficiency is small, as was already pointed out in section 30.4.2. Finite element modelling showed that changes up to 100% in elastic modulus of the bond line hardly change the repaired stress intensity factor. Changes in elastic modulus of the patch proved to be

more detrimental. However, looking at the rule of mixtures (where for the prepreg in used in Glare the following values can be used: $E_{FM94}\approx 2240$ MPa, $E_{fibres}\approx 86200$ MPa and $v_{fibres}=60\%$):

$$E_{prepreg} = v_{fibres} \cdot E_{fibres} + (1 - v_{fibres}) \cdot E_{matrix} \tag{30.1}$$

It can be calculated that the elastic modulus of the matrix (FM-94) accounts for only 1.7% of the elastic modulus of the prepreg in dry condition. Changes in the matrix properties due to moisture and temperature will thus hardly change the prepreg modulus, which in turn means that the Glare patch modulus is hardly affected.
The analytical model(s) discussed earlier also indicate that material property degradation in the form of modulus decrease has hardly any effect on patch efficiency. Because the entire bond line can be used to transfer the loads from the defect adherend into the patch, simply more adhesive length is used when the adhesive properties decrease. This is the so-called load transfer length (β^{-1}). Using the load transfer length equation from the analytical models [9], a decrease in adhesive shear modulus (meaning also a decrease in elastic modulus) can be related to a change in load transfer length, see Figure 30.17.

It can be seen that even for a 90% decrease in shear modulus the added load transfer length is only about 8 mm. The bond line in repaired structures is large enough to accommodate much larger changes in load transfer length than the above mentioned worst-case 8 mm.

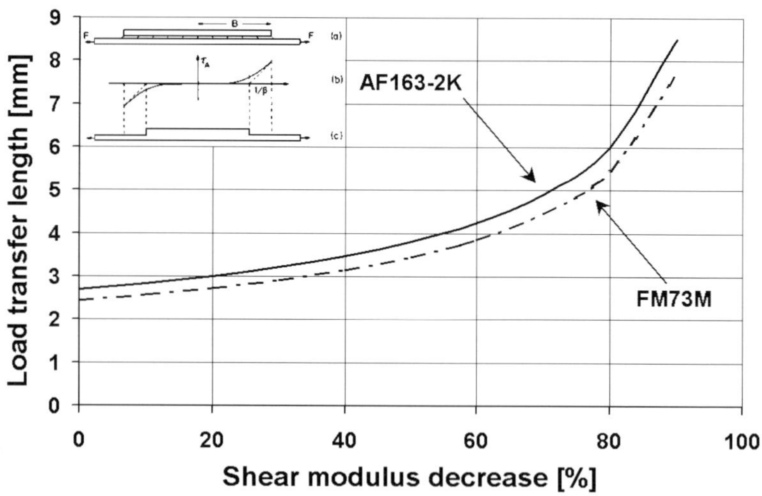

Figure 30.17 Load transfer length as function of shear modulus reduction. (Also shown the load transfer length for bonded reinforcements as used in the Rose model: (a) cross-section; (b) adhesive shear stress distribution; (c) equivalent. rigid-bond configuration [9]).

30.5.2 In-service effects

Even though epoxy/adhesive degradation only has a marginal influence on the bonded patch efficiency, in-service temperature and moisture can have an effect on the repaired crack growth. Temperature and the accompanying changes in relative humidity can affect the aluminium crack growth behaviour, especially in the low stress intensity factor regimes the crack experiences after repair. Moisture entering the crack and embrittlement of the bond line, possibly resulting in increased debonding, can change the stress state at the crack tip or the crack opening behaviour. To investigate these in-service effects, specimens with artificial debonds in the bond line have been tested [20]. The temperature/moisture effect has been looked at in isothermal low temperature tests [21].

Artificial debonds

The debond mechanism is not static, but caused by fatigue of the adhesive. C-scanning of bonded repairs during and after fatigue tests confirmed that debonds grow gradually in the wake of the repaired crack tip [21]. Increased debond sizes due to moisture and temperature influences can occur, therefore the influence of large debonds has been investigated using artificial debonds [20]. These debonds under Glare 2 patches were created by using Teflon inserts between adherend and adhesive. Tested debonds are shown in Figure 30.18. The Glare 2 patches were bonded to a 1 mm CCT aluminium adherend.

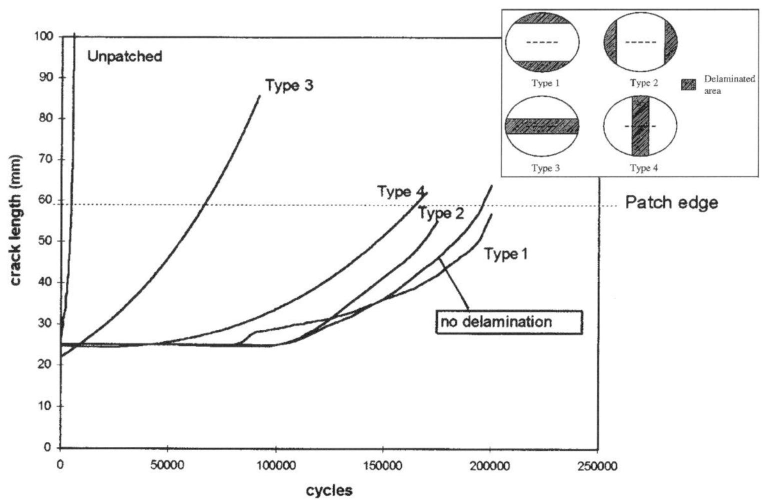

Figure 30.18 Effects of four types of artificial debonds under Glare 2 patches on the crack growth behaviour of a repaired crack [20].

Apart from the superior fatigue behaviour of bonded Glare repaired specimens compared to unpatched specimens (due to a dramatically lower stress intensity factor under the repair), it can be seen that only debond type 3 causes a severe decrease in fatigue life. The other debond types show fatigue lives comparable to a Glare patch without delaminations (i.e. only a natural debond growing in the wake of the crack), with a slightly shorter fatigue life for debond type 4. From a crack opening point of view this behaviour is understandable. Type 3 debonds result in more crack opening (less restraint by adhesive-patch combination). More crack opening means a higher stress intensity factor and hence increased crack growth rates. Debond type 4 will have slightly more crack opening because the debond lies over the crack. It will not be as severe as type 3 while only part of the crack is affected and there is still bonded patch width left for crack opening restraint. Debonds type 1 and 2 are entirely bonded around the crack, so there will be no effect on crack opening compared to specimens without delaminations.

Isothermal low temperature

Experiments were conducted at low temperatures comparable to cruise altitude conditions to see what the influence of these conditions was on the fatigue behaviour of bonded patch repairs [21]. Butkus [22] found a significant drop in fracture toughness and adhesive embrittlement at low temperatures, which would in itself cause degraded fatigue behaviour. Additional tensile thermal stresses at these temperatures can also cause increased crack growth rates of the repaired crack. Figure 30.19 shows the results of the isothermal low temperature tests performed.

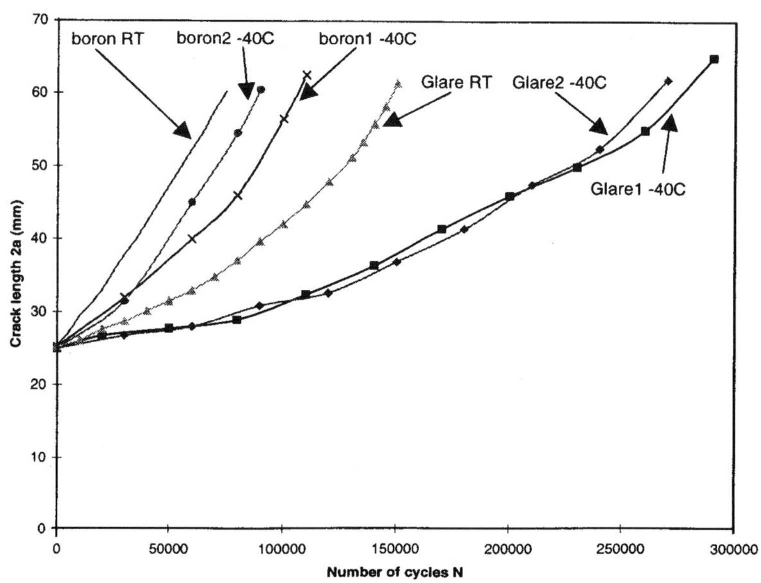

Figure 30.19 Crack growth curves for isothermal low temperature test specimens [21].

It can be clearly seen that both specimen types (boron and Glare patch) show improved fatigue performance at –40 °C compared to room temperature (RT). For the boron-repaired specimens the improvement is less pronounced because of the larger mismatch in coefficient of thermal expansion between boron and aluminium. The improved fatigue life at low temperatures can be explained by the fact that the adhesives becomes stiffer, resulting in decreased crack opening and hence, decreased crack growth rates. The shear modulus of FM-73 for example, increased from 503 MPa at room temperature to 791 MPa at –40 °C. Another favourable effect is the decrease in crack growth rate in the aluminium caused by the low relative humidity at low temperatures. Crack growth tests on Al2024-T3 sheets at –35 °C showed a reduction in fatigue crack growth rate of up to a factor six [23]. However, in view of the minor influence of adhesive property changes on repair effectiveness (as discussed in section 30.5.1) it can be concluded that the improved fatigue life found at low temperatures is largely caused by the aluminium crack growth behaviour.

30.6 Conclusions

The technological development of a new repair technique for aircraft structures showed that bonded Glare repairs are feasible, damage tolerant, safe and relatively easy to apply. In situations where thermal residual stresses have a pronounced influence on repair effectiveness, Glare repairs are superior to the fibre/epoxy repairs commonly used for thick structures. This is due to the lower CTE mismatch of Glare with the aluminium skin. Design guidelines have been developed for more complicated situations such as multiple patches in close proximity. Finite element modelling proved to be a useful additional research tool; not only for studying these complex configurations, but also to gain understanding of the physical behaviour of bonded repairs for a large variety of boundary conditions. The ability to accurately model bonded repairs analytically is increasing with every addition to or refinement of the analytical models available. Several in-service effects were considered and revealed no problems during the service life of the repair. Cruise altitude conditions even proved to be beneficial for bonded repairs because of the improved fatigue behaviour of the aluminium skin. Adequate control of the entire bonding process produces strong and durable bonded repairs; even when subsequent environmental exposure causes severe bond line and patch matrix degradation. This is because the amount of (degraded) adhesive needed for load transfer stays well below the amount of adhesive available.

References

[1] J.J. Denney and S. Mall, *Effect of Disbond on Fatigue Behaviour of Cracked Aluminium Panel with Bonded Composite Patch*. Department of Aeronautics and Astronautics. Wright-Patterson AFB Ohio: Air Force Institute of Technology.

[2] A.A. Baker, 'Crack Patching: Experimental Studies, Practical Applications'. In: A.A. Baker and R. Jones (eds), *Bonded Repair of Aircraft Structures*. Dordrecht (NL): Martinus Nijhoff Publishers, 1988, p.107-173.

[3] R.S. Fredell, *Damage Tolerant Repair Techniques for Pressurized Aircraft Fuselages*. PhD-thesis, Delft University of Technology, 1994.

[4] C.B. Guijt and R.S. Fredell, 'Delamination Effects in Fuselage Crack Patching'. In: *Proceedings of the 41st International SAMPE symposium*. 1996.

[5] The Online Materials Information Resource, *http://www.matweb.com*. Automation Creation Inc.

[6] M. Davis and D. Bond, 'Principles and Practices of Adhesive Bonded Structural Joints and Repairs'. In: *International Journal of Adhesion & Adhesives*. Volume 19. Oxford (UK): Elsevier Science Ltd.,1999, p.91-105.

[7] G. Ipenburg, *Analysis of Stresses in the Periphery of Bonded GLARE Repair Patches*. Master Thesis, Delft University of Technology, 1997.

[8] W. Mortier, *Crack Initiation in Glare and Aluminium Patches for One-Sided Repairs*. Master Thesis, Delft University of Technology, 1999.

[9] L.R.F. Rose, 'Theoretical Analysis of Crack Patching'. In: A.A. Baker and R. Jones (eds), *Bonded Repair of Aircraft Structures*. Dordrecht: Martinus Nijhoff Publishers, 1988, p.77-106.

[10] L.J. Hart-Smith, *Analysis and Design of Advanced Composite Bonded Joints*. NASA CR-2218. Hampton (USA): NASA, 1974.

[11] J.J.A. Massar, *Calculation of Stress Intensity Factors of a Bonded Repair by Means of a Finite Element Method.* Preliminary Thesis, Delft University of Technology, 1998.

[12] J.J.A. Massar, *Crack Growth Predictions of Bonded Repairs - by Means of Finite Element Models*. Master Thesis, Delft University of Technology, 1998.

[13] M.M. Ratwani, 'Cracked, Adhesively Bonded Laminate Structures'. In: *AIAA Journal 17*. Number 9. New York: AIAA, 1979, p.988-994.

[14] C.H. Wang et al., 'Analysis of Out-of-Plane Bending in One-Sided Bonded Repair', *International Journal of Solids and Structures* (35), Number 14, 1998, p.1653-1675.

[15] R.D. Mindlin, 'Influence of Rotary Inertia and Shear on Flexural Motions of Isotropic, Elastic Plates'. *Journal of Applied Mechanics*. Volume 18, 1951, p.31-38.

[16] E.F. Rybicki and M.F. Kanninen, 'A Finite Element Calculation of Stress Intensity Factors by a Modified Crack Closure Integral', *Engineering Fracture Mechanics* (9), 1977.

[17] M.J. Young and C.T. Sun, 'On the Strain Energy Release Rate for a Cracked Plate Subjected to Out-of-Plane Bending', *International Journal of Fracture* (60), 1993, p.227-247.

[18] P.J.M. Nijssen, P.J.M., *Qualitative Investigation of Disbond Mechanisms under Single-Sided Adhesively Bonded Repairs*. Master Thesis, Delft University of Technology, 1997.

[19] R. Brenninkmeijer, *The Effect of Moisture Ingress on the Bonded Repair Effectiveness*. Master Thesis, Delft University of Technology, 2000.

[20] J.J. Denney, *Fatigue Response of Cracked Aluminum Panel with Partially Bonded Composite Patch*. Thesis, AFIT/GAE/ENY/95D-7. Dayton (USA): Air Force Institute of Technology, 1995.

[21] S. Verhoeven, *In-Service Effects on Crack Growth under Bonded Composite Repairs*. Master Thesis, Delft University of Technology, 1998.

[22] L.M. Butkus, *Environmental Durability of Adhesively Bonded Joints*. PhD Thesis, Georgia Institute of Technology, 1997.

[23] D. Broek, *Residual Strength and Fatigue Crack Growth in Two Aluminium Alloy Sheets at Temperatures Down to -75°C*. NLR Report TR 72096. (NL): NLR, 1972.

31

Bonded repairs for C-5A fuselage crown cracking

C.B. Guijt, S. Verhoeven and J.M. Greer

Disclaimer: *The views expressed in this article are those of the authors and do not necessarily reflect the official policy or position of the United States Air Force, Department of Defense, or the US Government.*

Abstract - This chapter describes the application of Glare repairs to a USAF C-5A transport. One of the problems in the ageing C-5A fleet is the multiple small cracks in the upper aft-crown section of the fuselage skin. These cracks are possibly caused due to the usage of stress-corrosion sensitive 7079-T6 aluminium. The crown section experiences significant longitudinal tensile bending in addition to biaxial tension due to internal pressurisation. Multiple short cracks (25 to 50 mm long) were found, see Figure 31.1, in the crown of several aircraft. In the autumn of 1995 two bonded Glare repairs were applied at fuselage stations 1700 and 1784 by a team of Wright Laboratory, San Antonio-Air Logistics Center, and United States Air Force Academy/Center for Structural Life Extension (CAStLE) personnel [1].

The cracks are believed to have nucleated in the rivet holes of the 7079-T6 aluminium skins due to high fit-up stresses induced in manufacturing and the bending moment caused by the T-tail.

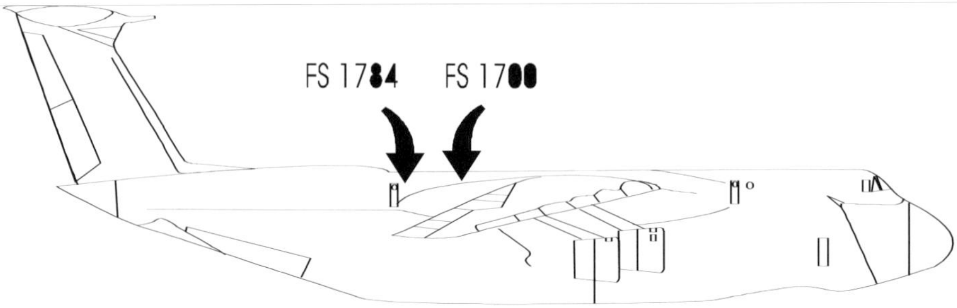

Figure 31.1 *Crack locations in crown section of C-5A.*

31.1 Introduction

For low temperature applications, analysis and experiments have shown that moderate Coefficient of Thermal Expansion (CTE) composite materials can perform better as fuselage skin repair materials than the more traditional low CTE composite patch materials. The moderate CTE composite material Glare is a viable alternative to traditional composite patch materials such as boron- and carbon/epoxy for pressurised fuselage skin repairs. The moderate CTE patch materials result in a lower $K_{repaired}$ near the crack tip and reduced adhesive shear strains due to the better CTE match with the aluminium skin material. The lower $K_{repaired}$ will result in slower, or no, crack growth. The lower adhesive shear strain is an important advantage for the durability of the bonded repair.

31.2 Damage tolerance analysis guidelines

This section describes some general guidelines for bonded repairs and will present details how these apply to the C-5A repairs.

31.2.1 General damage tolerance guidelines

General guidelines for USAF bonded repairs are based on the problem that no techniques are currently capable of predicting the durability of a bonded joint [2]. Process control and verified surface pre-treatment procedures are the only means of guaranteeing the durability of the bonded joint. For many applications this must be achieved in the field under less than ideal circumstances. These inspection issues lead to the problem that there is no guarantee regarding the durability of the patch. Since the durability is not predictable, the most conservative assumption is that the patch might become ineffective as soon as it is applied. This leads to the following application boundaries and guidelines.

When repaired, the following criteria must be satisfied assuming that the patch could be non-functional:

- *Residual strength*
  At 1.2*Limit Load (LL) for the component, the damage to be repaired cannot be critical without the repair. If the repair fails (or debonds) the unrepaired structure must be able to carry this load.
- *Inspection interval*
  The inspection interval must be based on the unrepaired crack-growth rate. The maximum inspection interval is half the time/flights the unrepaired damage grows to a critical defect at 1.2*LL.

When repaired, the following criteria must be met with the patch present:

- *Remaining life*
  The repaired damage must be able to provide double the (remaining) life of the structure. This means that the defect must not grow to critical size with the repair in place. Typically a safety factor of two is used; the repaired damage must be able to provide two (remaining) lifetimes of the structure.
- *Ultimate strength*
  The static ultimate strength (DUL) cannot be lowered after application of the bonded repair and the repair shall not create new critical areas under DUL.

31.2.2 C-5A Damage tolerance considerations

The cracks grow in the second rivet row of a 5-rivet row joint. This makes it obvious that the crack is not an ordinary fatigue crack. The direction in which the crack grows indicates that the longitudinal stress is the dominant stress for crack growth. The complicated sub-structure (frame and stiffeners) affects the stress field around the crack. A typical exceedance table of the fatigue spectrum of the C-5A, is given in Table 31.1. The total C-5A design life is 30,000 flight hours.

Table 31.1 Typical exceedance table.

Stress level (MPa)	Nr. of exceedances (per 1000 flight hours)
145	1
130	10
125	100
120	1,000
110	10,000
100	100,000
65	1,000,000

31.2.3 FEM DTA analysis

Finite element modelling was done to predict more realistic unpatched *K*-values for the crack. A Franc2D-L [3] model was made consisting of the substructure, rivets, and the skin. Due to the support of the substructure there is no bending and the 2D-model can be used. In order to simplify the model, the following modifications were made compared to the actual conditions:

1. Shear loads were not applied.
2. The crack was modelled as vertical, perpendicular to the longitudinal stress.
3. The elliptical patch was modelled as a circular patch.

The result for an applied stress of 120 MPa longitudinal and 64 MPa hoop stress is given in Figure 31 2. The lines in the figure indicate the location of the frame, stringer, and the patch. The rivets are modelled using rivet elements provided in Franc2D-L. These elements create a point connection between the layers. In this figure the patch is not bonded to the skin (unrepaired).

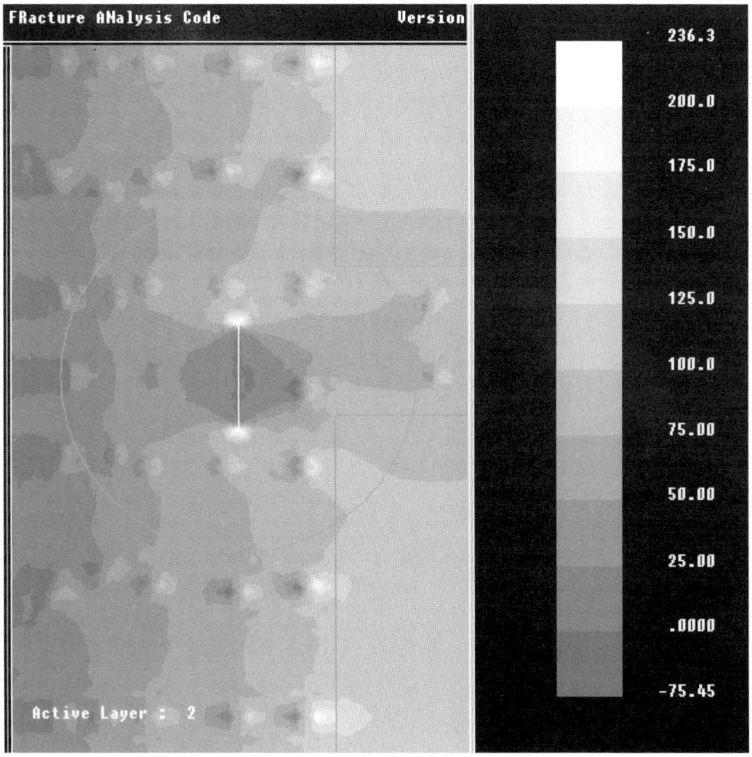

Figure 31 2 Stresses in the unrepaired skin at FS 1784 (scale in MPa).

Table 31.2 *Stress intensity factors.*

Applied stress (MPa)	long	80	120	145	160	227
	hoop	64	64	100	64	83
K (Franc2D-L) (MPa√m)		8.5	12.4	15.2	16.4	23.2

With the *K* result from the model and the applied far-field stress, the geometry factor β can be determined:

$$\beta = \frac{K}{\sigma\sqrt{\pi a}} \qquad (31.1)$$

with: K = Stress intensity factor from the FEM calculation
σ = The applied longitudinal stress
a = The half-crack length in the model (27.66/2 mm)

The result for β in the unrepaired case is approximately 0.50. This is not an exact solution due to the simplifications in the model listed above. In general, a biaxial load reduces the K and $K_{repaired}$ slightly.
It can be concluded that the substructure lowers the stress in the skin at the crack location by roughly a factor two. Table 31.2 gives the stress-intensity factors from the Franc-model at different biaxial stress levels.

At the time this analysis was performed, the guidelines for bonded repairs were evolving into the guidelines described earlier. Limit Load was the load at which the unrepaired defect should not be critical. Since then this load changed to a slightly higher (less likely to occur) load of 1.2*LL. Although the analysis does not address this load, it is obvious that the 20% increased load will not result in a critical crack. Limit load is 227 MPa for the longitudinal stress. The unrepaired stress-intensity factor is 23.2 MPa√m at limit load; this is well below the lowest reported fracture toughness for 7079-T6 in the Damage Tolerance Design Handbook [4], 75 MPa√m. Therefore, a bonded repair is allowed according to the current USAF guidelines: the crack is not critical at limit load without a patch.

31.3 Repair options

One possible alternative for bonded repairs for the C-5A's aft upper crown skin is to re-skin the area. However, the rivet pattern in the stringers and frames are hand-drilled and some holes may have edge-distance problems. Moreover, re-skinning could be very costly if some frames and stringers have to be replaced due to edge distance problems. Finally, skin replacement must be accomplished in a "fuselage level" condition.

Riveted repairs were attempted. In some cases, new skin cracks grew from the corner rivet holes of the patch. Considering these factors, the most viable repair alternative might be a bonded repair.

31.3.1 Stiffness of the patch material

In order to match patch stiffness to that of the repaired structure, a boron patch will be much thinner than a Glare patch. For very thick sections, or aerodynamic critical areas, the thickness of the patch will play an important role. On a relatively thin fuselage skin in an area where the boundary layer is large, the thickness of the patch will play a minor role; therefore both patch materials are candidates.

31.3.2 Coefficient of thermal expansion

The highest loads in a pressurised fuselage typically occur at low temperatures. The difference in CTE between the patch material and the fuselage plays an important role in patching effectiveness. For all patch materials a stress free state between patch and fuselage skin occurs between the cure temperature and the glass transition temperature, T_g, of the adhesive. The cure temperature of the adhesive used is typically 100 to 125 °C. After cooling down to room temperature, or to –50 °C at cruise altitude, in general a moderate- or high-CTE material will cause the crack to be in compression. On the other hand a low CTE material can actually put the crack in tension. Because of this effect, for this application a moderate- or high-CTE patch material is favourable over a low CTE patch material. Aluminium would be the best alternative from this point of view. However, since the patch is applied over a fatigue crack it is likely that the fatigue crack will propagate through the thickness of the patch and fail the aluminium patch. Composite patches will not suffer from this problem due to the intrinsic resistance against crack propagation (through the thickness). A Glare patch combines the advantages of composite materials with a higher CTE and is therefore a good candidate for this type of repair.

31.4 Design of the bonded repair

An analytical closed-form computer programme was developed, CalcuRep [5], which allows a quick design of bonded repairs. The 1995 version of this programme was used to design the patches and to study the effects of different patch materials, cruise temperature, and cure temperature on the effectiveness of the repair. The results of the predictions were verified using fatigue crack-growth specimens.

However, the programme can only handle limited, geometrically simple cases. Therefore, the results are also verified using a two-dimensional multi-layer finite element model in Franc2D-L [3].

31.4.1 CalcuRep calculations

To make a proper patch design the configuration must meet the guidelines for typical loads, limit load and ultimate load. At typical loads, the repair must show durable behaviour: the bonded repair should easily outlive the surrounding aircraft structure without causing additional problems. At limit load, no yielding of any component is allowed (except for the adhesive). At ultimate load (1.5 times limit load) the structure shall not fail; the stresses should be under the ultimate strength of the components.

CalcuRep calculates the five most important design parameters for a bonded repair:

- *K* repaired
- patch stress σ_{patch}
- skin stress at patch tips σ_{tip}
- shear strain of the adhesive $\gamma_{adhesive}$
- minimum patch length (P_l) vs. Load Transfer Length (LTL).

The Load Transfer Length (LTL) is defined as:

$$\frac{1}{\lambda} = \sqrt{\left(\frac{t_a}{G_a}\left[\frac{E_p \cdot t_p \cdot E_R \cdot t_R}{E_p \cdot t_p + E_R \cdot t_R}\right]\right)} \qquad (31.2)$$

with: E_p = Young's modulus of the structure (plate)
t_p = Thickness of the structure
E_R = Young's modulus of the patch (reinforcement)
t_R = Thickness of the patch
t_a = Thickness of the adhesive
G_a = Shear modulus of the adhesive

Note: In more recent versions of CalcuRep the length of the patch is determined using both the elastic and plastic transfer length. This method is different from that used here to design the Glare patches for the C-5A crown section.

Table 31.3 shows the design limits of the 1995 CalcuRep version of these variables for the different load cases.

Table 31.3 Design limits used for bonded repairs.

Variable	Typical loads	Limit load	Ultimate load
K (MPa√m)		$<K_c/1.5$	$<K_c$
σ_{patch} (MPa)	< 0.5 yield	< yield	< ultimate
σ_{tip} (MPa)	< 0.5 yield	< yield	< ultimate
$\gamma_{adhesive}$	$< 0.5\ \gamma_e$	$< 1.2\ \gamma_e$	$< 1.5\ \gamma_e$
P_l (mm)	40 × LTL	40 × LTL	40 × LTL

For the typical repair configuration on an aluminium 7079-T6 crown panel this results in the guidelines given in Table 31.4.

The design of the Glare patch must meet the design guidelines given in Table 31.4. The CalcuRep input data, used to design the patches is given in Table 31.5.

Additional inputs were:

- Heat blanket 275×275 mm
- Cure temperature 125 °C
- Cruising altitude: 10,000 m typical, 1000 m limit/ultimate
- Taper thickness 0.3 mm
- Patch length: 100 mm
- Patch width: 90 mm
- Ultimate load: 1.5 × Limit

The results of the calculations for FS 1700 are given in Table 31.6.

Table 31.4 Design limits for bonded Glare 2 patches on 7079-T6 skins.

Variable	Typical loads	Limit load	Ultimate load
K (MPa√m)		<50	<75
σ_{patch} (MPa)	<236	<472	<1100
σ_{tip} (MPa)	<236	<472	<1100
$\gamma_{adhesive}$	< 0.0446	<0.10704	< 0.1338
P_l (mm)	40 × LTL	40 × LTL	40 × LTL

Table 31.5 CalcuRep input file for patches @ FS 1700 & FS 1784, both 7079-T6.

Variable	FS 1700	FS 1784
Skin thickness (mm	1.3	1.3
Crack length (mm)	12	22
Frame spacing (mm)	628	508
Stringer spacing (mm)	155	155
Long stress typical* (MPa)	121	121
Hoop stress typical (MPa)	100	100
Long stress @ limit (MPa)	223	227
Hoop stress @ limit (MPa)	81	83
Patch thickness (mm)	Glare 2-3/2 1.4	Glare 2-4/3 1.55

* Load that occurs 500 times per 1000 flight hours (15000 per life), see Table 31.1

Table 31.6 *CalcuRep results for FS 1700.*

FS 1700	Typical	Limit	Ultimate
K (MPa√m)	3.03	7.13	19.6
σ_{patch} (MPa)	148	268	394
σ_{tip} (MPa)	195	352	517
$\gamma_{adhesive}$	0.0343	0.0805	0.1523
P_l (mm)	153	153	153

Table 31.7 *CalcuRep results for FS 1784.*

FS 1784	Typical	Limit	Ultimate
K (MPa√m)	2.67	6.98	20.7
σ_{patch} (MPa)	133	246	363
σ_{tip} (MPa)	199	364	533
$\gamma_{adhesive}$	0.0276	0.0723	0.1361
P_l (mm)	157	157	157

Comparing the calculated results with the table for the design limits, see Table 31.4, shows that all guidelines are met except for patch length (100 mm instead of 153 mm) and adhesive shear strain, γ, at ultimate (0.1523 instead of 0.1338). The patch length is 26 times the load transfer length, which should still provide sufficient resistance to creep of the adhesive. Also, the loads carried at the frame location are significantly lower than the skin loads in the middle of the bay, see the DTA section with the determination of β=0.5. At ultimate load, the adhesive is allowed to go to the maximum plastic-strain, γ_p; for the adhesive used (3M's AF163) this is typically a lot higher than 1.5 times γ_e. For this adhesive the strains remain within the plastic limit.

For FS 1784 the results are given in Table 31.7. All guidelines except patch length are met. Adhesive shear strain at ultimate is only slightly over the recommended value. The patch length used is, again, about 26 times the LTL, which should be sufficient to prevent creep.

31.4.2 FEM model of the patched crack

The patched configuration was modelled in Franc2D-L. The resulting stresses with the patch bonded to the structure are shown in Figure 31.3.
The effect of the repair on the crack is drastic, compared to the unpatched situation. A summary of the results obtained with Franc2D-L is given in Table 31.8.

Both analysis methods have weaknesses. CalcuRep does not model the substructure properly and therefore overestimates the applied stress in the repaired area. Franc does not model the thermal issues. A comparison of the results for $K_{repaired}$ between the two methods is given in Figure 31.4.

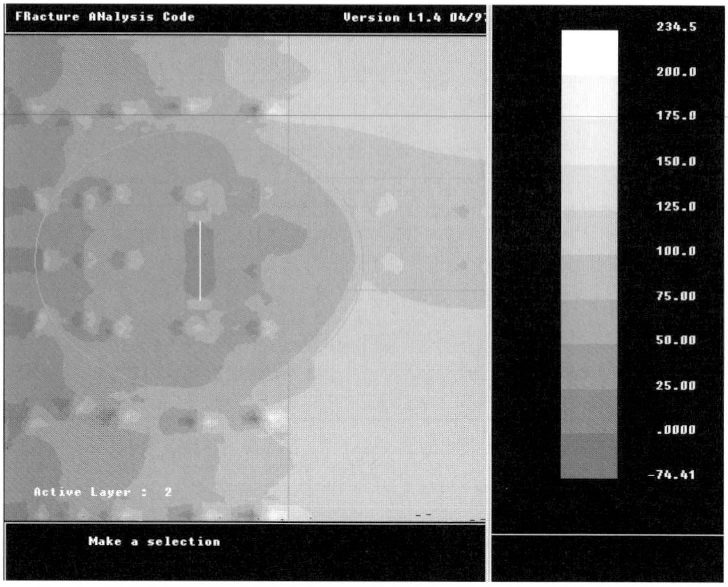

Figure 31.3 *Patched crack at FS 1784, applied far field stress is 120 MPa (scale in MPa).*

Table 31.8 *Results from Franc2D-L calculations.*

Applied long stress (MPa)	$K_{repaired}$	σ_{patch} (MPa)	σ_{tip} (MPa)	$\gamma_{adhesive}$
80	2.49	46	75	0.018
120	3.67	67	110	0.024
145	4.47	82	137	0.030
160	4.84	89	145	0.033
227	6.86	126	202	0.045

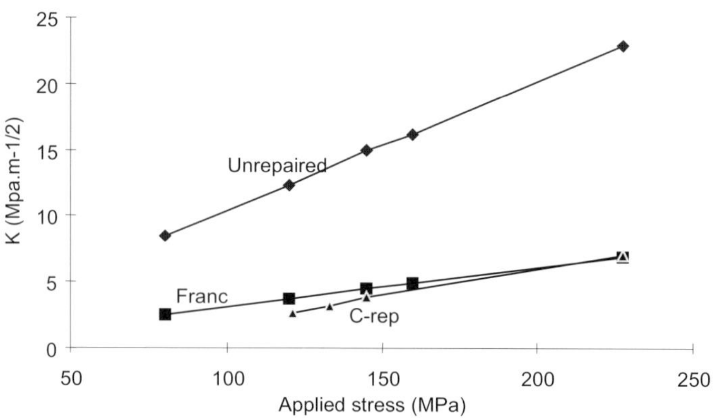

Figure 31.4 *$K_{repaired}$ for Franc and CalcuRep models.*

31.5 Patch application

CalcuRep can be helpful to determine an optimised patch configuration, but the programme cannot design the actual patch shape to fit the existing structure. CalcuRep will give the proper patch dimensions but the final outline largely depends on location of splices, doublers, stringers and frame configurations, and the existing rivet pattern.

Figure 31.5 shows the rivet pattern, substructure, and the patch location on the aircraft for FS 1784. In the critical areas at the patch tips, the patch covers all the rivets. By covering the rivet a double stress concentration due to the patch tip and the rivet is prevented. At the time when these patches were installed, there was only experience with elliptically shaped patches. A better fit to the structure can be achieved using a square patch with rounded corners. Current experience and research would have resulted in using such a patch for this repair.

31.5.1 Patch manufacturing

Once CalcuRep gives the rough dimensions of the patch, a paper tracing of the actual crack and the rivets can be used to fine-tune the final patch dimensions. A sheet of Glare, bigger than the patch dimensions is cold rolled to match the curvature of the fuselage. The outline of the ellipse is drawn on the Glare sheet and a shear cutter is used to cut it to the rough dimensions. A belt sander is used to trim the patch to the final outside dimensions.

In order to decrease secondary bending at the patch tips, the patch is tapered. For the taper a minimum of 0.2-0.3 mm is used, which is the thickness of the last aluminium layer, to prevent a knife-edge effect. The taper angle is normally 1/10 up to 1/20. Drawing a smaller ellipse on the patch, and filing the material from that line up to the last aluminium layer of the patch results in the desired taper angle. Due to the nature of applying the taper this way, Glare patches have a "wedding cake" taper. In order to handle the patch while making the taper, the patch is fastened to a wooden template with double-sided tape.

Big scratches in the taper are removed with a fine file and waterproof sandpaper. After degreasing the patch surface (pre-treated with Phosphoric Acid Anodisation (PAA) or Chromic Acid Anodisation (CAA) and primed (Cytec's BR-127) - the back surface is the bonding side and is not damaged by applying the taper) with a solvent, such as MEK or Acetone, the patch is ready to be bonded to the fuselage skin. The patch can be prepared any time during or before the process.

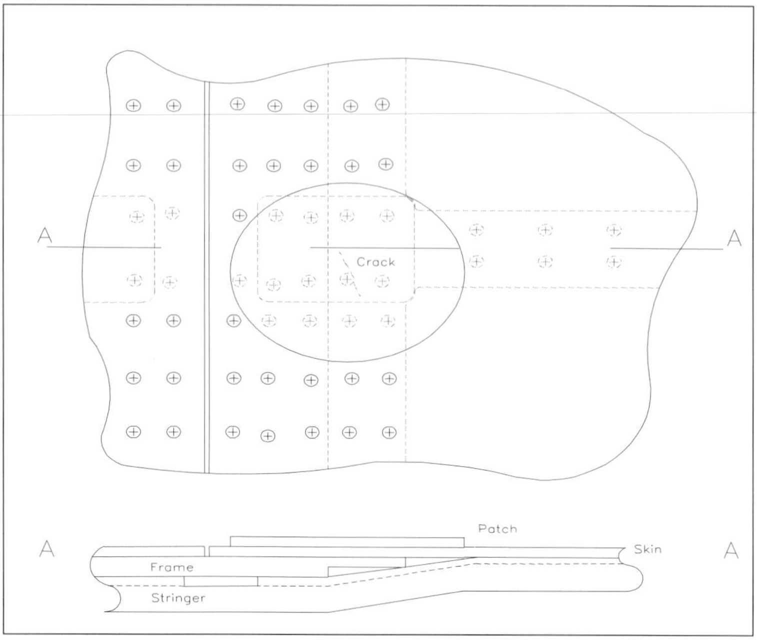

Figure 31.5 *Patch at FS 1784.*

31.5.2 Bonding procedure

Very few bonding procedures can deliver the durability required for structural repairs. Achieving this under field conditions is even more difficult. The procedure used for this repair is the Grit Blast Silane procedure (GBS) developed by the AFRL/ML division of the US Air Force [6]. The adhesive used was AF163-2M with a cure cycle of 125 ^{0}C.

31.6 Constant amplitude testing

In order to verify the patch design, several fatigue specimens were built and tested. Coupon tests were performed on boron and Glare patched panels. The configuration of the panels used was:

- panel width / length = 152 / 400 mm
- fatigue crack length 25 mm
- uniaxial loading, R=0.05 (Glare) and R=0.1 (Boron), maximum stress 120 MPa
- adhesive AF163-2, 0.13 mm thick
- patch size optimised per material

31.6.1 Repaired 2024-T3 panels

Initially experiments were performed using available aluminium panels to prove the concept of bonded repairs and to verify the possible advantages of the higher CTE Glare patches. Figure 31.6 shows the results for an unpatched panel and both Glare and Boron patched panels. Both patch materials extend the fatigue life of the panels considerably. The Glare repair results in a lower $K_{repaired}$. At a lower $K_{repaired}$ it takes more load cycles for the crack to reinitiate.

Due to the lower CTE of boron, this repair results in a higher $K_{repaired}$ than the Glare repair, and it takes fewer cycles for the crack to reinitiate. These tests were all run at room temperature. At lower temperatures these differences might even be more pronounced. On the other hand, the repairs were applied to unstiffened panels. A real-life structure (with substructure and only partly heated) will have a lower CTE than an unrestrained panel, and therefore the lower CTE of the boron will better match the effective CTE of the actual structure. The support of the structure prevents bending, therefore the crack-growth on the C-5 will actually be slower than for the unsupported specimens.

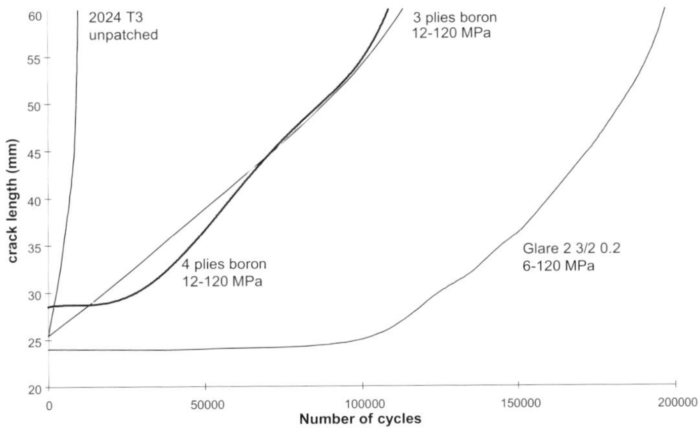

Figure 31.6 Performance of bonded repairs on 2024-T3 using Glare and Boron patches (respectively moderate and low CTE).

31.6.2 Repaired 7079-T6 panels

The next test was to evaluate the performance of the repairs on the actual fuselage skin, 7079-T6. Since this material is not available, some scrapped panels from a C-5A were used to prepare the specimens. The panels had multiple rivet holes from installation on the aircraft. All the holes outside the patched area were filled with rivets. The configuration of the specimens was similar to the 2024 specimens tested earlier. A centre-crack was cut and fatigued up to the initial starter length after which the patches were bonded to the panels. The results of these fatigue tests are given in Figure 31.7, the stiffness ratio of the repairs was 0.99.

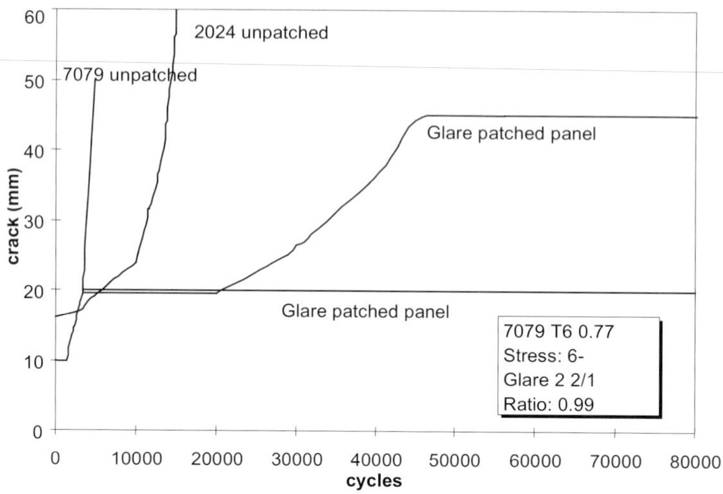

Figure 31.7 *Glare patched 7079-T6 panels.*

The first noticeable (and expected) difference between 2024-T3 and 7079-T6, is the higher crack-growth rate in 7079 at the same fatigue level. This indicates that an effective bonded repair is necessary. The first patched panel gave no fatigue crack growth after the bonded repair was applied. The second panel gave an initial retardation period, after which the crack slowly grew until it terminated in one of the existing rivet holes under the patch.

31.7 Spectrum testing

Before a decision can be made to repair the C-5A fleet using bonded Glare patches, it is necessary to perform spectrum tests to prove the effectiveness of bonded Glare repairs under loads that occur during the actual service life of the aircraft [7],[8]. This section will focus on the modification of analytical C-5A spectrum data, not compatible for fatigue testing and normally used in combination with a fatigue prediction code. It will also give the test results of a Glare repair under spectrum loading.

31.7.1 Supplied data

The C-5A spectrum data supplied by San Antonio ALC (Air Logistic Center) was used in an analytical fatigue programme to predict crack growth. Data in this particular form is not compatible for use in combination with the fatigue testing equipment. Table 31.9 gives an example of the typical spectrum data.

Table 31.9 *Format of spectrum data.*

Flight Type	DSA code	Partial cycles	Minimum stress (psi)	Maximum stress (psi)
X	Y	35.73	4881.302	5916.822
X	Y	35.73	4580.98	6217.142
X	Y	7.501	4263.112	6535.012
X	Y	1.0000	3861.912	6936.212
X	Y	0.4601	3297.695	7500.429
X	Y	3.8270E-02	2547.69	8250.429
X	Y	1.6042E-03	1797.69	9000.429

The data consists of five columns. The first column represents the flight type, followed by a DSA code (Damage Source Assignment, for example taxiing, take off, manoeuvre, etc.). The main incompatibility is the third column, representing partial cycles. A partial cycle can be non-integer, which cannot be present in a fatigue test (and real-life). The fourth and fifth columns represent the stress combinations, σ_{min} and σ_{max} (psi). Each flight data file consists of approximately 2500 rows. There are a total of 122 flights (21 different types) in one block of 508 flight hours. The total life of a C-5A being 30,000 hours, 60 blocks need to be constructed to simulate one lifetime. This will result in a life of 7320 flights.

31.7.2 Proposed spectrum building method

Because of the large amount of data, a computer programme was developed to generate input data suitable for the fatigue testing equipment. The programme, called Generated Spectrum Processor, adds partial cycles, within set limits.

Each row contains a number of partial cycles (non-integer, in third column). The programme starts reading the data and identifies for each row: σ_{min}, σ_{max} and the number of partial cycles. After a row is identified, the programme will check whether a particular min-max combination has occurred earlier. If not, this row is separately placed in a buffer where a so-called bucket is created to store the number of partial cycles encountered for that specific σ_{min}-σ_{max} combination. If the σ_{min}-σ_{max} combination occurred earlier, the (partial) number of cycles is added to the bucket already in the buffer (within the buffer tolerance, described later), as long as the cycles originate from the same flight type. This method only adds cycles that belong to the same flight type.

After combining the cycles, the programme will check whether the partial cycle amount exceeds the bucket size, which can be set to any value. If the partial cycle amount is more than the bucket size, the bucket will be emptied and the min-max combination for that row in the buffer will be placed in the spectrum. If the partial cycle amount is more than the bucket size, the method will keep subtracting and placing the load combinations in the spectrum until the partial cycle amount is smaller than the bucket size.

A reason for a bucket size higher than one is that high stresses tend to occur in groups caused by turbulence, for example. These high stress cycles will have larger retardation effects on crack growth when they occur with higher scatter (bucket size 1). If loads are grouped in sets of more than one load, they will provide conservative crack growth data.

One of the extra options of the programme is the adjustable buffer tolerance. This means that if the buffer tolerance is set at 100 psi, for example, the programme will check if there is a load combination in the buffer which differs no more than the buffer tolerance of 100 psi both on σ_{min} and σ_{max}. In this way the total $\Delta\sigma$ will differ no more than twice the buffer tolerance and only partial cycles are added together within the specified range. This means that the buffer tolerance can be seen as a floating bucket related to both σ_{min} and σ_{max}. If the buffer tolerance is set at zero then only partial cycles of load combinations that are exactly the same are added.

Using this method, no partial cycles with large differences in $\Delta\sigma$ are added into one bucket. The adjustable buffer tolerance is implemented so that by increasing this tolerance, more load combinations can reach a full bucket size depending on its value. If the buffer tolerance is set at zero then a load combination may never reach the bucket size even though many cycles might only be slightly different for both σ_{min} and σ_{max}. After a complete flight has been processed this way, the buffer is stored until the same flight type occurs again. The stored buffer is then used as a start. Therefore, the filling of the bucket continues from the previous flight of that type. This way, each flight of the same type will probably differ slightly. At the end of the construction of a block, all the buffers of the different flight types are saved and used in the next block (60 total). All of the 60 blocks will be slightly different.

31.7.3 Truncation of spectrum

Truncation of high loads may have a large effect on the fatigue life of a specimen. In a spectrum, it is possible that exceptionally high peak loads occur incidentally, reducing crack growth, even in patched specimens. Not all aircraft in service will encounter these high loads.

Due to the fact that high peak loads cause plasticity at the crack tip, this can cause a significant retardation in crack growth. To get conservative results when testing specimens, it is necessary to truncate infrequently occurring high loads.

In general it is suggested that in order to get conservative results, a spectrum should be truncated at a load level which will be exceeded fewer than an average of ten times in the life of the aircraft. The results of the Generated Spectrum Processor showed that an unfiltered fatigue life did not contain such exceptionally high loads as the highest loads occurring in the spectrum appear more frequently than ten times. For this reason there was no truncation.

Negative loads have to be avoided during testing to keep the unsupported specimen from buckling resulting in damage. In the full-scale aircraft, the structure will be stiffened and able to carry compressive loads. To avoid the (very rare and small) negative loads on the specimens during testing, negative loads will be set to 100 psi to retain as much of the original $\Delta\sigma$ as possible.

31.7.4 Description of spectrum specimens

Two specimens, one unpatched and one patched panel, were tested under a full C-5A spectrum (constructed as described in the previous sections). The spectrum had a bucket size of 1. The output was between 100 and 40000 psi. The buffer tolerance used was 10 psi. The two specimens tested under this spectrum are listed in Table 31.10.

The dimensions of the patched panel are given in Figure 31.8. The material is 7075-T6 clad, 1 mm thickness, to represent the C-5A skin. The panel was pre-cracked using a fatigue load of 60-6 MPa at a frequency of 10 Hz to prevent large plastic zones in comparison to the plastic zones caused by the low loads occurring in the spectrum.

The unpatched panel was also 7075 T6-Alclad with a width of 152.4 mm and a length of 400 mm. The pre-crack is 25 mm. The panel was pre-cracked at 60-3 MPa.

Table 31.10 *Specimens.*

Specimen #	Spectrum type	Pre-crack level (MPa)	Patch
06	Unfiltered	60-3	None
S6	Unfiltered	60-6	Glare 2

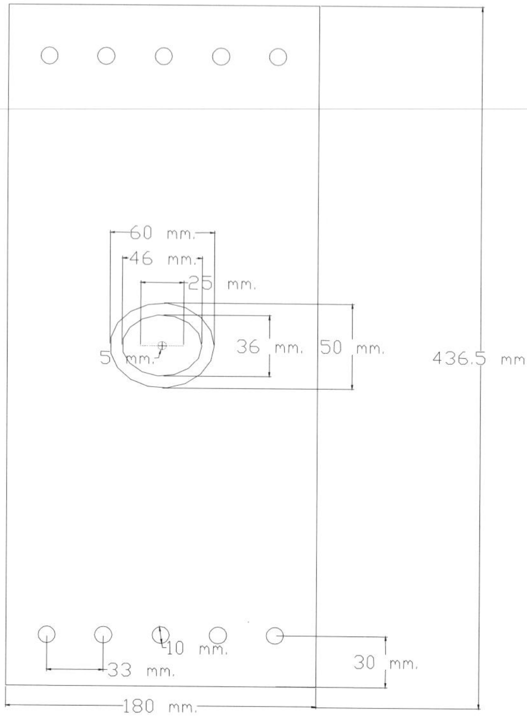

Figure 31.8 *Panel dimensions of patched spectrum specimen.*

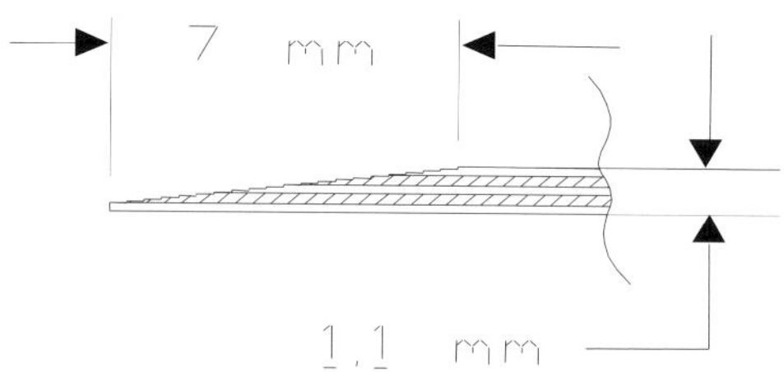

Figure 31.9 *Close-up of taper of Glare patch.*

Description of the Glare patch

The Glare patch was made of Glare 2 3/2 0.2, 3 layers of aluminium and 2 layers of glass-epoxy prepregs, with a total thickness of 1.1 mm.

The edges of the Glare patch were tapered to prevent high skin stresses and adhesive peel stresses. The taper length was 7 mm and the patch was tapered down to a minimum thickness of 0.2 mm, which is the thickness of the bottom aluminium layer. A close-up of the taper is given in Figure 31.9. The taper was CNC-milled in twenty equal steps. Glare 2 is a uni-directional material, i.e. the fibres are in one direction, the fibre direction of the patch was perpendicular to the crack.

Manufacturing of the specimens

Before patching, the panel was pre-treated using PAA and then primed using Cytec's BR-127. For bonding, AF-163-2M (0.29 kg/m^2) adhesive from the 3M company was used. Bonding was carried out in an autoclave of the Structures and Materials Laboratory, only using vacuum to simulate the on-aircraft process.

Fatigue machine

An 810 MTS hydraulic fatigue machine was used. The load frame has a capacity of 500 kN, the load cell a capacity of 250 kN. The machine is controlled by a Teststar II control system. The spectrum tests were load-rate controlled and the rate set at 90 kN/s. A higher rate resulted in inaccurate loads on the specimen, caused by overshoots and undershoots of the machine.

31.7.5 Spectrum tests results

Figure 31.10 shows the crack growth curves for the unpatched specimen 06 as well as for the patched specimen S6.

The C-scan of the patched specimen after completion of the test is shown in Figure 31.11. No debond is visible on the scan. The circular shape in the middle is a drilled hole from where the saw cut was made.

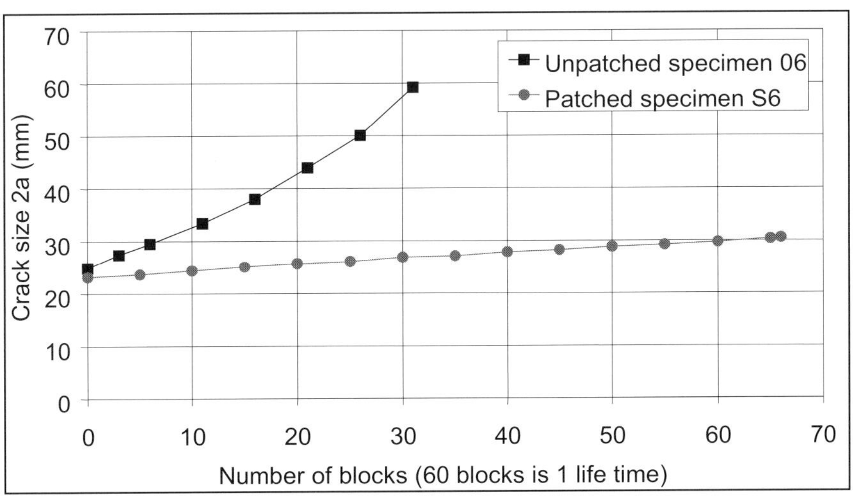

Figure 31.10 Crack growth curves for unpatched and patched 7075 specimens, under C-5A spectrum loading.

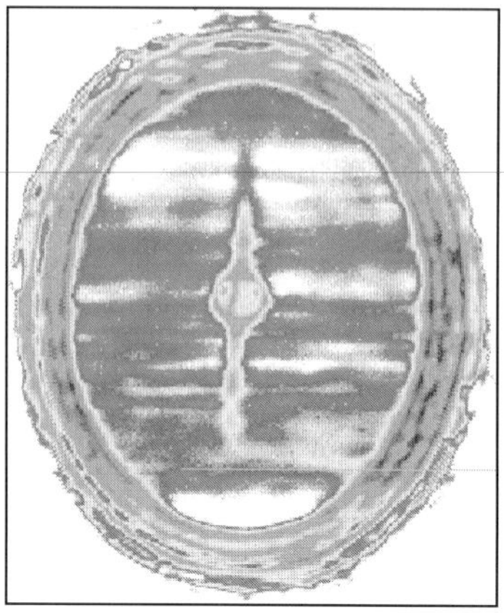

Figure 31.11 C-scan of specimen S6 after fatigue testing

31.7.6 Discussion of the spectrum test results

As can be seen in Figure 31.10, the unpatched panel showed significantly higher crack-growth rates than the patched specimen. The Glare repair is very effective in slowing crack growth under a realistic crown fuselage spectrum. As can be seen in Figure 31.10, the patched specimen showed hardly any crack growth under spectrum loading, even after testing more than one C-5A life.

The repair can withstand one full life without problems. If the patched structure is assumed to be non-inspectable, the minimum required life is twice the remaining life of the aircraft. Assuming that the C-5As are at 50% of their fatigue life, one lifetime has to be accomplished. The patched crack has this capability. However, the C-5A is probably closer to 75% of its lifetime and therefore the life of the patched crack is well within the requirement.

As can be seen in Figure 31.11, the C-scan of S6 shows no sign of debonding after being tested for more than one complete C-5A life.

31.8 Conclusions

Two bonded repairs were applied as a prototype repair on a C-5A. The patches have been in service since October 1995 and are stopping further crack growth in the skin of the aircraft. The repairs are a technically viable solution for the crown-cracking problem of the C-5A. With respect to the limited remaining service-life of the C-5A, the possibly higher initial cost of bonded repairs vs. the lower cost of mechanically fastened repairs might prohibit the technically superior and more durable (bonded) repair option.

This chapter described two different analysis methods: (1) CalcuRep addresses the thermal issues and helps to optimise and size the repair, while (2) the finite element programme Franc2D-L can model the substructure as well as the skin and the repair as a verification of the CalcuRep results. The finite element model was used to predict K in the unrepaired case and verify the load carrying capabilities of the cracked structure before a repair was applied. The analysis methods were verified using fatigue experiments. Both constant amplitude testing and spectrum testing provided verification of the performance of the bonded Glare repairs. The spectrum testing proved the fatigue durability of the bonded repair.

References

[1] R. Fredell, et al., 'Design Development of a Bonded Fuselage Repair for the C-5A'. In: *Proceedings of the 1995 USAF Structural Integrity Program Conference*, WL-TR-96-4094, August 1996.

[2] Dr J.W. Lincoln, ASC/Deputy for Engineering, Wright-Patterson AFB, Ohio, USA.

[3] Kansas State University, *http://www.cfg.cornell.edu/.*

[4] J. Gallagher, *Damage Tolerant Design Handbook*, MCIC-HB-01R.

[5] For more information see: *http://www.usafa.af.mil/dfem/castle.*

[6] R.J. Kuhbander and J.J. Mazza, 'Understanding the Australian Silane Surface Treatment'. In: *Proc 38th International SAMPE Symposium and Exhibition, Anaheim*, Calafornia, May 1993.

[7] Verhoeven, S., *Preparation of the Spectrum Testing of Bonded GLARE Repairs for the C-5A Galaxy, a study of the interaction effects of peak loads on bonded repairs and a proposal for the modification of C-5A aftcrown spectrum data.* TZ Report, Delft University of Technology, The Netherlands, June 1997.

[8] Verhoeven, S., *In-Service Effects on Crack Growth under Bonded Composite Repairs.* Master Thesis, Delft University of Technology, The Netherlands, July 1998.

32

Eddy current inspection

C. Borsboom

Abstract – The eddy current non-destructive inspection technique can be used effectively on fibre metal laminates. It uses the principle of electromagnetic induction, i.e. the interaction between a magnetic field with an electrically conductive object. With this method, the different Fibre Metal Laminate types can be easily discriminated and fatigue cracks in the metal layers of the laminate can reliably be detected in, e.g. plain material, lap joints and spliced laminates.

32.1 Basic principles of electromagnetic induction

The eddy current non-destructive test uses the physical phenomenon of electromagnetic induction. Electromagnetic induction is the interaction of a magnetic field with an electrically conductive object. It occurs when a magnetic field that changes in strength with time is present around a conductive object [1, 2, 3, 4].

The strength of the magnetic field can be changed by the relative movement of a magnetic object with respect to the conductive object, or by placing a coil near the object and applying an alternating current to the coil. If a current excites a coil then it will produce a magnetic field and this field will be present both inside and outside the coil. The direction and the strength of the magnetic field directly depend on the strength and the direction of the current flowing through the coil. The direction of the magnetic field can be determined by using the right hand rule [1, 2, 3, 4].

If the current through the coil is constant then it will induce a stationary magnetic field. However, if the current through the coil is alternating then the magnetic field will also alternate. In this situation the phenomenon of (self-)induction occurs. According to Lenz's law, the coil will try to maintain the magnetic field at a constant strength and will oppose any changes in strength. Due to a change in the magnetic field, there will be a potential difference across the coil. This potential difference is called the induction potential, and does not exist when the magnetic field is stationary. The induction potential depends on the rate of change of the magnetic field strength; the faster the field strength changes, the higher the induction potential will be. Thus, a high frequency current will induce a higher voltage than a low frequency current [1, 2, 3, 4].

If the excited coil is close to a conductive object, and the windings of the coil are parallel to the object's surface, then the alternating magnetic field penetrates the object. The degree of penetration depends on the material being inspected, as there is a difference between magnetic and nonmagnetic materials. This difference is expressed by the relative magnetic permeability. For nonmagnetic materials, such as aluminium, the relative magnetic permeability equals one, but for magnetic materials this number is greater than one [1, 2, 3, 4].

As a result of the existence of the magnetic field near the object, Lenz's law once again plays its role. Because of the alternating magnetic field, currents will flow in circular loops perpendicular to the magnetic field. These currents are called eddy currents. They produce their own magnetic field (following the right hand rule) that opposes the primary magnetic field from the coil, see Figure 32.1 [1, 2, 3, 4].

The opposition depends on the electrical characteristics of the object and the characteristics of the primary magnetic field itself. The primary and the secondary magnetic fields form a new magnetic field, which has its inductive effect on the coil. As a result, the impedance of the coil changes, causing a change in the voltage across the coil. The resulting voltage is different from that mentioned previously [1, 2, 3, 4].

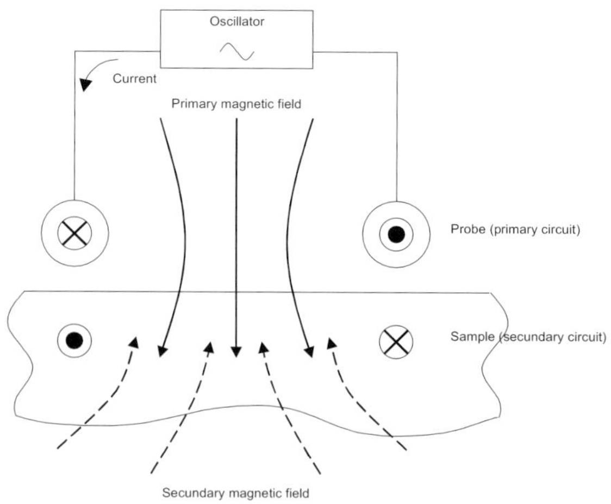

Figure 32.1 *Excited coil near a conductor* [2].

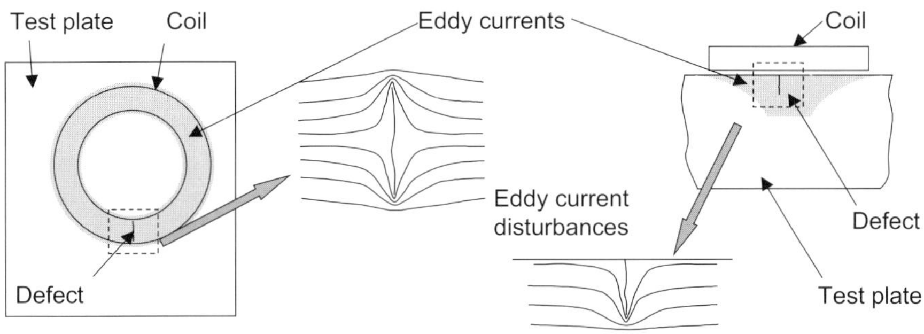

Figure 32.2 *Disturbance of eddy current flow.*

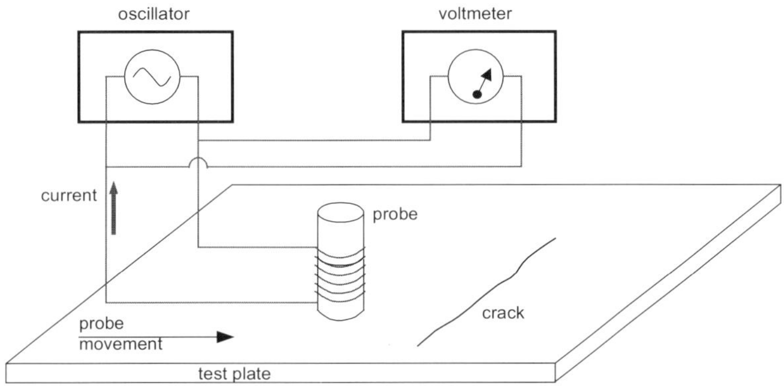

Figure 32.3 *Measurement of voltage* [2].

When the eddy currents are obstructed by the edges of the object or other obstacles (disruptions in the conductivity) inside the object, the circular loops change shape, see Figure 32.2 [2]. This causes a change in the secondary magnetic field and thus in the total magnetic field, resulting in a different impedance of the coil. The coil produces an electric potential, or tension, according to the change in impedance, which can be measured by a voltmeter, see Figure 32.3. The tension provides information about the electrical characteristics of the material [2].

32.2 Depth of penetration and eddy current phase lag

In eddy current testing, the phenomenon of 'skin effect' exists. This means that the eddy currents concentrate near the surface next to the excited coil, and that eddy current density decreases with increasing depth. The magnitude of the skin effect depends on the electrical conductivity and magnetic permeability of the test object and the (adjustable) frequency of the primary magnetic field. To get an idea of the penetration depth of the magnetic field, the concepts of the standard penetration depth (SPD) and the effective penetration depth (EPD) are introduced [2].

The decrease in eddy current density with depth is exponential, and is given by the following equation:

$$J_x = J_0 e^{-\left(\frac{x}{\delta}\right)} \qquad (32.1)$$

where:

$$\delta = 50\sqrt{\frac{1}{f \cdot \mu_r \cdot \sigma}}, \text{ [in millimetres]} \qquad (32.2)$$

with: J_x = eddy current density at depth x
J_0 = surface eddy current density (which remains unknown)
x = depth below the surface
δ = standard penetration depth
f = frequency, expressed in Hz
σ = conductivity in %IACS (International Annealed Copper Standard)
μ_r = relative magnetic permeability

The eddy current density is displayed graphically in Figure 32.4 [2]. The equations are only valid for infinite thickness and coil diameters greater than ten times the thickness of the inspection part. The standard depth of penetration is arbitrarily defined as the depth at which the eddy current density has decreased to 1/e (approximately 37%, e = 2.71828) of the surface density J_0. The eddy current density decreases 1/e per δ. At 3δ the eddy current density has decreased to $1/e^3$ (approximately 5%) of the surface eddy current density. This depth is called the *effective depth of penetration*. For practical coil sizes and materials, the eddy currents attenuate more rapidly than predicted with the equation for J_x. δ thus becomes more of a material or test parameter rather than a true measurement [2].

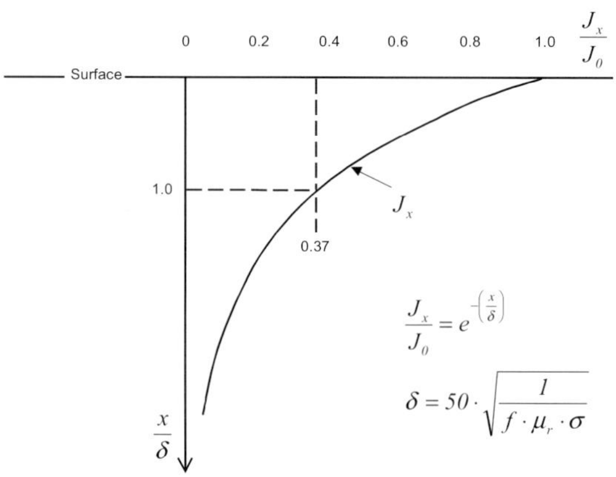

Figure 32.4 Eddy current density.

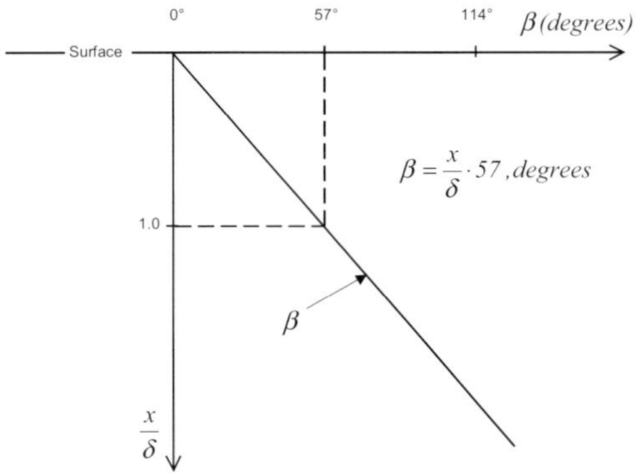

Figure 32.5 *Eddy current phase lag.*

Sensitivity to subsurface defects depends on the eddy current density at the defect depth. Therefore 3δ represents a practical limit of defect detection. Because of the lower eddy current intensity at greater depths, the signal amplitude of a deep subsurface defect will be smaller than the signal amplitude of a defect of the same size just below the surface. Defects located at different depths below the surface will change the probe impedance by different amounts. This also implies that a small defect at or just below the surface can produce the same signal amplitude as a larger defect at a greater depth. Therefore, it is not possible to rely on the signal amplitude alone to judge subsurface defects [2].

In addition to the decrease of the eddy current density with increasing depth, the deeper eddy currents lag in phase compared to those at the surface. This phase lag increases linearly with depth, see Figure 32.5, and is expressed by [2]:

$$\beta = \frac{x}{\delta}\,(rad) = \frac{x}{\delta} \cdot 57\ (degrees) \qquad (32.3)$$

with: β = phase angle in radians or in degrees
x = depth below the surface
δ = standard penetration depth

Responses in the magnetic field from defects at different depths will show this phase lag too, independent of the signal amplitude. The phase lag also affects the impedance; an additional phase angle will be introduced. With both amplitude and phase angle indication, it is possible to discriminate between defect depths and sizes. So, the phase lag of the deeper eddy currents offers an opportunity to estimate the depths of defects [2].

The influences of the electrical conductivity and the frequency of the primary magnetic field on the penetration depth are considerable. When the test object has a high electrical conductivity, strong eddy currents will be induced. The magnetic field inside the test object is attenuated quickly and the eddy currents concentrate near the surface. A low conductivity object will show weak eddy currents, which are less powerful in attenuating the magnetic field. The eddy currents show less concentration near the surface. Therefore, the penetration depth in a low conductivity object is greater than in a high conductivity object, see Figure 32.6 [2].

The relative magnetic permeability is unimportant for the inspection of Arall and Glare, because it is equal to one [1].

A higher coil frequency reduces the penetration depth of the eddy current, see Figure 32.7. However, a higher frequency is more sensitive to smaller defects. Thus, in real applications, the inspector may vary the inspection frequency depending on what sort of defect must be found.

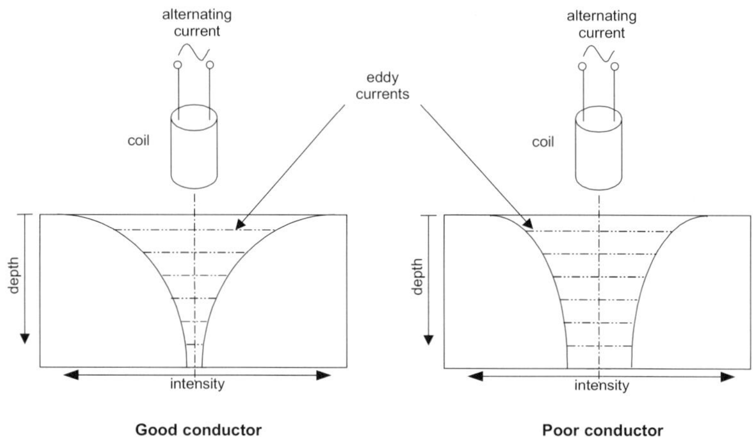

Figure 32.6 *Effect of conductivity on eddy currents* [2].

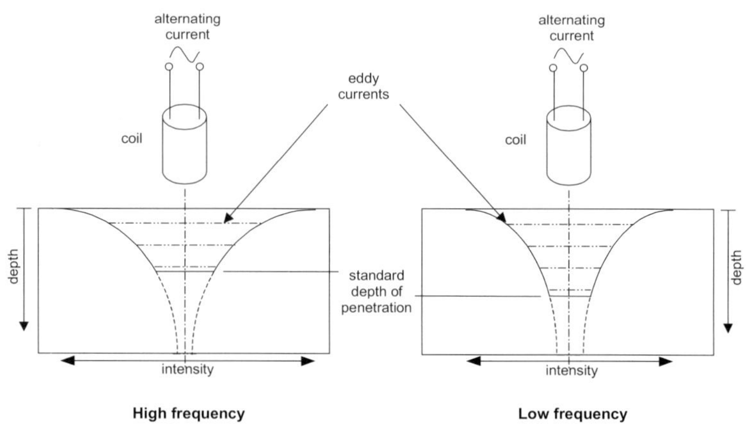

Figure 32.7 *Effect of frequency on eddy currents* [2].

32.3 Impedance diagram for signal interpretation

For a coil, the resistance and the inductive reactance are the two dominant parts, and the impedance for such a circuit, see Figure 32.8 (a), is expressed as [2]:

$$Z = R + j2\pi fL = R + jX_L \tag{32.4}$$

with: Z = impedance (Ω)
R = resistance (Ω)
f = frequency of the alternating current (Hz)
L = inductance (H)
X_L = inductive reactance (Ω)

This depends on the frequency. The voltage drop is expressed as:

$$V = I \cdot Z \tag{32.5}$$

with: I = (alternating) current through the coil

The impedance can be represented by complex analytic equations. However, for a clear, graphical representation of the signal from the probe (coil) the impedance diagram can be used. The advantages are that the two parts of the impedance, the resistance and the inductive reactance, are visible. Also, the phase angle between the excitation current and the total voltage drop can easily be obtained. In the diagram, shown in Figure 32.8 (b), the inductive reactance (X_L) is plotted on the vertical axis and the resistance (R) on the horizontal axis. With the impedance diagram, the changes in coil impedance resulting from the test object can be plotted as a function of the resistive and inductive reactance components [1, 2, 3]. A more extensive explanation on impedance analysis can be found in [2] and [3].

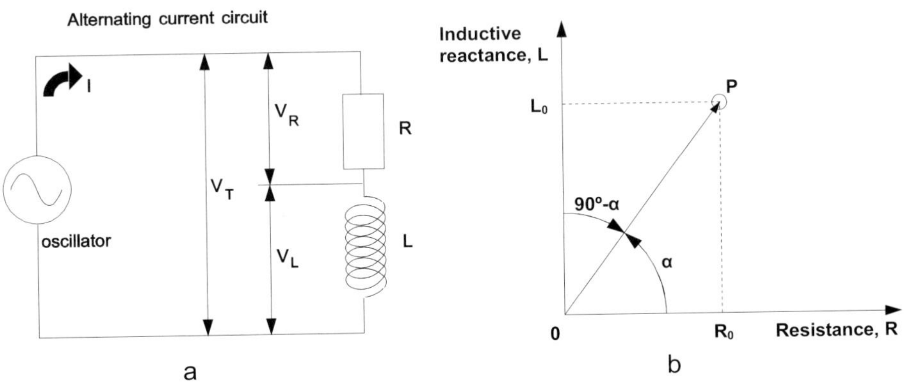

Figure 32.8 (a) The coil in an electric circuit and (b) the impedance diagram.

32.4 Application of eddy current non-destructive testing on fibre metal laminates

With Eddy Current NDI subsurface defects can be reliably found. Also, the different laminate types can be discriminated. For these tests the Nortec 19e Eddy Current Tester was used [1]. Figure 32.9 shows a typical impedance graph, and it shows the relationship between setting and displayed number. The figure also shows a response and how the data, with which the signals can be interpreted, are obtained from a response [1].

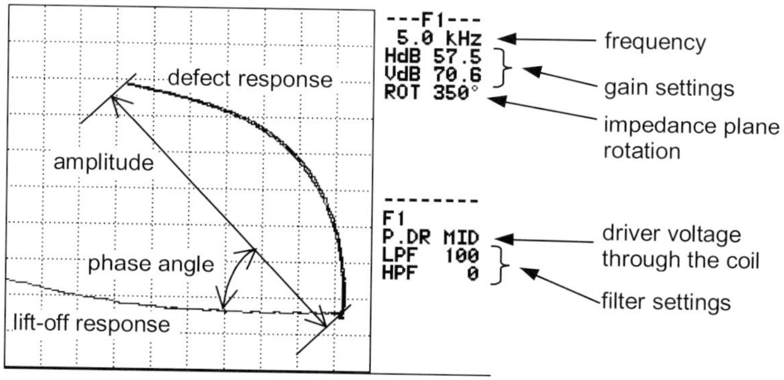

Figure 32.9 Typical display of the eddy current instrument.

32.4.1 Identification of fibre metal laminates

Each of the fibre metal laminates can easily be identified with the eddy current test [1]. Each material has its own unique impedance location on the screen. This is shown in Figure 32.10, where the aluminium alloys 2024-T3, 7475-T761P and 7075-T6 are displayed as reference; the air point is the impedance of the probe in air. The fibre metal laminates were 3/2 layups with an aluminium sheet thickness of 0.3 mm, although each unique layup will have its characteristic trace. The identification takes place by comparing the lift-off signals (moving the probe away from the sample). A small reference standard with samples of the various fibre metal laminates can be used to identify an unknown material. The frequency is not important for identification; it is a comparison. The difference between them is clear and makes identification easy. However, the difference between the material's impedance points depends on the frequency. These differences are often small. The fibre metal laminates have a conductivity of about 8-9% IACS [1].

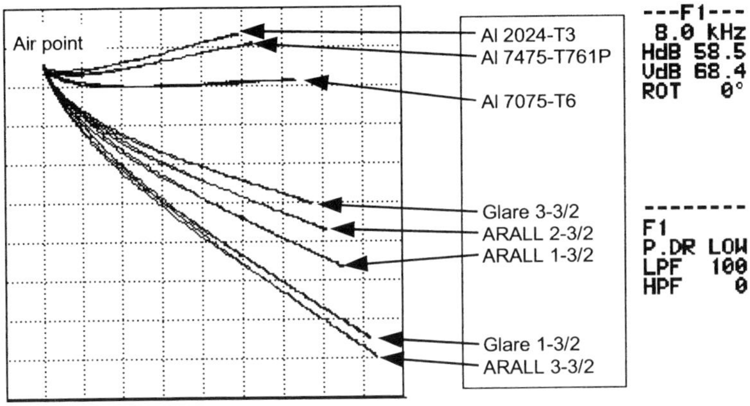

Figure 32.10 *Identification of several Fibre Metal Laminates* [1].

32.4.2 Lap joint inspection

Both a Glare 3-3/2 and an Arall 1-3/2 lap joint reference sample were made to investigate the applicability of this inspection method [1]. Two probes were used for the inspection of the samples. A sliding probe (Nortec SPO-3806) for the inspection of the facing aluminium sheets in the lap joints, defined as layer 1 of the top sheet and layer 1 of the bottom sheet, see Figure 32.1132, and a small pencil probe (Nortec) for the outer two layers of the material [1].

The choice of the frequency is a compromise between the penetration depth and sensitivity. With the inspection of the reference standards, a frequency of 9 kHz for Glare and 10 kHz for Arall was used. The frequencies can be used to inspect both central layers as these layers are sufficiently close to each other. These frequencies enable detection of all cracks, but the smallest cracks (1 mm) are difficult to find. The choice of the frequency did not appear to be very critical for finding a 2 mm notch (in both layers). Two typical responses from the sliding probe are shown in Figure 32.12 below. Figure 32.13 shows the responses from the outer two layers obtained by the pencil probe [1].

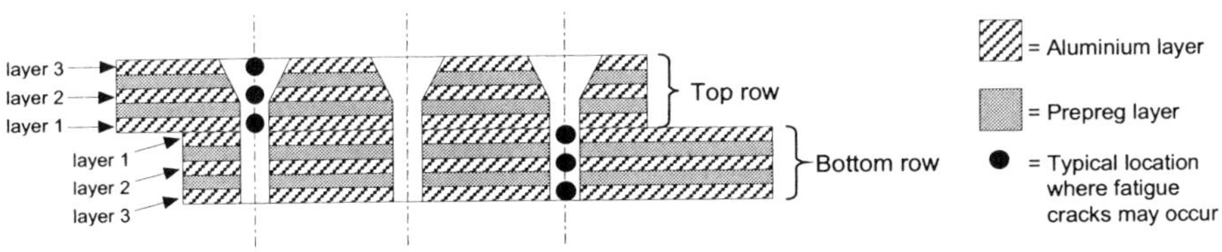

Figure 32.11 *Layer definition in a 3/2 lap joint and the typical locations where fatigue cracks may occur* [5].

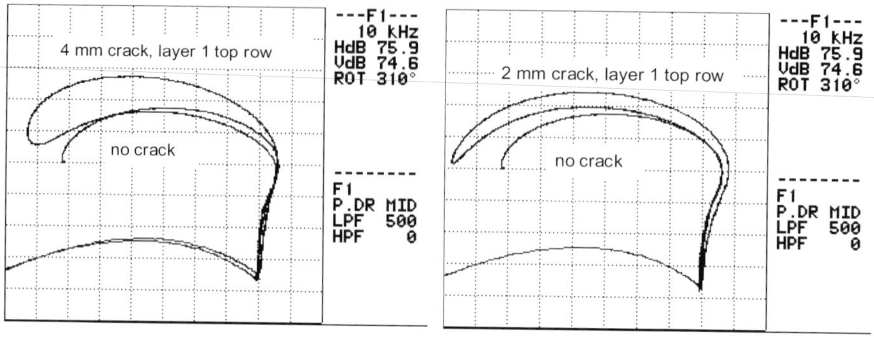

Figure 32.12 *Sliding probe responses from Arall reference sample* [1].

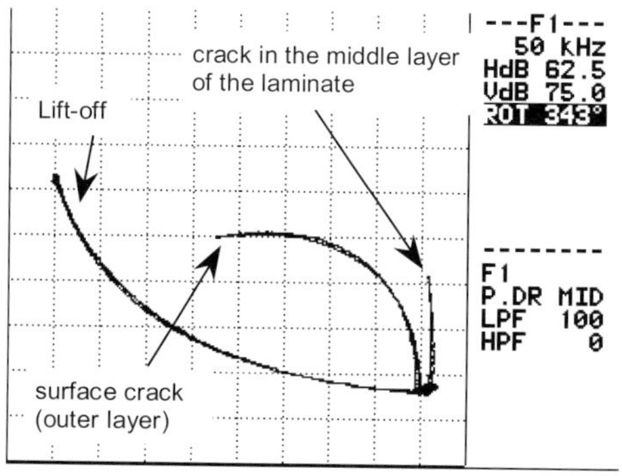

Figure 32.13 *Pencil probe responses from the outer two layers (layers 2 and 3) of the laminate.*

Using the sliding probe, very peculiar and spectacular roller-coaster type responses can be obtained from a severed lap joint, see Figure 32.14 [5]. This response originates from a Glare 3-3/2 lap joint that was fatigue tested in the reduced scale barrel test set-up of the Structures and Materials Laboratory of the Delft University of Technology [5, 6]. With such responses, crack length estimation with the sliding probe in the subsurface aluminium layers is extremely difficult if not impossible. Under these circumstances the surface probe and the pencil probe can be used, even when the crack tips are relatively close together (in the different layers when the cracks are on top of each other or in the same layer where cracks grow towards each other). However, inspection with the surface probes is more time consuming, especially when two different probes have to be used for an accurate crack length assessment [1].

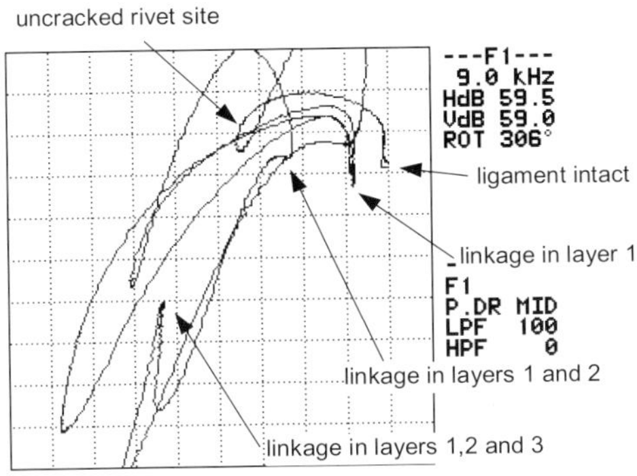

Figure 32.14 *Link-up responses from layers 1, 2 and 3 in the top row in the reduced scale barrel test* [5].

32.4.3 Probability of detection of fatigue cracks in fibre metal laminate lap joints

Eddy current inspections were performed on fatigued test coupons [1]. Following the eddy current inspection, the rivets were drilled out so that the cracks could be checked visually. This was done to check the accuracy and, more importantly, the probability of detection of cracks with the eddy current test.

The probability of detection and the minimum detectable crack size are important for determining inspection intervals. The minimum detectable crack size depends in part on the sensitivity of the inspection system, and may have been specified based on a damage tolerance analysis of the structure. The use of extremely slow crack growth materials, such as fibre metal laminates, might allow a relaxation in inspection techniques or frequency. A bigger crack is easier to find, and a repair action can be undertaken in plenty of time as the crack growth is slow. It could relieve inspection requirements and so reduce effort and costs, without any compromise of safety.

Therefore, some insight in the probability of detection (POD) and the smallest detectable crack size is required. The POD-value increases with increasing crack length. For the estimation of the probability of detection, inspection data from real (unknown) cracks combined with tear-down data were necessary. When these data were combined, an estimation of the POD could be given.

All cracks discussed here were found in facing aluminium layers, which were 0.3 mm thick. The data used were obtained with specimens fatigued under the influence of air, salt water or distilled water. The inspector was not specially training beyond simple instructions on eddy current inspection. The combination showed a very low POD-value for 1 mm cracks. A reliable value, based on the obtained data, cannot be given. For 2 mm cracks, the POD-value increased to about 55%. When the (subsurface) cracks were 3 mm long, the POD-value became about 90%. The POD-value for 4 mm cracks was 100%. The POD as a function of the crack length is displayed in Figure 32.15. The percentage of false indications was 8% [1].

The influence of the environment on the POD of cracks longer than 3 mm was negligible, but crack length estimation accuracy went down with more aggressive environments. Often, when the cracks were quite long (7-9 mm), a difference of 2 mm between eddy current indications and tear-down results occurred. The length estimation of the cracks that had grown under air conditions were the most accurate [1].

The POD of cracks in the second aluminium layer of the bottom sheet is not high, but the inspection frequency used showed an especially high sensitivity to cracks in the two central layers. In deeper layers, the eddy current density decreases rapidly. However, given the strong analytical and experimental indications that cracks will almost always initiate in the facing surfaces of Fibre Metal Laminate lap joints, this is not considered to be a shortcoming of the eddy current technique [1].

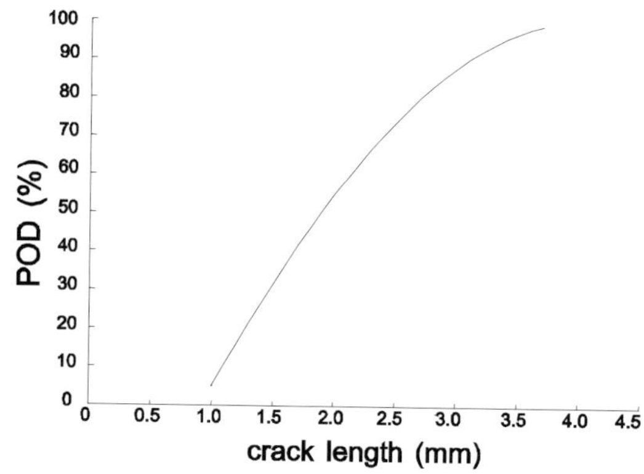

Figure 32.15 *The POD as a function of the crack length* [1].

The POD value may change with experience, mental state of the inspector or sensitivity of the equipment. The values presented here are only a first estimation, but can provide a basis for further work to establish inspection intervals for fibre metal laminates fuselage lap joints [1].

32.4.4 Inspection of spliced laminates

Eddy current inspection techniques have been developed for splices [1], which have a crack stopping effect in fibre metal laminates. The most sensitive probe for inspecting spliced fibre metal laminate sheets was the Nortec SR/.31 surface probe. The best results were obtained using a scanning path where the edge of the probe is about 1 mm from the splice. This scanning path keeps the influence of the splice small. Surface probe scanning needs guiding by a (non-conductive) ruler, as the probe is sensitive to deviations from the scanning path at this position. A Nortec pencil probe was used at a frequency of 50 kHz to inspect two cracks close to each other. At this frequency, the eddy currents only exist in the top two aluminium layers and discrimination between these layers is easy [1].

In a 5/4 Glare sheet, small cracks of 2.5 mm at a splice can be detected in both top and bottom aluminium layers at a frequency of 3.0 kHz. A typical response is shown in Figure 32.16 (a). Such small cracks did not occur in the scanned 3/2 Glare sheet, but similar or higher sensitivities were expected here. The responses were clear and easy to interpret. It was not difficult to identify the crack depths and lengths [1].

The Nortec SPO-3806 sliding probe can also be used. It was operated at a frequency of 10 kHz, which enabled inspection of all aluminium layers in one scan. The small cracks in the first and fifth layers in 5/4 Glare could not be detected at this frequency. A typical response obtained with the sliding probe is shown in Figure 32.16 (b) [1].

A three-layer subsurface crack that grew 4 mm beyond the splice gave a clear response. Some advantages of the sliding probe were that larger responses were obtained and that it is insensitive to deviations from the scanning path. It makes the sliding probe easier to operate than the surface probe [1].

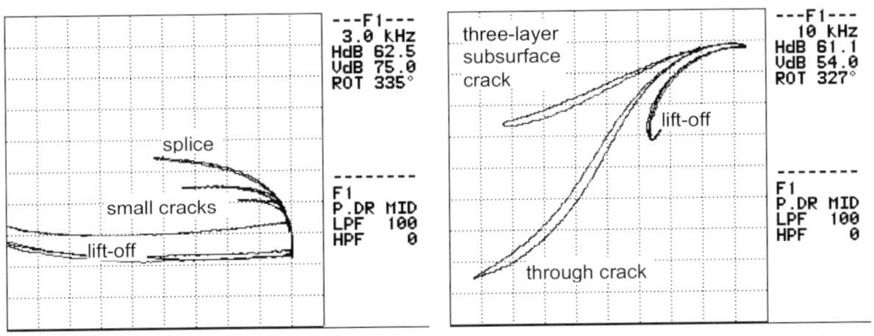

a) 2.5 and 4.5 mm cracks starting at splice, 5/4 Glare 3, surface probe reponse.

b) three-layer and through-crack at a splice, 5/4 Glare 3, sliding probe response.

Figure 32.16 Some typical responses from fatigue cracks near splices [1].

32.5 Conclusion

The eddy current inspection technique can be used quite effectively to inspect fibre metal laminates. The electrically non-conductive prepreg layers confine the eddy currents to the aluminium layers. This works in favour of the interpretation of eddy current responses: one cracked aluminium layer gives a discrete response and depth estimation is more accurate than with monolithic aluminium. Depth estimation is expressed in layer numbers, instead of mm.

The Fibre Metal Laminates have a conductivity of about 8-9 %IACS. Their identification is reasonably easy; each fibre metal laminate type has its own unique impedance point on the screen, which is often clearly recognisable. The maximum depth of penetration in fibre metal laminates is expected to be comparable with monolithic aluminium.

In lap joints, where the two central aluminium layers are in contact with each other, distinct phase rotations are still recognisable. The sliding probe is best used for rivet row inspections at structural joints. The best frequency for Glare 3/2 is 9 kHz, for Arall 3/2 it is 10 kHz.

The probability of detection: Cracks of 3 mm long can be detected reliably in a lap joint of 3/2 layups, the POD for these cracks is approximately 90%. For longer cracks, the POD is practically 100%. A 1 mm crack in the facing layers in a lap joint will not be found because of the influence of the rivet head on the magnetic field.

Inspection was performed successfully on spliced 3/2 and 5/4 sheets. Eddy current responses were clear. Cracks of 2.5 mm, starting at the splice in the bottom layer of a 5/4 layup, can be detected. The best probe for this type of inspection was the surface probe.

Eddy current inspection of fibre metal laminates can yield peculiar impedance responses. This is probably the result of the fact that one crack consists of several cracks, which will not always be directly above each other.

References

[1] C. Borsboom, *Eddy Current Testing of Fiber-Metal Laminates*, Structural Laminates Company, Technical Report TD-R-93008, Delft, The Netherlands, June 1993.

[2] V.S. Cecco, et al, *Eddy Current Testing*, Volume 1, Atomic Energy of Canada Limited, 1987, originally published under the title 'Eddy Current Manual', 1983, chapters 1 - 6, p. 164.

[3] H.L. Libby, *Introduction To Electromagnetic Nondestructive Test Methods*, Wiley-Interscience, 1971, Chapters 1 and 2.

[4] J. Blitz, *Electromagnetic And Magnetic Methods Of NDT* , Brunel University, Uxbridge, England, 1991, p. 13-15, 89-91, 116-125, 137-139, 141-147, 150-152.

[5] C. Borsboom, *Damage Tolerance of Glare Lap Joints in Pressurized Fuselage Structures*, Master's Thesis, Delft University of Technology, Faculty of Aerospace Engineering, Structure and Materials Laboratory, Delft, August 1994.

[6] R.G.P. Müller, An Experimental and Analytical Investigation on the Fatigue Behaviour of Fuselage Riveted Lap Joints, PhD Thesis, Delft University of Technology, Delft, 1995.

33

Glare as part of Sustainable and Environmentally Sound Engineering

A.R.C. de Haan

Abstract – The difference between the terms "Environmental Friendly" and "Sustainable Development" is their space and time frame. Sustainable Development has a much wider scope in terms of time and space.

Glare may not be completely environmental friendly (although it is completely harmless for humans when proper precautions are taken) during its life cycle, but, more important, it is a very promising material with respect to Sustainable Development. Recycling of Glare can not create a completely closed loop of energy and material without any landfill. But, this limited environmental friendliness does not make a contribution to Sustainable Development impossible.

The IPAT theory shows that for a Sustainable Society, an enhancement of the efficiency of technology by a factor of 5 to 50 is needed in the next 50 years. The promising aspect of Glare is that it contributes to this required factor to make a 20% higher efficiency possible, of the new generation of aircraft, because of reduced fuel consumption.

Weight savings, for instance due to the use of Glare, are usually used to study how much energy, i.e. fuel can be saved. However, the question should be raised whether this saving will create more transport as a result of lower prices. This is called the 'rebound effect'. But it is also possible that more payload is transported on each flight and thus the number of flights is reduced. The effect of the application of Glare is therefore not certain.

33.1 Introduction

During the last three decades of the 20th century a lot of articles have been written about 'The Environment'. Governments tried to force people to use less energy because the general idea in the late seventies was that the oil fields contents were decreasing rapidly. Another cause for criticism was the accident at the Union Carbide firm in India, which contaminated a large area and killed lots of people. Society asked for better security and better standards of living in densely populated

areas. Waste pipes were expanded, filters were placed, some insecticides were banded, frogs returned to the pools near factories. A lot of troubles seemed to be solved by what we now call "end of pipe solutions", e.g. not changing the process itself, but making its negative side effects less harmful.
In the eighties this all seemed to be not enough. Scientists warned of the enormous rise in cancer casualties if the decrease of the ozone layer would not be stopped. Other scientists warned for an even bigger problem of climate change and all its hazardous effects. After the nuclear accident at Tsjernobyl near Kiev in 1986 society deeply realised that effects of its handling on earth did not only effect a local area but could also lead to global deterioration of the earth [1].
'Sustainable Development' as a concept was introduced in the 1987 United Nations report of the Brundtland committee. The committee defined Sustainable Development as follow [2]:

> *Sustainable Development is a development that meets the needs of the present without compromising the ability of future generations to meet their own needs, It contains within it two concepts: (1) the concept of needs, in particular the essential needs of the world's poor, to which overriding priority should be given; and (2) the idea of limitations imposed by the state of technology and social organisation on the environment's ability to meet present and future needs.*

The popularity of the concept Sustainable Development is enormous and ever increasing. Sustainable Development has similar aspects to the concept 'Democracy': Nobody will argue its importance. At the same time the concept, at least in this definition, is open and can have different implications. In that respect there is a big distinction between Environment and Sustainable Development [1]. Everybody knows what his or her environment is. It is local, both in time and space and it is something you can see, touch, feel, hear, smell etc., but what does the open term Sustainable Development mean?
My experience as a university professor is that engineering students like to have a more direct definition, and although maybe not completely accurate, I often define Sustainable Development for my students as the following short description [1]:

> "*Sustainable Development is creating capacity for continuity*".

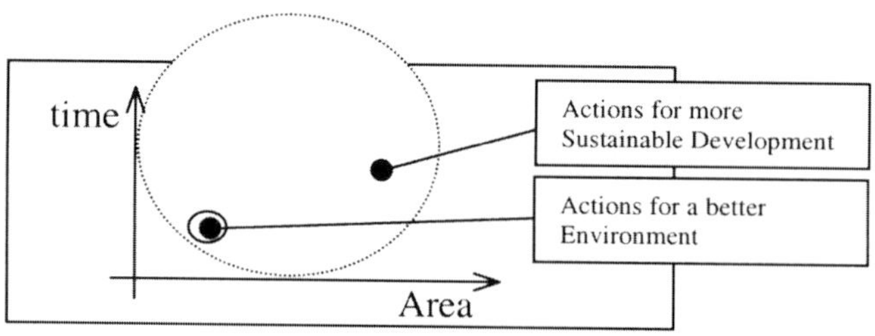

Figure 33.1 Graphical distinction between Environment and Sustainable Development.

This description clearly states Sustainable Development as a long term view on activities and therefore also concerns a geographically wide area. A good and healthy environment is therefore part of a Sustainable Development. The distinction of both terms is plotted in Figure 33.1.

This chapter is divided into two main parts: Environmental Engineering and Sustainable Development. In the introduction the distinction was made between the two. Results of two case studies in the environmental impact of Glare are presented in the second part. The third part deals with the implication of Glare for more sustainable transport. Finally some conclusions are drawn.

33.2 Glare and the environment: two case studies

Using the distinction outlined in the prior section, this section deals with two studies about the impact of Glare on the Environment during processing and the end of life of the product.

33.2.1 Processing Glare

Chromates

Glare is a laminate made up of traditional aluminium sheets and glass fibre/adhesive sheets bonded together. During surface preparation for the bonding and curing cycle both sides of the aluminium sheets are sprayed with primer containing chromate substances. Most of this primer stays inside the laminate, as it is not exposed to the outside environment during the lifetime of the Glare. By far the most important moment when chromate particles leave the laminate is when Glare is being processed, e.g. drilling, milling, sawing etc.
Amongst the human body organs sensitive to chromates are the skin, the lungs and reproductive organs. All chromium compounds are considered carcinogenic in human bodies.

Figure 33.2 Drilling in Glare.

The Dutch limit of acceptable concentration is 10 μg/m^3. This is the strictest acceptable level in Europe. Analysing samples taken during experimental conditions according to standardised methods measured that the maximum chromium level in the air is reached for Glare during sanding and drilling, see Figure 33.2. Still, these levels are at least 24 times below maximum allowed levels according to the Dutch regulations [3], and therefore there is no reason to expect negative effects.

Fibres

When processing any kind of material, dust will always in some form. Most of this dust is harmless, but some particles with certain specific dimension can get into the most inner part of the lungs. This may cause damage and in a later stage lung cancer. Such particles are called to be "respirable".

As an important part of the Glare material consists of glass fibres, special attention is paid to the dimensions of the fibre particles resulting from processing. Samples of dust left over from sawing, drilling, sanding and other kinds of processes were collected and studied under a scanning electronic microscope (SEM).

The original fibre diameter of glass fibres is larger than 3 microns, which is the size that is considered to be respirable and therefore can cause damage to the lungs. Breaking up of these fibres in shorter fragments, will not reduce its diameter, so these non-respirable fibres will never become respirable. Analysis of the samples taken show no signs of fibre fragments with diameters smaller or equal to 3 microns, see Figure 33.3.

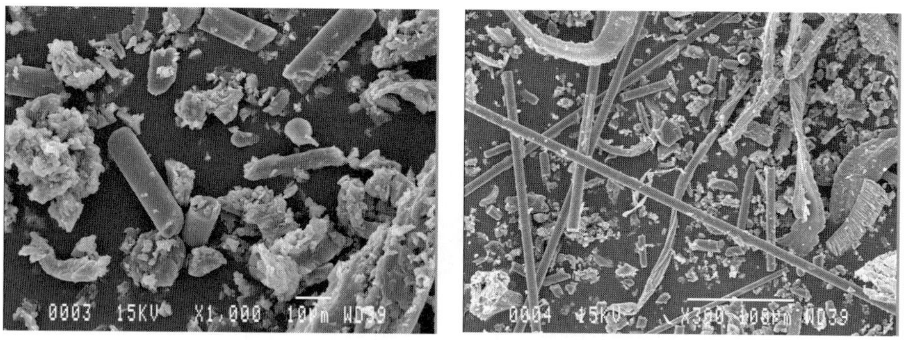

Figure 33.3 *Samples taken from drilling dust show no harmful fibres of less than 3μm diameter.*

33.2.2 Recycling of Glare

Traditional aluminium is a valuable material, because a lot of kWh is invested to produce it through the process of electrolyses. After the use phase, when the product becomes waste, aluminium products therefore usually find their way to the recycling industry. Glare meets specific problems when it must be recycled. And, next to that, it is not always clear whether recycling is enhancing the Environment.

Is recycling environmental friendly?

Recycling appears to be a wizards word in society. It is correct to use it frequently and no one is arguing it. However, is recycling always the most favourable thing to do when products are at the end of their life?
Recycling is done in different ways. The best thing to do, at least the thing that costs least processing energy is re-using the product for the same purposes it was originally used for. E.g. Glass recycling preferable takes place as reusing the bottles instead of crushing them and making new bottles out of them.
As soon as the material of the recycled product is used in it's next life on a lower level than it was originally used for; there is a loss. Think of plastic coffee cups recycled and used in composite window frames instead of new coffee cups because of hygienic reasons. To create the original products, new material of the same type is needed. There is no closed cycle that can be used infinitely.
A typical example of the use of the 'wizards word' recycling for purposes where it does not suit an environmental friendly approach, is the new European requirement for car manufacturers. A new car must be at least recyclable for 95% of its components in the next few years. For lightweight materials such as glass fibre, carbon fibre and Glare, the costs for recycling are substantially higher than for the traditional steel used by car manufacturers. The obligation to recycle cars for 95% of their components will result in using the traditional heavier metals in the car industry. This results in heavier cars burning up more fossil fuels per kilometre and thus exhausting more harmful gasses. The negative effect of the burning of this extra fossil fuel is usually substantial when compared to the negative effects of burning up the non-recyclable lightweight materials and using the heat or even when just landfilling them [4].

Typical problems when recycling Glare

Whereas aluminium is easily melted in an oven and deformed into new shapes suiting new products, Glare can not be melted to reuse it. It is a laminate product and therefore it needs specific attention when wasted.
If Glare is burned, the aluminium will melt quickly, but it will pick up all kinds of unwanted components, which are around 32% of the total volume. Therefore this form of recycled aluminium has a very low market value. The epoxy will burn, leaving relatively harmless fumes, and the glass fibres will be left in the slag, usually in the form of a sand-like residue. Both the epoxy and the fibres do not have any market value left [4].

Recycling Glare and getting well separated materials at the end of the process, means separating Glare in a non pre-heated environment. A promising technology is cooling down Glare until it reaches a temperature of minus 180 degrees Celsius. The epoxy then becomes brittle and some separation between the aluminium sheets and the glass fibre sheets is possible. The process gets more efficient when the Glare scrap is cut into pieces, see Figure 33.4. Although the separation is not perfect, the aluminium components can be sold for a higher price than the costs of the recycling process [4], making it an economical process. The question must be raised what the environmental advantage is of recycling Glare, because a substantial amount of energy is needed for the cooling down process.

Figure 33.4 Aluminium flakes obtained by processing Glare for recycling of its components.

Is their a need for recycling Glare?

Remembering the section about recycling, an interesting question is whether we really need to recycle Glare. Is it worth the effort?
Even if recycling of Glare is not advantageous from an environmental point of view, a study by Tempelman [4] shows that when using Glare for civil aviation purposes the vast majority (>97%) of the environmental effects is caused by burning fossil fuel to fly the aircraft. Putting much effort in developing techniques for recycling Glare is then a matter of debate, as these efforts can also be spent in making aviation more sustainable in other ways. Even landfilling Glare after its use is tolerable if flying is made more environmentally friendly. However, the recycling of the aluminium content of Glare is shown to be possible and economically attractive for a certain minimum amount of Glare scrap.

33.3 Glare and Sustainable Development

Sustainable Development in flying is a development that makes it possible for current and future generations (in principle infinite future generations to come) to choose whether to fly or not. If flying keeps on being a growing market as it is now (rates over 5% per year) and nothing is found to make flying less inconvenient than it is today, the sector will inevitably run into crises within a considerable timeframe.
In current international treaties about global sustainability problems such as climate change, the aviation sector is left out. However, the aviation sector is expected to come up with solutions to these problems or it will be treated hard in the next international treaty. Although questions may be raised about what the international community can do to sanction the aviation industry, the treaty is strict enough to not leave it unnoticed. Among the fields of aviation technology where experts expect most sustainable improvement are Air Traffic Control and Aircraft Manufacturing. We focus our attention to the materials used in the manufacturing industry.

33.2.1 Improvement by factors necessary

A simple way of calculating the needed efficiency of aircraft over a period of, say, fifty years, compared with the efficiency today is the IPAT model [5, 6].
First of all, a baseline must be chosen at which the inconvenience of aviation is acceptable. Being completely sustainable does not automatically mean not flying at all. What is needed is operation in the buffer zone of the earth system, i.e. operating in such a way that the existing equilibrium is not disturbed. When knowing the dimensions of the buffer zone, society must make a choice about the amount of the buffer zone that can be spent on flying.
Let's assume for sake of simplicity that the inconvenience of flying today is smaller than the Earth's buffer zone, and acceptable for society, e.g. society accepts that this part of the buffer zone can not be used for anything else. The challenge for the engineer is now to keep the technological efficiency in pace with additional growth of consumption, welfare etc.

For calculation the following is assumed:

- Impact (I) stays the same, e.g. a level 1 relative to the situation today.
- Population growth (P) has a factor of 2 by 2050.
- Rate of adverse effects on sustainability per unit of consumption (A) increases with factor 5, as developing countries will raise to the level of consumption the western world has today.
- Technology efficiency (1/T) must increase to keep pace with growing P and A, to maintain I equal to 1.

The relation between the factors is given by:

$$\left.\begin{array}{l} \mathrm{I} = \mathrm{P}\cdot\mathrm{A}\cdot\mathrm{T} \\ 1 = 2\cdot 5\cdot \mathrm{T} \end{array}\right\} \rightarrow \quad \mathrm{T} = \tfrac{1}{10} \qquad (33.1)$$

This IPAT model is very simple, and there is a lot of discussion about the precise numbers. What this model correctly shows is that the technological answer to the question for more sustainability, must be an increase in efficiency with an order of magnitude. Just 15 % or even 50% is by far not enough. Therefore aviation should explore and develop ways to keep on fulfilling its service while in the meantime raising its energy and material efficiency by a factor 5 to 50 [5]. The question is which materials can contribute to this required factor.

33.3.2 Sustainable Development and technology

Through technology human beings were brought a lot of welfare. The use of technology also gave mankind some major local and global problems to solve. Still, the belief in a technology that can solve every occurring problem is substantial. Especially governmental agencies appear to think that it is worth more to have advanced technologies solving problems, than to try to change people's behaviour causing those problems.
Some people see technology as something inevitably leading mankind to disaster. For instance the French philosopher Jacque Ellul, see Figure 33.5, who died in 1994. He shows in his famous works that technology can only solve a problem by introducing more new problems. Unfortunately he is only diagnosing and not giving any clues to break through this vicious spiral process.

Ellul's interesting work forces engineers and other people to think about the role we would like technology to play in our society. As engineers we are warned, but are not that sceptical. We see it as our responsibility to use technology to give our part of the answer for a more sustainable society. Nevertheless we have to remember Ellul's warning and use the concept of Sustainable Development to solve problems not only locally or for short periods of time or not only by technical solutions. Sustainable Development is about creating visions for the long term and working out scenarios to make sustainability a reality.

Figure 33.5 The French philosopher Jacques Ellul (1912-1994).

33.3.3 Sustainable transport and advanced materials

As Glare is especially used in the air transport sector we will narrow our focus of sustainability and Glare. The main point of this section is finding out the inconveniences of the transport sector for society in general. The next section will focus on the contribution of Glare in solving these inconveniences.

In the PhD-thesis of Tempelman [4] he distinguishes five categories of drawbacks as a direct result of society's wish for transport:

1. harmful emissions for health and ecology
2. land use and, linked with that, noise hindrance
3. consumption of material, energy consumption and waste generation
4. costs, including discarding, congestion, ecological degeneration, etc.
5. safety, i.e. direct, through accidents, and indirect, through emission, etc.

The choice for a specific material will influence all these 5 categories, either in a direct or indirect way. The manufacturing of Glare sheets, as we have seen in earlier sections, does not cause harmful situations for human beings. That does not prove that Glare production does not have any environmental harmful effects. It is an energy intensive process, consuming non reproducible energy and materials like aluminium and different kinds of chemical components. Using Glare in products also means producing waste that is not easily recycled.
Other reasons explain why this new material is so promising and why so much effort is spent for its development. Glare gives prospects for lighter constructions, which will have the same function and at least keep other properties equal. Results can be vehicles using less energy and therefore producing less harmful emissions. In the case of safety, Glare can substantially extend the time that passengers have to leave a burning aircraft, as it has excellent fire resistance.
An important issue illustrated here is that overall sustainability is the result of all effects together. Materials can not be called Sustainable or Environmentally Friendly upon one quality alone, e.g. recycling. Materials can also not be called Sustainable or Environmentally Friendly as a label that is valid under all circumstances.

33.3.4 Glare's contribution to sustainable air transport

Modern civil aircraft are almost completely made of aluminium skin panels, stiffeners and frames. In the traditional world of aviation, with its high investment risks, advanced materials like composites have been promising for a long time. At this very moment one of these advanced materials, Glare, appears to be making a breakthrough after twenty years of development.
In aircraft design, a dominant drive is reducing weight. In many cases fatigue characteristics are an important design criteria for aircraft. Therefore, in producing lighter aircraft, most effort is put in avoiding the occurrence of metal fatigue.

Glare could change this, because it has excellent fatigue properties. Next to that, it makes fewer joints necessary, due to its possibility of splices. Overall Glare makes it possible to reduce the weight of an average aircraft by at least 20%.
How can this 20% weight saving be used? In a study by Tempelman [4], a section of the Airbus A-320 was redesigned in Glare. A Life Cycle Analysis showed that the 20% weight saving led to a 27% less harmful impact on the environment. A number larger than the weight savings, due to secondary weight effects. This effect has both local and global benefits for the humans and their environment, so one can speak of a contribution to Sustainable Development.

However, the question is whether using a lighter structure for savings of consumed transport energy is the best way to reach sustainability. A negative comment is that weight savings are used to transport more, and therefore, the total effect for sustainability could even be negative. That is only the case if transporting more tons of payload results in cheaper transport, and with it creating more transport due to lower prices. Using the weight saving for transporting more, could also lead to less flights. This could be preferable to using less fuel in the same amount of flights, because the vehicle empty weight does not need to be transported [7].

Clearly, Glare has good opportunities to contribute to Sustainable Development. By itself Glare can not be called Sustainable or Environmentally Friendly. It totally depends on the way it is used and under what circumstances. Whether society will use this engineering part of the sustainability puzzle for either short term optimisations or long term optimisations, e.g. Sustainable Development, is both a necessity and a matter of debate [1, 2, 4, 6].

33.4 Conclusions

- There is a distinction between the often used terms "Environmentally Friendly" and "Sustainable Development". The first one is related to local inconvenience due to all kinds of activities and usually its timeframe is also shorter. A good environment is definitely part of Sustainable Development, though, sometimes, a contribution to a sustainable development can be made with environmentally unfriendly activities.
- Processing Glare causes locally unfriendly environmental effects, that are not directly harmful for humans when proper precautions are taken. Its effects of metallic particles in the air and the effect of fibres on the lungs are for humans not harmful and factors below the limits set by European governments.
- Recycling Glare is economically possible in theory. However, recycling asks for investing a lot of energy, so questions may be raised whether recycling is an undoubtedly beneficial process for the environment. Most people see recycling as a wizards word that can not be argued against.
- For reaching sustainability the simple IPAT model shows that efficiency of technologies should increase in the order of 5 to 50. Glare contributes to

Sustainable Development by an efficiency raise of around 20%. It is therefore a promising development in the process towards more sustainability.
- Questions may be raised whether weight savings can contribute to sustainability by reducing the amount of energy the vehicle needs to move. Other possibilities, such as transporting more payload in one time, therefore reducing the amount of flights, should be considered as well.

References

[1] Haan, A.R.C. de, *LR in Duurzame Ontwikkeling*, Delft, Netherlands: faculty of Aerospace Engineering, 2000.

[2] Brundtland, G.H., et al., *Our common future*, Oxford, UK: Oxford University Press, 1987.

[3] Jalving, N.H., *Health, Safety and Environmental Aspects of the manufacturing and processing of Glare*, report B2v-00-02, Delft University of Technology, 2000.

[4] Tempelman, E., *Sustainable Transport and Advanced Materials*, Haarlem, Netherlands: Eburon, 1999.

[5] Jansen, J.L.A. et al., *Duurzaam: zo gezegd, zo gedaan*, Delft, Netherlands: Delft University of Technology, juni 1999. Emirital Speech.

[6] Weizsäcker, E. von, et al., *Factor Four,* London, UK: Earthscan Publications Limited, 1998.

[7] Haan, A.R.C. de, et al., *Optimising weight-savings for sustainability in aircraft structures,* submitted for publication.

Authors and Affiliations

R.C. Alderliesten	Delft University of Technology
A. Auffret	Delft University of Technology
Th. Beumler	Delft University of Technology
T. de Boer	Delft University of Technology
B. Borgonje	Delft University of Technology
C. Borsboom	Fokker Aerostructures B.V.
R. de Borst	Delft University of Technology
O.J. Bosker	Fokker Aerostructures B.V.
P. Broest	Delft University of Technology
R.A.M. Coenen	Delft University of Technology / ING Groep N.V.
J.W. Evancho	Aviation Equipment, Inc.
A. Gennai	Delft University of Technology / Pisa University, Italy
J.M. Greer	United States Air Force Academy, Colorado Springs, USA.
C.B. Guijt	United States Air Force Academy, Colorado Springs, USA
J.W. Gunnink	Delft University of Technology / Fibre Metal Laminates Centre of Competence
C. Haack	EADS-Airbus-Germany
A.R.C. de Haan	Delft University of Technology
M. Hagenbeek	Delft University of Technology
W.G.J. 't Hart	National Aerospace Laboratory NLR
F. Hashagen	EADS-Airbus-Germany
C. van Hengel	Fibre Metal Laminates Centre of Competence
W. van der Hoeven	National Aerospace Laboratory NLR
J.J. Homan	Delft University of Technology
P.A. Hooijmeijer	Delft University of Technology
M.S. IJpma	Delft University of Technology
O.C. van der Jagt	Advanced Lightweight Engineering BV, Delft
N. Jalving	Delft University of Technology
E.L. Jansen	Delft University of Technology
T.W. de Jong	Delft University of Technology
J.L.C.G. de Kanter	Delft University of Technology
E. Kroon	Fibre Metal Laminates Centre of Competence
S.C.H. van Meer	Delft University of Technology
W.J. Mortier	Delft University of Technology
R.P.G. Müller	Urenco, Almelo, the Netherlands
B.C.L. Out	Detabas, Rotterdam, the Netherlands
F. Pellenkoft	EADS-Airbus-Germany
J.J.C. Remmers	Delft University of Technology
J.J.M. de Rijck	Netherlands Institute for Metals Research / Delft University of Technology
G.H.J.J. Roebroeks	Fibre Metal Laminates Centre of Competence
J. Sinke	Delft University of Technology

F.O. Taddey	EADS-Airbus-Germany
S. Verhoeven	United States Air Force Academy, Colorado Springs, USA
A. Vlot	Delft University of Technology
T.J. de Vries	EADS-Airbus-Germany / Delft University of Technology
M. Wiedemann	EADS-Airbus-Germany
G.P. Wit	Fokker Aerostructures B.V.
T.C. Wittenberg	Delft University of Technology
H.J.M. Woerden	Delft University of Technology

Sponsors

Main sponsors

- Delft University of Technology
- Airbus
- Fokker Aerostructures B.V.
- Aviation Equipment, Inc.
- FMLC - Fibre Metal Laminates Centre of Competence
- European Office of Aerospace Research and Development, Air Force Office of Scientific Research, United States Air Force Research Laboratory

TU Delft
Delft University of Technology

AIRBUS

STORK®
Fokker

AVIATION EQUIPMENT, INC.

Co-sponsors

- KLM Royal Dutch Airlines
- Advanced Glassfiber Yarns LLC
- SP aerospace & vehicle systems
- Embraer - Empresa Brasileira de Aeronáutica S.A.
- Advanced Lightweight Engineering BV
- National Aerospace Laboratory NLR
- Netherlands Agency for Aerospace Programmes NIVR
- Kluwer Academic Publishers
- TNO
- ASCO Industries NV
- Pechiney Rhenalu
- NAG - Netherlands Aerospace Group

FMLC
Fibre Metal Laminates Centre of Competence

EUROPEAN OFFICE AEROSPACE R&D

(updated until 6 July 2001)

FMLC

Fibre Metal Laminates
Centre of Competence

Kluyverweg 6
2629 HT DELFT
The Netherlands

http://www.fmlc.nl

phone +31 15 251 47 40
fax +31 15 257 07 86
e-mail info@fmlc.nl

PO Box 6050
2600 JA DELFT
The Netherlands

http://www.tno.nl/homepage.html

phone +31 15 269 69 69
fax +31 15 261 24 03
e-mail infodesk@tno.nl